. . as surely as the person born with six
gers or the calf with two heads, Isaac
wton was a mutant, seeming, as often
not, more a phenomenon than a man
he was hardly a conventional figure in
y sense of the word. His intellect was
profound; his capacity for rage too
at; his desire for seclusion from the out-
e world too obsessive; his passion for
thing not having to do with original
ought and scholarship too subdued. He
s, in truth, the incarnation of the ab-
acted thinking machine."

n the first major popular biography of
Isaac Newton in 50 years, historian
le E. Christianson paints a compelling
trait of this seminal thinker—a tower-
genius who, in the words of Albert Ein-
in, "stands before us, strong, certain,
l alone."

Drawing on the full body of Newton pa-
s (nearly four million words), this majes-
work details Newton's life in its en-
ty: from an introspective boyhood in
al Lincolnshire, to Cambridge, where
came to question the very order of
igs, to the heretical religious ideas that
uld ultimately absorb him more 'than
nce itself, to celebrity as leonine Mas-
of the Mint and President of the Royal
iety. Throughout, Newton emerges as
assionate recluse, given to sleepless
hts working alone with little more nour-
nent than bread and wine.

s the legend unfolds, so, too, do New-
's epoch-making discoveries in mathe-
ics, physics, optics, and astronomy. At
ie had already established the elements
ifferential calculus. Soon after he cre-
l the reflecting telescope and described
properties of light. At 45 Newton seed his reputation by publishing the
cipia Mathematica, a treatise on uni-
al gravitation that would alter forever
's vision of the cosmos.

Yet Newton's own assessment of his
life—one as devoted to human understanding as any in history—was surprisingly self-deprecating. "I do not know
what I may appear to the world," he wrote,
"but to myself I seem to have been only
like a boy, playing on the sea-shore, and diverting myself, in now and then finding a
smoother pebble or prettier shell than ordinary, whilst the great ocean of truth lay
all undiscovered before me."

In the Presence of the Creator traces
Newton's tireless struggle to find a grand
design in the universe. And it sets his brilliant intellectual feats against a fascinating
backdrop of scientific, political, and religious turmoil—when, suddenly, nearly all
he assumptions of the old order were
alled violently into question.

In the Presence
of the Creator

In the Presence
of the Creator
Isaac Newton and His Times

GALE E. CHRISTIANSON

THE FREE PRESS
A Division of Macmillan, Inc.
NEW YORK

Collier Macmillan Publishers
LONDON

The Free Press
A Division of Macmillan, Inc.
866 Third Avenue, New York, N. Y. 10022

Collier Macmillan Canada, Inc.

Printed in the United States of America

printing number
1   2   3   4   5   6   7   8   9   10

Library of Congress Cataloging in Publication Data

Christianson, Gale E.
      In the presence of the Creator.

      Bibliography: p.
      Includes index.
      1. Newton, Isaac, Sir, 1642–1727.   2. Physicists—
Great Britain—Biography. 3. Mathematicians—Great
Britain—Biography. I. Title.
QC16.N7C49   1984   509′.24 [B]   83–49211
ISBN 0–02–905190–8

To Pelly,
who alone knows the cost

Contents

Illustrations

*[H]e gives one the notion of
belonging to a bigger race than
ours with a bigger future opening
before it.*

Roger Fry

Preface

I cannot say with certainty when I first heard the name Isaac Newton; no doubt it was as a child attending elementary school. I have vague recollections of a drawing in my science textbook depicting a youth in knee breeches seated under an old apple tree, his eyes uplifted in mock ecstacy, ripened fruit scattered at his feet. But I can also recall a then more compelling drawing of the legendary Swiss patriot William Tell, who, for refusing to remove his hat in deference to one placed on public display by the black-hearted bailiff, Gessler, was forced to shoot an apple off his small son's head. In those days of innocence the riddle of universal gravitation proved no match in my mind for the derring-do of the man of action. Many years were to pass before I came to the realization that there are other, more subtle kinds of heroism capable of rendering old worlds meaningless while erecting new ones upon the ruins; that an apple can serve not only as a metaphor for good and evil but as the symbol of an elegant scientific law binding every particle of matter in the universe to every other particle, thus substituting mathematical order for seeming disunity.

Creativity and genius—the primary forces which continually revitalize and enrich the human experience—have defied the persistent attempts of scholars to track them to their mysterious source. They remain, in the words of philosopher–educator John Dewey, "set in the invisible." Nor, it must be made clear at the outset, have I been able to illuminate "the invisible" in a manner wholly satisfactory to myself. At best, I have shed some light upon the preeminent thinker of modern science—perhaps of the age itself—and upon the shadows at the margin where cultural development and the truly extraordinary mind inevitably meet.

Historians have tended increasingly to interpret Newton and his intellectual achievements not in seventeenth-century terms but in the light of our own times. In doing so we have been made ever more conscious of his limitations and ever less appreciative of the revolutionary nature of his many accomplishments. Moreover, the twentieth century has made out of Newton something that he was not—an Enlightenment figure whose dedication to the principle of a mechanical universe became his reason for being and his single most important legacy to posterity. That Newton did adhere to a philosophy of mechanistic causation in the physical world is undeniable; but to argue, as did Voltaire, that this is the whole Newton, or even the essential Newton, is erroneous. Isaac Newton held tight the conviction that science (or natural philosophy, as it was known in his day) must be employed to demonstrate the continuing presence of the Creator in the world of nature. Indeed, he believed that the earth is gradually running down. Unless God chooses to restore its motion at some point in the forseeable future, its fate, and that of every living thing, is sealed. And, since Newton felt that God's intervention is by no means certain, he can hardly be counted among the eternal optimists weaned on the very scientific discoveries associated with his name.

It is frequently the tragedy of the true genius that he strikes fear into the hearts of more ordinary mortals. While this is true in all fields of creativity, it seems especially so in the case of the great scientists. The intellectual waters in which Newton lived are admittedly deep. My intent, however, is not to plunge the reader into depths known only to scholars, but to lead him on a voyage through the universal as well as the particular aspects of scientific discovery and creativity.

In this account of Newton's life and times his own voice has been invoked whenever possible: It would be a poor biographer indeed who did not make use of his brilliant observations, keen insights, and striking epithets. I have also been candid and unreserved when dealing with the more difficult parts of Newton's life, but never, I hope, for the sake of sensationalism. In sum, my purpose has been to bring Isaac Newton to life rather than to embalm him, and one can neither know nor sympathize with the man unless one shares his frailties as well as his strengths, his failures as well as his triumphs.

A Note to the Reader

Only a very few changes have been made in the grammar, spelling, and punctuation of quotations taken from original sources. Operating on the theory that the evolution of language is an important part of historical development, I have permitted the figures quoted to speak for themselves just as they spoke some three centuries ago.

Chapter One

Inside a Quart Pot

A star danced, and under that I was born.

Shakespeare

I

The years 1641 and 1642 were not kind to the Newtons of Woolsthorpe, a tiny hamlet situated on the gentle River Witham in the county of Lincoln. In the nearby village of Colsterworth, where the family regularly attended church, the parish register contains two brief but sobering entries which stand as lasting testimony to the tragedy and sorrow visited upon the Newton household:

> 1641: Robert Newton buried Sept. 20.
> 1642: Isaac Newton buried Oct. 6.[1]

Robert Newton, a yeoman or independent farmer, was born about 1570, and as the eldest son of Richard and Isabel Newton succeeded to the recently acquired family property at Woolsthorpe. Apparently a thrifty and industrious sort, Robert added to his patrimony by purchasing Woolsthorpe Manor, where Isaac Newton, his illustrious grandson, would be born some fifteen months after his death. The name of Robert's wife, grandmother of Isaac Newton, is not known. No record of her burial has been found, although it appears that she preceded her husband to the grave. In terms of offspring, theirs was a fertile union, producing eleven children—seven daughters and four sons—six of whom survived infancy.[2]

Of Isaac Newton, the second son of Robert and father of the soon-to-be-born genius, precious little is known. The Colsterworth parish register records the date of his baptism as September 21, 1606, which means he had

Isaac Jr.

1

just turned thirty-six when death struck. As the oldest surviving son, he had been given Woolsthorpe Manor some three years earlier under a settlement in 1639, perhaps in anticipation of his marriage.[3] In April 1642, only five months prior to his death, Isaac wed Hannah Ayscough, daughter of James Ayscough, a gentleman from nearby Market Overton in county Rutland.

Nothing of the character or temperament of Isaac the father has been recorded, and exactly how much his posthumous son knew of him by questioning the widowed Hannah in later years is anyone's guess. Newton, always reluctant to discuss his personal life, would on occasion speak of his mother with a relative or long-time acquaintance, but there is nothing in the writings of those who kept a record of these conversations about the father, whom he never saw. According to a frequently repeated story, the elder Newton was a "wild, extravagent, and weak man," but closer scrutiny indicates this harsh judgment to have been misplaced, for it was made regarding the conduct of a relative and falsely applied to Isaac's father.[4] Fortunately, the will and property inventory of the senior Newton have survived, and they provide some interesting insights into the family circumstances, both personal and financial.[5]

Isaac's death did not come without warning. His will begins: "Inn the name of God amen the first day of October (anno Dom 1642) I Izacke Newton of Woolsthorpe in the parish of Coulsterworth in the country of Lincoln yeaman sicke of body but of good and perfect memorie...." How long he lingered and of what he succumbed cannot be said, but he was dead within a matter of days after putting his affairs in order. More revealingly, and in glaring contrast to his yet unborn child, he, like all the clan, remained illiterate to the point of not even being able to sign his name: Isaac's mark was the traditional X.

There were a few bequests to relatives and friends of between five and seven pounds. Forty shillings were also bequeathed "unto the poore of the parrish" and ten shillings to repair the bridge between Colsterworth and Woolsthorpe. The remainder of the goods, "moveable and unmoveable I give unto my loveinge wife." These consisted, in addition to the land and manor house, of some £460 worth of livestock, grain, timber, and household furnishings. No mention is made of the unborn child. Hannah and her relatives could be trusted to provide for it adequately—that is, if it lived. Infant mortality being what it was, the possibility loomed large that the baby would be carried to the churchyard at Colsterworth, its name recorded as the third Newton to be buried there in less than two years. Death in the seventeenth century cast a very long shadow.

II

The date of Hannah's birth is unknown. According to the Bishop's transcripts taken from the parish register of the nearby village of Stroxton, her parents, "James Aiscough and Margery Blyth," were wed on December 24, 1609.[6] The Ayscough marriage produced at least two other children, William and James. Both brothers became clergymen, and both took an interest in the rearing and education of their sister's son, Isaac. Since it would appear that the boys were born before Hannah, her age at the time of Isaac's birth was probably about twenty-five.

In contrast to the Newtons, who had been yeomen for several generations, the Ayscoughs, at least until recently, had been a family of greater wealth and property. Among their ranks were gentlemen farmers, members of the clergy, architects, and lawyers. The fortunes of Hannah's branch of the family had undergone a gradual reversal of late, however, as is evidenced by her marriage to a yeoman rather than a man, like her father, of gentle birth. It seems that the Ayscoughs of Market Overton were related to the Askews (or Ayscoughs) of Harlaxton, with whom the Newtons were also connected. In any case, the Newtons of Woolsthorpe were gradually improving their economic and social position while that of the Ayscoughs was in slow but perceptible decline.

In addition to the livestock, crops, and other material goods left Hannah by her husband, there was the "manor" itself. The rather poor land consisted of more than 100 acres of fields, pasture, and woods. The manor house, which still stands for the visitor to see, is a modest two-story structure of gray stone, facing west and overlooking the garden, which contains a reputed descendant of the famous apple tree. The rooms are fairly large, with low ceilings and stone floors. Even on a sunny day relatively little natural light reaches the interior, creating a melancholy atmosphere. The setting was even more somber during Newton's childhood, when the feeble glow of a candle barely held its own against the encroaching shadows of day's end.

At the top of the stairs and to the left is the room where Isaac Newton was probably conceived and born. Hannah apparently went into labor some time on the twenty-fourth of December 1642. Who attended her and what special difficulties, if any, she may have experienced are not known. It was the time of the full moon, however, and an hour or two after midnight the child was born. It was Christmas morning.*

What the average size may have been of infants born during the seventeenth century is impossible to say; but, considering what is known of the general state of health and diet during the period, it seems reasonable to conclude that they were considerably smaller than today. This precarious circumstance took on added meaning with the birth of Hannah's son, for the evidence suggests that the child arrived several weeks prematurely. We

---

*On the Continent, where the Gregorian calendar was in use, the date of birth was ten days later: January 4, 1643. Besides this ten-day difference in dating between England, which employed the Old Style (O.S.), and the Continent, which employed the New Style (N.S.), the calendar year in England started on March 25 rather than on January 1. With few exceptions, which are duly noted, I have employed the modern style in order to avoid confusion.

can imagine the women attending the birth, upon seeing the tiny baby for the first time, exchanging knowing glances solemnly underscored by silent resignation. So little hope was entertained for Isaac's survival that two women who had been sent to the home of Lady Pakenham, a member of the local gentry, to obtain some medicine for the baby "sate down on a stile by the way and said there was no occasion for making haste, for they were sure the child would be dead before they could get back." Many years later Newton told John Conduitt, husband of his half-niece Catherine Barton, that "when he was born he was so little," according to Hannah, "they could put him into a quart pot." Furthermore, because he was too weak to hold his head upright for proper feeding and breathing, he had to wear a bolster around his neck to prevent serious self-injury. And as sometimes occurs in such cases, he remained "very much below the usual size of children" his own age.[7] Yet, despite these life-threatening adversities, the tiny infant was endowed with a constitution to match his intellectual capacity. His survival proved to be the first of many instances were Isaac Newton would confound the predictions of those who underestimated his extraordinary powers.

Despite the risks associated with exposing mother and child to the winter elements, tradition holds that a week after the infant's birth Hannah and her tiny baby were taken to the family church at Colsterworth. If so, it must have been a touching scene to the relatives and friends in attendance; the young widow, still pale and drawn from the trauma of giving birth, bringing her dangerously premature child to receive the sacrament of baptism before the very alter where the obsequies had been delivered over her husband only three months before.

In point of fact, it is quite likely that the child, owing to his delicate condition, was baptized by the Rector of Colsterworth in the room where he was born. In either case, the Rector dutifully recorded the event in the parish register:

Isaac sonne of Isaac and Hanna Newton Baptized Jan. 1.

So for the third time in less than two years the name of a Newton of Woolsthorpe appeared, this time a record not of death but of a precious new life. All three entries are to be found on the very last page of the parish register, as if predestined to stand out in relief.

III

It was natural that Hannah Newton should become the central figure in the life of this frail, fatherless only child. Virtually all that is known about her derives from a meager character-portrait drawn decades later by John Conduitt from conversations he had with a Mrs. Hutton (nee Ayscough), an

elderly family relative: "She was a woman of so extraordinary an understanding & virtue that those who think that a soul like Sr Isaac Newton's
could be formed by any thing less than the immediate operation of a divine
Creator might be apt to ascribe to her many of those extraordinary qualities
with wch it was endowed."[8] Needless to say, such exaggeration offers little
insight into the true relationship between mother and son.

Among the voluminous papers left by Newton at his death is a tattered,
barely literate but illuminating note written to him by Hannah while he
studied at Cambridge. Dated "woolstrup may the 6. 1665" it reads:

> Isack
> received your letter and I perceive you letter from mee with your cloth
> but none to you your sisters[*] present thai love to you with my motherly lov
> you and prayers to god for you I your loving mother
>
>   Hannah[9]

One cannot help but be impressed by Hannah's generous use of the words
"love" and "loving."

Conduitt also saw evidence of their close relationship in Newton's conduct once he had left home. "He made frequent journeys from Cambridge
to visit [Hannah] and even at that time when he was in the warmest pursuit
of those enchanting discoveries wch made him forget his food & his rest &
seemed to transport his imagination above all sublunary things, broke loose
to pay his duty to her."[10] While it is true that Newton spent considerable
time at Woolsthorpe during the plague years of 1665 and 1666 and that he
ministered to Hannah during her terminal illness in 1679, his other visits
can hardly be characterized as frequent or prolonged, a possible indication
of a later strain in their relations over Isaac's choice of careers. Nevertheless, when Hannah died Isaac became heir to the bulk of her considerable estate.[11]

The widow Newton, most probably not yet thirty, had caught the eye of
one Barnabas Smith, Rector of North Witham, a tiny village little more than
a mile southeast of Woolsthorpe. The differences between the Reverend
Smith and Isaac's father could not have been more pronounced. Smith was
about sixty-four when he made his proposal of marriage to Hannah, a ripe
old age by seventeenth-century standards. A well-educated man of much
property and capital, Smith matriculated at Lincoln College, Oxford, in
1597, where he took a B.A. and an M.A. He became Rector of North
Witham, a sinecure purchased for him by his clergyman father, in January
1610, only a month after Hannah's parents were married.[12] According to
Conduitt, Smith was a bachelor, but the parish register of North Witham
indicates otherwise. At the very bottom of the badly effaced page listing
deaths for the year 1645 is the following entry, the first word of which is il-

---

*Mary (Marie) and Hannah Smith, Isaac's stepsisters by Hannah Newton's second mar-
riage.

legible: "[?] Smith, wife of Mr Barnabas Smith, North Witham."[13] Apparently this union, of whatever duration, was childless, for Smith's will names only the offspring born of his marriage to Hannah Newton.

If Conduitt's account of the premarital negotiations, which he also obtained from Mrs. Hutton, is to be believed, the clergyman was either too retiring or too fearful of being rejected to ask in person for Hannah's hand. (Perhaps he was not a little embarrassed as well, since his first wife had died only a few months earlier.) Instead he dispatched a willing neighbor—who we are told was paid a day's wages for his trouble—to make the proposal. "Tho she had no reason to distrust her own strength and prudence she would not treat with him personally upon a concern in wch passion generally bears so great a sway."[14] Hence the matchmaker, happy no doubt to garner a reported second day's wages for such easy "work," dutifully carried Smith's proposal to Hannah's brother William Ayscough, Rector of Burton Coggles.

Though Smith's age was almost certainly a negative factor, it was far outweighed by his education, social standing, and, most important of all to the Ayscoughs, his money and land.[15] However, before accepting the proposal of marriage Hannah, no doubt acting on the advice of her brother, insisted that Smith "settle some of the property upon her son."[16] The elderly suitor complied, and a piece of land near Sewstern worth about £50 per annum became Isaac's when he attained his majority in 1663. Smith seems to have agreed to another request as well, namely, that he refurbish the manor house at Woolsthorpe, which apparently suffered from considerable disrepair.[17] These matters settled, the marriage took place on January 27, 1646, a month after Isaac's third birthday.

All might have gone well had not the single major object of the child's affection literally been wrenched from his naturally possessive grasp, for the resulting trauma played a significant if not decisive role in the development of his sensitive temperament and always enigmatic character. Hannah moved to the North Witham home of her new husband, leaving the bewildered Isaac at Woolsthorpe in the care of his maternal grandmother, Margery Ayscough. Considering the fact that Smith agreed to finance repairs to the manor house, it appears that this decision was also an integral part of the marriage agreement. Perhaps the child was an unpleasant reminder to the Rector of Hannah's former union and a possible source of disruption in her new one. On the other hand, this arrangement has all the markings of a carefully conceived Ayscough plan to restore some of the luster lost during their recent period of financial reversal. Isaac was to remain on the estate to establish his position as future "lord of the manor" and to protect his eventual inheritance against the claim of any Newton with a covetous eye. In time, Hannah would have access to the aging Smith's money, with which she could ensure her son's elevation to the gentry. Though he could not now understand, the day would come when Isaac would thank his mother for her judicious protection of his financial interests—or so Hannah must have thought.

Margery Ayscough had married in 1609 and so was probably in her late fifties or early sixties when she came to live with Isaac. There is no evidence that James Ayscough, Isaac's grandfather, also resided at Woolsthorpe manor during this period.[18] Unless one chooses to explain the Ayscoughs' long separation as the seventeenth-century alternative to divorce, which at their relatively advanced age seems highly questionable, the most plausible interpretation must be the economic one. Otherwise Isaac could have lived with the Smiths or, failing that, gone to the home of his grandparents at Market Overton.

Of the relationship between grandmother and child, which lasted for nearly eight years, nothing is known. Newton's silence on the matter, as in the case of his dead father, was absolute. Perhaps the boy respected and, in his own way, cared for his mother-substitute, but he seems never to have felt any deep personal affection for her. When Isaac grew a little older, he learned just how agonizingly close his mother remained; by climbing a tree or wiggling his way to the top of a haystack, he found the steeple of North Witham's church clearly visible against the southern skyline. There was Hannah, but there too was the mysterious and feared interloper who had stolen her away.

The loss, at such an early age, of Hannah to another man was a deeply traumatic experience, which ate at Newton over the years like an emotional cancer. In 1662, at the age of nineteen, he underwent a period of intense religious awareness, during which he compiled a list (in shorthand) of fifty-eight "sins," most of them innocuous, which he hoped to expiate through the act of self-confession. Number thirteen from the catalog is particularly revealing: "Threatening my father and mother Smith to burne them and the house over them."[19] If indeed he did utter such a threat, it was a transgression of very long standing, because by this time Barnabas Smith had been dead for nine years, and Hannah had long since returned a wealthy widow to the family home at Woolsthorpe.

The next entry on the list also invokes the ultimate sanction: "Wishing death and hoping it to some." Whether this brief statement is simply a reassertion of the previous threat combined with a death wish of his own is not clear; but for Newton to have harbored for so long a time these bitter thoughts repugnant to his Puritan conscience says much about his childhood sense of betrayal. In agrarian societies fire traditionally has been the method whereby the weak settle their grievances against the strong. The thought of burning their house over the heads of his supposed tormentors was a child's way of seeking release from the emotional conflagration that raged within. This proclivity for bearing a grudge and waiting years, if need be, to gain revenge on his enemies, both real and imagined, became an indelible character trait of the adult Newton.

Perhaps, in addition to the absentee parents, the unspecified "some" for whom Isaac wished death were the half-brother and half-sisters fathered by Smith in his late sixties. By Hannah he had three children: Mary (Marie), baptized May 6, 1647; Benjamin, baptized in August 1651; and Hannah, baptized September 7, 1652. From Isaac's point of view, each must have been additional rivals for his mother's love and attention. However, by the time of Smith's death in 1653, when Hannah returned to Woolsthorpe, bringing with her the children of her second marriage, other less emotional concerns were beginning to occupy her first-born's attention. Isaac had already discovered that he could insulate himself from human contact by withdrawing into the caverns of the mind, becoming lost for hours among the inner labyrinths unknown to and untouchable by the outside world.

IV

Careful examination by historians of Newton's ancestry has provided no significant clues to the source of his genius. The Newtons themselves hardly seem to have been distinguished by an special intellectual gifts, while the branch of the Ayscough family from which his mother came, though better educated and more prominent socially, produced no men or women of clearly extraordinary merit. If the late historian of science Alexandre Koyré is taken at his word, any attempt to comprehend Isaac Newton—the man and his work—through a careful examination of English society in his time is as doomed as the genetic approach: "The social structure of England in the 17th century cannot explain Newton."[20] But Koyré also knew full well that the historian is not interested in single causal explanations. Thus while it is quite true that no complete understanding of Newton can be derived from an examination of the powerful social and political forces of his day, such factors are not without relevance.

The year of Isaac's birth saw the outbreak of military conflict of the worst kind in England—civil war—the wrenching, bloody struggle that turns cherished neighbors into implacable enemies and tears close-knit families to shreds. A number of early battles between Royalists and Roundheads were fought within less than a day's walk of the Newton home. On May 13, 1643, when Isaac was an infant of less than six months, Cromwell's troops, greatly outnumbered, won an important if little-known victory over the Royalists near Grantham, a few miles to the north of Woolsthorpe. The Cavaliers lost more than one hundred soldiers, while only two of the future Lord Protector's ill-equipped men were killed. Cromwell, as many another Puritan was prone to do, took this as a sign of God's benevolent intervention. "With this handful it pleased God to cast the scale," he later wrote, thus providing the underlying rationale for all of his subsequent actions.

During the period between Cromwell's first victory and the establishment of the Commonwealth, much of Lincolnshire fell prey to the excesses of rival armies as they sought to maintain an adequate supply of foodstuffs and other provisions. Local property-holders stood by in fear and frustration while their farms were stripped of everything that could be carried away by men described in a bitter petition to the government as having "lost the naturall softness of Englishmen and Christians" and "degenerated into almost a Turkish inhumanity." Countless buildings were put to the torch, crops went unplanted, and additional bolts were affixed to the heavy oak doors of the homes occupied by the once prosperous yeomenry. One can imagine that Margery Ayscough, mindful of the constant danger, frequently cautioned her grandson to stay within the close unless accompanied by her or one of the servants. What political position, if any, the Newtons and Asycoughs took during the Civil War is unknown, nor is there any information to indicate in what manner the conflict may have shaped young Isaac's political and religious outlook. The child had just turned six when the startling news arrived of Charles I's execution in front of the royal palace at Whitehall in January 1649. At this early age Isaac could have had little knowledge of political affairs, but the memory of such great events and their subsequent influence in shaping English history became prominent in his thinking later on.

Because of his background, education, and wealth, Barnabas Smith was likely to have identified with the Royalist cause rather than with the common and often crude soldiers responsible for Cromwell's early victories. The same can be said of Newton's uncle William Ayscough, Rector of Burton Coggles. But whatever the nature of their private thoughts in these matters, both men adopted a public position that kept them in favor with the Parliamentary ecclesiastical authorities. Once the anti-Royalist forces triumphed and established the Commonwealth, a number of suspect divines in the region lost their positions, but Smith and Ayscough were not among them. The score was evened during the Restoration, however, for many alleged Dissenters were in turn purged from the ranks of the practicing clergy. By this time Barnabas Smith was dead, but Ayscough emerged unscathed. They, like the father of the great lexicographer Samuel Johnson, had apparently reconciled themselves "by casuistical arguments of expediency and necessity, to take oaths imposed by the prevailing power." This ability to compromise in religious matters seems to have made a lasting impression upon Isaac, for though he remained a member of the Church of England all his life, as was required by the influential positions he held, in private he disagreed with certain of its most fundamental doctrines.

As with so many facets of Newton's early life, the exact nature of his boyhood religious training is unknown. On special occasions, when he was allowed to visit Hannah at her North Witham home, he would have listened to sermons delivered by the stepfather he so despised. But on most Sundays he rode in the company of his grandmother to Colsterworth to attend services in the fourteenth-century church where he had been baptized, and on whose north side his father and grandfather were buried.[21] One of the few clues pointing to a possible sectarian religious influence in Newton's youth comes from the period of his adolesence, when he attended grammar school at Grantham. John Angell, a noted Puritan divine and lecturer at the school, was the community's leading religious figure until his death in 1655. Angell deeply influenced other Puritans associated with the church and the adjoining school, two of whom were removed for nonconformity during the Restoration. Furthermore Isaac probably had access to the Grantham library, housed above the church porch, which was donated by Francis Trigge, another eminent Puritan. Not surprisingly, the collection contained many volumes that reflected the theology of its donor.[22]

Though no airtight case for a Puritan upbringing can be woven from such admittedly tenuous thread, the fact remains that by the time Newton matriculated at Cambridge, in 1661, his conduct, temperament, and outlook on life all bore a marked resemblance to Puritanism in its broadest religious and social sense. There was the characteristic pessimism and the passion for work, the profit motive and the constant desire to get ahead, the suspicion of art, the devotion to family, the hatred of oaths, and the drive to strip formal religion of all but the most fundamental of its rites and doctrines. His was a creed that taught that one must be wary of all men, struggle against their evil ways, and be even harder on oneself than on others. As for the stirrings of the heart, they are not to be trusted; any show of emotion must be reserved for members of the family alone, the flesh of one's flesh. Ceaseless labor is the only proper way to beautify the earth, which has been entrusted by God to His children, so that they might improve it by the sweat of their brows and the unremitting application of their minds. Success in itself is not enough, however, because its wages are often the twin evils of pride and vanity.

In addition to the other radical Protestant traits identifiable in Newton's behavior, he, like Cromwell, was seemingly possessed of that regal Puritan notion of special election. In Newton's case, however, the evidence is admittedly circumstantial, whereas the messianic Lord Protector openly and without embarrassment represented himself as God's chosen Englishman. Dr. William Stukeley, a friend of Newton in the natural philosopher's later years and the man responsible for collecting virtually all of the surviving information about his childhood activities and education, reported that a Christmas birth was widely thought of as an omen of future success. John Conduitt also noted the special circumstances surrounding Newton's birth by alluding to the equally popular notion that posthumous children were thought to be endowed with extraordinary powers which destined them to success and good fortune.[23] These folk beliefs would hardly have gone unnoticed by so sensitive and curious a boy as Isaac, especially when coupled with Hannah's account of his seemingly miraculous survival against over-whelming odds.

Such considerations aside, very early on Newton became aware of significant differences between himself and other children. As he grew to intellectual maturity, those differences widened until he became largely isolated from the outside world. Unlike Francis Bacon, who envisioned the new science as the most promising method of improving humanity's lot in a harsh and often brutal natural order, Newton viewed the pursuit of mathematics, physics, astronomy, and alchemy—not to mention history and Biblical exegesis—as the proper means by which the truly devout Christian might reveal the meaning of God's kingdom, thereby proving himself worthy of acceptance into it.

V

As heir to the Woolsthorpe estate and recipient of the land at Sewstern ob-tained for him by Hannah before her second marriage, Isaac seemed des-tined from the first to become a yeoman, as his father and grandfather had been. With a little luck he might even join the ranks of the gentry. From what we know of the Ayscoughs, it would have been out of character had they not insisted that Isaac receive some degree of formal education before settling into a reasonably comfortable agrarian existence. Even the Reverend Smith, who took little personal interest in his stepson, un-doubtedly favored at least a limited amount of schooling for him. Illiteracy was no virtue in a Puritan society, especially for a young man whose for-tunes were generally improving—tiller of the soil or no.

Isaac's introduction to education came by attending two little dame schools (named for the mistresses who ran them) in Skillington and Stoke Rochford, villages near enough to Woolsthorpe for him to walk back and forth each day. Who his teachers were and what he learned from them we do not know, nor is there any record of how long this experience lasted or how well he took to his early studies. His powers of observation being what they were, the boy could have easily acquired as much knowledge by trudg-ing the dusty country roads and cutting across the green fields on their borders as in the classroom proper. Self-education was Newton's strong suit.

Isaac was eleven when his twice-widowed mother again took up residence in the home she had left some eight years earlier. While he may have welcomed Hannah's return, he now found himself cast in the un-familiar role of an older brother to children whose father he detested and whose infancy and early development he had not witnessed. The only solid clue to his feelings on the matter is a brief entry, born of exasperation, in his

Latin exercise book: "I have my brother to entreate."[24] Whatever else Isaac thought of the new arrivals was kept secreted within his heart.

A year later, at the age of twelve, he was enrolled in King's School at Grantham, about seven miles to the north of Woolsthorpe. The stone building in which he attended classes dates from the 1520s or 1530s and opened directly on to the broad churchyard, which served as the student's playground. Next to the school were the house and offices of the master, who in Newton's day was Henry Stokes. The patents given by Edward VI to the school expressly provided for the instruction of the students in Latin and Greek, and it was here that the foundations of Newton's primarily classical education were laid. Bible studies also formed an integral part of the curriculum, and Isaac became familiar with Hebrew script, as is demonstrated by a surviving notebook dated 1659.[25] While most of his formal education at Grantham fell into the grammatical and literary categories, he also received a limited amount of instruction in arithmetic. Whether or not his keen interest in geometry was awakened by an introduction to the subject in the Grantham classroom remains a matter of some disagreement among scholars, though most are inclined to believe Newton did not study it before entering Cambridge.[26]

Of his schoolmaster, Henry Stokes, little is known. The son of a blacksmith, he was born at Melton Mowbray in October, 1619. On December 15, 1638 Stokes was admitted to Pembroke Hall (College) in Cambridge. He received his B.A. in 1643 and later became master of his old school at Melton. In February 1650 he was appointed master of Grantham school where he remained until late 1663, two years after Newton had left for Trinity College. Thus the proud, teary-eyed "old man" whom Conduitt describes as having made of Newton an example to be emulated by his fellow classmates was in truth only forty-one when his greatest pupil entered Cambridge University. Stokes, a potentially rich source of information concerning Newton's adolescence, died at the age of fifty-three, a decade before his pupil's reputation as a natural philosopher was solidly established.[27]

It was under Henry Stokes' perceptive eye that Newton's intellectual interests were gradually awakened, as is evidenced by a growing fondness for books and the construction of numerous ingenious mechanical toys and models. However, Stokes was not the only positive influence on Newton during his Grantham school days. Since Grantham was too great a distance to travel every day from Woolsthorpe, arrangements were made to board Isaac at the house of Mr. Clark, the town apothecary. Clark's second wife was a close friend of Isaac's mother, and his brother Joseph, a physician, held the position of usher (assistant teacher) at King's School. Clark's grandson, Ralf, also became an apothecary, and in 1727 the latter supplied Stukeley with some important information (via the family grapevine) regarding Newton's life in the Clark household.

The Clarks lived on High Street next to the George Inn, whose expansion in 1711 resulted in the complete alteration of the garret quarters inhabited by Newton in his youth. Mr. and Mrs. Clark, affable landlords, both believed that children should be given as much freedom in their daily activities as good judgment would allow. The stimulating yet relaxed environment proved ideal for a child of Isaac's reserved and contemplative demeanor, and he soon began to undertake various projects that prefigured the intellectual achievements of his later life. According to Stukeley:

> Every one that knew Sr Isaac, or have heard of him, recount the pregnancy of his parts when a boy, his strange inventions and extraordinary inclination for mechanics. That instead of playing among the other boys, when from school, he always busyed himself in making knicknacks and models of wood in many kinds: for which purpose he had got little saws, hatchets, hammers and a whole shop of tools, which he would use with great dexterity.[28]

Among the things that most fascinated him were the forces generated by moving air and rushing water. Isaac frequently visited the construction site of a new windmill on the road to Gunnerby near Grantham to observe the operations as they were carried out by the workmen. Imitating their efforts on a smaller scale, he built a working model of his own, which he would carry to the top of the Clark house so the wind could turn its handmade sails. "But what was most extraordinary in its composition," according to Stukeley, "was that he put a mouse into it which he called the miller, and that the mouse made the mill turn around when he pleased, and he would joke too upon the miller eating the corn that was put in." Accounts vary as to just how the diminutive captive propelled the mill: "Some say he tyd a string to the mouses tail, which was put into a wheel like that of turnspit dogs, so that pulling the string made the mouse go forward by way of resistance, and this turned the mill. Others suppose there was some corn placed above the wheel, this the mouse endeavoring to get to, made it turn."[29]

Isaac's thoughts soared far higher than windmills, however. Disdainful of what Stukeley calls the "trifling sports" of his schoolfellows, Newton sought to "teach them . . . to play philosophically" so that he—smaller in stature and physically weaker—"might willingly bear a part; and he was particularly ingenious at inventing diversions for them above the vulgar kind. As for instance in making paper kites, which he first introduced here." Driven by an extraordinary curiosity to know not only how but more importantly why things function as they do, Isaac soon transformed such a simple thing as kiting into a minor theater for continuous experimentation. The boy made kites of many shapes and sizes, varying the positions of their strings in an effort to discover the design most suitable for sustained flight. About the same time he learned to fashion lanterns from crumpled paper into which candles were placed to light the way to school in the dark mornings of deepest winter. These he sometimes tied to the tails of kites at night "which at first affrighted the country people accordingly, thinking they were comets."[30]

Newton's childhood foray into the field now known as aerodynamics did not end with his experiments with kites and windmills. When Oliver Cromwell died in September 1658, a great storm swept over England, causing widespread property damage and giving rise among the superstitious to the belief that it was Satan riding the whirlwind to claim the Lord Protector's lost soul. Isaac, so the story goes, took part on that day in a rare competition with his more athletic schoolfellows to see who could jump the greatest distance. By carefully timing the gusts of wind, he "took so proper an advantage of them as surprisingly to outleap the rest of the boys," wrote Stukeley. Many years later Newton told Conduitt that this was one of his first experiments.[31] We might legitimately ask exactly what Newton meant when he used the word "experiment" in this instance. Was he simply referring to a test of the extra distance he could jump by cleverly timing a powerful natural force? Or did he see in his feat a more revealing connection with many later experiments, whereby he demonstrated his superior grasp of the solution to a problem at the expense of all who dared challenge him, and demonstrated as well a compulsive drive to be in absolute control of every situation? Failing that, he would withdraw from the field, leaving the game to others.

Newton's early preoccupation with space and the movement of objects through it was matched by an equally precocious interest in time. "They tell us how diligent he was in observing the motion of the sun," wrote Stukeley, "especially in the yard of the house where he lived." The youth carefully tracked the shadows cast upon the walls and roof of the building and then drove in pegs "to mark the hours and half hours ... which by degrees from some years observations, he made very exact and anybody knew what o'clock it was by Isaacs dyal, as they ordinarily called it."[32]

It would seem that the sundial of which Stukeley wrote was on the side of the Clark house at Grantham, but even before entering school there Isaac designed and carved a number of smaller sundials, at least two of which were to be found on the south wall of his Woolsthorpe home. This acute sense of the interplay between light and shadow proved of great benefit to Newton when be began his optical experiments in the darkened rooms of his student quarters at Cambridge. Nor did his unusual powers of observation fade over the years, if John Conduitt's testimony is to be believed: "To the time of his death he retained this custom of making constant observations in the rooms he chiefly used where the shade of the sun fell; and I have often known him ... when anyone asked what o'clock it was, [to] tell immediately by looking where the shadow of the sun touched as exactly as he could have by his watch."[33]

Of the many toys and mechanical models built by Newton while at Grantham, the most frequently mentioned is his clepsydra or water clock, made out of a box he begged from his landlord's brother-in-law. Not having seen it himself, Stukeley described as best he could how it looked and the principle by which it worked:

> It resembled pretty much our common clocks and clockcases, but less: for it was not above four feet in height and of a proportionable breadth. There was a dyalplate at the top with figures of the hours. The index was turned by a piece of wood, which either fell or rose by water dropping. This stood in the room where he lay, and he took care every morning to supply it with its proper quantity of water. And the family upon occasion would go to see what was the hour by it, and it was left in the house long after he went away to the University.[34]

Writing a description of the clock brought to Stukeley's mind a discussion he had once witnessed between Newton and the astronomer Edmond Halley concerning homemade timepieces. "S[r] Isaac talk'd of those kind of instruments. That he observ'd the chief inconvenience in them was that the hole through which the water is transmitted being necessarily small, was subject to be furr'd up by impurity in the water, as those made with sand will wear bigger; which at length causes an inquality in time."[35] Newton had obviously learned much from his youthful experience as a clockmaker.

Like such other gifted men as Leonardo da Vinci, Benjamin Franklin, and Robert Hooke, Newton was not simply an aimless childhood tinkerer but a tinkerer playing with ideas and with mechanisms. Not only was he possessed of an unusual mechanical aptitude and the physical dexterity with which at a later time to grind and polish his own lenses, construct unlimited varieties of experimental apparatus, and fashion the alchemical furnaces that at certain times of the year burned round the clock, he also manifested innate geometrical powers coupled with a considerable, if not extraordinary, degree of artistic deftness. He was tutored in writing and in drawing by one "Old Barley," who, so far as Stukeley could learn, was not remembered as having had any artistic ability himself. Isaac covered the walls of his garret with sketches made from prints as well as from life. "They mention particularly several of the kings heads, D[r] [John] Donne and likewise his master Stokes." Below a picture of King Charles I were the following lines of verse quoted to Stukeley from memory by an aged Mrs. Vincent, one of Isaac's few close childhood companions.

> *A secret art my soul requires to try,*
> *If prayers can give me, what the wars deny.*
> *Three crowns distinguish'd here in order do*
> *Present their objects to my knowing view.*
> *Earth's crown, thus at my feet, I can disdain,*
> *Which heavy is, and, at the best, but vain.*
> *But now a crown of thorns I gladly greet,*
> *Sharp is this crown, but not so sharp as sweet.*
> *The crown of glory that I yonder see*
> *Is full of bliss and of eternity.*[36]

Thought by the old woman and by earlier biographers to have been composed by Newton himself, the lines (and several of the drawings as well) were actually copied by him from the widely read *Eikon Basilike: The Portraiture of His Sacred Majesty in His Solitudes and Sufferings.* Louis T. More took this poetry as conclusive proof that the boy was brought up in a family devoted to the Church of England and the Royalist cause, a conclusion open to question now that we know the origin of the verse.[37] Newton did tell Conduitt, however, that as a boy he "excelled particularly in making verses." If so, none of his poems have survived the centuries. Moreover, whatever poetic talent he may have had was apparently repressed at an early age, perhaps for leading too far into the unpredictable and therefore suspect realm of imagination. Aesthetic sensitivity—one of the qualities which so enrich the human spirit—was conspicuously absent from virtually all facets of Newton's later life.

Isaac's artistic efforts were not limited to the portraits drawn on sheets of paper and hung on the walls of his attic room. Ralf Clark told Stukeley that the entire wall of the room where he lodged was full of drawings he had made with charcoal. There were birds, beasts, men, ships, and mathematical schemes, all of which have long since perished.[38] Still, some inkling of Newton's childhood artistic endeavors may yet survive. During the 1940s the National Trust undertook the renovation of Woolsthorpe manor house, which involved the unbricking of windows closed two and one-half centuries earlier as a way of evading taxes. When the bricks were removed from a window in the room of Isaac's birth, several geometrical figures were revealed, all carved on plaster and stone. The subsequent removal from the walls of nineteenth-century newsprint led to the discovery of additional drawings in the hallways and dining rooms. The designs occur about four feet above the floor, suggesting the work of a child perhaps eleven or twelve years of age.[39] Though they have little if any scientific significance, these few abstract glyphs lend credence to Stukeley's account of the youth's early love of drawing. Nor was he able to resist the seemingly universal boyhood compulsion to carve his name in a public building. On a window ledge of King's School, with the aid of a penknife, he left a simple, permanent record to be gazed upon by later generations of curious admirers: "I. Newton."

In addition to the pursuit of his numerous boyhood interests, Isaac passed many an intriguing hour in the company of his landlord as the apothecary concocted remedies for the illnesses of Grantham's citizenry. Isaac's natural curiosity and the rare gratification of being in the company of a friendly and understanding adult male were sufficient in themselves to sustain his budding interest in the still imprecise sciences of medicine and chemistry. Yet it is difficult not to suspect that at least one other significant factor was at work. All his adult life Newton tended to be something of a hypochondriac, a quite common condition among creative, high-strung individuals. In his case the neurosis may well have manifested itself at a rather early age, and for reasons that at first had nothing to do with the rigors of subsequent intellectual pursuits. The Freudian hypothesis notwithstanding, it can be clearly demonstrated that the mental anguish suffered by an isolated and seemingly unwanted child (no less than an adult) is rather easily transformed into physical complaints and bodily ailments. This would have been all the easier in Newton's case, for his strong sense of maternal rejection was reinforced by a keen awareness of his perilously close brush with death as a tiny baby. Add to that his childhood introversion, substantially due to his being smaller than normal boys his own age. Taken together, these factors suggest a personality haunted from early youth by the prospect of an untimely demise, a subject much discussed in Puritan circles.

Though he enjoyed remarkably good physical health throughout most of his eighty-four years, Newton mixed his own medicines to alleviate the various symptoms of his imagined illnesses. At a very early age he became an inveterate collector of all manner of recipes and nostrums. The first of several books, other than the Bible, that seem to have captured his attention as a child was *The Mysteries of Nature and Art* by John Bate. Its third edition was published in 1654, when Isaac was eleven. It is clear from comparisons made between it and entries in the Morgan notebook of 1659 that he copied extensively from the popular book. In fact, many of Newton's childhood "inventions," such as his water clock, kites, paper lanterns, and windmills, were more or less derived from Bate's instructions. Also copied from Bate were recipes for mixing paints and for effecting cures of common ailments.[40] Another book from which Newton collected information on diseases and their cures was Francis Gregory's *Nomenclatura*. Many of the illnesses described by Gregory appear under the heading "Of Diseases" in the Morgan notebook, along with remedies from Bate and still others that seem to have resulted from Newton's experiences with the apothecary Clark. A somewhat later manuscript notebook contains reputed cures from everything from a swollen hand and toothache to falling sickness and fistulas—the latter disorder to be healed by "drinking twice or thrice a day a . . . small portion of mint & wormwood & 300 Millipedes well beaten (when their heads are pulled off) in a mortar . . . & suspended in 4 gallons of Ale in its fermentation."[41] Unedifying (and unpalatable) as this first tentative exploration into the realm of medicine and alchemy may seem, it would soon broaden to encompass Newton's crucial theory of matter with which he hoped to find nothing less than the underlying cause of universal gravitation and to rediscover the lost wisdom of the ancients, the elusive *prisca sapientia*.

VI

The precocity demonstrated by Newton while a boarder in the Clark household contrasted sharply with his performance in the classroom, at least in the beginning. Isaac's meager preparation for grammar school left much to be desired in his master's eyes, for Stokes put him in the lowest form. There he ranked second to last out of eighty-odd students, hardly an auspicious beginning. Nor did he immediately offer any evidence of exceeding Stokes's limited expectations: As Conduitt reported, Newton often told him that he "continued very negligent" in his studies at Grantham School. Scant though our knowledge of his behavior during this period is, it seems safe to say that his indifference toward education was anything but the result of laziness, that character defect so repugnant to the Puritan conscience. Rather, it sprang from a natural diffidence in combination with an intellectual craving for the pursuit of ideas and projects little discussed in the classroom.

An unexpected turning point was reached one morning as Isaac was on his way to school. The boy who ranked immediately above him administered a painful kick to Isaac's stomach, which provoked the normally timid youth to thoughts of revenge. As soon as school was over for the day, he challenged the aggressor to a fight, and they went into the neighboring churchyard to settle their differences. An older youth supposedly happened on the scene and, according to Conduitt, encouraged the rivals by "clapping one on the back" while "winking at the other." Though smaller than his opponent, Isaac "had so much more spirit and resolution that he beat him till he declared he would fight no more." Then, egged on by the older boy, Isaac grabbed his opponent by the ears and dragged him to the church, where he pushed his face against the stone wall. Dissatisfied with a mere physical victory, "he could not rest till he got above him in the school, & tho before he never minded his book ... he from that time began to follow it with great application."[42] Not only did Isaac pass the boy he had bested in the churchyard, he rose to become the first student in the school! Never had revenge been more complete or tasted sweeter; the taunts and insults suffered at the hands of his fellow students were at last repaid in full.

Whether partly apochryphal or not, Conduitt's story has a decidedly prophetic ring of truth about it. All during his life Newton required an external stimulus to arouse his latent powers and to exert himself to complete his work and to make public the fruits of his thought.[43] And, as in the case of his childhood adversary, he was often most easily aroused by those whom he disliked and sought to humiliate rather than by the encouragement of admirers, who wanted him to reveal his discoveries to a waiting scholarly world. Only if some aspect of his work was seriously questioned or an unannounced discovery was innocently claimed by another did he become emotionally involved. These occasional great storms were separated by broad stretches of quiet solitude. Once stirred, however, Newton's temper became a force almost impossible to quell; it fed upon itself, rising in feverish crescendo, then leveling whatever—or whoever—chanced to be in its path. A ferocious aggression lurked beneath a deceptively placid exterior—a strange and awesome combination of fire and ice.

How long it took Newton to establish his intellectual preeminence at school is a matter of conjecture, for no record of his scholarly attainments at Grantham has survived. Nor do we know what thoughts crossed Hannah's mind when she received word of her son's unexpected progress from a watchful and admiring Henry Stokes. But by the time he was preparing to leave school for good, Isaac had convinced the headmaster that his talents were too considerable for squandering in the isolated countryside of parochial Lincolnshire. Stokes must have revealed his feelings to the youth, whetting Isaac's appetite for additional formal education but, at the same time, setting the stage for a conflict of wills.

He was probably in his sixteenth year when his mother decided that the time had come for him to master the operation and management of the Woolsthorpe estate. Neither Stukeley nor Conduitt relates whether the youth overtly protested Hannah's decision or quietly, if superficially, acceded to her wishes, as would have been expected of a dutiful son. In either case, it was a plan destined to fail from the outset. Perhaps as early as his enrollment in dame school at the age of six or seven, Newton began to see himself as different from the simple country folk that were his people; inside, he gradually ceased to share their lives. Had he remained on the family farm he would have been stifled and might have grown more bitter by the year—perhaps to end as the kind of recluse whose bizarre behavior is the stuff of local legend.

Of the many sins to which Isaac confessed in the Fitzwilliam notebook of 1662, only three refer specifically to his mother. In addition to the previously mentioned threat of burning the house over the heads of Hannah and Barnabas Smith, Newton cites his "refusal to go to the close at his mother's command" and "peevishness with my mother."[44] Whether these rather minor transgressions were a product of the period after he had been called home is impossible to say, but they are suggestive of the generally contentious, if not openly rebellious, conduct attributed to Newton at this time. Stukeley comments:

> When at home if his mother ordered him into the fields to look after the sheep, the corn, or upon any rural employment, it went on very heavily through his manage.* His chief delight was to sit under a tree, with a book in his hands, or to busy himself with his knife in cutting wood for models of somewhat or other that struck his fancy, or he would go to a running stream, and make little millwheels to put into the water.... The dams, sluices and other hydrostatic experiments were his care without regarding the sheep, corn, or such matters under his charge, or even remembering dinnertime.[45]

Frustrated, uncomprehending, and more than a little annoyed by this turn of events, Hannah tried to remedy the situation by placing her son under the supervision of a trusted elderly servant. Together they went to Grantham on Saturday to purchase the needed staples for the farm and to

---

*Meaning he did not accomplish the tasks assigned to him.

sell the corn and other crops raised in the Woolsthorpe fields. But no sooner did they arrive at their inn, the Saracen's Head in Westgate, then Isaac, perfectly content to leave the mundane affairs of business to his older companion, would head for the Clark home in an attempt to recapture the spirit of the old world in which he found any true contentment. Returning to his old room above the apothecary shop, he passed the day reading books on various scientific subjects stored there by his former landlord's brother. On other occasions he found no need to "waste" time by going all the way into Grantham; he simply found a comfortable spot under a roadside hedge where he read until the servant collected him on the way home. "No doubt," observed Stukeley, "the man made remonstrances of this to his mother."[46] No doubt indeed! Yet the complaints did nothing to alter the lad's disconcerting behavior.

To what degree Isaac's conduct was influenced by a subconscious desire to prove Hannah wrong in her decision to tie him to the soil is moot. It is worth noting, however, that his slipping away to read books in the Clark attic, letting the sheep and cattle go unattended, and missing meals while he experimented with models all went unlisted as sins in his confession of 1662. Perhaps, in spite of frequent parental admonitions, this is because in reflecting back, he saw nothing in his behavior to be ashamed of, even though Hannah and the servants viewed it as damning evidence of a gradual moral breakdown. In any case, Isaac was still too inexperienced in the ways of the world to appreciate his mother's marriage of convenience, a union undertaken in large measure to secure his future as a substantial landowner and potential gentleman. She had left him when he most needed her; now she would hold him back from the wider world that beckoned in the distance. Hannah, on the other hand, was too poorly educated to understand the irresistible attraction to the life of the mind being experienced by her enigmatic son. Both were justified in their feelings, yet nothing could change the simple fact that Isaac could walk but a single path. If, as seems likely, his determination had come from Hannah's seed, something was deeply moved during this wrenching conflict of wills. The ground between them shifted and would never be the same again.

The abstracted behavior so characteristic of Newton during this period was not solely a product of his discontent with agrarian life. This pattern of withdrawal, which had its genesis in a lonely childhood, continued throughout his life as a natural philosopher. Many tales of his legendary absent-mindedness have been told by those who knew him, the first of which dates from the period under discussion. Isaac had gone to Grantham for some unspecified purpose and was returning home late in the day. On the south edge of the town he came to a steep natural incline known in the seventeenth century as Spittlegate Hill. Riders often dismounted at the bottom, led their horses up the grade, then remounted at the summit. While in this process, Isaac became so lost in thought he failed to notice that his horse had slipped away. Only when he reached the top and was ready to remount the animal did he notice an empty bridle in his hand.[47]

Whether exaggerated or not, such frequent periods of boyhood detachment are early intimations of a coming intellectual eruption. Very much alone and left to his own devices after Hannah's second marriage, Isaac yielded to the pleasure of creative self-expression, and it quickly became the chief solace in his life. He lived among the creations of his own mind, and in many ways they were more real to him than anything he encountered in the external world. The energy he might otherwise have put into cultivating relationships with the members of his family and children his own age was turned inward upon itself, producing an inexhaustible ambition to understand intellectually a world that remained mute to his emotional advances.

VII

A significant feature of Newton's early development was his ability to attract and to hold the attention of older scholars, who seemed to sense promise in him. The behavior that seemed so bizarre to Hannah and the servants she employed was taken by Henry Stokes as further evidence of the brilliance Isaac had ultimately shown as a student. Neither Stokes nor anyone else had any way of knowing then the magnitude of the boy's genius, but the headmaster knew that here were gifts beyond the ordinary. Whether, during his trips to Grantham, Isaac secretly visited Stokes to protest his depressing fate is unknown. Even if he did not, the headmaster must have learned about the solitary reading sessions in the Clark attic, for the apothecary's brother had been usher at Grantham School, and it was the latter's books Isaac was reading. Unable earlier to persuade Hannah Smith that her son should receive additional education, Stokes decided to renew his efforts in light of the problems she was having with the boy: "He told her it was a great loss to the world, as well as a vain attempt, to bury so promising a genius in rustic employment, which was notoriously opposite to his temper; that the only way whereby he could preserve or raise his fortune must be by fitting him for the University." In an attempt to ease the imagined financial burden—for Hannah, though very well off for a widow with four children, was ever concerned about money—Stokes offered to remit the annual forty-shilling fee he received from the parents of all boys born more than a mile away from Grantham, no small sacrifice for a man of modest means.[48]

But Hannah could be almost as difficult to budge as her eldest son: Stokes had to argue his case again and again. In her defense we must remember that her long-standing plans for Isaac were falling to pieces, and with them an important part of a mother's dreams. Where could the

University lead her son? For what purpose? Land was the thing that truly mattered—the only thing! Still, the land did not matter enough to Isaac, at least not in terms of its unremitting demand for daily care, a fact which Hannah finally admitted to herself. As in the past, she turned to William, her trusted brother, for advice. Already aware of the situation, perhaps from having discussed the matter with Stokes himself, Ayscough, too, supported the idea of Isaac's returning to school. Her options closed, Hannah acceded. For better or for worse, the boy would go on to the University.

At least one other person besides the jubilant headmaster experienced great joy at the thought of Isaac's return to Grantham, albeit for reasons having nothing to do with the life of the mind. Among the members of the Clark household was a young woman identified by Stukeley as "Miss Storey." Three years Newton's junior, she was a child of Clark's second wife, an intimate of Hannah Newton. Twice widowed and in her early eighties when interviewed by Stukeley, "Miss Storey"—by then referred to by her married name, Mrs. Vincent—was the main source of the doctor's information about Newton's Grantham days. Her description of Isaac's activities and character serve to reinforce the portrait already drawn of him: a sensitive child who usually played abstract games with fragile toys he dared not share with the more raucous children of the street:

> She says Sr Isaac was always a sober, silent, thinking lad and never was known to play with the boys abroad at their silly amusements, but would rather chuse to be at home even among the girls, and would frequently make little tables, cupboards, and other utensils for her and her playfellows, to set their babys and trinkets on. She mentions likewise a cart he made with four wheels wherein he would fit, and by turning a windlas about, he could make it carry him around the house wither he pleased.[49]

But what is even more intriguing about Mrs. Vincent's recollections, particularly in light of Newton's extremely limited dealings with women, is that she revealed to Stukeley the existence of a romantic attachment between them:

> Sr Isaac and she being thus brought up together, tis said that he entertained a love for her, nor does she deny it. But her portion being not considerable, and he being [a] fellow of a College, it was incompatible with his fortunes to marry, perhaps his studys too. Tis certain he always had a kindness for her, visited her whenever in the country, in both her husbands days, and gave her forty shillings upon a time, when it was of service to her. She is a little woman, but we may with ease discern that she has been very handsom.[50]

If Mrs. Vincent and Newton were once pledged to each other, he could never have married and retained the fellowship he so coveted at Trinity College, for with few exceptions Cambridge Fellows were barred from taking wives, a regulation that remained in effect until the nineteenth century. Perhaps even more important is the fact that the range of Newton's sensibilities, rather than following the usual pattern of broadening and deepening with age, grew ever more narrow. He could remember and visit his childhood sweetheart, even give her money in a period of distress, but a deeper emotional commitment to a member of the opposite sex was out of the question. His prolonged immersion in the great intellectual struggle to come would leave no room for either the pleasures or the responsibilities of the hearth.

Exactly what form Isaac's final months of preparation at Grantham took is unknown. Since Latin was still very much the language of scholars it seems almost certain that Stokes grilled his star pupil in the classics one last time. Two of the many extant books from Newton's personal library—a Pindar and Ovid's *Metamorphoses*—are inscribed and dated 1659; both were standard texts in the grammar schools of the period. Though he had inherited nothing under the terms of his stepfather's will, Isaac also had the use of Barnabas Smith's library of some two or three hundred volumes, mostly theological in content, which was transferred to Woolsthorpe and subsequently given to him by his mother. It was an act of generosity which Hannah, upon deeper reflection, may have come to rue, for the books could have only contributed to her son's discontent with life at the manor. To preserve the collection and insulate himself from the constant intrusions of a younger brother and sisters, Isaac set aside a corner of his upstairs bedroom as a makeshift study. There, in the lengthening shadows of evening, the aggravating distractions of the outside world could be kept at bay for hour after solitary hour. Stukeley described the room as it appeared during an October 1721 visit, when Newton was seventy-eight and living in London:

> They led me up stairs, and showed me Sr Isaacs study, where I suppose, he study'd in the country in his younger days, as perhaps when he visited his mother from the University. I observed the shelves were of his own making, being pieces of deal boxes, which probably he sent his books and clothes down in, upon those occasions. There were some years agoe 2 or 300 books in it of . . . Mr Smiths, which Sr Isaac gave to Dr Newton of our town.[51]

It was at least partly within the confines of these obscure walls that Newton, as a student refugee from the Great Plague of 1665, laid the foundations of the fluxional method, conducted critical experiments with prisms, and formulated, in general outline, the inverse square law of planetary attraction, a crucial step in the direction of the greatest among his many discoveries—universal gravitation.

From the time it was decided that Isaac should receive a university education, there was little question about which institution he would attend. With the exception of Barnabas Smith, who was an Oxford man, almost all of the better-educated individuals with whom the youth had contact were former students at Cambridge University, some sixty miles south

of Woolsthorpe. Newton's uncle, William Ayscough, had attended Trinity
College, while Henry Stokes, the most important influence on Newton during this period, had completed his studies at Pembroke. Joseph Clark, usher
at the school, was a pupil of Henry More, the famous Platonist and a Grantham native, at Cambridge's Christ's College. Given such connections,
Isaac could have become a student of any one of several colleges, depending, of course, on the amount of influence that could be exerted on his
behalf.

Though the details remain sketchy, Ayscough appears to have been
determined that his nephew undertake his advanced studies at Trinity College, the rector's alma mater. He found a willing ally in the person of a
fellow clergyman. Less than three miles to the north of Ayscough's tiny
village of Burton Coggles is the equally small community of Boothby
Pagnell, whose rector was Humphrey Babington, a Fellow of Trinity College. Equally important is the fact that Babington was none other than the
brother of Mrs. Clark, the apothecary's wife, and the uncle of Catherine
Storey (Mrs. Vincent), Newton's first and only "love." Stukeley related that
Babington "had a particular kindness for [Isaac] which was probably owing
to his own ingenuity." This, Stukeley conjectured, was why Newton
became a student of Trinity College in 1661.[52] There is little reason to question this account, for the name of Dr. Babington shows up time and again
in Newton's student handbooks, as does that of Newton in Babington's
daybook records, strong evidence of an enduring relationship.

If William Stukeley, in recounting the story of Newton's childhood, exaggerated somewhat in writing, "Every one that knew Sr Isaac . . . recount
the pregnancy of his parts when a boy," the good doctor should be forgiven.
For what may not have been exactly true of Newton's youth had become
unassailable fact long before he died, the doyen of eighteenth-century
natural philosophers. Stukeley had not only been privileged to become the
friend of one of the great intellects produced by any age, he had witnessed
the rare apotheosis of a still living thinker. Under the circumstances, the
physician had done well to resist the temptation of turning complete
mythographer, as so frequently happens with those who belong to the inner
circle of a seminal intellect.

Unlike Stukeley, few of those who have carefully sifted through the
meager surviving record of Newton's youth are ready to maintain that the
seeds of genius were clearly in evidence, waiting only to be isolated and
identified. One twentieth-century biographer put the matter quite bluntly:
"There seems to have been nothing of the prodigy about the young
Newton."[53] The majority of contemporary Newton scholars have made
similar, if somewhat less pointed, observations.

Perhaps this view, by virtue of its being so widely shared, is the correct
interpretation of the facts as we know them; perhaps Newton, in order to
demonstrate the presence of an early and active genius, should have ac-

complished intellectual feats of profounder substance than those recounted by Stukeley and Conduitt. Still, it is difficult to support the conclusion that Newton, except for an inordinate shyness and a disposition to
withdraw into a private inner world of his own making, differed little from
most other boys of his day. Both the range of his interests and the number
of his largely unaided accomplishments, however childish many of them
may now seem, suggest that powerful forces were astir beneath that
taciturn exterior. Newton's constant observing, reading, recording,
building, experimenting, and, yes, even his daydreaming were all part of an
inexorable movement in the direction of his still undetermined lifework.
That this was no normal intellect Stokes, Ayscough, and Babington well
knew. But what they did not know, indeed, could not have known at that
time, is that the youth's ceaselessly active mind was starved for lack of
guidance in the direction of its greatest potential, scientific and
mathematical discovery. The boy's continual construction of ingenious
mechanical devices was as close as he could come to indulging his gift for
science, a subject to which he as yet had no formal introduction.

We are told that on the eve of Newton's final departure from Grantham
Henry Stokes "with the pride of a father put him in the most conspicuous
place in the school & with tears in his eyes made a speech in his praise to excite the other boys to follow his example."[54] Whatever the truth of this account, and whatever the actual level of Newton's intellectual development
as a child, there is no disagreement among historians concerning one
supremely important fact: At Cambridge, Isaac Newton found his genius.

Chapter Two

Of Giants and Dwarfs

What a chimera, then, is man! what a novelty, what a monster, what a chaos, what a subject of contradiction, what a prodigy! A judge of all things, feeble worm of the earth, depository of the truth, cloaca of uncertainty and error, the glory and the shame of the universe!

Pascal

I

The first two and one-half decades of Newton's life saw a great political and intellectual revolution, culminating in the execution of Charles I and the proclamation of a republic. Censorship broke down after Parliament's victory, and every area of thought, from religious expression to scientific inquiry, blossomed as never before. When John Conduitt looked back on these special times, he rightly singled out freedom of expression as the critical factor in spawning the most vigorous activity in the sciences since the days of classical Greece:

> Sr Isaac had the happiness of being born in a land of liberty & in an age where he could speak his mind—not afraid of The Inquisition as Galileo was . . . his works not in danger of being expunged as Des Carte's was nor [was] he obliged to go into another country as Des Cartes was in to Holland to vent his opinions, nor reduced to the miserable shifts as Des Cartes was of saying his Philosophy was the Philosophy of Moses & that he could prove Transubstantiation mathematically.[1]

Historical development, like biological change, is an organic process, no part of which exists in isolation. Not only was the English political and social climate conducive to the expansion of human thought on a broad

26

front, the age produced an abundance of keen and penetrating intellects, a rare and nearly perfect marriage of men and circumstances. For at the very time the condemnation of Galileo marked the climax of the scientific revolution in astronomy in Italy, England was giving rise to a constellation of distinguished scientific thinkers not often seen in any similar span of time or geographical region: Robert Boyle, Robert Hooke, Isaac Barrow, Edmond Halley, John Flamsteed, Christopher Wren, William Petty, John Wilkins, John Wallis, Seth Ward, John Ray, William Harvey, and Jeremiah Horrocks. Theirs and Newton's was the illustrious age of intellectual achievement which witnessed the demise of classical science and the cometary rise of the modern Promethean order.

Because Newton's impact on science has been so profound and long lasting, much of what he learned from others is often thought of as completely his own. Yet the great feats of scientific innovation and synthesis, like those in the arts, music, and literature, are never accomplished in a vacuum. The political, social, and religious influences, the important figures, and the scholarly resources that surround a young genius are of critical importance if one would understand how his or her intellectual gifts are shaped. Newton was heir to a rich and dynamic mathematical-experimental tradition, deriving from the innovative work of Copernicus, Kepler, Galileo, Bacon, Gilbert, and Boyle. And from France the mechanical model of the universe constructed by René Descartes became well known to him. Moreover, he was soon to fall under the spell of the natural philosophers—the virtuosi—of his own country, gifted men, who, in spite of their many differences, believed that a rational approach to knowledge held the greatest promise for those who would truly comprehend the works of the Creator. It was this invigorating world of thought—shaken by civil war, attacked by the proud defenders of the ancients, and impeded in its evolution by the encrusted scholasticism of the universities—which Newton entered as a callow youth from Lincolnshire in the summer of 1661.

II

Exactly how Cambridge first appeared to Isaac Newton is not known, but we can speculate with some assurance that a pleasing image greeted the eager but unworldly yeoman's son, for he had not yet visited London or even set foot in a town with a population of more than a few thousand souls. As his coach approached Cambridge from the north via the ancient road constructed by Roman slave labor, his pulse must have quickened when the spires of the town's numerous churches first came into view. Little did the youth imagine, as he strained for a better glimpse of his new home through the vehicle's undulating windows, that the peaceful community in the hazy distance would lay claim to the next thirty-five years of his life.

It is not very probable that Newton, eighteen years of age and educated mainly in the classics, was intimately acquainted with the historical development of the University of Cambridge or of Trinity College, in which he was about to enroll. Still, he may have been aware of the fact that little more than a century earlier the continued existence of England's unique college system had been in doubt. At the center of this political and intellectual crisis stood the ponderous, quick-tempered Henry VIII, whose combination of concupiscence, desire for a male heir to the English throne, and general contentiousness had led to an irreparable break with the Church of Rome and the permanent establishment of the Anglican Church under the deft guidance of his shrewd daughter, Elizabeth I. In 1544, three years before Henry's death, an Act of Parliament gave the King the power to dissolve any college, chantry, or hospital in the realm and to expropriate its possessions. Now that most of the monastery properties had been either given or sold at bargain prices to English nobles to ensure their adherence to the King's controversial religious policy, several favorites at court, their materialistic appetites unfulfilled, urged upon Henry a similar program of dissolution of the country's wealthier educational institutions.

The authorities at Oxford and Cambridge mounted an intensive campaign at court to save their colleges, and they luckily found a friend and influential advocate in the Queen, Katherine Parr, Henry's sixth and last wife—the famous "survivor." The historian George M. Trevelyan called her a wise woman, deserving of a statue in Cambridge and most of all in Trinity.[2] Katherine, who better mastered the art of defusing Henry's mercurial termperament than any of her hapless predecessors, succeeded in persuading the King that the colleges should be spared for the primary purpose of preparing a loyal clergy in the newly established state religion. Moreover, Henry, who sensed that death was near and, like seventeenth-century France's morally errant Sun King, Louis XIV, wanted to depart the world in a manner pleasing in the sight of God, acceded to Katherine's wish that he found a royal college at Cambridge of "unprecedented size and magnificence." Hence, instead of dissolving the colleges and dividing up the spoils, as he had done with the monasteries and convents, the King, a learned man in his own right, established a college to serve as the foundation of the new order, even as it was endowed with money and land wrenched by naked force from the old.

A younger, more egocentric Henry might have decreed that the new institution carry his came, but the reality of life's ephemeral quality had left its indelible mark during his pain-racked and often irrational final years. Instead, the King decided that the new college should be named Trinity, for the "Holy and Undivided Trinity" of the Christian religion. Originally made up of one Master and sixty Fellows and Scholars, the college was required to fly the royal standard. Its first Master, John Redman, was appointed by the Crown, as each of his many successors has been, a practice which at Cambridge University applies to the head of Trinity alone.

Both the studies and the conduct of students at Cambridge were governed by the Elizabethan Statutes, which went into effect in 1571 and remained in force until the reign of Queen Victoria in the nineteenth century. These same regulations, supplemented by statutes of the various colleges, also set forth the professional obligations and standard of behavior required of the Master and Fellows. Indeed, seemingly very little, from the election of University officers to the proper type of academic dress, was left to the discretion of the individuals concerned, although in practice the force and application of the rules was significantly altered over three centuries of evolving custom and tradition. In and of themselves, the Elizabethan Statutes are an incomplete guide to the diversity and quality of contemporary academic life. Yet any serious consideration of academic life, especially as regards its formative influence on Newton, must begin with some consideration of them.

Though the intellectual pulse of seventeenth-century England beat more vigorously with each passing decade, the new spirit of learning ushered in by the Puritan Revolution was hardly reflected in the formal curriculum of the nation's two leading institutions of higher learning. The prescribed course of study at both Oxford and Cambridge harkened back to the scholastic pattern of medieval education established in England during the thirteenth and fourteenth centuries.

The Statutes required that the scholar preparing for the B.A. at Cambridge reside in the University for a minimum of twleve terms (four academic years) and attend all public lectures given by the members of his college faculty. During the first year rhetoric, the art of eloquent oral and written communication, was his primary subject of study. It encompassed drama, poetry, classical history and geography, epistolary prose, and readings from Scripture, ever with an eye toward establishing unimpeachable sources of ideas and examples of the phraseology thought indispensable to the educated man. Eloquence of expression by itself was not deemed sufficient, however; every argument, no matter how compellingly presented, had to be underpinned by a rigorously logical thought process, the very art of reasoning. The treatises of Aristotle and his commentators were employed to teach the syllogistic method of deductive logic in the belief that an intellect trained to think in a systematic manner in one field would be prepared to attack any problem presented by another with equally controlled rigor. Latin, the language of instruction, received much scholarly attention, as did Greek and Hebrew. The entire second and third years of study were devoted to dialectics, the fourth year to philosophy alone. The latter subject was subdivided into several distinct categories, including metaphysics, physics, mathematics (arithmetic, geometry, astronomy, music, optics), and ethics (economics, politics, military and diplomatic history). Each had a strictly classical orientation, so that little of what men of the day called "the new natural philosophy"—modern physical science—was offered on the baccalaureate level until late in the seventeenth century.[3] (Only after Newton published the *Principia*, in 1687, did arithmetic, geometry, physics, and astronomy begin to receive the attention they deserved.[4] For those who, like Newton, went on to take the M.A., intensive study in natural philosophy, as expounded by the ancients, was mandated by the Statutes.

Most professors were required to lecture four days a week, Monday through Thursday, for periods of about forty-five minutes. A lecturer who failed to put in an appearance without providing a substitute was subject to a fine. A similar penalty was exacted from students for any unexcused absences from lectures. Enforcement of these regulations, however, hardly hewed to the letter of the law.[5] Judging by the miserable turnout at his own lectures after he became Lucasian Professor of Mathematics, Newton was as much a victim as anyone of the student disaffection and apathy that characterized English university life in the seventeenth century.

As one would imagine, the limitations imposed by the Statutes gave professors little flexibility in terms of introducing their students to new authors or of disseminating recently acquired knowledge, even if they wished to do so, which was not often the case. The professor of philosophy, for example, had to teach the *Problemata, Moralia,* and *Politics* of Aristotle, as well as the work of Pliny and Plato, while the professor of mathematics, if teaching cosmology, was required to interpret Mela, Pliny, Strabo, or Plato; if arithmetic, Tonstall or Cardanus; if geometry, Euclid; if astronomy, Ptolemy.[6] But we must remember that, the Statutes notwithstanding, curricular flexibility coupled with the freedom to increase the inherited body of knowledge and wisdom, while primary concerns of university education today, were not what most professors demanded or most students expected. In the Tudor and Stuart period Cambridge, like its sister insitution Oxford, was concerned primarily with maintaining the *status quo*.[7] Thus at the time Newton entered Trinity College, the wisdom of the ancients still reigned supreme, while the advocates of substantive change, comparatively few in number and just beginning to communicate with one another, carried on an arduous and quite lonely struggle for reform.

During the rule of Elizabeth Trinity adopted the custom, established in the Middle Ages, of requiring every member of the college below the degree of Master of Arts to have a tutor. In Newton's day these were ordinary Fellows, either chosen by the student according to the subject he wished to read or selected for him by the Master. Since the tutor was responsible for both the educational development and general deportment of his charges, he, rather than the college as a whole, normally had the greatest influence on the students.[8]

With few exceptions, Fellows were required to take holy orders and could wed only if they gave up their coveted positions, a price many were unwilling to pay, and with good reason. Once obtained, a fellowship was virtually a sinecure for life. Even the Masters, who were permitted to wed, often chose celibacy as an example to those who served under them. It was largely this consideration that caused Newton to break his youthful pledge of marriage to "Miss Storey." Indeed, he would not resign the fellowship to which he was elected in 1667 until 1701, some five years after leaving Cambridge for London. Even then he seems to have been under no official pressure to do so. Once a Fellow always a Fellow became the operative dictum of the period.

A primary weakness of the tutorial system lay in the assumption that a single individual was capable of giving each of his students all the instruction and guidance needed; rarely, even in the seventeenth century, an era of nonspecialization, was this the case. In truth, few Fellows aspired to the acquisition of universal knowledge, and fewer still attained it. The number of good tutors was always limited, and the general inefficiency and lack of accountability in the system spawned gross incompetence. The mathematician John Wallis, one of the distinguished founders of the Royal Society and a leading advocate of educational reform, wrote his friend Dr. Thomas Smith that as a Cambridge undergraduate in the 1630s, "I had none to direct me, what books to read, or what to seek, or in which method to proceed. . . . And amongst more than two hundred students (at that time) in our college [Emmanuel], I do not know of any two (perhaps not any) who had more mathematics than I . . . which was but little."[9] Of course, good tutors could and did do much for energetic students by judiciously selecting their reading matter and requiring them to comment in detail upon what they had learned, both orally and in writing. And those tutors who broke with tradition by mastering new theories on their own were instrumental in raising the intellectual level of a younger generation. But for the most part tutors were no more interested in the new natural philosophy than were the majority of university professors, something that should hardly surprise us, for they simply mirrored the educational values of the professional class to which they often aspired. Many of them even neglected the brilliant, if outdated, achievements of ancient science whose influence on the still functioning scholastic system was immense.[10]

A general laxity of student discipline further plagued the University throughout the seventeenth century and beyond. So serious had matters become by the 1630s that Charles I felt constrained to draft a series of ordinances aimed at curbing the more flagrant aspects of student carousing. Among other things, the ordinances included prohibitions against student liaisons with "women of mean estate and of no good fame" (whore was too indelicate a word) and the excessive frequenting of taverns "to eat or drink or play or to take tobacco." But the problem was compounded by the fact that the students frequently took as their examples Senior Fellows and members of the faculty, who not infrequently indulged in the proscribed conduct. The latter relied upon their traditional immunity from prosecution by secular authorities when they ran afoul of the law, which only exacerbated the traditionally strained relations between "Town and Gown."

Further regulations, handed down in 1636 by a Consistory composed of the Vice-Chancellor and the Masters of the several colleges, forbid the wearing of long hair over the forehead or ears, absence outside the college walls after eight o'clock in the evening, or "venturing into town or country" without the express permission of a tutor or other college official. Finally, in an attempt to curb sexual promiscuity within the University itself, female bedmakers under the age of fifty were no longer to be employed.[11]

Whatever the short-term effects of the new regulations, their long-range impact was considerably less positive than had been hoped by the authorities. Even Newton, who was far more circumspect when it came to his private conduct than most young men of his day, occasionally succumbed to the newly experienced temptations of student life. He accompanied his friends to the local taverns, where he drank ale and lost small amounts of money gambling at cards, hardly hanging offenses. Yet never did he allow himself the luxury of regularly indulging in such minor vices. And once he attained the position of Major Fellow, he seldom if ever took part in the more gregarious social activities so popular among his Cambridge peers.

Of the less than positive factors that contributed to a feeling of profound moral and intellectual depression at Cambridge on the eve of Newton's matriculation, none was more critical or far-reaching in its impact than the Civil War. The majority of townsmen had sided with Parliament, while most of the students and Fellows favored the Crown. Both sides were arming themselves by the spring of 1642, and there were scattered reports that some local citizens had already attempted to settle old grievances by firing into the windows of particularly "obnoxious students."[12] Fearing that Cambridge might be seized by the enemy, Cromwell sent his volunteer army—variously estimated at between 15 thousand and 30 thousand men—into the town, thus deciding the balance of power in the region for the remainder of the war.

Unlike Oxford, which took up arms and surrendered only after a siege by the parliamentary forces, Cambridge saw nothing of military conflict and suffered relatively little material damage as a result of its occupation. The academic cost of the Civil War, while less tangible, proved much higher. Many scholars, intimidated by the none-too-friendly troops quarterd in their colleges, quietly left Cambridge of their own accord. Others were sent away either because of their open resentment toward the new order or for refusing to pledge allegiance to the Oath and Covenant as decreed by Parliament in February 1644. Of the sixteen heads of colleges, ten were ultimately ejected for nonconformity, the same number removed from Oxford. Trinity, Newton's college, was among the foundations that suffered most. The majority of its Fellows were purged, including the poet Abraham Cowley and Humphrey Babington, the man primarily responsible for opening the way to Newton's acceptance at Cambridge after the Restoration. It is estimated that some 254 ejections from senior positions (Fellows and Masters) occurred at Cambridge while it was under Puritan rule. Oxford, which suffered even more because of its armed resistance to Cromwell, lost 370 of its senior members.[13] All appointments to fill the large number of forced vacancies came under the purview of the parliamentary commissioners, who had no set criteria to guide them save absolute loyalty to the political and religious principles of the new regime. Hence a broad range of Puritan divines found themselves in academic positions, many for the first time. The liberal arts curriculum of neoscholastic design continued in vogue, as a majority of educators showed an abiding conservatism during the 1650s and 1660s. With the Restoration came further purges, principally from among those very scholars who had taken the positions of the "uncovenanted" Fellows several years earlier. And yet in spite of this political turmoil, academic rigidity, and student ennui, there were those special men within the intellectual community whose minds reached beyond the finite world of classical scholarship to embrace a seemingly infinite universe of human knowledge, irrespective of their individual political or religious convictions. These were the thinkers who made up what Roger North, a Cambridge student of the 1660s, characterized as "the brisker part of the university."

III

Until now we have stressed the fact that seventeenth-century Cambridge accorded precious little attention to a growing body of postmedieval scholarship, whether in the sciences or in the many other areas of human thought. This is especially so if one sees the University as bound by its statutory regulations. But the term "university" can also be applied to the wider, less formal activities of the community of scholars associated with the institution. Thus the intellectually acute individual of the 1600s, bound as he was by institutional constraints, almost invariably played the part of the dutiful traditionalist by day; but by night, he not infrequently exchanged this role for that of the questing savant, reaching out for *terra incognita*.

William Oughtred, author of the important textbook *Clavis Mathematica* (1631), a work studied independently by Newton at Trinity, was just such an individual. Oughtred recalled his difficult, solitary struggle to master mathematics while a Cambridge undergraduate: "I redeemed night by night from my naturall sleep, defrauding my body and inuring it to watching, cold, and labour, while most others tooke their rest."[14] Newton was to follow an identical path: somber and secretive, he moved about his shuttered rooms like some gray disheveled ghost, his previous night's supper littering the table, untouched.

Oughtred later established an enviable reputation as a mathematician and friend of young scholars who found the neoscholastic curriculum stifling. They came to his home in a small but steady stream and often stayed with their congenial tutor for months on end. Upon returning to Oxford and Cambridge they passed their newly acquired skills on to fellow students. Important among their number was Seth Ward, a Fellow of Sidney Sussex, who lived and studied with Oughtred for a period of six months. So taken was Ward by his new master that he sent his friend, Charles Scarborough, then a Fellow of Caius College, to study under Oughtred. When Ward and Scarborough returned to their respective institutions in Cambridge they read the *Clavis Mathematica* to their undergraduate students, the first occasion, so far as is known, on which the work was used for instruction in a university.[15] Ward was later appointed to the Savilian Chair in astronomy at Oxford during the 1640s, and with him the revolutionary cosmology of Copernicus formally crossed the threshold of an English university for the first time.

Another of Oughtred's brilliant pupils was the young Puritan, John Wallis, whom we have already met. Educated at Emmanuel College, Cambridge, Wallis was chosen Savilian Professor of Geometry in 1649. At Oxford he became a member of the august group of natural philosophers who later founded the prestigious Royal Society in London. Christopher Wren, who possessed one of the finest minds of his time, also turned to Oughtred for advanced instruction in mathematics while pursuing his early studies at Oxford. Indeed, there is every reason to believe that Wren, who followed in Ward's footsteps as Savilian Professor of Astronomy, would have made his name as a natural philosopher had not the city of London, devastated by the Great Fire of 1666, wisely made him its chief architect for life. Such were the men of precocious intellect who undertook to challenge the established order even as they mastered the astronomy of Ptolemy, immersed themselves in the vast medical literature of Galen and his commentators, and entered into statutory disputations employing the logic of Plato and Aristotle.

While the new interest in experimental science and mathematics was never the monopoly of any single group, there is no denying that the Puritans stood at the cutting edge of seventeenth-century educational reform. Looking forward to the impending millennium, the more active Puritan educators sought to prepare the way for the final golden age, when the last obstacles to knowledge would be overcome and the spirit of man would stand forth exalted in the unblemished light of divine wisdom. Toward this end, they encouraged higher standards of individual achievement and the expansion of education to encompass a broader range of society. Even before the establishment of the Commonwealth they had gained the upper hand in a number of colleges at both Oxford and Cambridge. So entrenched had they become that even during the purges that accompanied the Restoration their grip could not be broken. Furthermore, their cause was aided by a growing body of intellectuals who had become alienated by the stultifying restrictions handed down during Elizabethan rule.[16]

The new pedagogy championed by the Puritans aimed at fulfilling the spiritual and intellectual needs of the individual from earliest childhood through advanced studies at the University. Its primary purpose was to prepare each member of the community for a vocational life of public service, a mission that derived from a deeply ingrained sense of religious obligation. Such a strong vocational orientation led to a deemphasis of humanistic philosophy and an embrace of the empirical sciences, which were underpinned by the new experimental method. Jan Amos Comenius, the much-admired Czech commentator on seventeenth-century educational reform, wrote:

> Children must be formed to be men by handling humane things; and by having all manner of occurences of this life represented both to their notice, and while they are in Schooles. Yes, all Philosophy in generall must be so ordered, that it may be a lively image of things, a secret fitting and dressing of mens minds for the business of this life.[17]

The program advocated by Comenius seems remarkably similar in spirit to the one pursued independently by the adolescent Newton while he lived in the Clark household at Grantham.

The man who more than any other both expressed and gave direction to the new age ushered in by the English intelligentsia was Sir Francis Bacon. Remembered today primarily for his eloquence and persuasive power as a speaker in Parliament and for the incisive wisdom of his beautifully written essays, Bacon claimed nothing less than all knowledge as his natural province. He attended Cambridge's Trinity College in the 1570s, but soon became impatient with the scholastic curriculum then dominant within the University. He was convinced that Aristotelianism had for centuries barred the way to the discovery of new and more valuable knowledge, especially in the realm of natural philosophy. Though he withdrew from Cambridge to undertake the study of law in London, Bacon never abandoned the idea that mankind's most pressing need was a revolutionary method for uncovering the knowledge denied to bright minds by the inflexible regimentation of the Schoolmen. Eventually he formulated an alternative manner of investigating the phenomena of nature and by so doing became the leading public advocate of the new experimental philosophy—what we now refer to as the inductive method of science: the gathering of data by observation, the formulation of a hypothesis, and the testing of the hypothesis by further observation and experimentation. Above all else, the use of Bacon's method provided English science with a strong utilitarian motive, for it turned attention away from the ultimate metaphysical explanations pursued by the medievalists toward an analytical, descriptive knowledge of natural phenomena. If pursued along Baconian lines, natural philosophy held out the promise of bestowing upon man the ability to control nature for his own edification and material wellbeing, a restoration of the power he had lost with the fall of Adam.

As devoted Christians, Bacon and his followers among the virtuosi fervently believed that the pursuit of the experimental method would lead them to new knowledge of profound religious significance. There was no question in their minds that God had fashioned a rational, orderly universe. He had given to man alone the capacity to discover and to put to use its most intimate secrets. Thus while the power of a single Divinity sustained the stage, the drama, and the actors, the pageant of life could be infinitely enriched through the critical examination of both the self and the wider world of which the individual is only a part. The ancient cyclical conception of the pagan world had grown wearisome and banal, its gods emasculated, its promise turned to pedantic stone.

Nothing better illustrates the spirit of the new age than the founding of the Royal Society, which received its formal charter from an admiring Charles II on July 15, 1662. On its coat of arms was engraved the motto *Nullius in Verba*—"at the dictation of no one"—to signify to the world that the institution's purpose was to discover scientific facts rather than to facilitate the exchange of private religious views or political opinions. Bacon, by virtue of his advocacy of the experimental method, became the idol of the original Fellows. It has been calculated that some 60 percent of the scientific problems dealt with by the Royal Society during its first thirty years of operation were prompted by practical considerations, while only 40 percent related to pure or theoretical science. Robert Hooke, who at twenty-seven was appointed the first Curator of Experiments, sometimes presented as many as six weekly demonstrations of new discoveries to members, an extremely demanding schedule even for a man of Hooke's considerable talent. Still, the appetite of the Fellows could not be satiated, for they were intrigued with any and all types of gadgetry and mechanical devices, no matter how bizarre. It was through this very avenue of practical demonstration that Newton grained admittance to the Society, for he successfully constructed the first working model of a reflecting telescope and shortly thereafter became a member by unanimous vote.

Thus did Francis Bacon—the great synthesizer—alter the outlook of future generations. The first among time-conscious moderns to sense the winds of the oncoming future, he often expressed himself in the language of the great voyagers of his time. Bacon believed the circumnavigation of the world to be the privilege and honor of his age; he compared it with "the new continent" of scientific thought that awaited human discovery. The title page of his *Instauratio Magna* pictures a ship passing through the Pillars of Hercules, the symbolic limit of classical learning, into a boundless yet beckoning sea beyond. History had arrived at a great divide: The ancient motto *non ultra*, "no further", must yield precedence to the new and more suitable vision of the times, *plus ultra*—"further yet!"

The experimental philosophy of Bacon provided Restoration thinkers with the inspiration and a *modus operandi* for exploring nature in a more systematic and detailed fashion than ever before. What they now required was a promising hypothesis with which to explain the many puzzling phenomena uncovered during their investigations. For this they turned in growing numbers to Europe and the work of René Descartes, the French mathematician and philosopher.

Born in 1596, the son of a prominent lawyer and a mother who died giving birth a year later, Descartes was enrolled in the Jesuit Royal College in La Fleche at the age of eight. As had Bacon a generation earlier, Descartes bridled against a degenerating curriculum centered on the teachings of medieval scholastics. Although his teachers considered him a brilliant student, Descartes considered himself ignorant, for he had acquired no certainty about anything. This attitude persisted even after he took a law degree in 1616 from the University of Poitiers. During his early years he frequently commented that he was sustained only by a deep religious faith, which he retained to his dying day.

Though the details remain obscure, Descartes tells us that on the night of November 10, 1619, at the age of twenty-three, he experienced what can only be described as an illuminating flash of mystical insight. During a few brief but emotionally overpowering moments, he became convinced that through the unlimited potential of mathematics human knowledge of the world could be synthesized into an all-encompassing wisdom. So profound was his sense of revelation that he embarked upon a pilgrimage to the shrine of the Lady of Loreto in Italy. There he gave thanks to God, who alone could have given him such inspiration. In the wake of this illuminating experience the young Frenchman immersed himself in the study of mathematics, and within a few months his efforts were richly rewarded by the discovery of a crucial mathematical tool—analytical geometry—which proved much easier to wield than the ancient geometry of Euclid. Without it Newton could not have formulated the law of universal gravitation or written the revolutionary *Principia*.

Having rejected the use of the senses as a reliable method of probing nature, Descartes took upon himself the task of describing the occurrences of the physical world in a strictly mathematical language. He knew, of course, that mathematics is nothing more than a complex system of symbols, but he also knew that it is the only symbolism that defies the attempts of the mind and the senses to shift or transform its meaning. Hence it is exact and by being exact is self-correcting. It does not confuse; it does not lie. To the dedicated Cartesian, the only truly scientific knowledge is mathematical knowledge.

According to Descartes, all bodies in the universe have moved in perfect order since their creation by God and will continue to do so time out of mind. There is no spontaneous motion; everything functions in strict accordance with the Creator's fixed principles. Descartes envisioned space as an infinite plenum filled with an all-pervading fluid, or ether, which has unlimited divisibility. Within this subtle, infinite matter there is a denser matter of globular clusters, which combine to form the physical bodies with which we are familiar—the sun, planets, comets, and stars. At the moment of creation, God fashioned the pervasive "first matter" into countless whirlpools or vortices, providing each planet, comet, and star with its own eddy in which it is spun round, much as a leaf captured by the current of a swift-flowing stream. The larger the body the vaster the vortex required to move it. The sun, for example, is propelled by a far larger vortex than are any of its smaller companion planets. The only forces operating between the individual bodies are those produced by the spinning vortices of matter in which they ride. "Attraction" and "repulsion" are merely appearances, the result of a continuous series of impacts. One body, on moving, fills the space vacated by another body, which in turn occupies the space vacated by a third, and so on, like celestial billiard balls in perpetual motion.

The world and universe, viewed from the Cartesian perspective, become amenable to rigorous mathematical analysis. Each body yields objective data in terms of its size, shape, and location in relation to other bodies. Of the quantitative attributes, Descartes was most interested in spatial dimensions, or what he called the "extension" of bodies. By "extension" he meant the amount of space occupied by a given object based on the computation of its volume. In addition to being extended, or occupying a calculable amount of space, all bodies in the Cartesian universe are constantly in motion. And as Galileo and Kepler so clearly demonstrated, motion is also subject to quantification. Indeed, time itself is nothing more than the measurement of objects in motion. Descartes fervently maintained that only through the mathematization of the universe can man grasp its principles of operation in any objective sense.

It should be obvious at this point that the universe of Descartes and of those who embraced his philosophy is totally mechanistic in character, a vast and beautifully synchronized machine whose movements are predictable and self-sustaining. Spontaneity has been banished by a God who employs no physical principles other than those of mechanics. Once He created the innumerable parts of the universe, God set those parts in motion relative to each other, and for all time. And because God is perfect His creation is also perfect, without need of further refinement or periodic repair. The basic laws that govern this magnificent construct simply await discovery by man, the most rational of Gods' creatures.

However simplistic Descartes's mechanized universe may seem more than three centuries after his death, his was the first comprehensive attempt to formulate a picture of the whole external world in a manner fundamentally different from the ancient model of Aristotle and Ptolemy. The new and infinite cosmos he envisioned was no longer spun round by the right hand of God, who in Descartes's system is relegated to the role of Creator alone. The happenings of the physical world are self-perpetuating and eternal; nature is a perfect engine whose every movement is necessary and predictable. If ever the cosmology of a single individual was deserving of the title "world-shaping," Descartes's most certainly is. Archimedes had said that with a giant lever and a place to stand he would move the earth. Descartes, even more boldly and just as confidently, declared: "Give me extension and motion and I will remake the world." As a result of his radical break with the static view of the universe, which gave comfort to ancients and medievals alike, the scientist of modern times has been gradually elevated to the position once held by the Creator alone, that of absolute lawgiver.

IV

The Cartesian hope of formulating a rational scheme of a universe governed by immutable mathematical laws became fruitful only when merged with the spirit of empiricism fostered by Galileo and Bacon. The crucial synthesis of these methods was accomplished in England during the mid-seventeenth century by a group of natural philosophers collectively known as the virtuosi. Since the eighteenth century the "virtuoso" has been thought of as a master musician whose extraordinary ability and exquisite technique set him apart as a brilliant individual performer. During the seventeenth century the term was applied not so much to the performing artist as to the practicing scientist or, as the chemist Robert Boyle put it, to "those that understand and cultivate experimental philosophy." Unlike scientists of today, who enjoy a status at least equal to that of other professionals, the virtuosi usually answered to other titles. Many, like Isaac Barrow and John Wilkins, were both practicing clergymen and teachers, while others were physicians, entrepreneurs, or employees of the government. Science was yet to become the institutional activity so commonplace today, and only a small number of men, usually of independent means, could devote their full time to its pursuit.

Just as their religious, educational, and professional backgrounds differed, so too did their comprehension of the natural world. While it is true that the mechanical philosophy, formulated by Descartes and championed by the materialist Thomas Hobbes, was viewed as a promising hypothesis by most of the virtuosi during the Restoration, its truth was only gradually being established through the systematic application of Bacon's method of experimental verification. Furthermore, there were almost as many mechanical conceptions of nature as there were virtuosi, no two of which presented exactly the same problem. On balance, however, there was fairly general agreement that Cartesian principles offered the most fruitful path of scientific exploration: the virtuosi tried to construct a theory picturing nature as a machine running without external aids, a machine that human beings could study and comprehend.[18] Indeed, the whole development of modern science from Copernicus to Newton has been rightly characterized as the mechanization of the world picture.

Implicit in the idea that nature's secrets can be unlocked through the systematic application of controlled experimental techniques is the belief that design must also be an integral part of God's master plan. The virtuosi employed a hierarchical concept of natural order borrowed from the Middle Ages called the Chain of Being, according to which nature is likened to a giant flight of stairs. At the very bottom one encounters insensate matter in the form of minerals and the more complex inorganic materials. As the observer begins his ascent, life forms make their appearance, with each arranged in strict ascending order of function and complexity. Finally, atop the first landing, stands man, the pinnacle of terrestrial creation, the only know sentient being capable of shedding light on the benign darkness and dumbness of the cosmos. From man, a second flight of stairs reaches into the more perfect spiritual realm above, where, at the apex of creation, stands God. The more the virtuosi explored the world about them the more they were dumbfounded by the plenitude and richness of nature's incomparable gifts. Each new discovery brought further evidence of the eternal presence of God, who was more and more worshiped through the bounty of His creation than in the words of Holy Scripture. Primary matters sustains the growth of vegetative life; vegetables reach upward to nourish the sensitive or animal species; and sensitive life supports the intellectual order embodied in man, who, of all earthly creatures, most nearly mirrors the Creator. Never has this symbiotic relationship been more aptly described than by the eighteenth-century English poet Alexander Pope in his stirring *Essay on Man*:

> *Vast chain of being! which from God began,*
> *Natures aethereal, human, angel, man,*
> *Beast, bird, fish, insect, what no eye can see,*
> *No glass can reach; from Infinite to thee,*
> *From thee to nothing.—On superior pow'rs*
> *Were we to press, inferior might on ours;*
> *Or in the full creation leave a void,*
> *Where, one step broken, the great scale's destroy'd;*
> *From Nature's link whatever link you strike,*
> *Tenth, or ten thousandth, breaks the chain alike.*

The search for design in nature became an uncontrollable passion through which everything was made to appear as though created for the single purpose of serving humanity and enriching the experience of life. Mountains were made to enhance man's visual experience; large animals moved about on four feet to make them subservient beasts of burden, and vast oceans proved not to be impenetrable barriers after all, but challenging pathways to uncharted continents and lost civilizations. Thus did a natural religion based upon the empirical examination of physical phenomena arise to challenge the institutional theology of the past. And so, also, did a dominant and subtly arrogant theme of Western thought begin to crystallize, namely, that humankind is meant to have control over the earth and all life forms, because it is the loftiest product of a preordained creative process. Though many temperamental virtuosi, not least Isaac Newton, chose isolation from their fellow men, they were never at rest in their secluded worlds, never the contemplative seer beneath the secred Bô tree of the Buddha. Like Spengler's Faustian man, they became the very embodiment of the restless and exploring ego now so familiar in the scientific civilization of the West, aggressive spokesmen of the will.

On this scene of increasing cosmic order, with its heightened interest in nature as a tangible manifestation of divinity, the idea also emerged that there is a double revelation of God: the one contained in His words found in Scripture, the other to be found in nature and its general laws. Having established this dualistic framework, the virtuosi faced the delicate problem of working out an acceptable compromise between a universe whose laws can be demonstrated mathematically and God's Biblical revelation, which contains many passages that are seemingly incompatible with the discoveries of Copernicus, Galileo, and Kepler. As a way out of this dilemma, they advocated the acceptance of two very different languages: the scientific-mathematical language of primary qualities with its attendant precision and rigor, and the everyday, emotional language of secondary qualities with its inherent inconsistencies and inexactitude. Being knowledgeable students of ancient history, the virtuosi contended that when God gave the Scriptures to man He necessarily employed the language of the common people. Biblical language had to conform to the daily experience of a pre-scientific Hebrew culture and hence contains certain unscientific propositions that support such beliefs as an earth-centered universe. But now that the penetration of the Divine mind had been made possible through the application of mathematics to natural phenomena, communication between God and man could no longer rely upon so undependable a vehicle as words alone. The virtuosi advocated that man, in his striving to know the physical nature of the universe, begin with God's *work* rather than his *word*. The teachings of the Bible must remain paramount in questions of human conduct and morality, but in dealing with nature the impersonal language of science must take precedence over the common tongue. This was the position taken by Galileo when confronted by clerical antagonism toward the new science in Italy, and never was it more forcefully presented or defended than in his famous letter to Christina, the Grand Duchess of Tuscany:

> I think that in discussions of physical problems we ought to begin not from the authority of scriptural passages. . . . It is necessary for the Bible, in order to be accommodated to the understanding of every man, to speak many things which appear to differ from the absolute truth so far as the bare meaning of the words is concerned. But Nature, on the other hand, is inexorable and immutable; she never transgresses the laws imposed upon her, or cares a whit whether her abstruse reasons and methods of operation are understandable to men. For that reason it appears that nothing physical which sense-experience sets before our eyes, or which necessary demonstrations prove to us, ought to be called in question (much less condemned) upon the testimony of biblical passages which may have some different meaning beneath their words. For the Bible is not chained in every expression to conditions as strict as those which govern all physical effects; nor is God any less excellently revealed in Nature's actions than in the sacred testaments of the Bible.[19]

The religious mystery previously associated with the physical world had suddenly vanished; the only remaining mysteries were those which natural philosophers would eventually solve through experimentation and mathematical deduction—or so it seemed. During the second half of the seventeenth century nature was reverently spoken of as the second book of God's revelation. Some considered it the most direct communication of all, untrammeled by words, less clouded by human contention. There had begun, by degree, the unremitting examination that led from innocent optimism and absolute certainty to the numbing reality of vast and unreachable galaxies composed of countless billions of stars, to time stretched across trackless space by millions of light-years, or read backward in the bleached bones of long-extinct animal species and the variegated strata of primordial stone.

Still, the Baconian belief in illimitable progress was slow to gain a foothold in the thinking of a learned world whose inexhaustible reverence for antiquity was matched only by an equivalent capacity for self-deprecation. And no single factor was more responsible for the feeling of modern inferiority than the belief that nature was gradually decaying.[20] This argument, though simple enough, was profound in its implications. For it was generally held that just as man through his sinful fall from Eden brought death upon himself, so God had imposed a death sentence upon all that He had created. Thus the farther anything is removed from the moment of its inception, the more precipitous its irreversible decline. Civilization, this whirling planet inhabited by mankind, indeed, the very cosmos itself must pass through the inescapable downward cycle from birth, to youth, to old age, and finally to a natural death. Since the seminal thinkers of antiquity—Aristotle, Plato, Herodotus, Pliny, Galen, and Ptolemy—had lived in an age closer to the moment of creation, it was reasoned by analogy that they were the last intellectual giants to have walked an increasingly corrupt earth. Never again would the human species match the level of learning attained by a long-vanished classical world, let alone surpass it.

Alexander Ross, one of the more dedicated supporters of the concept of ancient superiority, responded to what he perceived as a hubristic and dangerous challenge to the intellectual legacy of established authority:

> The Dictates and Opinions of the ancient Champions of Learning are slighted and misconstrued by some modern Innovators; whereas we are but children in understanding, and ought to be directed by those Fathers of knowledge: We are but Dwarfs and Pigmies compared to those Giants of Wisdom, on whose shoulders we stand, yet we cannot see so far as they without them. I deny not but we may and ought to strive for further knowledge, which we shall hardly reach without their supportation. I disswade no man from inventing new; but I would not have him therefore to forget the old, nor lose the substance whilst he catches the shadow.[21]

John Donne, in what has been characterized as "the poetry of skepticism," was more sardonic than Ross in his assessment of the shattering effect of the new learning. In his *Anatomy of the World* Donne pictured the "new philosophy" of Copernicus and his followers as putting "all in doubt" by destroying the natural order and harmony implicit in the hierarchical universe constructed by the ancients:

> *And new Philosophy calls all in doubt*
> *The Element of fire is quite put out;*
> *The Sun is lost, and th' earth, and no mans wit*
> *Can well direct him where to looke for it.*
> *And freely men confesse that this world's spent,*
> *When in the Planets, and the Firmament*
> *They seeke so many new; then see that this*
> *Is crumbled out again to his Atomies.*
> *'Tis all in peeces, all cohaerence gone;*
> *All just supply, and all Relation.*

Not only, from Donne's perspective, did the new philosophy destroy the harmony of a finite, geocentric universe by seeking new planets and scattering the stars through the seemingly boundless abyss of the firmament, but the once strong bond between family members and society was rapidly dissolving as the new individualism that accompanied the new learning opened the way to social anarchy, a premonition of our own age of relativity:

> *Prince, Subject, Father, Sonne, are things forgot,*
> *For every man alone thinkes he hath got*
> *To be a Phoenix, and that then can bee*
> *None of that kinde, for which he is, but hee.*

The advocates of modern thought felt constrained to respond in kind to the attacks mounted by the traditionalists. If moderns are inferior, they argued, it is not because of any diminished intellectual capacity but the result of sloth and negligence. The person who claims to be a dwarf perched atop the shoulders of ancient giants is himself a potentially creative individual who is unwilling to face the challenges of a world no less, but certainly no more, corrupt than that of the classical age. Moderns never tired of contrasting their methods of acquiring knowledge with those of the ancients: nature versus books, works versus words, experimental laboratories versus libraries, and industry versus indolence. The very physical effort required by experimental research became a virtue in itself, while the inactivity of prolonged reading was characterized as a form of sloth. The efforts of the ancients were not to be depreciated, but neither must they stand in the way of a more activist mode of investigating the phenomena of nature. As Agricola Carpenter so eloquently expressed the position of the moderns: "Truth is not ingrossed by aged Parents; there is an *America* of knowledge yet unfound out, discoverable by the endeavours of some wiser *Columbus* and the promised fertility of succeeding Ages."[22]

The moderns had their champions among the poets as well. John Milton expressed a genius's contempt for the very astronomical system of Ptolemy, which, to Donne's disappointment, yielded to the new Copernican astronomy. In Book VIII of *Paradise Lost* Milton declares:

> From man or angel the great Architect
> Did wisely to conceal and not divulge
> His secrets to be scann'd by them who ought
> Rather admire; or, if they list to try
> Conjecture, he his fabric of the heavens
> Hath left to their disputes, perhaps to move
> His laughter at their quaint opinions wide
> Hereafter; when they come to model heaven
> And calculate the stars, how they will wield
> The mighty frame, how build, unbuild, contrive
> To save appearances; how gird the sphere
> With centric and eccentric scribbled o'er,
> Cycle and epicycle, orb in orb.

Such, then, is the background of the challenging age in which Isaac Newton was destined to reach his intellectual maturity. At the time of his death, in 1727, Alexander Pope penned two eloquent lines of epitaph that forever silenced those who had argued the case for the dwarflike nature of modern man. Carved above the fireplace in the room of Newton's birth at Woolsthorpe, they read:

> Nature and nature's laws lay hid in night:
> God said, let Newton be! and all was light.

Chapter Three

The Transit of Genius

Towering in the confidence of twenty-one.
Samuel Johnson

I

Trinity, bordered on one side by St. John's and on the other by Caius, was, in the words of the seventeenth-century scholar Thomas Fuller, "the stateliest and most uniform Colledge in Christendom."[1] Though a Cambridge man himself, Fuller can hardly be accused of bias, for he took his degrees at Queens' and Sidney Sussex rather than at the institution whose architectural praises he so unstintingly sang. The Great Gate, the most common avenue of access to the College, was erected between 1490 and 1535. On its massive exterior, facing Trinity Street, is a statue of the founder, Henry VIII. From the interior steps of this towering entrance the visitor gains a commanding view of the sweeping court enclosed by Tudor Gothic façades, which connect the Chapel, Master's Lodge, and magnificent Dining Hall, with its hammer-and-beam ceiling, minstrels' gallery, and dominating portrait of Henry. Even in our own age of massive architecture it is no exaggeration to characterize the initial impact as overpowering. It was from this vantage point that the young Newton had his first good look at his new residence. To one of his limited background the sense of wonder must have touched on the miraculous, only to be replaced by nagging intimations of self-doubt.

Like many other young men of his social rank and religious persuasion, Newton kept strict accounts of his income and expenditures. Among the most important of these is a tiny notebook, now in the possession of Trinity

College Library, begun even before he left Woolsthorpe for Cambridge. It contains a partial record of his financial affairs during the first two years of college. Under the heading "Impensa propria" (special expense) he provides a vivid glimpse of his very first student outlays:[2]

	£	s.	d.
Sewsterne	0	1	0
Stilton	0	2	0
Cambridge white lion	0	2	6
Carriage to ye Collodge	0	0	8
A Chamberpot	0	2	2
A table to jot down ye number of my cloathes in the wash	0	1	0
A paper booke	0	0	8
For a quairte bottle & inke to fill it	0	1	7
Income for a glasse and other things to my Chamberfellow	0	0	9
To the butlers	0	2	6
To Wolfe	0	0	6
A lock on my deske	0	1	4
A pound of Candles	0	0	6

From this list we learn that during his initial journey to Cambridge Newton stopped off at Sewstern, perhaps to have his first long look at the property deeded to him by Barnabas Smith, the income from which would be his on turning twenty-one. He spent the following night in Stilton, arriving in Cambridge on the fourth of June. He took accommodations that evening at the White Lion Inn and the next day hired a carriage to take him to Trinity, where he presented himself to the appropriate officials. Then came the task of settling in, which required the purchase of such basic necessities as a chamberpot, ink, paper, and candles. Interestingly, the last item is the most frequently recorded purchase in all of Newton's accounts, testimony to the fact that he often worked through the better part of the night while others took their rest. Some idea of his taste in food can also be gained by examining this notebook. Among the items he purchased to supplement the fare served up in the dining hall were: "Marmelot, Custords, Apples, paires, Raisons, Whitewine, Wafors & sugar." Fruit resisted spoilage, an important consideration in an age when many fell ill and died as the result of food poisoning.

Newton was admitted to Trinity on June 5, 1661, as a subsizar and matriculated in the University as a sizar the following month. In the original meaning of the word, which still obtained at the time, a sizar was one whose financial circumstances dictated that he maintain himself by performing menial tasks, which included waiting on tables and running errands for the Fellows and his tutor. Sizars also served as valets to their classmates of wealth and privilege, waking them for morning chapel before dawn, fetching special orders of food, dusting their rooms, polishing their boots, and dressing their hair. At the top of the undergraduate hierarchy were the fellow-commoners, scions of England's most influential families, who paid for additional privileges such as dining at high table. Below them were the pensioners, offspring of the prosperous but less wealthy. Sizars and sub-sizars (there was little if any practical distinction between the two), like the servitors of Oxford, were at the very bottom of the university social structure, yet grateful for the rare opportunity to improve their economic and social standing through the avenue of advanced education. The following entry from the *Conclusion Book* of Trinity College offers both an example of the sizar's duties and a revealing glimpse of the prevailing lack of discipline at Restoration Trinity:

> Jan. 16. 1660–1. Ordered that no bachelor . . . nor any undergraduate, come into the upper butteries [pantries], save only a sizar that is sent to see his Tutor's quantum [share], and then to stay no longer than is requisite for that purpose, under penalty of 6*d.* for every time; but if any shall leap over the hatch or strike a butler or his servant, upon his account of being hindered to come into the butteries, he shall undergo the censure of the Master and Seniors.[3]

The pattern was as old as class distinction itself: The fellow-commoners and pensioners made sizars the object of their scorn, and they in turn vented their frustrations on the even more hapless employees of the College.

Though the university system of the seventeenth century was deeply flawed, Cambridge, like Oxford, at least provided a route, albeit capricious, for boys of humble origin to achieve greatness not only in learning but also in the world. Augustus De Morgan, Newton's nineteenth-century biographer, viewed the passing of the sizarship in his own day as a tragic sacrifice to the stilted concept of Victorian gentility. "Those who know the old constitution of the universities see nothing in it except the loss to the laboring man and the destitute man of his inheritance in those splendid foundations." The weakness in De Morgan's argument lies in the example he offered to support an otherwise sound thesis. "If sizarships with personal services had not existed, Newton could not have gone to Cambridge; and the *Principia* might never have been written."[4] By no stretch of the imagination can Newton be considered a child of humble circumstance, let alone of need. His mother, by virtue of the substantial inheritance from her second marriage, was about as well fixed as any woman in Lincolnshire. Moreover, Barnabas Smith had deeded property to Isaac worth some £50 a year. How is it, then, that the youth entered a highly class-conscious institution with so little status; that he waited tables and dined on the leavings of young men, many of whose fathers possessed considerably less money and land than the twice-widowed Hannah?

We know that Newton's mother was ever conservative in financial matters, and she may simply have seen no good reason to pay for something that could be earned through honest labor, a typically Puritan trait. Apparently strong of will, it is also possible, indeed more probable, that she had not yet become reconciled to Isaac's rejection of the secure agrarian life planned for him since childhood, a life virtually guaranteed by her less than idyllic union with a wealthy clergyman over twice her age. If her son wanted to attend the University so be it, but she would do little, at least in the beginning, to smooth the way. His education would have to be earned. Only after he attained his majority do we find entries in Newton's notebooks of fairly substantial sums obtained via his mother, money now legally his to do with as he pleased.[5]

Thus while others seem to have entered Trinity on a social and economic footing similar to Newton's, in truth his ambiguous status set him apart to a considerable degree. He was certainly not, as De Morgan and other biographers have suggested, the child of an impoverished home, at least in the pecuniary sense; yet neither had he been prepared for Cambridge by attendance at one of the great public schools in London—St. Paul's, Merchant Taylors', or Westminster. Waiting upon fellow students and silently suffering their indignities could have come as nothing less than a shock and an embarrassment to one who, though not raised in luxury, was used to having his own basic needs tended to by servants. At the same time the rural environment from which Newton came placed little value on the social graces cultivated by the wealthy, least of all the art of polite conversation and repartee. Rather, it stressed the Puritan virtues of humility before God, a respect for daily toil, and an abiding sense of moral duty. The notoriously loose ethical standards of Restoration Trinity were ever locked in mortal conflict on the plain of Newton's Calvinistic conscience with an unshakable disgust for any and all forms of moral laxity. He repelled the forces of evil on a daily basis but at the same time, judging by the contents of the Fitzwilliam notebook, harbored a powerful desire to yield to the agonizing temptations of the flesh.

The ambivalent and personally confusing nature of Isaac's position was compounded by the fact that he was well into his nineteenth year in the summer of 1661, and therefore somewhat older than most of his peers. One biographer has conjectured that this may have been an advantage in terms of his mental maturation and seemingly latent physical development.[6] How it affected his ability to relate to others is an entirely different matter. Since Newton was already introspective and aloof, the difference of age, which for the most well-adjusted young men would have counted for relatively little, only served to intensify a deep sense of alienation present since childhood. When, in his mid-fifties, he chose to vacate his Cambridge professorship for the wider world of London, it was to accept the prestigious appointment as Warden and subsequently Master of the Mint. This decision was based in no small part on a desire to overcome the parvenu's sense of social inferiority that, like a lingering shadow, had haunted him since his earliest days at Trinity: at last to join the ranks of the powerful, whom he long and secretly admired from afar, and to reinforce in his own mind his claim to the title of "gentleman" made during the Herald's visitation to Grantham in 1666. Given this background, it is hardly surprising that during the last quarter-century of his life Newton thoroughly relished his lionesque role as President of the Royal Society, the undisputed autocrat of English science.

II

By the time Isaac left Woolsthorpe for Cambridge, he had carved out a local reputation for himself as a scholar of considerable promise. As is so often the case, however, there were those who harbored little but contempt for intellectual accomplishment, and they made known their feelings on the matter. Hannah's servants, wrote Stukeley, "rejoyc'd at parting with him, declaring, he was fit for nothing but the 'Versity." At least they were consistent, for Stukeley had also learned that "His mother, as well as the servants, were somewhat offended at this bookishness of his: the latter would say the lad is foolish, and will never be fit for business."[7] Nevertheless, the more enlightened citizens of Woolsthorpe, Colsterworth, and Grantham could be proud of "their Isaac," though they might have been surprised to learn how comparatively little he knew of formal scholarship. While Henry Stokes had done the best within his limited resources to lay the foundations for his pupil's future education, the youth was not particularly well, or even systematically, trained.

Fortunately, he was solidly grounded in Latin, the international language of learning. Isaac had also acquired some Greek, but he did not master the subject while at Grantham. His acquaintance with French was at best cursory, and when he read the language it was always with dictionary in hand. (In later life the letters written to him in the vernacular by French natural philosophers were turned over to young and trusted followers for translation.) Newton seems not to have understood German at all. Hebrew was in fact considered the most crucial language after Latin, and early entries in the Morgan notebook show that he had a more than passing acquaintance with Hebrew characters. His eventual mastery of the language proved indispensable in advancing the Biblical scholarship that claimed as much of his time as natural philosophy and mathematics combined. From his Grantham teachers he also received lessons in ancient history, grammar, and Christian exegesis. His preparation in the fields of science and mathematics, in which his lasting discoveries were made, was pitifully meager. The mathematical problems of his childhood, which some have cited as a foreshadowing of his latent genius, were in truth nothing more than simple arithmetical calculations incidental to his mechanical inven-

tions. At best he had learned basic arithmetic and no more than a smattering of geometry, if that. Grantham School provided no formal introduction
to Euclid. The undeniable skill and passion for experimentation and the
construction of all manner of mechanical models found no outlet in the
classrooms of Newton's childhood, nor did they find a place in the neoscholastic curriculum of Cambridge. Newton suddenly found himself in competition with forty other young men whose knowledge, acquired in many
cases from the public schools or private tutors, was considerably in advance
of his own. Cambridge, to put the matter bluntly, was understandably quite
unmoved by his arrival. Less comprehensible is the fact that the intellectual
gifts of her supreme genius, destined to reach full fruition within three
short years of his matriculation, would go unrecognized until after
Newton's Master's degree was well in hand.

Newton, like all undergraduates of his day, was assigned a tutor who
assumed the dual responsibilities of teacher and of surrogate parent (*in loco
parentis*). The meager records of the period inform us that this task fell to
one Benjamin Pulleyn, an obscure figure about whom we would like to
know more. Pulleyn, a native of Leicestershire, was admitted sizar at Trinity in June 1650; he earned a B.A. in 1654, became a Fellow in 1656, and
took an M.A. in 1657. He also took holy orders, as was the statutory requirement of Major Fellows, and during the last four years of his life served as
Rector of Southoe in Huntingdon. Pulleyn's most important appointment
at Cambridge was as Regius Professor of Greek, a position he held from
1674 to 1686.[8]

Though Newton was hardly unconcerned with the study of classical
languages and literature, he was soon to be captivated by intellectual interests that ran counter to those of his tutor and of Cambridge in general.
Oxford and London's Gresham College were the homes of what modern
scientific study there was at this time; it was Newton's own achievements
that brought this legacy to Cambridge several decades later. Whether or not
Pulleyn, whose scholarly field looked to the past instead of the future,
sensed that he had a student of extraordinary ability on his hands is difficult
to say. Only an oblique reference to his attitude toward Newton survives in
the Conduitt papers, and many scholars consider it of dubious value. Still, it
seems worth recounting, given the dearth of information on this point:

> Mr Aiscough had given Sr Isaac before he sett out for Cambridge Sanderson's
> logick & told him that was the first book his tutor would read to him, this Sr I.
> read over by himself & when he came to hear his tutor's lectures upon it found
> he knew more of it than his tutor, who finding hir.ı so forward told him he was
> going to read Kepler's Opticks to some gentlemen commoners & that he might
> come to those lectures. Sr I. immediately read it at home & when his tutour gave
> him notice of the lectures he told him he had already read that book throu.[9]

If Conduitt's account is true, it could only have been related by an aging
giant who, in a rare and unguarded moment, was considerably more willing
to boast of his youthful exploits than he normally was. It would also indicate

that Newton's tutor had a greater intellectual grasp of the new science than other sources have granted him. We may also legitimately question Conduitt's assertion that the ability to master an elementary textbook on logic should lead to an invitation to attend lectures on an advanced optical treatise, to be delivered, surprisingly, by a classical scholar. Whatever the character of their relationship, Pulleyn carried his thoughts on Newton to the grave. Newton, for his part, was never one to give even partial credit to others for his successes. Pulleyn, like poor Henry Stokes, is not even damned by faint praise in the voluminous writings of his most gifted and famous pupil. Still, in fairness to Newton, there is a distinct possibility that his tutor was deserving of none.

Nothing was more prized by Newton than his personal privacy, which he guarded with the same ferocity sporadically unleashed against those he accused of stealing his work. In this regard he had chosen his new home well. Trinity College, even in its present-day setting of central Cambridge, remains essentially impervious to encroachment from the outside world. Whether one enters from the front, by way of the Great Gate, or from the rear, via the footpath across the broad green affectionately known to generations of Cambridge scholars as "the Backs," inside the court the atmosphere becomes unmistakably medieval—sequestered, serene, otherworldly, the perfect refuge for an intellectual and emotional monastic like Newton.

Where he resided during this period is a mystery. In earlier times most students lived either with or in close proximity to their tutors. In the case of sizars this arrangement facilitated the running of errands and the doing of odd jobs. But by the 1660s the practice had become less common, a casualty of the looser moral and academic standards that accompanied the Restoration. Thus while it is possible that Newton roomed with Pulleyn for a term or two after coming up to Cambridge, it is unlikely. Moreover, Pulleyn was known as a "pupil monger," a term of derision applied to those Fellows who took on large numbers of pupils in order to expand their incomes. Upwards of fifty students were assigned to him during Newton's undergraduate years, further reducing the chances that the two shared the same living quarters or forged a strong intellectual bond.

However much Newton may have wished to live alone, he did not have the money, nor Trinity sufficient space, to permit it. The best he could hope for was to find a roommate of similar outlook and temperament. Some eighteen months after Isaac's arrival John Wickins, son of the Master of Manchester Grammar School, was admitted a pensioner. Wickins died in 1719, and his son Nicholas later gave the following account of his father's chance meeting with Newton and the relationship that grew out of it:

> My Father's intimacy with Him came by meer áccident. Father's first Chamberfellow being very disagreeable to him he retired one day into ye walks where he found Mr Newton solitary & dejected. Upon entering into discourse they found their cause of Retirement ye same & thereupon agreed to shake off their present

disorderly Companions & Chum together, wch they did as soon as conveniently they could & so continued as long as my Father staid at College.[10]

For whatever reason, John Wickins never committed to paper his invaluable memories and observations, which spanned upwards of two decades of Newton's life at Trinity, a critical period about which so little information of a personal nature is available that it has been characterized as the least known of all the Newtonian dark ages.[11] Son Nicholas wrote of his father's lamentable silence: "He went no farther than ye transcribing three short Lett^{rs} he received from Him [Newton] & a Common Place of His part of wch I find under Sr Isaacs own hand, ye rest . . . is lost."[12]

As a further measure of the distance between Newton and his fellow students, we may ponder this: Even though his place in history was assured decades before he died, none among Newton's Trinity classmates left even the most meager testimony to their having known him; and this in a highly class-conscious society where the slightest contact with the great and the powerful was an honor to be cherished and a privilege to be envied. The fact is that while Newton did major work, he led a socially minor existence, touching precious few lives along the way. Given these circumstances, it was next to impossible for contemporaries to form detailed memories of him.

Of the few important sources of illumination from the early undergraduate period, the most significant is a self-revealing document compiled by Newton toward the middle of 1662. For reasons that are not stated, he underwent a crisis of conscience on or near Whitsunday (Pentecost), the seventh Sabbath after Easter, commemorating the descent of the Holy Ghost upon Christ's disciples. In 1662 Whitsunday fell on May 18, and the catechism from the Book of Common Prayer poses this question: "What is required of them who come to the Lord's Supper? Answer: To examine themselves whether they repent of their former sins, steadfastly purporting to lead a new life."[13]

In the interests of economy, but more importantly, one suspects, to conceal thoughts of the most private kind, Newton employed shorthand to compile two lists of sins, which are labeled "Before Whitsunday 1662" and "Since Whitsunday 1662."[14] The former contains some forty-five transgressions, the latter only nine. No reason for this imbalance is supplied. The shorter time elapsed in which to accumulate the later list seems the most obvious explanation, but he also may have wearied of searching out ordinary acts of human frailty and elevating them to the level of damnable breaches of the Decalogue. In both lists Newton's failure in his relations with God predominates. He had used the word "God" openly, eaten an apple "at Thy house," failed to turn "nearer to Thee for my affections," and cared "for worldly things more than God." Violations of the Sabbath are among the most common lapses of duty, and he recalled many from his boyhood and adolescence: "Making a feather [pen] while on Thy day; Making a mousetrap on Thy day; Squirting water on Thy day; Swimming in a kimnel [tub] on Thy day." Minor sins of the flesh are also confessed in number. Gluttony appears no less than three times, while a childhood craving for sweets resulted in an occasional breach of the Eighth Commandment: "Robbing my mothers box of plums and sugar" and "Stealing cherry cobs from Eduard Storer," a "crime" compounded by "Denying that I did so." John Wickins had not yet arrived at Trinity, but already Newton was deceiving his boorish first chamberfellow "of the knowledge of him that took him for a sot."[15]

For one of Newton's puritanical caste, thought virtually equaled deed. The youthful zealot was as much plagued by the possibility of sin—lust, idleness, dissipation—as with specific acts of commission. "Having uncleane thoughts words and actions and dreamese," he wrote. It was with unspeakably immoral fantasies rather than intellectual doubts that the dark powers from beyond the pale would tempt him. Until the very moment of his death, Newton feared that an entire lifetime devoted to moral rectitude and Christian piety would go for naught. Should one give in to temptation, no matter if only temporarily and in the abstract, the fall from grace would surely be unmercifully swift and irreversible.

In surveying Newton's conduct during this crucial stage of his religious development, one is struck by the close parallel between the shaping of the Protestant temperament in pre-Restoration England and that in colonial America. At puberty the child of rigid Protestant upbringing commonly faced a crisis of self. Emergent sexuality confirmed his own sinfulness, producing feelings of guilt, which merged with a rebellious desire for independence from parents and their tyrannical God. Each of these elements is amply present in Newton's list of sins: "Having uncleane thoughts ... and dreamese; Peevishness with my mother, Threatening my father and mother Smith to burne them and the house over them; Not loving Thee for Thyself; Not desiring Thy ordinances." This heady and temporarily liberating experience of protest served as a necessary prelude to a personal and often highly emotional conversion to Christ. Afterward, for having experienced what he had been taught to look upon as sin, the young adult feared damnation. He reacted, as did Newton, by throwing himself on God's mercy and promising the impossible—to sin no more.[16] Though now a selfless child of the Creator, the newly regenerate Christian found that guilt was his unshakable companion, that evil lurked not only in the shadows but in the hearts of men.

III

There is no reason to suppose that Newton's formal education deviated in any significant respect from the straight and the narrow. Indeed, the fact that Benjamin Pulleyn, a Regius Professor of Greek in the making, had

been named his tutor would suggest that the youth was as thoroughly in-
doctrinated in the mysteries of peripatetic philosophy as any of his fellow
students, if not more so. The most convincing evidence of this can be found
in a commonplace book begun by Newton in his first year at Trinity and
continued until his graduation in 1665. In this small volume, bound in worn
brown leather and containing 140 leaves, Newton entered his student notes
at both ends, a widespread custom among scholars in an age when writing
paper of quality was a dear commodity. He recorded his ownership with the
inscription: "Isaac Newton, Trin: Coll. Cant. 1661."[17]

Both ends of the commonplace book begin with incomplete notes in
Greek from Aristotle's works, the *Organon* and the *Nicomachean Ethics*.
The beautiful script has obviously been entered with great care, for a
number of the folios contain the handsomest examples of Newton's
classical hand to be found anywhere. Pulleyn was undoubtedly influential
in this regard; yet the careful attention to detail, especially in a notebook
hardly meant for his tutor's scutiny, suggests something more profound: a
genuine warmth and reverence for the works of the preeminent thinker of
the ancient world, perhaps of all history. Even though Newton's discoveries
in natural philosophy would at last seal the doom of Aristotle's seemingly
deathless cosmological system, one searches his voluminous works in vain
for any signs of disrespect for the philosopher, a startling contrast to
Galileo's frequently contemptuous and pointed assaults.

Then, at some point in 1663, Newton ceased taking copious notes on
the works of Aristotle and his commentators. In a manner wholly uncharacteristic of one who so hated to waste paper that his thoughts and equations
were frequently jotted on the unfolded envelopes of letters from his later
correspondents, he left dozens of pages blank before beginning a new,
radically different series of entries. In this instance the most plausible explanation is also the most obvious. Newton had begun to experience the
most revolutionary chapter of his intellectual development, the significance of which he consciously underscored by the simple act of spatial
separation in his notebook. He titled the new section, appropriately
enough, *Quaestiones quaedam Philosophicae*. And at the top of the first
page, in somewhat fainter ink, he penned a line that not only set the tone of
the philosophical questions he postulated and sought to answer but also
summarized to perfection his unshakable commitment to the new order:
"Amicus Plato amicus Aristoteles magis amica veritas." ("I am a friend of
Plato, I am a friend of Aristotle, but truth is my greater friend.")

He wrote out an index of the various subjects to be investigated. These
range from "Aer [Air]," "Earth," and "Matter" to "Time and Eternity,"
"Soule," and "Sleepe." Under a number of the headings he made no entries
at all, a foreshadowing of his later aversion to the formulation of insupportable hypotheses. Under others he wrote from a few sentences to several
pages. As in the Morgan notebook of 1659, Newton had taken much of the
information from the works of contemporary thinkers. Yet the *Quaestiones*

themselves were most certainly formulated by him, thus providing the earliest extant evidence that his nascent genius was on the verge of full flower. What fascinated him about nature at this early stage of his scientific studies was its variety, its infinite adaptability, the fitness and yet the individuality of its many parts. At the same time, one observes the origins of a far more profound struggle to achieve a principle of unity, a singleness, a model in which the disparate pieces can be synthesized and made whole.

This quest is reflected in the beginning series of entries concerning the nature of primal matter. Following the example of the natural philosopher Walter Charleton, Newton recorded several possibilities regarding matter's composition: whether it be points, points and parts, a simple entity before division indistinct, or atoms.[18] He immediately dismissed the first two alternatives and attacked the third, which had been postulated by Descartes. For if matter were an infinitely extended plenum, as the French philosopher had argued, motion would be impossible, because the space required for the displacement of one particle or body by another would not exist. "It remains therefore y^t ye first matter must be attomes and y^t Matter may be so small as to be indiscernible."[19] This idea of matter, which Newton would come to embrace even more fervently in later years, had been revived from the ancient Greek atomists by Pierre Gassendi, a seventeenth-century mechanist. Gassendi and his fellow atomists held that matter occupies only a part of the great void. Between these invisible particles are equally invisible pockets or pores, which contain nothing of a corporeal nature. Moreover, all bodies are composed of an indefinitely large number of unseen atoms of various shapes and sizes. Physical objects differ from one another because of the number, density, and movement of these minute constituent parts. But even the most solid of bodies, the planets included, are honeycombed by minuscule interstices of empty space. By choosing the atoms of Democritus and Gassendi over the plenum of Descartes, Newton set his course down a particular path from which he did not return. At the same time, however, he continued to embrace Descartes's mechanical conception of nature as the only feasible basis for scientific research and discussion. Invisible, yet discernible, mechanisms must stand behind the attractions and repulsions generated by discrete particles in ever continuous motion. Newton even employed a modified form of the atomic or corpuscular theory of matter to account for the very life process itself:

Vegetables

Suppose ab ye pore of a vegetable [be] filled with fluid mater and y^t ye Globule c doth hitt away ye particle, b, y^n ye rest of subtile matter in ye pore riseth from a towards b and by this means juices continually arise from ye rootes of trees upward: wch juices leaving dreggs in ye pores and y^n wanting passage stretch ye pores to make y^m as wide as before they were clogged wch makes ye plant bigger untill ye pores are too narrow for ye juice to arise through ye pores and y^n ye plant ceaseth to grow any more.[20]

Though introduced to natural philosophy through the systematic study of Aristotelian science, at the very outset of his scientific career Newton parted company with the Stagirite, albeit on amicable terms. He rejected the classical qualitative conception of matter for a quantitative model, which held that natural phenomena are the product of shifting atomic particles controlled by fixed mechanical laws. The world of self-willed elements—earth, water, air, and fire—had already given way in Newton's mind to the universe of machine.

It is apparent from his childhood preoccupation with all things mechanical that Newton harbored a strong predilection for Bacon's new experimental method long before he came to know it by name. By 1664 he clearly looked upon it as the indispensable tool for overcoming the inherent weaknesses of a classical science based upon sense experience and *a priori* reasoning alone. "The senses of divers men are affected by y^r same objects according to ye diversity of their constitution," he wrote in the philosophical notebook. "The nature of things is more securely and naturally deduced from their operacons out upon another y [than] upon ye senses. And when by ye former experiments wee have found ye nature of bodies ... wee may more clearly find ye nature of ye senses." Since "wee are ignorant of ye nature of both soule and body," our exploration of nature must not be colored by either.[21]

Under the heading of "Water & Salt," Newton presented an excellent example of the type of experimental problem he had in mind. Descartes offered as his explanation for the ancient riddle of the tides an extension of the vortex hypothesis. He reasoned that since the gigantic whirlpool of matter surrounding the earth must continuously pass between the planet and the moon, it is compressed into a smaller than normal space. This in turn exerts a greater pressure, which pushes down upon the oceans causing a low tide. Would not this same pressure, Newton queried, affect a barometer in a like manner; indeed, do elevated barometric readings occur in regular cycles which coincide with low tides?

> To try whither ye Moone passing ye Atmosphere cause ye flux and ye reflux of ye sea. Take a tube ... with water which is larger ... y^n 30 inches ... ye top being stopped ye liquor will sink 3 or 4 inches below it leaving a vacuum (perhaps) then as ye air is more pressed without by ☽ [moon] so will ye water rise or fall as it doth in a witherglasse by heate or cold. ... Observe if ye sea water rise not in day & fall at nights by reason of ye earth pressing from ☉ [sun] upon ye night water etc. Try also whither ye water is higher in mornings or evenings to know whither ☉ or its vortex press forward most in its annual motion.[22]

This experiment, earlier designed by Christopher Wren and written up by Boyle, in whose work Newton had probably discovered it, did not support Descartes's ingenious but erroneous hypothesis. A little more than two decades later Newton, with a sweep and finality that stunned his incredulous contemporaries, announced his own elegant mathematical solution to the ebb and flow of the seas.

A thorough reading of the philosophical notebook indicates that Newton carried his program of independent study far beyond the works of Descartes, Charleton, and Gassendi. Extracts and interpretations from the writings of other virtuosi abound. Among the most prominent are Boyle, Kenelm Digby, Henry More, Thomas Hobbes, Galileo, and Joseph Glanville. Many of their books and those of Copernicus, Thomas Digges, Tycho Brahe, and Kepler were to be found in the steadily expanding collection of Trinity College Library. The principal exceptions were Galileo's two most important works, *Dialogue Concerning the Two Chief World Systems* and *Discourses Concerning Two New Sciences*. It would appear that the Italian's open hostility toward Aristotle made his books anathema to the authorities of Aristotelian Cambridge.[23]

In Newton's day the College library was not open for extended periods on a regular basis. Fellows had their own keys and could come and go as they pleased, a privilege denied their students. Nor could students use the collection without being accompanied by a Fellow. Such restrictions, though irritating and inconvenient, raised no insuperable barriers to one, like Newton, obsessed by the desire to know. He continued his boyhood practice of taking extensive notes when given access to stimulating books; neither would it have been unusual for his tutor to borrow volumes on his behalf, especially during the last year of his studies, when more freedom from the narrow scholastic curriculum was tolerated. Humphrey Babington was another logical source of access to printed knowledge. He took Newton under his wing and, being a Major Fellow, could have obtained almost any book the young man desired. And what Trinity or the University library did not have could be borrowed from Fellows at St. John's, Christ's, or Queens'. Moreover, Babington, like others among Newton's Trinity contacts, had an extensive personal library. When forced to leave Cambridge by the plague in 1665, Newton visited Babington's home in Boothby Pagnell, near Woolsthorpe. There he used the amiable Rector's books and undertook advanced mathematical calculations based on the pioneering work of John Wallis, Johann Hudde, and Franz van Schooten.[24]

Despite his heavy reliance on the innovative thinkers of his day, there was much in the philosophical notebook that Newton did not extract from the works of others. For example, he had become intensely fascinated by celestial phenomena, which included the regular tracking of comets across the starry firmament. The first was observed on December 10, 1664; a second appeared a week later at 4:30 in the morning. He made regular observations of the celestial visitor until it finally disappeared from sight a month later. Still not satisfied, Newton continued to scan the night skies of winter while the rest of the University slept. These solitary efforts were rewarded a third time when he spied yet another comet streaking through the constellation Andromeda in early April 1665, concerning which the last datable entries in the notebook are made.[25] How is it, he wondered, that these brilliant objects could move with such speed and seeming impunity through the massive celestial vortices of Cartesian space?

Not only did Newton court exhaustion and illness during this intense period of independent study, but he also put his eyesight in jeopardy. Desirous of experiencing first hand the effects of highly concentrated light on the eye, he fixed his gaze on a looking glass in whose reflection he captured the image of the sun. Many years later he vividly recounted the experiment and its result in a letter to his friend the political philosopher John Locke. "I looked a very little while upon ye sun in a looking-glass with my right eye & then turned my eyes into a dark corner of my chamber & winked to observe the impression made & the circles of colours wch encompassed it & how they decayed by degrees & at last vanished." Growing bolder, he repeated the procedure a second and third time. At length the brilliant, multicolored imprint of the sun overlay almost any object upon which Newton gazed, from an open book to a moving cloud. A period of crisis ensued:

> And now in a few hours time I had brought my eyes to such a pass that I could look upon no bright object . . . but I saw the sun before me, so that I durst neither write nor read but to recover ye use of my eyes shut myself up in my chamber made dark for three days together & used all means to direct my imagination from ye Sun. For if I thought upon him I presently saw his picture though I was in ye dark. But by keeping in ye dark & imploying my mind about other things I began in three or four days to have some use of my eyes again & by forbearing a few days longer to look upon bright objects recovered them pretty well, tho not so well but that for some months after the spectrum of the sun began to return as often as I began to meditate upon ye phaenomenon, even tho I lay in bed at midnight with my curtains drawn.[26]

If Newton temporarily yielded to a sense of rising panic, there is no hint of it among his notes and recollections. He seems to have viewed the experience as simply another opportunity to learn, unpleasant and terrifying though the consequences must have been. When he had sufficiently recovered his sight to take pen in hand, the various phases of the lingering affliction were duly written up in the philosophical notebook in the form of ten carefully worded steps.[27] They stand as a lasting reminder of Newton's ability to exert an extraordinary degree of self-control and detachment under the most strenuously adverse of physical circumstances.

However ill-advised the experiment with sunlight may have been, it was undertaken for a purpose that transcended idle curiosity. The first Newton scholar to draw attention to the *Quaestiones quaedam Philosophicae* wrote: "The most profound as well as the most original writing in this commonplace book is undoubtedly that on the optics."[28] Indeed, the first major discoveries Newton considered worthy of announcement to other virtuosi were made in this science, even though he had simultaneously founded the calculus and roughed out the inverse square law, a most significant step in the direction of universal gravitation.

The dates of the notes on light and color, like so many of the other entries, are not given. "In August 1665," according to Conduitt, "Sr I bought a prism at Sturbridge fair to try some experiments upon Descartes's book of colours."[29] Conduitt's chronology is incorrect, however, for Newton had already left Cambridge to escape the plague.[30] Furthermore, the King had issued a proclamation "that Sturbridge fayre should not this yeare [1665] be kept because of the great Plague at London." It has been suggested that the prism purchase actually occurred a year earlier than Conduitt had supposed and that the early optical experiments were recorded between August 1664 and the end of 1665.[31] Whatever the case, Newton's first serious scientific thoughts on the questions of light and color are contained in the philosophical notebook. From these notes it is clear that at this time he had only scattered elements of the far deeper knowledge gained a year or two later by further research and experimentation. Yet his early findings were by no means insignificant, and they established a solid working foundation on which to build a revolutionary theory. He had already discovered that the apparent color of an object is related both to the nature of its surface and the composition of the light striking it. He conjectured that the color of the refracted ray was dependent upon its velocity and apparently believed already that light was the product of corpuscular motion. A prism simply acted as a filter to separate slow-moving corpuscles from the fast. Most important, he had discovered the varying refrangibility of light—that those rays which produce blue are more sharply bent than those which produce red.[32]

If the philosophical notebook reveals anything, it is this: By 1664 Newton had determined what his life's method of scientific investigation would be. He had already fused the two major strands of modern science, the rational and the empirical. The logical outlook of Descartes is joined with the experimental passion of Bacon and Galileo. A substratum of yet undefined particles and mathematically determinable laws has replaced the classical-medieval world of Aristotelian hierarchies. At the same time, however, the clockwork precision with which the universe functions is for Newton anything but the result of blind circumstance. For behind it all he sensed the presence of intelligible planning and purposeful direction. Under the heading "Of God" he wrote' "Were men and beasts & c made by fortuitous jumblings of the attomes there would be many parts useless in them, here a lumpe of flesh there a member too much. Some kinds of beasts might have had but one eye some more y^n two."[33] This was more than a passing observation; it was a reminder to himself that the underlying unity in nature, revealed to man through rational inquiry and observation, is a product of the Divine Mind. Atoms there are, and mechanical laws, but when measured against the wisdom and knowledge of the Creator they are as nothing. Hence from the very beginning Newton, like Boyle, held that no matter how rational the world may appear to us, there is nothing below a certain depth that is truly explicable in human terms. Few things would have angered or dismayed him more than the Enlightenment belief that the *Principia* contained the framework of a universe in which God was no longer a vital, or even necessary, part.

IV

A further indication of Newton's rapid intellectual maturity was his decision, made toward the beginning of 1664, to cultivate a new style of handwriting. The rather ornate script of his adolescence has yielded to a simpler, somewhat less pretentious form, which, with occasional modifications, was to remain Newton's for the rest of his life. This conscious sacrifice of a degree of classical elegance must have seemed of little consequence when considering the utilitarian benefits conferred by the more natural hand. For in addition to his published works, he left thousands of pages and many millions of words in manuscript at his death. In the heightened consciousness of genius the mind insists on expressing itself; it will not be silenced. It will speak, even if it must speak on paper to itself alone. It is little wonder, then, that as a young man Newton confessed to violating the Sabbath by fashioning pens from goose quills and was constantly in search of better recipes for making the untold gallons of ink that flowed, like a dark, flood-swollen river, from their tips. He had literally taken to heart the words of the ancient prophet Isaiah: "Precept upon precept, precept upon precept; line upon line, line upon line; here a little, and there a little."

As the contents of the philosophical notebook testify, Newton pursued many disparate thoughts at once, moving in peripatetic fashion from one concept to another as his fevered mind sought to encompass and assimilate a newly discovered embarrassment of intellectual riches. Though possessed of an iron will and a physical constitution to match, Newton, too, had his limits, and these he apparently overstepped for the first time in 1664. He became so fascinated with celestial phenomena that his nightly vigils deprived him of the rest needed to sustain his health. Newton informed Conduitt in the course of their conversations that "He sate up so long in the year 1664 to observe a comet that appeared then that he found himself much disordered & learned from thence to go to bed betimes."[34] What form this affliction took is not stated, nor is there any way of determining its duration. We may surmise with some measure of confidence, however, that it was similar to, though less severe in degree, the nervous exhaustion suffered by Newton in 1693, the so-called Black Year. In the latter instance his mental equilibrium became so disturbed that he wrote to accuse John

Locke of a plot "to embroil me with women" and astounded the diarist Samuel Pepys with the assertion "that I must withdraw from your acquaintance, and see neither you nor the rest of my friends any more." Both men, deeply concerned for Newton's sanity, responded to his bizarre outbursts with admirable equanimity. Newton later apologized and explained himself to Locke: "When I wrote to you, I had not slept for an hour a night for a fortnight together, & for 5 nights together not a wink."[35] While he hardly seems to have suffered from such extreme physical deprivation in 1664, the essential cause of his illness would appear to be the same as that of the later period—a prolonged lack of rest and excessive mental strain. Newton's work habits, both in early life and in middle age, were pathological in their addictiveness. He willingly starved himself socially, sexually, and, not least, nutritionally. He courted sickness, failure, even insanity in order to test whether the established world would break him or whether he would triumph by displacing certain of its time-worn fundaments with new ones of his own. John North, Master of Trinity from 1677 to 1683, once remarked to his nephew that "if Sir Isaac Newton had not wrought with his hands in making experiments, he had killed himself with study. A man may so engage his mind as almost to forget he hath a body which must be waited upon and served."[36] As North's observation clearly illustrates, it is not enough to say that Newton was a man of dogged determination. His unquenchable thirst for work combined with an almost total abstinence from life's everyday pleasures add up to something far more compelling. Though one hesitates to use the word, he seems to have been nothing less than a man possessed. In this instance, however, no witch, warlock, or demon had laid claim to a human soul. Newton's possession was self-induced, the result of genius turned sorcerer. Because of his absolute refusal to accept anything approaching reasonable limits of physical and mental endurance, a large part of the most creative period of his life was spent perilously close to that indefinable boundary which separates the rational and the orderly from the frenzied abyss of psychic chaos.

V

If Newton was to pursue his scientific studies more deeply, it was essential that he develop a mastery of existing mathematical techniques. In the absence of such knowledge large stretches of the work of Galileo, Kepler, and Descates—not to mention many lesser lights—would be denied him. Even though Newton the mathematician makes no appearance in the philosophical notebook, there is every reason to believe that the origins of his studies in the field date from this very period. During the late summer of 1664 he composed his first mathematical essays,[37] and shortly thereafter he began the quest that would lead to the discovery of the fluxions (calculus) in the plague year of 1665. Fortunately a partial record of his earliest forays into the subtleties of advanced mathematics survives. Having been given the library of his deceased stepfather by Hannah, Newton found among its contents Barnabas Smith's commonplace book. He appropriated the remaining blank pages and generous margins for his own use, filling them with problems in the analytical geometry of Descartes and, more important, his own revolutionary fluxional calculations.[38] It is indeed ironic that he should have laid the foundations of one of his most profound intellectual achievements on pages interspersed with those containing the thoughts and scholarly notations of a man he had hated.

Like any other student new to the field, Newton did not begin his mathematical studies at the advanced level. And, as with the many other facets of his intellectual development, a truly definitive chronology of his first steps in mathematics has never been established. Of the scattered accounts of his mathematical baptism, the most frequently cited is that compiled by the French mathematician Abraham de Moivre, one of a coterie of privileged young men who were allowed access to Newton's papers once his creative period had come to an end. The following is de Moivre's low-key yet remarkable account of genius in the throes of metamorphosis:

> In 63 [Newton] being at Sturbridge fair bought a book of Astrology to see what there was in it. Read it till he came to a figure of the heavens which he could not understand for want of being acquainted with Trigonometry.
>
> Bought a book of Trigonometry, but was not able to understand the Demonstrations.
>
> Got Euclid to fit himself for understanding the ground of Trigonometry.
>
> Read only the titles of the propositions, which he found so easy to understand that he wondered how any body would amuse themselves to write any demonstrations of them.

But he soon realized that he had underestimated the work, especially after encountering the Pythagorean theorem that the sum of the squares of the lengths of the sides of a right triangle is equal to the square of the length of the hypotenuse.

> Began again to read Euclid with more attention than he had done before and went through it.
>
> Read Outhtreds [*Clavis Mathematica*] which he understood tho' not entirely. . . . Took Descartes's Geometry in hand, tho' he had been told it would be very difficult, read some ten pages in it, then stop't, began again, went a little farther than the first time, stop't again, went back again to the beginning, read on till by degrees he made himself master of the whole, to that degree that he understood Descartes's Geometry better than he had done Euclid.
>
> Read Euclid again, & then Descartes's Geometry for a second time. Read next Dr Wallis's Arithmetica Infinitorum. . . .
>
> In 65 & 66 began to find the method of Fluxions, and writt several curious problems relating to that method bearing that date which were seen by me above 25 years ago.[39]

In addition to the de Moivre memorandum, we have from Newton's pen a brief but valuable firsthand account of his early mathematical studies. In 1699 his claim to priority concerning the calculus was under serious question, and to protect his position he undertook a reexamination of his student papers. On the facing page of a notebook containing, among other things, annotations from the work of John Wallis, he wrote:

> July 4th 1699. By consulting an accompt of my expenses at Cambridge in the years 1663 & 1664 I find that in ye year 1664 a little before Christmas I being then senior Sophister [undergraduate] I bought Schooteen's Miscellanies & Cartes's Geometry (having read this geometry & Oughtred's Clavis above half a year before) & borrowed Wallis's works and by consequence made these Annotations out of Schooten & Wallis in winter between the years 1664 & 1665. At wch time I found the method of Infinite series. And in summer 1665 being forced from Cambridge by the Plague, I computed ye area of ye Hyperbola at Boothby in Lincolnshire to two & fifty figures by the same method.
>
> Is. Newton[40]

In comparing these two generally compatible accounts, it is interesting to note that de Moivre strongly emphasizes Newton's early, though somewhat disjointed, use of Euclid, while Newton himself makes no mention of the ancient mathematician. Henry Pemberton, a second-generation Newton disciple, addressed himself to this issue near the time of Newton's death: "I have heard him even censure himself ... and speak with regret of his mistake at the beginning of his mathematical studies, in applying himself to the works of Des Cartes and other algebraic writers, before he had considered the elements of Euclide with that attention, which so excellent a writer deserves."[41] Newton's early mathematical notes bear out Pemberton's recollections. For the time being Euclidian geometry, in which he had no formal instruction, was of secondary concern to one whose interest in modern analysis was quickly transformed into an all-consuming passion.

In neither account is any mention made of a guiding hand, thus leaving the distinct impression that Newton alone was responsible for his mathematical education. His tutor, Benjamin Pulleyn, a classical scholar, could have provided at best some measure of encouragement but little, if any, of the expertise to advance his pupil's mathematical studies. Moreover, on at least two occasions in later life Newton commented on self-education in mathematics and natural philosophy in a manner too suggestive of his own learning experience to be considered the result of coincidence alone.

In 1682 Edward Paget, a Fellow of Trinity College for whom Newton seems to have harbored a rare personal affection, became a successful candidate for the post of Master of the Mathematical School at London's Christ's Hospital. Newton wrote a strong letter of recommendation to the Board of Governors on Paget's behalf, which contains this passage: "He understands ye several parts of Mathematicks ... & wch is ye surest character of a true Mathematicall Genius, *learned these of his owne inclination, & by his owne industry without a Teacher.*"[42] A decade later, Newton compiled a set of instructions for the scholar and critic Richard Bentley to aid him in understanding the *Principia*. Before turning a page of the formidable treatise, Bentley was encouraged to undertake extensive background reading, beginning with Euclid and continuing through the works of Jan de Witt, Phillippe de la Hire, Schooten, and Gassendi. Then, "At ye first perusal of my Book it's enough if you understand ye Propositions with some of ye Demonstrationes wch are easier than the rest. For when you understand ye easier they will afterwards give you light into ye harder."[43] In other words, if you would comprehend what I have created you must learn, as I too learned, by building, in natural progression, on a solid foundation. In such a manner was Newton rapidly borne to the limits of existing mathematical knowledge, and beyond.

A specter has haunted virtually every account of Newton's early years of study at Cambridge. William Stukeley, who seems to have given life to this remarkably tenacious phantom, wrote that "The famous Dr Barrow, [later] Master of Trinity, was Sr Isaacs tutor & tis likely he took a byass in favor of mathematical studys from him." This is not an unreasonable conjecture for one who had no access to Newton's early papers. Aware that Barrow later resigned his professorship and favored Newton as his successor, Stukeley could not resist the temptation to gild the lily. "The D^r had a vast opinion of his pupil, & would frequently say that he himself truly knew somewhat of the mathematics, still he reckon'd himself but a child in comparison of Newton."[44]

Isaac Barrow was the first to hold the Lucasian Chair of Mathematics endowed in 1663 by Henry Lucas, Fellow of St. John's and one of the University's members of Parliament. It was Lucas's wish to give Cambridge a lectureship equal in status to those founded at Oxford in geometry and astronomy by Sir Henry Savile in 1619.[45] In a letter of January 18, 1664, confirming royal acceptance of the Lucasian statutes, Charles II expressly forbade the chair's holder to tutor anyone but a fellow-commoner, which Newton never was.[46] In this same letter the King ordered "all Undergraduates after their 2nd year, and all Bachelors of Arts" to attend the Lucasian Professor's lectures, which Barrow inaugurated on March 14, 1664. Newton, by his own account, was among the auditors on at least two occasions. Though he was vague about their content, Barrow's ideas on motion do seem to have made a positive, though by no means profound, impression at the very time he was beginning the mathematical studies that would lead him to the calculus. "Dr Barrow then read his Lectures about motion that might [have] put me upon taking these things [fluxions] into consideration," he wrote many years later.[47] What else, if anything, Barrow did at this time beyond stimulating Newton's interest in mathematics will never be known. Even the good Doctor's biographer was forced to admit that Barrow's beginning lectures would probably never have found their way into print had his wishes been consulted, for they were far more elementary than he would have liked.[48]

Stukeley's account is further belied by an anecdote contained in the de Moivre memorandum and repeated in somewhat embellished form by Conduitt. If Newton entertained any hopes of winning a fellowship, so that he could continue his studies indefinitely at Trinity after graduating B.A., he first had to be elected to one of the sixty undergraduate scholarships supported by the College. The scholarship examinations were held in April 1664:

> When he stood to be scholar of the house his tutour sent him to Dr Barrow then Mathematical professor to be examined, the D^r examined him in Euclid wch Sr I. had neglected & knew little or nothing of, never asked him about Descartes's Geometry wch he was master of. Sr I. was too modest to mention it himself & Dr Barrow could not imagine that one could have read that book without first [be-ing] master of Euclid, so that Dr Barrow conceived then but an indifferent opinion of him but however he was made scholar of the house.[49]

The puzzling thing about this story is why Newton should have been examined by Barrow at all, for the mastery of mathematics was deemed of little importance compared with the other studies that formed Trinity's neoscholastic curriculum. One can only surmise that Pulleyn, who was by now aware of Newton's keen interest in the subject, sought to aid his tutee by enlisting Barrow's help. If so, Newton, through no fault of his own, embarrassed not only himself but also his well-meaning tutor. Cut to the quick, he found that the only way to assuage his deep sense of humiliation was to attack and master Euclid's geometry, even as he had attacked and mastered his childhood tormentor in the schoolyard and classroom at Grantham several years before. Few if any of the surviving books from Newton's library are more thumb-worn or more thoroughly annotated than the *Euclidis Elementorum,* a geometry textbook authored by none other than Isaac Barrow, whose "indifferent opinion" of Newton was soon to undergo a marked change for the better.[50]

Thus, Newton had began his Promethean quest alone, and so too would he end it—locked up in his private world, haunted by the accumulated images of a lifetime. It is true that a person too early cut off from the common interests of other individuals is exposed to a degree of inner impoverishment. But it is equally true that the rare and the beautiful sometimes survive only in protective isolation. Observers and participants alike have often spoken of the loneliness of this inspiring process. As a young man, Charles Darwin took his greatest enjoyment in tramping across the Andean highlands in the silent company of illiterate natives: "The whole of my pleasure was derived from what passed in my mind," he wrote in his journal. And when, after he became famous, Newton was asked how he had made his discoveries, he remarked that truth is "the offspring of silence and unbroken meditation."[51]

Such is the transit of genius.

VI

Though the study of nature and mathematics became Newton's primary preoccupation while a Trinity undergraduate, he also evidenced a constant and at times troubled concern for all matters financial. Included among the list of his sins compiled in 1662 was the admission to "setting my heart on money" more than God. And this was followed by a further confession that he had suffered "a relapse."[52] However guilty Newton may have felt after yielding to his materialistic desires, he was ready and willing to take advantage of the extravagances of his peers. The Trinity College notebook reveals that Newton the budding natural philosopher was also Newton the student moneylender. Sizars and pensioners alike took advantage of his services, and by the time he was in his second year of college business was thriving.[53]

	£	s.	d.
Lent Pollard	0—	2—	0
Lent Bigg	0—	7—	6
Lent Pollard	0—	1—	0
Lent Agatha	0—	1—	0
Lent Andrews	0—	11—	5
Lent Oliver	0—	1—	0
Lent Wilford	0—	6—	0
Lent Gosh	1—	0—	0
Lent Guy	0—	10—	1

Puritan that he was, he lent as much as a pound to only one person during this period, and with considerable trepidation, noting that the sum was "to bee payed on fryday."

When a loan was repaid, he drew a line through the borrower's name and further marked the entry with an X. Occasionally no such marks appear, proof that Newton, as all creditors eventually must, had come across a bad debtor. What rate of interest he charged his youthful borrowers is not recorded, but given the considerable number of loans made and the fact that occasional losses were incurred, we can surmise that this service was not performed gratis. Besides, the frequent granting of favors, financial or otherwise, for the sake of friendship alone was scarcely Newton's way. And since the moneylender has never been a popular figure among those forced to rely on his goodwill, Newton's decision to take up the calling of Shylock may have only served to increase the already considerable distance between himself and his fellow students. His account book for 1663–64 is lost, and with it any hope of gaining more direct knowledge of his social life as a Cambridge undergraduate.

A product of the parochial existence of an isolated village, Newton found himself in a community awash with the moral license that accompanied the Restoration. There was no more significant sign that the University had returned to its traditional life and customs than the replacement of the pileus, the round student cap of the Commonwealth, by the mortarboard. The circle, as one wag put it, had been squared and with it, we might add, old scores settled. But even when Puritanism had reigned supreme, the authorities had found it extremely difficult to impose discipline on the high-spirited student body. With the Restoration such attempts became little more than face-saving pretense. Drinking, gambling, and whoring once more became commonplace, and the student who refused to join in could easily find himself a pariah. Thus was Newton, the outsider, driven from his quarters by a boisterous roommate and his equally loutish friends, perhaps even made the butt of their ribald humor. Wandering disconsolately along the walks, he counted himself fortunate to encounter a like spirit in the form of Wickins, who shared his distressing pangs of moral and intellectual alienation. Together they would weather the hedonistic storm that swirled round them, secure and content in their private quarters, an island of self-sufficient order.

Having faced and survived the scholarship examinations of 1664, Barrow's allegedly indifferent opinion of his preparation in classical mathematics notwithstanding, Newton no longer bore the to him odious title of "sizar." As a scholar, he was now entitled to receive meals (commons) from the College, a small livery allowance, and a stipend of equal amount. Of greatest importance, however, was the guarantee that he could remain at Trinity to take his M.A. and, if all went well, extend his residence at Cambridge indefinitely by eventually obtaining a fellowship.

It was at this point that any remaining hope Catherine Storer had of wedding her childhood sweetheart vanished forever, sparing her the heartache of playing Penelope to a young Odysseus of the mind. Having learned of Isaac's plans, she married Francis Bakon, a Grantham attorney, in a service performed by her uncle Humphrey Babington at Boothby Pagnall in 1665. Still, Catherine's affection for Newton did not die, as is evidenced by the touching manner in which—more than sixty years later—she recounted to Stukeley from memory the verses from the *Eikon Basilike* that Isaac had copied on the walls of his room in the attic of her stepfather's Grantham home. Here she may have lingered in a state of melancholy after his departure, seeking solace from the scattered drawings, poetry, and books he left behind. Neither, in his final years, did Newton forget the only woman in his life for whom he entertained the slightest romantic interest. When Stukeley informed him in 1726 that he had decided to establish a medical practice in Grantham, Newton asked the physician to inquire as to whether the house to the east of the church could be purchased at a reasonable price, for "his old acquaintance Mrs Vincent [once] lived in the place."[54]

Within less than a year of being made scholar Newton, along with his fellow candidates for the B.A., stood *in quadragesima,* the outmoded medieval disputations which provided the last hurdle on the path to the degree. As a student at Cambridge a generation later, Stukeley heard "that when S^r Is. stood for his bachelor of Arts degree, he was put in second posing, or lost his groats, as they call it, which is looked upon as disgraceful." Having no way to verify the story, Stukeley nevertheless found it quite in character: "it seems no strange thing, notwithstanding Sr Isaacs great parts, for we may well suppose him too busy in the solid parts of learning, to allow much time to be master of words only, or the trifling nicetys of logic, which the universitys still make the chieftest qualification for a degree."[55] In this instance Stukeley's conclusion would appear to match the known facts, for Newton had abandoned any serious pursuit of the traditional curriculum by early 1664. He may well have relied on the general state of intellectual torpor that plagued the University to carry him through the final disputations, not to mention the aid of an influential protector like Humphrey Babington and the support of his tutor, Pulleyn, who served on the examination committee.[56] In this he was not disappointed. Together with twenty-five other Trinity men, Newton took his B.A. in the spring of 1665. What his final ranking was we do not know, because the "Ordo Senioritatis" for this year is annoyingly omitted from the Grace Book. Yet, whether first or last, as a young man of twenty-two Isaac Newton had already recorded in his notebooks brilliant flashes of that mysterious power which would soon compel not only men but the modern world to move to his ideas.

Chapter Four

A Movable Feast

The young men's vision, and the old men's dream.

Dryden

I

With few exceptions, Englishmen of the seventeenth century were as yet unaccustomed to seeking natural causes for the great disasters that periodically ravaged their island nation. They turned instead to Scripture and the chilling Revelation of John, the mercurial evangelist of Patmos:

> And I looked, and behold a pale horse: and his name that sat on him was Death, and Hell followed with him. And power was given unto them over the fourth part of the earth, to kill with sword, and with hunger, and with pestilence, and with the beasts of the earth.[1]

Thanks to civil war, England had known much of the sword in its recent past, while famine, though less a problem than in earlier times, remained endemic until the ninteenth century. Disease, of course, was an ever present threat to life, but never more so than when manifested in the form of plague, whose sporadic visitations more radically altered the course of European history before 1700 than any other series of disasters, whether natural or man-made.

The most virulent and dread form of plague was the infamous Black Death, which first appeared in Europe during the late summer of 1347. Carried to Italian ports amid cargo from western Asia, its primary host was none other than the flea which infested *Rattus rattus*, the black house or ship rat, diminutive and unsuspected companion beast of Revelation's Four Horsemen.

In the absence of modern antibiotics there was no cure for the disease, so in late medieval times the professional gravedigger came into his own. Since Death could not be denied, he became a dominant figure in European art, for he was king, leading in alternately rowdy or stately dance the population of the earth—lord, merchant, and begger alike. In some representations he abandoned the dance and appeared as a sudden, unexpected visitor who came to take the peasant from his field, the banker from his wealth, or the lover from his mistress. Finally Death became the plague incarnate, a grinning skeleton or a stooped pilgrim ravaged of body but ever in human form—a hideous, mocking caricature of man himself. "As I am now, so you soon shall be," he whispered menacingly from the artist's macabre canvas.

Some forty years had passed since the last serious outbreak of bubonic plague in London during the accession of Charles I in 1625. As the decades accumulated, men and women tended to forget the horrors that their parents and grandparents had endured in silence and resignation. In the meantime the city's population rapidly increased, reaching an estimated half a million by 1660. Many of the poor were crowded outside the city walls into the slum districts of the "Liberties" beyond—Cripplegate, Whitechapel, St. Giles-in-the-Fields, Westminster, and Stepney—where their numbers multiplied geometrically despite an appalling rate of infant mortality. Here the Great Plague erupted in 1665, spreading outward from this rat-infested epicenter to engulf all of Greater London and countless towns and villages beyond.

On April 30 Samuel Pepys, future Secretary to the Admiralty, made the first of many entries concerning the plague in his invaluable diary: "Great fears of the Sickenesse here in the City, it being said that two or three houses are already shut up. God preserve us all." Some two months later the disease had spread so rapidly that Pepys resolved to "put all my affairs in the world in good order, the season growing so sickly that it is much to be feared how a man can scape having a share with others in it."[2] By now every Londoner knew the ominous symptoms and the inexorable chain of progression associated with the disease. Thomas Vincent, a clergyman who, like so many others, saw the plague's coming as a manifestation of Divine judgment, recorded this vivid description of its onset in his widely read book, *God's Terrible Voice in the City*:

> [It] first began with a pain and a diziness in the head, then trembling in other members; when they have felt boiles to arise under their arms, and in their groins, and seen blains to come forth in other parts: when the disease hath wrought in them to that height, as to send forth those spots which (most think) are certain Tokens of near approaching death; and now they have received the

sentence of death within themselves, and have certainly concluded, that within a few hours they must go down into the dust.[3]

Go down into the dust they did. So sustained was the virulence of bubonic plague over the course of four centuries, the mortality rate of those infected probably never fell below 60 percent and may have approached the 90 percent level in the initial stages of an epidemic.[4] By September 1665, when the contagion reached its peak, upwards of 8,000 were dying every week in London alone. And by the time the epidemic had run its course a year later some 100,000 souls had met an agonizing and senseless death. Little wonder that it was viewed by evangelical millennialists like Vincent as a consummating act, an Armageddon.

Quarantine was considered the most effective method of controlling contagious disease, and in 1665 the College of Physicians reissued old orders mandating the sequestration of the sick: "That to every infested house there be appointed two watchmen, one for every day, and the other for the night; and that these watchmen have a special care that no person go in or out . . . upon pain of severe punishment." Nurses were also hired at public expense to care for the ill; and every parish was mandated to appoint women searchers to perform the grisly task of determining whether those who died were victims of "the infection" or of some other affliction.[5] Cleanliness was encouraged, and all forms of public entertainment and feasting were strictly prohibited. "Disorderly tippling in taverns, ale-houses, coffeehouses, and cellars [must] be severely looked unto, as the common sin of this time and the greatest occasion for dispersing the plague." It was further ordered that no domestic animal, including the cat(!), be kept in any part of the city, thus indirectly encouraging the propagation of the deadly house rat through the destruction of its natural predator.

Every man, rich and poor, learned and ignorant, became his own doctor, praising amulets, patronizing the astrologer, applying powders and ointments, imbibing purgatives, sweating over summer fires, and regularly sniffing the essences of rue, wormwood, thyme, or juniper supplied to the populace by harried but prospering apothecaries. Even tobacco was thought to offer some measure of protection, and in early June, not long after his first sighting of marked houses, Pepys bought liberal quantities "to smell and chaw." Yet in the final analysis the only effective method of dealing with the Black Death lay, as Charles II, the members of his Court, and Parliament well knew, in adhering to the proverbial antidote born of fourteenth-century Europe: *cito, longe, tarde*—fly quickly, go far, return slowly. Thus did the Government temporarily abandon London for the security of bubolic Oxford. And so would Newton, when Cambridge closed its doors pending the plague's abatement, return to the isolation of Woolsthorpe and the yeoman's house that rested at the beginning of his memory.

II

Even though many lives were spared when the citizens of London responded to the overwhelming urge of self-preservation by taking flight, their retreat into the countryside produced the inevitable result of spreading the plague to the eastern counties, and from there to the Midlands. By August 1665 the approaching pestilence had made it necessary to cancel Sturbridge Fair, annually held on the outskirts of Cambridge, and soon after all public meetings, whether of town or University, were prohibited by the corporation governing the city. In October the University Senate passed a grace for the discontinuance of sermons at St. Mary's Church and of exercises in the public schools. But by this time the scholars and faculty of the various colleges had already been sent away, only a select few remaining behind to handle administrative affairs and protect valuable property. Those who stayed were aptly characterized as defenders of a beleagured fortress. At dinner they ritualistically laced their wine with a preservative powder, while the gatehouses of their deathly quiet institutions were bathed in the reddish glow of fires fed by charcoal, pitch, and brimstone, a last desperate attempt to ward off the invisible enemy.[6]

According to tradition, Newton left Cambridge in the summer of 1665 and stayed away in Lincolnshire without returning for upwards of two years. The supposed time of his departure clearly matches the known facts, but the belief that he remained at Woolsthorpe until the spring of 1667 does not. He was still at Trinity in May 1665, for on the twenty-third he paid his tutor, Pulleyn, five pounds.[7] There is also Hannah's only surviving letter to him, dated May 6. The College was dismissed on August 8, 1665, but Newton must have left Cambridge considerably before that, since he did not claim the extra commons paid to those few who, risking infection, were in residence the previous six and one-half weeks.[8] The evidence would therefore indicate that he returned to Woolsthorpe sometime in June or early July.

For eight months the University was virtually deserted. The fourteen parishes of Cambridge, following the example of London, issued weekly bulletins of mortality in which the deaths resulting from the plague and those attributable to other causes were listed over the signatures of the mayor and vice chancellor. At the top of each report, a number of which are now in the possession of Clare College, the following announcement appeared: "All the Colledges (God be praised) are and have continued without any Infection of the Plague."[9] The authorities, lulled into a false sense of security when the death rate declined to zero while the rat-flea passed the winter months in a state of dormancy, invited the students to return in March 1666. Newton, by his own account, joined the intellectual migration back to Cambridge on March 20, where he remained until Trinity was again dismissed on June 22, when the plague reappeared. He retreated to Woolsthorpe a second time, not returning to Cambridge until the following April.[10] It now seems clear that Newton was at Trinity until June or July 1665, again between March 20 and late June 1666, and from April 22, 1667, thus belying the time-honored myth that the brilliant accomplishments of the so-called *annus mirabilis*—the miraculous year—were wholly carried out in the secluded fasts of rural Lincolnshire.

An equally durable component of the Newtonian myth would have us overlook the record of his studies between 1664 and 1665. Because Newton made no ripple in the stagnant pond of the academy, his emergence as a natural philosopher and mathematician, though admittedly swift and profound, has too frequently been characterized as a development whose origins are compactly rooted in the period of exile from Cambridge. However the *Questiones quaedam Philosophicae*, his undergraduate forays into advanced mathematics, and his equally bold experiments with light and color stand as incontrovertible proof that the Newtonian intellectual miracle was of an earlier genesis. If such a term must be employed, it would be more accurate to call that period the *anni mirabiles*, because it encompassed the better part of three years rather than one. Yet this in no way detracts from the magnitude of Newton's achievements, for the real miracle lay in the fact that any human being should have accomplished what he did, let alone a withdrawn young autodidact.

True creativity in the sciences, as in music, art, and literature, comes when the individual is in an aesthetically sensitive mood. Relieved of the student's usual concerns, Newton was now free to let his mind wander where it would, to put together seemingly disparate thoughts whose pieces were still little more than promising fragments of a gigantic intellectual puzzle. He meditated upon a mathematical method that would enable one to calculate accurately both the position and the speed of moving objects; became attentive to the apparently insignificant phenomenon of ripened apples falling from the branches of a wind-blown tree in his mother's garden; and formulated thought experiments with which to test the theory of light. At his death he left the partial draft of a letter to the Huguenot scholar Pierre Des Maizeaux concerning the months of discovery during the plague period. From reading it one might reasonably conclude that what was arguably the most strongly ideational period in the life of any scientist, before or since, was for him an experience little out of the ordinary:

> In the beginning of the year 1665 I found the Method of approximating series & the Rule for reducing any dignity of any Binomial into such a series. The same year in May I found the method of Tangets of Gregory & Slusius, & in November had the direct method of fluxions & the next year in January had the Theory of Colours & in May following I had entrance into ye inverse method of fluxions. And the same year I began to think of gravity extending to ye orb of the Moon & having found out how to estimate the force with w^{ch} [a] globe revolving

within a sphere presses the surface of the sphere from Keplers rule of the periodical times of the Planets being in sesquialterate proportion of their distances from the centres of their Orbs, I deduced that the forces wch keep the Planets in their Orbs must [be] reciprocally as the squares of their distances from the centres about wch they revolve: & thereby compared the force requisite to keep the Moon in her Orb with the force of gravity at the surface of the earth, & found them answer pretty nearly. All this was in the two plague years of 1665 & 1666. For in those days I was in the prime of my age of invention & minded Mathematics & Philosophy more than at any time since.[11]

This memoir, composed some fifty years after the events it describes, has been questioned by a number of contemporary Newton scholars, though not because of any doubts about the author's claims to paternity. It is simply that distant memories of major scientific advances, no less than in other fields of human endeavor, tend to become foreshortened in the retelling, thus enfolding them in a mantle of deceptive simplicity. Of the three major achievements associated with the period from mid-1664 to early 1667, only the steps in the development of the calculus have been reconstructed in reasonably satisfying detail. The experimental findings on light and, most especially, the depth of Newton's first penetration into the mysteries of gravitation continue to be matters of intense debate among specialists, who have less clear-cut documentation to go on and whose divergent interpretations of the surviving papers will never produce a truly definitive account. Yet this much can be said with certainty from the outset: The irrepressible mental powers that had recently surfaced maintained their control no matter where Newton spent his time during the mid-sixties—Cambridge, Woolsthorpe, or the Lincolnshire home of his congenial protector Humphrey Babington. This was indeed a movable feast, a magnificent intellectual banquet at which Newton dined—alone.

III

Though arithmetical prowess is of unquestionable value to the theoretician, other intellectual gifts sometimes weigh even more heavily in the balance. In Newton's case, one thinks of the untold hours of a solitary childhood spent in the fashioning of kites, sundials, water clocks, and various mechanical models, a foreshadowing of his innate geometrical powers, which reached full maturity in the university days. Clearly this pattern was shaped by some intuitive sense of what the true world must be like. When a certain critical point in his creative processes had been reached he somehow "knew" that he must be right and waited confidently for the results of experimentation to bear him out. Only rarely—and almost never for long—was he frustrated by subsequent developments. William Whiston, who succeeded to the Lucasian Professorship on Newton's resignation, wrote that "Sir Isaac, in Mathematics, could sometimes see almost by Intuition, even without Demonstration . . . and when he did propose Conjectures in Natural Philosophy, he almost always knew them to be true at the same Time."[12] John Maynard Keynes, to whom posterity owes a special debt for acquiring many of Newton's papers at auction some two centuries later, echoed Whiston's sentiments: Newton's "experiments were always, I suspect, a means, not of discovery, but always of verifying what he already knew."[13]

While he was not a brilliant calculator,[14] one sure measure of Newton's versatile genius was his ability consistently to devise methods for solving almost any scientific or mathematical problem. Indeed, he had to devise and perfect the very mathematical instrument required for testing his physical theories. But because he withdrew ever deeper into himself, Newton seldom communicated his advances to the world—an irritating habit that, as in the case of Robert Hooke, frequently led to unnecessary and acrimonious controversy. A similar and just as ugly dispute erupted later between Newton and Gottfried Wilhelm von Leibniz, the gifted German philosopher-mathematician. Their notorious quarrel centered on one of the key discoveries attributable to the *anni mirabiles*, the method of fluxions, or what is now known as the differential and integral calculus.

By late 1664 Newton, having given free rein to a voracious intellectual appetite, was master of the most advanced mathematics of his day and stood poised on the brink of original discovery. Promising avenues of investigation had already been opened by a number of seventeenth-century mathematicians, including Wallis, Descartes, Hudde, Gregory, and Fermat. Together they foreshadowed a method by which infinitesimal quantities might be computed and infinitesimal differences in varying quantities measured. Their attention was directed to several classes of problems such as the finding of maxima and minima, the computation of quadrature, and the construction of tangets to curves. The essential difficulty lay in the fact that while various mathematicians had been able to solve individual problems, their methods of analysis were simply too unwieldy for general use. Even more critical was the inability of anyone to formulate a general method of analysis that would apply to all problems of a given type. Fermat and Descartes, in their development of analytic geometry, had envisioned such a method, but they were destined to fall short of realizing that goal.

Drawn to the Cartesian method while an undergraduate, Newton began taking notes on analytical geometry in the autumn of 1664. Within six months he had crossed over into the uncharted realm of original investigation; within a year, he would become the most advanced mathematical thinker history had yet known. Newton composed his first major paper at Cambridge in May 1665, shortly before the University closed on account of the plague. It dealt with the summation of infinitesimal arcs of curves. He completed a second paper on fluxions and their applications to tangents and the curvature of curves at Woolsthorpe in November of the same year, and three others during the *"annus mirabilis"* of 1666, culminating in the brilliant October tract in which he revised and condensed the product of two years' feverish research and analysis.[15] Ironically, it was the formulation of the calculus that ultimately led Newton to reject the Cartesian belief, reaching back to the mystical Pythagorean Brotherhood, that the universe is fundamentally geometrical. Perhaps partly because he had little background in classical mathematics, Newton realized that concealed behind the geometrical forms that so appealed to Copernicus, Kepler, and Descartes were such dynamic, though not yet fully articulated, concepts as mass, force, and acceleration, none of which could be fully represented geometrically.

Unlike algebra, a tool for calculating specific numerical values, calculus is based upon the idea of considering quantities and motions not as definite and unchanging but as in the process of originating, fluctuating, or disappearing. It is employed, for example, to calculate the most subtle variations in the acceleration of a body falling through space, to compute the precise arc of a revolving planet, or to measure the exact rate of deceleration of a ball as it gradually comes to a stop after being rolled across the ground. In other words, the calculus is a most effective tool both for resolving problems concerning infinitesimal variations in the rates of motion and for determining the path of a given body through space. It is hard, though admittedly not impossible, to think of any changes in natural motion that are not mathematically reducible in Newtonian terms. And though in his demonstrations and proofs he did employ certain geometric constructs of finite dimensions—triangles, lines, and rectangles—he always thought of them as nothing more than abstract representations of far more complex and continually changing forms. During a visit to the home of Humphrey Babington at Boothby Pagnell in the summer of 1665, Newton became one of the first to compute accurately the area under a hyperbola. So thrilled was he by his discovery that he carried the calculations to fifty-two places, an operation that for most would have turned exultation into abject drudgery.[16] Thus while Descartes's conception of the universe is pictorial and general, Newton's is rigorously mathematical and precise. Recognizing the need for a more simplified and comprehensive method, Newton found it by drawing all the traditional questions together in a clean and harmonious manner, a *tour de force* of originality and synthesis.

Discontinuity is alien to nature and equally so to the history of science. No major discovery or significant innovation is absolutely new, removed from the past and the spirit of its own age. One might argue, therefore, that since the germs of the calculus were very much in the air, it was a discovery whose "time had come," as was shown when Leibniz, a decade later, developed it independently of Newton. Yet the latter's priority testifies to his greater sensitivity and penetration when compared with other mathematicians of the day. Newton was quicker than others to grasp the essentials of a problem and to develop its ultimate consequences with an unequaled mastery, which is further evidenced by his formulation of the binomial theorem even as the calculus was being born. In a letter to the well-known scientific intermediary John Collins, dated November 8, 1676, Newton wrote that there is no curved line expressed by an equation of three terms for which he could not, within "less than half a quarter of an hower," tell whether it may be squared or determine the simplest geometric figure with which it may be compared. "This may be a bold assertion . . . but it's plain to me by ye foundation I draw it from, though I will not undertake to prove it to others."[17] As always, proof for Newton was sufficient unto itself; he knew what he knew. The rest of the world did not matter, that is, not until another, namely Leibniz, innocent of the unmarked path already trodden by a solitary predecessor, laid claim to what Newton believed to be his and his alone.

IV

Although some two millennia and more had passed since the death of Aristotle, Newton, as we have seen, was born into a world still very much influenced by the thought of the ancient master. And if an increasing number of learned men no longer looked upon the efforts of those who relied on experiment as impudent or foolish, many still looked on the new method as a questionable expenditure of time and energy. The resolve of Marlowe's Faustus still mirrored the goal of many a seventeenth-century student:

> Having commenc'd, be a divine in shew,
> Yet level at the end of every art,
> And live and die in Aristotle's works.

This Newton conciously chose not to do. He wrote of his intellectual friendship for Plato and Aristotle in the philosophical notebook but of an even higher regard for the pursuit of truth, meaning the new mechanical philosophy. This quest, in the minds of most who know anything of his scientific accomplishments, was supposedly triggered by the fall of an apple from a tree in Hannah's garden, leading almost immediately, in a flash of numinous insight, to the discovery of universal gravitation. Like many genuine myths, it is an intriguing tale built upon a number of half-truths, which at their center contain a small but solid kernel of documented fact.

On April 15, 1726, a year before Newton's death, Stukeley visited him at his lodgings in Kensington near London. They dined together and afterward went into the garden to drink tea "under the shade of some appletrees, only he and myself. Admidst other discourse, he told me, he was just in the same situation, as when formerly, the notion of gravitation came into his mind. It was occasion'd by the fall of an apple, as he sat in the contemplative mood."[18] Newton evidently told much the same story to his niece, who, in turn, passed it on to an admiring Voltaire, through whom Newtonian thought was popularized in eighteenth-century France. In the English translation of his *Elémens de la Philosophie de Newton*, the *philosophe* wrote: "One day in the year 1666, Newton, having returned to the country and seeing the fruits of a tree fall, fell, according to what his niece, Mrs. Conduitt, has told me, into a deep meditation about the cause that thus attracts all bodies in a line which, if produced, would pass nearly through the center of the earth."[19] If one accepts the dubious psychoanalytic hypothesis that Newton's creativity sprang from spending the plague years "in the protective bosom of his mother,"[20] it would seem to follow that the apple—tainted fruit of Judeo-Christian religious tradition—once again became the symbolic source of knowledge offered up by a latter-day Eve in her isolated country garden.

More detailed accounts of the incident survive. Like Stukeley, Henry Pemberton also talked with Newton of gravitation toward the end of the natural philosopher's life. Pemberton's summary of Newton's recollections retains the garden but omits any reference to falling fruit:

> The first thoughts, which gave rise to his Principia, he had, when he retired from Cambridge in 1666 on account of the plague. As he sat alone in a garden, he fell into a speculation on the power of gravity: that as this power is not found sensibly diminished at the remotest distance from the center of the earth, to which we can rise, neither at the tops of the loftiest buildings, nor even on the summits of the highest mountains; it appeared to him reasonable to conclude, that this power must extend much farther than was usually thought; why not as high as the moon, said he to himself? and if so, her motion must be influenced by it; perhaps she is retained in her orbit thereby. However, though the power of gravity is not sensibly weakened in the little change of distance, at which we can place our selves from the center of the earth; yet it is very possible that so high as the moon this power may differ much in strength from what it is here. To make an estimate, what might be the degree of this diminution, he considered with himself, that if the moon be retained in her orbit by the force of gravity, no doubt the primary planets are carried round the sun by the like power. And by comparing the periods of the several planets with their distances from the sun, he found, that if any power like gravity held them in their courses, its strength must decrease in the duplicate proportion of the increase of distance. This he concluded by supposing them to move in perfect circles concentrical to the sun, from which the orbits of the greatest part of them do not differ. Supposing therefore the power of gravity, when extended to the moon, to decrease in the same manner, he computed whether that force would be sufficient to keep the moon in her orbit. In this computation, being absent from books, he took the common estimate in use among geographers and our seamen, before Norwood had measured the earth; that 60 English miles were contained in one degree of latitude on the surface of the earth. But as this is a very faulty supposition, each

degree containing about $69\frac{1}{2}$ of our miles, his computation did not answer expectation; whence he concluded, that some other cause must at least join with the action of the power of gravity on the moon.[21]

Finally, William Whiston presented a similar account based on a discussion with Newton not long after they became acquainted in 1694:

An Inclination came into Sir Isaac's Mind to try, whether the same Power did not keep the Moon in her Orbit, notwithstanding her projectile Velocity, which he knew always tended to go along a straight Line the Tangent of that Orbit, which makes Stone and all heavy Bodies with us fall downward, and which we call *Gravity?* Taking this *Postulatum,* which had been thought of before, that such power might decrease in a duplicate Proportion of the Distances from the Earth's Center. Upon Sir Isaac's first Trial, When he took a Degree of a great Circle on the Earth's Surface, whence a Degree at the Distance of the moon was to be determined also, to be 60 measured Miles only, according to the gross Measures then in Use. He was, in some degree, disappointed, and the Power that restrained the Moon in her Orbit . . . appeared not to be quite the same that was to be expected, had it been the Power of Gravity alone, by which the Moon was there influenc'd. Upon this Disappointment, which made Sir Isaac suspect that this Power was partly that of Gravity, and partly that of Cartesius's Vortices, he threw aside the Paper of his Calculation, and went on to other studies.[22]

Before any comment can be made about the accuracy of the above accounts, it is well that the reader have some grasp of the many technical obstacles Newton had to surmount before he could even think to ask the deceptively simple question relating the fall of an apple to the orbit of the moon. We must turn once again to the contents of the *Waste Book*, where Newton's early thought on dynamics is first merged with the analytical geometry of Descartes and, subsequently, with the calculus of his own design.

Of the multiple problems that seem to have simultaneously occupied his thoughts, none was more critical than the question of uniform circular motion. As the historian of science John Herivel points out, the only other relatively simple kinds of movement, besides motion in a circle, were rectilinear, parabolic, and elliptical.[23] The kinematics (the study of motion exclusive of the influences of mass and force) of rectilinear and parabolic motion, which occur under the influence of gravity, had been examined by Galileo in *Dialogue Concerning the Two Chief World Systems.* Newton, who had access to Tome I of Thomas Salusbury's *Mathematical Collections* (1661), an English translation containing the brilliant pro-Copernican polemic, greatly benefited from its contents, as he soon would from the Italian's masterpiece, *Discourses Concerning the Two New Sciences,* published by Salusbury in Tome II (1665). Of particular importance to Newton was Galileo's principle of uniformly accelerated rectilinear motion, the paradigm case for all other more complicated motions.[24] On the other hand, Galileo dealt not at all with the concept of force, nor did he take under consideration Kepler's discovery of elliptical planetary orbits derived from the key astronomical data left to the German by the imperious Tycho Brahe.

In the beginning, the complexity of coming to grips with Kepler's elliptical orbits was too great even for Newton, but the problem of uniform circular motion was not. By late 1664 he was already familiar with the principle of inertia, derived in no small measure from his having read and assimilated Descartes's *Principia Philosophiae*.[25] Newton enunciated his interpretation of the concept in the first two Axioms of the *Waste Book*.

(1) If a quantity once move it will never rest unlesse hindered by some external cause.
(2) A quantity will always move on in ye same steight line (not changing ye determination nor celerity of its motion) unless some externall cause divert it.[26]

He now conceived of a crucial "thought experiment" in which a ball is moving along the interior surface of a hollow sphere. Drawing upon the principle of inertia, he was able to demonstrate a constant tendency for the ball to continue in a straight line; in other words, along a tangent to a circle. But because it is constrained by the spherical wall along which it moves, the ball describes a circular as opposed to a rectilinear path. This could only be the result of a force continually acting upon it, and such a force can come into existence only because of a pressure exerted between the sphere's surface and the ball. The recognition of this reciprocal action pointed Newton toward a single acceptable conclusion: that "all bodys moved Olarly [circularly] have an endeavour from ye center about wch they move."[27] The "endeavour" to which Newton refers is nothing less than what Christiaan Huygens called "centrifugal force"—the movement away from a center or axis—or what Newton would soon come to refer to as *conatus recedendi*.*

To accomplish his analysis of circular motion in terms of a centrifugal tendency, Newton applied quantitative methods to another thought experiment similar to the one just discussed. In this instance, however, he ingeniously conceived of a ball or "globe" moving along a square inscribed in a circle (Figure 4–1). He reasoned that if the ball moves along the sides of the square and is "reflected" from the circle at the angle points (A, B, C, D), the combined force with which it strikes the circle at each of the four points will be its change in momentum. As Herivel has shown, Newton proceeded to formulate the following mathematical relationship concerning the magnitude of the "shocks" (impulse) encountered by the ball:

---

*Conversely, he termed "centripetal force," or the "endeavour of approach," *conatus accedendi*. In the *Principia* (1687), the Latin terminology has generally been replaced by the modern usage. It is in this, his masterwork, that Newton himself coined the phrase "centripetal force."

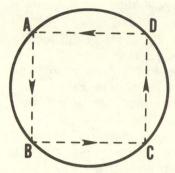

Figure 4-1. Newton's analysis of circular motion.

$$\frac{\text{total sum of shocks at 4 corners}}{\text{force of movement of ball}} = \frac{\text{sum of sides of square}}{\text{radius of circle}}$$

When any other regular polygon, no matter the number of its sides, is substituted for the square the formula also applies. This limiting process was the first major clear-cut contribution of a modern thinker to the study of dynamics. With it Newton derived his version of the law of centrifugal force, whose effects are identical if the ball, rather than being constrained to a spherical surface, is allowed to proceed on a plane.[28]

As with the calculus, Newton was not the only person of his day to direct his attention to questions of circular motion. Huygens, one of the most gifted of the continental virtuosi, had accomplished a similar result several years earlier, in 1659. Then only thirty and at the height of his powers, the Dutch genius employed the methods of ancient geometers and minute kinematical analysis, rather than the modern analytical methods and dynamics known to Newton alone. Neither man was aware of the other's discovery, for Huygens did not publish his findings until 1673, when they appeared in an appendix to the *Horologium Oscillatorium*. By that time the principle had also been Newton's for seven years.

Ever sensitive when another "encroached" upon intellectual territory he thought of as his own, Newton never forgot what Huygens had "done" to him. Many years later he ruefully observed: "What Mr. Huygens has published . . . about centrifugal forces I suppose he had before me."[29] And this grudging recognition came only while he himself basked, like an aging god, in the glory surrounding the publication of the *Principia*. Still, there can be no doubt that Newton was deeply impressed by Huygens's facility in the classical mathematics whose value he had so rashly underestimated during the early stages of his own self-instruction. Prodded, in old age, by Pemberton to reveal his feelings about the relative worth of the work done by prominent mathematicians of his day, "Sir Isaac Newton has several times particularly recommended to me Huygens's stile and manner. He thought him the most elegant of any mathematical writer of modern times, and the most just imitator of the ancients."[30] Seldom did Newton speak more highly of any man.

Having worked out the law of centrifugal force, Newton began to speculate that the same principle that defines the motion of a circling ball also causes the planets to move away from the sun, which, as Kepler demonstrated in his first law, rests at one focus of their elliptical orbits. But since no spherical wall prevents the planets from shooting outward into space, some invisible and inward-pulling force must counteract this powerful tendency. If so, the same must also be true of the earth's moon and of the moons orbiting Jupiter, then the only known planetary satellites in the solar system. Like Kepler and Huygens, Newton entertained the idea of a power emanating from celestial bodies by which they attract other bodies and influence their motion. He conceived that when acted upon by such a force a body (like the moon) would deviate from its rectilinear path and be constrained to revolve in a closed orbit, its centrifugal force counterbalanced by the inward pull of "gravity." If this concept could be demonstrated mathematically, then the moon could be thought of as a giant apple perpetually falling around the earth. It would thus appear that this original insight was indeed triggered by the fall of an apple witnessed by Newton while in a deeply contemplative state. This seemingly inconsequential occurrence may well have provided the very physical demonstration of a principle which, until now, was for him only loosely rooted in the abstract.

Suppose, as Pemberton stated, that Newton, while sitting in the garden, was led to ponder the question of how far the earth's attraction might extend. Since from the bottom of the deepest crater to the summit of the highest mountain "this power is not found sensibly diminished," might not its action be reasonably extended to the moon? Does the moon, in fact, obey the same law as a moving body (a falling apple) on earth? Perhaps so, but the path of a falling apple is essentially perpendicular to the earth while that of the moon round the planet is roughly circular. For Newton, the most promising explanation lay, as Pemberton noted, in the premise that the moon is so high "this power may be much different in strength from what it is here." Put more simply, gravitational attraction must decrease at a distance, thus preventing the orbiting moon from crashing into the earth, but also from hurtling outward into the unplumbed celestial abyss. And though it is not mentioned as such by either Whiston or Pemberton in their respective accounts, Newton, unless he misinformed both Stukeley and his niece, must have considered the case of the apple as well. For if one could somehow place such a small object at the appropriate distance from earth it, too, would become a satellite like the moon, its centrifugal tendency or outward pull exactly counterbalanced by its inward fall. Perhaps at this point even Newton's seemingly unflappable composure slipped a bit. The creative probing of his mind's eye appeared to have resulted in a concrete application after all. Material objects must behave in a manner comparable to the circling balls of this thought experiments. He was indeed dealing with the real world, a world of mathematically expressible relationships!

Whether in fact the mental tumblers fell into place in this manner will never be known. Chances are they did not. Yet whatever the precise nature of Newton's inspiration in that secluded Woolsthorpe garden, an experience he remembered until the day he died, it proved essential in leading him to that most critical stage of scientific inquiry, the point at which a promising hypothesis is ready for testing.

In his memoir of the *anni mirabiles* quoted from above, Newton tells us of the method he used to solve this complex problem:

> I began to think of gravity extending to ye orb of the Moon & having found out how to estimate the force with w^{ch} [a] globe revolving within a sphere presses the surface of the sphere from Keplers rule of the periodical times of the Planets being in sesquialterate proportion of their distances from the centres of their Orbs, I deduced the forces wch keep the Planets in their Orbs must [be] reciprocally as the squares of their distances from the centres about wch they revolve.

Restated in the language of today, Newton discovered the law of centrifugal force and then proceeded to combine it with Kepler's third law, which states: The square of the periods of the revolutions of the planets are to one another as the cubes of their mean distances from the sun. From the resulting formula he was able to demonstrate that the attraction of the planets to the sun must decrease inversely as the squares of the distance separating them from it. Hence if the attraction between the sun and a planet amounts to a certain force when the bodies are at a given distance from each other, at twice that distance the force will not be twice as great, but one-fourth as great; at three times the distance, not one-third but one-ninth as great, and so on, even smaller in proportion as the distance grows larger.

Other than his own tersely written reflections and the valuable but less than definitive oral accounts recorded by Pemberton and Whiston, until quite recently scholars had little additional information to go on in their efforts to determine just how successful Newton's early forays into planetary dynamics had been. But in 1953 two invaluable holograph manuscripts, both of which have subsequently shed much light on this intriguing question, were discovered in the Cambridge University Library.[31] The first is a direct outgrowth of Newton's mechanical investigations in the *Waste Book*. Written in his newly developed utilitarian hand and sprinkled with data taken from Salusbury's translation of Galileo's *Dialogue*, the document is thought to date from late 1665 or, more likely, 1666. Willing, as ever, to work with any readily available writing material, Newton recorded his calculations on the ratios of the force of gravity to the centrifugal forces generated by the earth's motions on the reverse side of a torn parchment once used by Hannah as a lease, additional evidence—along with the story of the apple and Pemberton's account of his "being absent from books"—that this chapter of the *anni mirabiles* was indeed written at Woolsthorpe.

The second manuscript, also undated, builds directly upon the conclusions of the first. In it is to be found the very inverse square relationship between the sun and the planets and the earth and the moon that Newton later claimed to have discovered in 1666. An exact dating of this manuscript is also impossible, but 1666 would appear to be the most likely guess, not only because of Newton's later testimony, but also because we have the statement of the astronomer David Gregory, one of his gifted protégés. Gregory visited Newton at Cambridge in May 1694 and, as rarely happened, was allowed to see a number of his host's early papers. Among them was one whose contents, though zealously overappraised by Gregory, almost certainly identify it as the document in question:

> I saw a manuscript written *before the year 1669* (at the time when its author Mr. Newton was made Lucasian professor of mathematics) where all the foundations of his philosophy are laid: namely the gravity of the Moon to the Earth, and for the planets to the Sun: and in fact all this even then is subject to calculation. I also saw in that manuscript the principle of equal times of a pendulum suspended between cycloids, before the publication of Huygens' *Horologium Oscillatorium.*[32]

Equally convincing is the fact that by the end of 1666 Newton had quite literally exhausted his experimental interest in the study of dynamics. The reams of extant papers for the next decade and more contain no evidence about new developments along these lines. Optics and alchemy would lay claim to almost all of his seemingly limitless creative energy until 1679. Only then would he return to the subject with a passion put to use and forge the law of universal gravitation, the brightest jewel in his intellectual crown.[33]

For now, Newton was faced with the daunting task of putting his highly ambitious but less sweeping theories of lunar attraction and planetary motion to the test. From his own calculations and those of other astronomer-mathematicians, he knew that the period of the moon—the time required by the satellite to complete one revolution about the earth—is approximately 27 days 8 hours. He knew also that the moon's distance from the earth is sixty times the radius of the planet. In other words, the center of the moon is sixty times as far from the center of the earth as is a body on the earth's surface. But just how far, exactly, is this? The figures then available told of the distance not always in miles but in terms of the earth's size, which immediately raises another question: How big is the earth?

Newton had previously turned to Salusbury's translation of Galileo's *Dialogue* for answers to these questions, and there is nothing to suggest that he would not have done the same in this instance.[34] Employing the measurement commonly used by the seamen of his day, Galileo took one degree of latitude as equaling 60 miles, the very figure cited in the accounts of both Whiston and Pemberton. However, Pemberton speaks of "English miles"—meaning the statute mile of 5,280 feet—while Whiston refers only to "60 measured Miles." Relying once again upon our knowledge of his use of Galileo's figures, it seems reasonable that Newton adopted the Italian mile of 5,000 feet rather than the somewhat longer English mile.[35] With this information, he calculated the earth's circumference as follows: $360° \times 60 = 21,600$ miles. Drawing upon this figure, he then computed a radius for the planet of some 3,500 miles, about one-eighth too small.

The only additional measurement now required was the rate of fall of a body at the earth's surface in a given period of time. Disappointed earlier when he used Galileo's grossly underestimated value, Newton arrived at a far more accurate one of his own. Experimenting with a conical pendulum inclined at an angle of 45°, he discovered that an object starting from rest will fall slightly more then 16 feet in a second, almost identical to the modern value and not quite double the one contained in the *Dialogue*.[36] If the moon's center were indeed sixty times as far from the center of the earth as an object at the planet's surface—and if the inverse square relationship (60^2) held good—the force of gravity acting upon the satellite would be only 1/3,600th as great, and it would fall toward earth only 1/3,600th as fast. Put another way, at a disctance of sixty earth radii the moon, or any other object, will fall the same distance in a minute as it would in a second if near the earth, that is, 16.1 feet.

Working through the problem, Newton came up with an answer characterized by Whiston as disappointing "in some degree" and "not quite the same that was to be expected." For based on the above figures, the fall of the moon in its orbit in one minute would have worked out to approximately 13.2 feet. If, as is possible, Newton then chose to substitute the English statute mile for its Italian counterpart, a figure of about 13.9 feet per minute would have resulted, still significantly below expectation.

Quite clearly, the erroneous value taken for a single degree of latitude had proved to be the principal stumbling block. Pemberton undoubtedly reflected Newton's thoughts on the matter when, with the wisdom of hindsight, he observed that "tis a very faulty supposition, each degree containing about $69\frac{1}{2}$ of our miles" rather than the 60 Italian miles of the youthful virtuoso's calculations. A thorough reexamination of his figures, if undertaken, proved futile. Newton simply could not account for the discrepancy which, in Whiston's words, "made Sir Isaac suspect that this Power was partly that of Gravity, and partly that of Cartesius's Vortices." He thus "threw aside the Paper of his Calculation, and went on to other studies." Pemberton concurred: When "his computation did not answer . . . expectation he concluded that some other cause must at least join with the action of the power of gravity on the moon." By the time these accounts were drafted Newton, for reasons both scientific and theological, had long since put as much distance between himself and Cartesianism as was possible for one who remained committed to the basic precepts of mechanical philosophy. That he should have so openly discussed his early attachment to Descartes's ideas thus seems most significant. It is a clear indication that the principle of attraction at a distance—not to mention universal gravitation—was approached with grave misgiving by the young virtuoso. Better, until some more suitable mechanical explanation could be worked out, that the French philosopher's exotic vortices of invisible material particles drive the planetary machinery through the firmament, than to contemplate the profane possibility that mysterious ghostlike fingers, of origins unknown, might be working their fathomless will across the dark and trackless abyss.

In 1670 the French astronomer Jean Picard completed a series of triangulation measurements near Paris. With the publication of the results, still another chapter was added to the Newtonian myth.[37] Picard quite accurately determined that one degree of the earth's meridian equals 69.1 English statute miles. This and certain others of his measurements found their way into the Royal Society's *Philosophical Transactions* for 1675. Newton, upon seeing them, is said to have rushed to modify the somewhat disappointing findings of his earlier work. So agitated did he allegedly become that his normally steady hand began to tremble with anticipation. Unable to complete the calculations because of this uncontrollable nervous palsey, he had to seek the aid of an unidentified friend to do it for him.

It is difficult to believe that anyone familiar with the true character of the man could accept the idea that Newton's composure literally disintegrated on being faced with a relatively simple mathematical problem. He had already handled others whose results were equally, if not more, important, and he would do so countless times more afterward. Yet even if we were to suppose that in this particular instance Newton temporarily lost control of his emotions, the question remains: Is this the kind of human failing one who continually took extraordinary pains to conceal his innermost feelings from the world would openly admit to another, whoever he might have been? After all, even the minor sins of his childhood and adolescence were kept from prying eyes through the ruse of short-writing. Equally important is the fact that Newton was preoccupied by more immediate concerns in 1675, as evidenced by the following communication from the mathematician John Collins to David Gregory: "I have nott writt to or seene [Newton] this 11 or 12 Months, not troubling him as being intent upon Chemicall Studies and practices, and both he and Dr. Barrow &c beginning to thinke mathcall Speculations to grow at least nice and dry, if not somewhat barren."[38] This is indeed a fitting description of the way Newton's mind worked, focusing on one field of study or series of problems to the virtual exclusion of all others. "I keep the subject constantly before me," he once stated, "till the first dawnings open slowly, by little and little, into a full and clear light."[39] As far as 1675 is concerned, Newton was fully in the grip of his alchemical studies. Most of gravity's complex mysteries remained shrouded in darkness, as yet impervious to the brilliance of the coming intellectual dawn.

The specter of the trembling hand notwithstanding, Newton did avail himself of Picard's measurements when his attention was once again captured by unresolved questions of dynamics during the late 1670s and early 1680s. The memorandum Abraham de Moivre drafted for Conduitt supports the basic contention of Pemberton and Whiston that Newton "found himself disappointed for awhile, the reason of which was that he took it for granted that a degree of the earth did contain 60 miles exactly, which made it that his calculation did not agree with Theory, but he entertained a notion that with the force of gravity, there might be a mixture of that force which the Moon would have [if] it was carried along in a vortex."[40] Later, however (de Moivre does not say when), Newton employed Picard's measure of the earth and began his calculations anew. This time he "found it perfectly agreable to the Theory." If nothing else, we have here a plausible explanation as to why Whiston, like de Moivre, characterized the early result as "disappointing" while Newton, who may have been thinking of his later success, rightly described it as "answer[ing] pretty nearly."

Having completed his early test of the force of gravity at the earth's surface and finding it to be upwards of four thousand "times greater than the endeavour of the moon to recede from the centre of the Earth," Newton was now in a position to take still another important step. By combining Kepler's third law with his own formula for centrifugal force, he discovered that "the endeavours of [the primary planets] receding from the Sun will be reciprocally as the squares of the distances from the Sun."[41] Here was the very inverse square relationship discussed in his much quoted memoir.

Important as the recently discovered documents are, they contain no discussion of gravity as operating anywhere beyond the immediate area of the earth's surface. Newton is ever careful to refer to the forces ("endeavour") of the planets *from* the sun and of the moon *from* the earth, but nowhere does he mention a *mutual attraction,* whatever his private thoughts on the subject. What creative people do not say is frequently as important as what they choose to communicate, and Newton, whatever the thrust of the insight experienced in his mother's garden, was twenty years away from solving the great riddle of universal gravitation. Reflecting on his early thoughts in a letter to Edmond Halley dated June 20, 1686, he made no claims beyond what was contained in his surviving papers: Gravity "was but an Hypothesis & so to be looked upon only as one of my guesses which I did not rely on."[42] Thus when reading the more general later accounts of a Whiston or a Pemberton—indeed of Newton himself—one must always keep in mind that the references to the idea of gravity in 1666 were references to a promising hypothesis or, as de Moivre called it, "a Theory," nothing less and assuredly nothing more.

"To be looked upon as one of my guesses," wrote Newton. Ah, but what a "guess" it was!

V

During the first week of September 1666, while Newton, ensconced at Woolsthorpe, quietly probed questions of mathematics and natural philosophy, a curious scene was unfolding more than 100 miles to the southeast. On the morning of the fourth, Samuel Pepys, who as future President of the Royal Society would live to see his name grace the title page of the *Principia*, was deeply occupied by considerations of a far less lofty nature. He watched intently as Sir William Batten, Surveyor of the Navy, dug a pit in his garden large enough to hold his considerable supply of wine. In the remaining space Pepys "took the opportunity of laying all the papers of my office that I could not otherwise dispose of." That evening, in the company of Sir William Penn, grizzled Commissioner of the Navy and father of the Quaker leader, "I did dig another [pit] and put our wine in it, and I parmazan cheese as well . . . and some other things."[43] About two the next morning Pepys was roused from a sound sleep by his worried wife. Weary and so sore-footed he could hardly stand, Pepys saw at a glance through the window a sight that caused him to forget immediately his temporary discomforts. He now realized that the outcome of the intense four-day battle in which he and his fellow citizens had been engaged was no longer in doubt. The inhuman enemy was advancing unchecked. Grabbing his sacks of gold, Pepys fled the endangered premises along with his wife; Will Hewer, one of his clerks in the Navy Office; and the Pepyses' maidservant, Jane Birch. Quickly they made their way to the protection of the river, unhampered by the accustomed darkness of a late summer's night, for on this unforgettable occasion the skies above the city were aglow from one end to the other: London was burning to the ground!

Pepys had been among the first to see the fire before it was transformed by strong winds and acres of dry timber shops and houses into a conflagration beyond the imagination of a Dante. The night of Septermber 1, a Saturday, some of the housemaids stayed up late preparing for the usual Sunday feast. About three in the morning "Jane called us up . . . to tell us of a great fire they saw in the City. So I rose, and slipped on my nightgown and went to her window . . . but being used to such fires as fallowed, I thought it far enough off, and so went back to bed again and to sleep."[44] It was the last full night of rest he would know for a week. Addicted to creature comforts though he was, Pepys might have sacrificed this night's sleep as well had he but known that no man or woman would ever again look upon the city as it stood at that deceptively quiet hour before dawn. The London of Chaucer, of Elizabeth I, of Marlowe, and of Shakespeare was about to disappear forever.

About the time Pepys caught his first distant glimpse of the flames, Sir Thomas Bludworth, the Lord Mayor, arrived on the scene. Already annoyed at having been dragged from a warm bed at such an ungodly hour, Bludworth became even more disgusted when he saw how little damage had occurred to that point. Like other Londoners, the Lord Mayor had witnessed scores of fires in his time, and so far as he could tell there was nothing about this one to warrant special concern. When asked whether several houses should be razed to contain the flames, Bludworth, unwilling to subject the city to unnecessary financial damages, uttered the tragicomic remark that would haunt him to his grave: "Pish!" he replied, "a woman might piss it out."[45]

When he awoke on Sunday morning, Pepys was told that more than three hundred houses had burned and that the wharves by London Bridge were ablaze. A considerable wind had risen during the night, and the outlook was now omnious. Pepys decided to investigate personally and "in an hour's time [saw] the fire rage every way, and nobody to my sight endeavouring to quench it, but to remove their goods and leave all to the fire." On a whim, he decided to make his way to Whitehall where, much to Pepys's surprise, the Court seemed to know little of what was happening. After discussing what he had observed with the King, Pepys was commanded to locate Bludworth and order him to "spare no houses, but to pull down before the fire every way." By the time he finally found the Lord Mayor the razing of buidings was already in progress, but the populace was starting to panic. The critical moment for effective countermeasures had slipped away with the dawn. "Lord, what can I do?" Pepys quoted the frantic Bludworth as saying. "I am spent. People will not obey me. I have been pulling down houses. But the fire overtakes us faster than we can do it."[46] And as if the beseiged official did not have problems enough, he was soon to see his own beautiful residence on Gracechurch Street reduced to a pile of white ash.

The rare conjunction of the three sixes in the numeral 1666 figured significantly in contemporary interpretations of prophecy, both Judaic and Christian. In the book of Revelation (13:6), for example, the Antichrist, emerging in the form of a beast, bears "the number of a man; and his number is Six hundred threescore and six." Little wonder it was widely believed that something unusual and quite terrible would mark that fateful year. Many a self-proclaimed seer had conjured up horrible visions of London's destruction, while in conventicle and Quaker meeting houses the devout had railed against the godless new Babylon—the city of blood. Not even the suffering and death wrought by the Great Plague had been catastrophic enough to silence the most zealous heralds of doom. As early as 1658 one Walter Gostelo had issued a prophetic warning: "If fire make not ashes of the city, and thy bones also, conclude me a liar for ever. Oh, London! London! sinful as Sodom and Gommorah! the decree is gone out, Repent, or burn, as Sodom, as Gomorrah!" A year later Daniel Baker, a divine, revealed his equally dark vision of London's future: "Yea, a geat effusion of blood, fire, and smoke shall encrease up in the dark habitations of cruelty; howling and great wailing shall be on every hand in all her streets."[47] And when fire came at last, there was no question in the minds of many that it had been unleased by a vengeful God. Speaking, no doubt, for the majority of his fellow citizens, Thomas Vincent simply observed: "It was the 2^d of September 1666, that the anger of the Lord was kindled against London, and the Fire began."[48]

The Great Fire, as it was soon to be known, burned four days and four nights. When it finally subsided on Thursday, September 6, the displaced and exhausted population had lost all track of time—not that it really mattered. The remains of the burned-out city had taken on the aspect of an Armageddon with no majesty among the ruins, only a deep and unbroken sea of ashes, soot, and dirt. The flames consumed an area roughly oblong in configuration, a mile and a half long and half a mile wide, between the Tower and the Temple. On the 436 acres destroyed—including more than 75 percent of the walled city—had stood 13,200 houses and 87 churches, among them beautiful old St. Paul's, rebuilt in 1628 by the brilliant Stuart architect Inigo Jones after a previous fire. If, as many believed, the fire, like the Great Plague, was indeed an act of Divine retribution, then God had also laid waste His own house, making it difficult to find a place of worship where, too late, the people of London could seek forgiveness of their wrath-provoking sins. There was, however, at least one important blessing to be counted in the fire's aftermath. Compared to the tens of thousands who succumbed from the plague, the death toll was inconsequential. The Bills of Mortality, though surely too low, listed only six as having perished in the flames.[49]

The plight of the refugee population was an altogether different matter. The masses that fled London during the plague at least had their dwellings to return to, ramshackle and squalid though they may have been. Many of the thousands routed by the fire were made permanently homeless, their plight movingly captured by John Dryden, a Trinity College graduate (1654) and future Poet Laureate, in quite another kind of "Annus Mirabilis":

> The most in fields like herded beasts lie down,
> The dews obnoxious on the grassy floor;
> And while their babes in sleep their sorrows drown,
> Sad parents watch the remnants of their store.

How or when the news of London's destruction reached Woolsthorpe we do not know. Nor is there any hint of Newton's thoughts on learning of the tragedy. The magnitude of what had occurred was beyond the comprehension of even those intimately acquainted with the great city. Newton, who would not make his first visit to the capital until 1668, was in no position to appreciate the consequences fully. Still, he had survived the furious charge of the Four Horsemen—the civil war of his childhood and, now, the plague and the fire of a world turned upside down. While others predicted the coming of the Last Judgment or spoke in exalted tones of the imminent appearance of the Messiah, modern science had at long last found the great lawgiver it had been waiting for, a new Moses capable of dispelling once and for all the scholastic myths of the medieval Baals. For even as the physical world seemed to be coming apart at the seams, a secretive young natural philosopher was beginning to put it back together in a more harmonious fashion than anyone since Pythagoras had dared dream.

VI

Isaac Newton died a wealthy man. Except for his land and a list of financial holdings, the inventory of his modest possessions commissioned by John and Catherine Conduitt in April 1727 gives no grounds for suspecting this.[50] Nothing that he owned had great intrinsic value; other than the library and voluminous manuscripts, there is little of abiding historical interest. Indeed, the most striking thing about his earthly goods is not their monetary worth but the fact that the normally staid Newton surrounded himself in a veritable sea of crimson. In the room where he slept was "a crimson mohair bed compleat with case curtains of crimson Harrateen"; the windows were hung with crimson drapes topped by matching valances, while crimson mohair hangings covered the walls. In the dining room a "crimson sattee" awaited the occasional guest, and in the back parlor a crimson easy chair and six crimson cushions promised welcome comfort to the aging Master of the Mint upon a late afternoon's return from the Tower.

Newton's interest in color can be traced at least as far back as his student days at Grantham. There he knew a rustic society of homespun fabrics—woolens and linens—of dull monotones: grays, browns, tans, and yellowish whites. Commercial dyes were generally unavailable, and only on special occasions were bright berries and roots gathered for the painstaking process of bringing a little color into the otherwise drab wardrobes of staunchly conservative farmers. One can imagine Newton as a boy accompanying a household servant to the woods, where he helped collect the needed ingredients for the dyeing pot, then standing transfixed while the brilliant colors—now lighter, now darker—bubbled forth, as if by magic, from the depths of the steaming cauldron. Perhaps he was permitted a small container of the precious liquid in which to dip his pen and trace marvelously vivid images on equally precious scraps of parchment scrounged from the papers of Barnabas Smith.

In any case, we perceive in his earliest notebooks a strong prediliction for colors and color mixtures. The Morgan notebook, dated 1659, contains some three dozen recipes for preparing various painters' colors. Interestingly, there are more formulas for the making of red than for any other color on the list. For example: "Take some of ye clearest blood of a sheepe & put it into a bladder & wth a needle prick holes in ye bottom of it y^n hang it up to dry in ye sunne, & disolve it in allum water according as you haue need." Among the many other recipes relating to drawing and painting are five for the preparation of red, blue, yellow, green, and black printing ink.[51] Scarcely any were of Isaac's invention; they were copied from the third edition of Bates's book on natural magic, *The Mysteries of Nature and Art.*

The aesthetic chords nature touched in the youth were eventually silenced by an adult obsession with the abstract world of symbol, equation, and geometrical lines. For unlike Kepler and Leonardo, Newton seems not to have paused to consider the snowflake on the frost-glazed windowpane, the tulip or the spring rose, the soaring bird, the course of the river as it flows quickly in the sunlight or seeps slowly through the darkness toward the hidden roots of giant oaks. In all the millions of words he left behind, one of the very few references to nature's beauty is contained in his famous letter to Henry Oldenburg of February 6, 1672, and even then Newton would not permit the aesthetic reverie that overcame him to linger beyond the passing moment:

> [I]n the beginning of the Year 1666 . . . I procured me a Triangular glass-Prisme, to try therewith the celebrated *Phaenomena* of *Colors.* And in order thereto having darkened my chamber, and made a small hole in my window-shuts, to let in a convenient quantity of the Suns light, I placed my Prisme at its entrance, that it might be thereby refracted to the opposite wall. It was at first a very pleasing divertisement, to view the vivid and intense colours produced thereby; but after a while applying my self to consider them more circumspectly, I became surprised to see them in an *oblong* form; which according to the received laws of Refraction, I expected should have been *circular.*[52]

Perhaps Newton indulged his senses more than he allowed on paper, but there is every reason to doubt it. After all, this is the same young man who risked permanent visual impairment, even blindness, by staring into the sun's image, and who later took a darning needle "and put it betwixt my eye and ye bone as neare to ye backside of my eye as I could." Then, in a test which sickens one at the thought, he pushed against the eyeball time and again until "severall white, darke & coloured cirlces" appeared. "Which circles were plainest when I continued to rub my eye wth ye point of ye bodkin, but if I held my eye & ye bodkin still, though I continued to press my eye wth it yet ye circles would grow faint & often disappeare until I renewed y^m by moving my eye or ye bodkin."[53]

As before, Newton dispassionately related the experiment as if it were nothing but an ordinary happening, but in fact he suffered intense pain and prolonged discomfort in order to test the theory that colors result from pressure exerted on the back of the eye. Galileo, as much as anyone, invented the world of primary qualities—of uncompromising mathematical rigor and emotional detachment, as against the subjective realm of everyday experience—but it was Newton who lived in it. All we are about to learn of his singular discoveries regarding light and color would not have been possible had he treated the eye as anything other than the coldest of nature's shutters.

VII

The mixing of colors and the construction of sundials were pleasant enough diversions for the clever, introspective lad from Lincolnshire, but neither activity, strictly speaking, can be said to have fitted within the context of contemporary natural philosophy. Precisely when Newton crossed this divide is impossible to say. As observed in the previous chapter, the chronology of the early optical research, like the dynamics, is at best clouded. Still, most signs point to late 1664 or early 1665, for it was then that Newton carried out the first of his many prismatic experiments, the results of which he duly recorded in the *Quaestiones*. There, in a seemingly innocuous passage, is found the first real challenge to the doctrine of colors since Aristotle:

> That ye rays wch make blew are refracted more y^n ye rays wch make red appeares from this experim^{nt}. If one hafe of ye thred abc be blew & ye other red & a shade or black body be put behind it y^n lookeing on ye thred through a prism one halfe of ye thred shall appeare higher y^n ye other & not both in one direct line, by reason of unequall refractions in ye 2 differing colours.[54]

In other words, by observing light refracted from differently colored surfaces through a single prism, Newton came to the previously unthinkable, albeit still tentative, conclusion that colors are *not* produced by complex modifications of white light, but that each color is singular and has its own degree of refrangibility. It is the ordinary white light of the sun that is a complex mixture of colors. This discovery, resting upon the bed rock of experimentation, supported all that was to follow.

Revealing though they were, these early findings could be considered little more than a promising beginning, as Newton realized. However naive the young scholar may have been about the ways of the world, his grasp of the fundamental operational precepts of modern science was nothing short of profound, especially at a time when the nascent scientific method had a decidedly limited intellectual appeal. Above all, Newton possessed an instinctive aversion to the dissemination of any hypothesis, no matter how potentifully fruitful, unless tested time and again by its formulator. When in 1672, at the age of twenty-nine, he presented his findings on light and color to the Royal Society, they had already taken the form of law. Theory simply would not suffice. As a prelude to a discussion of his discoveries, Newton made his position, from which he deviated only rarely, absolutely clear. His words might well have served as his epitaph and as the touchstone for the scientific age he, above all others, ushered in: "For what I tell . . . is not an Hypothesis but the most rigid consequence, not conjectured by barely inferring tis thus because not otherwise or because it satisfies all phaenomena . . . but evinced by ye meditation of experiments concluding directly & without any suspicion of doubt."[55] And to allay any suspicion of doubt Newton devoted as much of his time to research in optics, both during the *anni mirabiles* and in the years immediately after, as to the new mathematics or the many intriguing questions of dynamics, gravitation included.

Since before Aristotle the nature of color and light had been the subject of almost continuous philosophical speculation. The Stagirite believed that colors are a mixture of light and dark, or of black and white, a view that, significantly modified and refined, survived Newton and appeared in the eighteenth-century writings of Goethe. The first modern to break really new ground was Descartes, who faced the problem of explaining the phenomena in terms compatible with his revolutionary mechanical philosophy. In Descartes's view, light was a pressure transmitted through the transparent medium of ether that filled all space, and sight results when that pressure impinges on the optic nerve. He further conjected that the ethereal globules rotate at different speeds and thus give rise to different colors, which are not distinct substances in and of themselves.

The prism had been known since the time of Seneca in the first century, but it was Descartes who in *Discours* VIII of *Les météores* (1637), established it as the primary tool of seventeenth-century optical studies, and Boyle who called it "the usefullest instrument men have yet employed about the contemplation of colours."[56] The most significant result of Descartes's experimentation with the prism was his discovery of the sine law of refraction, with which he could calculate exactly how much a ray of sunlight is bent when it passes through a prism at a given angle. However, Descartes focused the spectral band on a screen only a few inches from the prism and did not confront the problem of dispersion encountered by Newton, who projected his spectrum on a wall many feet distant. Descartes's prism made a nearly perfect circle of colors from which he successfully calculated the sine for mean refrangible rays of light. But when Newton repeated the experiment Descartes's law became temporarily unhinged: "I became surprised to see them [the colors] in an *oblong* form," he wrote to Oldenberg, "which, according to the received laws of Refraction, I expected should have been *circular*." This revelation, coming as it did on the heels of Newton's earlier prismatic discoveries, persuaded him to test more fully his own tentative theory of colors and subsequently to reformulate the sine law in a manner consistent with the new experimental data.

Moreover, Descartes's break with the ancient doctrine of colors was not nearly so complete as the French virtuoso believed, for he had clung to the peripatetic notion of colors as modifications of pure light. He thought red, as did Aristotle, to be the nearest to white, produced when strong light undergoes modifications by a dark medium or is reflected from a dark surface. As light weakens and darkness increases, green emerges followed by violet, the last stage before total blackness. And, like Aristotle, Descartes held that all colors fitted on a scale between the absolute extremes of black and white, of darkness and purity. This doctrine of modifcation, redolent of Christian moral overtones, has been aptly characterized as the seemingly embodiment of common sense, inviting no questions and requiring no alterations.[57]

Despite the inherent limitations of Descartes's quasi-analytical approach, the doctrine of colors quite rightly acquired a new significance in the eyes of those who believed the mechanical philosophy held the key to nature's most closely guarded secrets. Among them was Robert Hooke, who in 1665 published the *Micrographia,* one of the true masterpieces of seventeenth-century science. With Leeuwenhook, it established him as the joint father of the science of microscopy. Judging from the contents of his early notebooks, there is every reason to think it was Hooke's theory of colors that stimulated Newton to intensify and broaden his own investigations into the celebrated phenomenon.[58]

Though an advocate of the mechanical philosophy of nature, Hooke rejected Descartes's idea that light is a pressure transmitted instantaneously through space from a luminous vortex. Instead, Hooke held that light consists of pulses or very quick vibrations propagated with a finite velocity, a hypothesis that makes him one of the forebears of the wave theory of light. He further theorized that light travels faster in a solid medium, such as glass, then in a more fluid or elastic medium, such as water. Because the velocity of light varies in different media, the resulting refraction renders the front of the pulse oblique to the direction of the beam; and it was with this "obliquity" that he connected the sensations of color: "That Blue is an impression of the Retina of an oblique and confus'd [mixed] pulse of light, whose weakest part precedes, and whose stongest follows. And that Red is an impression on the Retina of an oblique and confus'd pulse of light, whose strongest part precedes and whose weakest follows."[59]

Hooke had simply substituted his own mechanism of modification, personally more satisfying, for that of Descartes. Like Descartes, however, he continued to embrace the ancient scale of colors running from red—the strongest, least modified, and closest to pure white—to blue—the weakest, most modified, and closest to black. Thus while the mechanism believed responsible for producing colors may have changed, the underlying doctrine had not. The spectral colors—red, orange, yellow, green, blue, indigo, and violet—remained, in the eyes of the virtuosi, modifications of pure white light, that is, in the eyes of all the virtuosi save one. In a single devastating sentence in the *Quaestiones* Newton rejected the two principal tenets of Hooke's hypothesis, that light consists of swift vibrating motions or pulses and that colors result from a "confus'd" mixture of impressions: The more uniformly the globules move ye optick nerves ye more bodys seem to be coloured red yellow blew greene &c but ye more variously they move them the more body's appeare white black or greys."[60] So much for common sense.

Admittedly, when taken out of context, even Newton's earliest statements concerning the heterogeneity of white light seem unequivocally self-assured. And yet he first approached the concept, as he did the principle of attraction at a distance, on cat's feet. For just as he was reluctant to cast aside altogether Descartes's vortices as the explanation behind planetary motion, he cautiously retained certain elements of the mechanical philosophy to support his infant theory of colors. Such concepts as "slow and swift rays," "elastic power," and "corpuscles" unmistakably harken to the mechanists' tune. Newton could not have agreed more with Dr. John Arbuthnot's statement that "Those who talk mechanically talk most intelligibly."

It appears that at some point in 1665, perhaps with the onset of the plague, Newton temporarily set aside his optical studies. But if his own testimony is taken at face value the hiatus was of short duration, for in his letter to Oldenburg he claims to have undertaken further prism experiments in the beginning of 1666, probably during his three-month stay in Cambridge from March through June. Of equal interest is his parenthetical reference to the fact that at the same time "I applyed my selfe to the grinding of Optick glasses of other figures than *Spherical*."[61]

Ever since childhood Newton derived deep satisfaction from working with his hands, his only significant diversion from purely intellectual pursuits. Even before retreating to Woolsthorpe during the plague he had taken up the art of grinding and polishing lenses with machines of his own construction and design.[62] Like others before, Newton observed that every lens at its edges is a small prism. The resulting spectral effect creates, as a frustrated Galileo well knew, the poor definition and visual distortion so common to the early refracting telescopes. This phenomenon, known as chromatic aberration, occurs because rays of light of different colors, that is, of different wavelengths, on passing through a lens are not brought to the same focus. In the *Dioptrique,* from which Newton learned so much, Descartes employed his mathematical knowledge of conics to determine the proper shape for the surface of lenses in an attempt to remedy this condition. The hyperbolic and elliptical lenses Descartes designed were so difficult to grind that no one, not even the indefatigable and manually adroit Newton, could produce them. What Descartes did not know, of course, is that each ray of colored light has its own index of refraction. Thus no single convex lens, its shape notwithstanding, will eliminate all of the distorting effects of chromatic aberration. The problem can be overcome only by using pieces of glass of differing refrangibility in combination to create an achromatic lens, a masterful technical feat first accomplished in the eighteenth century by John Dolland and Chester Moor Hall.

It was against this background that Newton doubtless began to reflect on the fundamental meaning of his earlier experiments with the prism, which contradicted Descartes's assumption of the homogeneity of light. For even if he were to succeed in grinding elliptical and hyperbolic lenses, a perfect focus could not be obtained because some light rays, like blue, are refracted more than others, like red. Realizing this, he quit working on nonspherical lenses, never to return to them.[63] Instead, he purchased another of several prisms "to try [once again] therewith the celebrated *Phaenomena* of *Colours.*" Convinced that there was far more to learn about them than anyone had previously suspected, Newton now marshaled all his formidable powers to solve, experimentally, a riddle as ancient as the oracle of Apollo.

As often happened when he was about to tackle a fresh problem or, as in this instance, to renew his assault on a familiar one, Newton began a new section in one of his notebooks. As if to emphasize the significance of what he was now about, he took the additional step of beginning an entirely new volume. He titled the opening essay "Of Colours," as he had the corresponding section of the earlier *Quaestiones.* In sixty-four carefully drafted steps, he set forth the basic ideas that later became the substance of his first Lucasian lectures, the famous paper to Oldenburg of 1672, and ultimately Book One of the *Opticks.*[64]

Drawing upon the intellectual momentum generated during his most recent period of thought on light, it took Newton no more than a page to penetrate directly to the heart of the most critical issue at hand: the validity of the sine law of refraction. For if his own theory of refraction were to stand, Descartes's law must inevitably fall or else undergo significant revision.

Once again Newton turned to the prism, which, as an instrument of scientific study, was never more creatively or deftly handled by any human being before or since. If, as Descartes believed, a beam of white light is homogeneous, then, on passing through a prism, all parts of the beam will be equally refracted, producing a circle of colors on whatever surface it chances to fall. Indeed, when Descartes focused the spectrum on a screen only inches from his prism he found this to be so and presented the results as confirmation of the sine law and as additional proof of the principle of modification. Unlike Descartes, Hooke did not employ glass prisms in his most important experiments on the spectrum. Instead, like Grimaldi, he refracted a beam of light off the surface of a deep container of water and focused the image on a screen no more than 2 feet distant. Boyle, whose *Experimental History of Colours* deeply influenced Newton's earliest optical studies, did much the same. In each instance the refracted image was more or less circular. Any perceived irregularities were either discounted as too minor to be of concern or attributed to various causes, such as dispersion from the apparent diameter of the sun, which were seemingly consistent with the unshakable doctrine of modification.

In contrast to his fellow virtuosi, Newton conceived of a novel variation of the oft-repeated experiment (see Figure 4–2). Darkening his chambers, he drilled a one-eighth-inch hole in a shutter and intercepted the beam of incoming light with a prism. He then ingeniously projected the spectrum on a wall some 260 inches (22 feet) distant. Newton observed, "The Colours should have been in a round circle were all y^e rays alike refracted," but instead "their forme was oblong terminated at their sides wth streight lines; their bredth being $2\frac{1}{3}$ inches, their length about seven or eight inches, & y^e

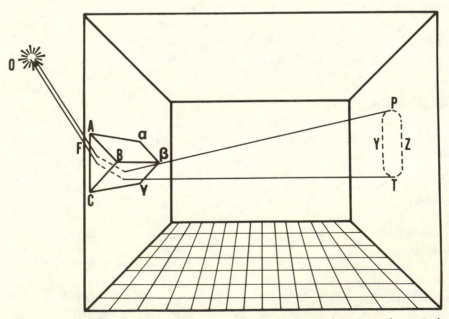

Figure 4–2.   The oblong spectrum. Redrawn from Newton's diagram in the optical lectures (University Library Cambridge, Additional Manuscript 4002, p. 3).

centers of y^e red & blew being distant about $2\frac{3}{4}$ or 3 inches."[65] Clearly, if Descartes's law were valid, the top part of the beam had been bent far more than it ought to have been, the bottom part less, which explains its elongated form. There was something fundamentally wrong with the "precise" law of refraction.

When Newton later recalled this critical moment in his paper to Oldenburg, he spoke of his "surprise" on first seeing the oblong spectrum. That he was deeply pleased, there is no doubt. But surprised—not really. By allowing sufficient distance for the spectrum to spread out, he found exactly what he expected to find: White light is heterogeneous, and the different rays that give rise to colors are refracted at different angles. To think otherwise—that this demonstration was nothing more than the result of fortuitous circumstance—is to misunderstand completely the essence of Newton's genius.

While convinced of the validity of his findings, Newton, like all great experimentalists, played devil's advocate to himself. He began to consider a number of contingencies, however remote, that might otherwise explain what he now attributed to the principle of heterogeneity.

> I could scarce think, that the various *Thickness* of the glass, or the termination with shadow or darkness, could have any Influence on light to produce such an effect; yet I thought it not amiss to examine first these circumstances, and so tryed, what would happen by transmitting light through parts of the glass of divers thicknesses, or through holes in the window of divers bignesses, or by setting the Prisme without [outside the shutter] so that the light might pass through it, and be refracted before it was terminated [confined] by the hole: But I found none of those circumstances material. The fashion of the colours was in all cases the same.[66]

Not yet fully satisfied, he placed a second prism, this one inverted, between the first and the wall of his room (Figure 4–3). If the oblong spectrum had resulted from imperfections in the glass, a second prism, with similar imperfections, would compound them. On the other hand, if the prisms were properly ground, a round beam "with as much regularity as when it did not at all pass through them" should result. This is exactly what occurred, further convincing him that he was not dealing with any contingent irregularity.[67]

To Newton, experimentation and quantification were merely different sides of the same coin; and he now brought mathematics to bear directly on the problem of light. "A naturalist would scearce expect to see ye science of [colors] become mathematicall," he wrote to Oldenburg, "& yet I dare affirm that there is as much certainty in it as in any other part of Opticks."[68] Aware that a beam of sunlight striking a prism is made up of nonparallel rays, Newton devised an experiment to eliminate any possibility of dispersion. He knew that if the angles of incidence and refraction of rays of colored light could be accurately calculated, a new law, according to which

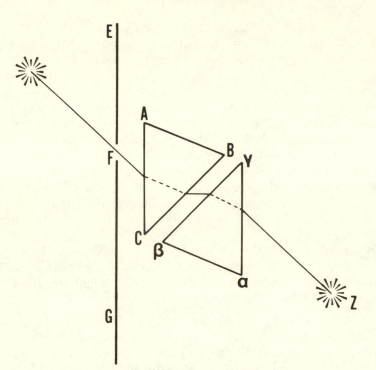

FIGURE 4-3. Testing the prisms. Redrawn from Newton's diagram in the optical lectures (University Library Cambridge, Additional Manuscript 4002, p. 17).

each color has its own proportion of sines, would result. Placing a board with a hole in it several feet from the shutter, he narrowed the incident beam from a visual angle of 31 inches to less than 7 inches. Although the length and breadth of the oblong spectrum decreased by more than half, the colors were now equally refracted by both faces of the prism. As expected, their relationships to one another were not radically changed. Finally, Newton rotated the prism at his window and found, as he had again expected, that this variation in the incidence of parallel rays did not alter their pattern on the wall. He could now calculate the sines for each color, something Descartes had been able to do for the middle of a refracted beam of white light only. Newton had again found vindication in the grand vision he shared with Pythagoras: that in the flow of things there is a perceivable fixity, that number holds sway above the flux.

In the *Novum Organum* Francis Bacon encouraged those who would seek out the underlying truths of natural phenomena to employ "luciferous experiments" rather than judge "solely by probable reasons," thus stressing the importance of what he called the *instantia crucis*, or crucial instance.

Robert Hooke, a fervent advocate of Bacon's experimental method, later coined the phrase *experimentum crucis* in his *Micrographia*, "serving as a Guide or Land-mark, by which to direct our course in the search after the true cause."[69] Newton, who knew the *Micrographia* well (indeed, too well from Hooke's point of view), borrowed the phrase, making it famous among the virtuosi through his 1672 paper on light.

Contrary to legend, the experiment that Newton thought more demonstrative of his revolutionary theory of colors than any other did not arise full-blown, like a mature phoenix, from the ashes of the disintegrating doctrine of modification. The phrase *experimentum crucis* is nowhere to be found in the notebook of 1666, for only on deeper reflection did he recognize the scientific value of expanding upon two relatively simple demonstrations in the essay "Of Colours." As before, he placed a prism near the hole in a shutter of his darkened chambers: "And holding another Prisme about 5 or 6 yards from ye former to refract ye rays againe I found first y^t ye blew rays did suffer a greater Refraction by ye second Prisme than ye Red ones. And secondly y^t ye purely Red rays refracted by ye second Prisme made noe other colours but Red & ye purely blew ones noe other colours but blew ones."[70] If colors were nothing but modifications of white light by the medium through which it passes, as the conventional wisdom dictated, the blue rays should not have been refracted more by the second prism than the red. Furthermore, the second prism should have produced other colors by turning red to orange or blue to indigo, but it did not. When fully developed a few years later, the *experimentum crucis* so effectively refuted the many arguments posed by its critics that it became the supreme symbol—a kind of scientific Holy Grail—of modern empirical analysis.

Newton, as one would expect, had concerned himself primarily with questions of refraction and dispersion up to this point. He was well aware, however, that he must also be able to demonstrate an ability to recombine individual rays of colored light to form white (Figure 4–4). It was hardly a new insight, for the second page of his essay on colors in the *Quaestiones* begin with this statement: "Try if two prisms ye one casting blew upon ye others red doe not produce white."[71] Now ready to pursue this question experimentally, he placed three prisms side by side so that their spectra overlapped. At the point in the center where the colors "are blended together there appears a white." He next fastened a piece of paper containing four slits to one side of a prism, making certain that the slits were parallel to the prism edges (Figure 4–5). Holding a second piece of paper close to the prism, he observed that the light passing through the slits produced four colored lines. He gradually moved the paper back until the colors combined to form white. As the paper was moved still farther away the area of white narrowed, finally yielding to the full spectrum.[72] Though the

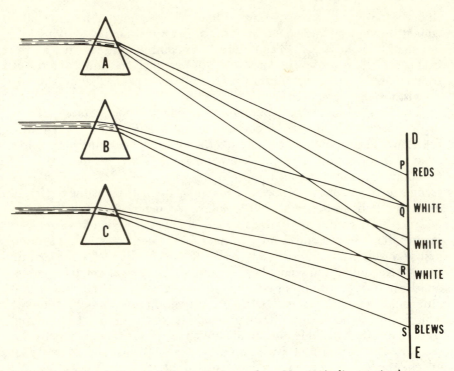

Figure 4-4.   Producing white light. Redrawn from Newton's diagram in the essay "Of Colours," p. 12.

recomposition of white light could hardly be put forward as an ironclad scientific principle based on these two experiments alone, further demonstrations in the coming years would lend full credence to the poet's claim that Newton had "untwisted the shining robe of day" and deftly put it back together again. It is little wonder that of all his discoveries he was most enthusiastic over the results of the experiments with light. No predecessors had clearly paved the way as for the yet to be refined calculus, and the evidence was far more conclusive than any he could now muster in support of the still highly questionable inverse square principle. Through skillful experimentation, dogged determination, and, not least, the gift of genius, Newton had singlehandedly reversed the relationship of colored and white light, thus standing a commonsense tradition spanning some two millennia on its head. Yet even after living intimately with this knowledge for nearly seven years and finding himself engaged in an intellectual battle with Hooke and others when he sought to establish it—a battle in which no quarter was given and none asked—Newton felt constrained to observe: "I perswade my selfe that this assertion above the rest appears *Paradoxicall, &* is with most difficulty admitted."[73]

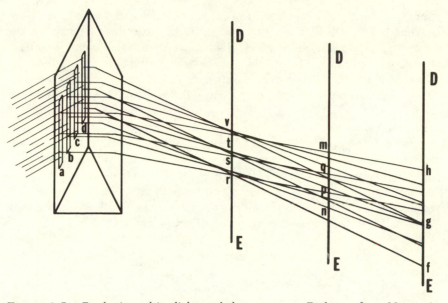

FIGURE 4-5.   Producing white light and the spectrum. Redrawn from Newton's diagram in the essay "Of Colours," p. 12.

VIII

Formidable though they were, the individual questions of natural philosophy faced by Newton during the *anni mirabiles* and after did not represent the whole of his scientific concerns. Beginning with the systematic composition of the *Quaestiones* in 1664 and continuing almost until his death, he labored under an extreme, albeit productive, tension wrought by the two great opposing forces of his genius. On the one hand was the rigorous self-imposed requirement that he proceed by methodically rational analysis—the familiar Newton of the public face. On the other was the secret, nonmystical visionary, fired by the barely controllable urge to rush forward in the hope of effecting the grand synthesis, no matter how intuitive. The recent examination of previously neglected papers lends much support to the thesis that certain of his most illustrious discoveries owe their paternity to heuristic principles, which themselves became untenable by virtue of the very truths to which they gave life. Among these principles was his commitment to the corpuscular conception of light.

From the very beginning of his concern with optics in the *Quaestiones*, Newton thought of light in mechanistic terms, as streams of globuli or particles. He assumed that the different colors resulted from the many sizes and varying speeds of these corpuscles being impressed on the optic nerves.

Blue was identified with small, slow-moving particles, red with large, swift-moving ones. Each of the other colors of the spectrum had its place on the scale between these two extremes. Hooke, in contrast, had opted for pulses, to which Newton strenuously objected, arguing that if light took the form of waves it would deviate from its linear path in the same manner as sound. The more deeply the concept of immutably colored rays penetrated Newton's thinking, the more he was convinced that the sweeping corpuscular hypothesis must be correct. Moreover, he found additional if inconclusive support for this view by experimenting with thin films in 1666.

Working with a lens and a piece of glass, he learned that the rings of colors produced in the layer of air trapped between their surfaces grew in diameter the more oblique the angle at which they were observed. He equated the size of the rings with the force of the ray's "blow" on the film. The stronger the blow the more readily the ray would pass through; therefore, the smaller the circle.[74] Buoyed by this "discovery," he immediately sought to quantify the phenomenon by working out a formula in which the diameter of the circle varies inversely as the perpendicular motion of the light to the film. Later, Newton would expand the corpuscular conception to embrace the principle of gravitation, which he sought to explain as the result of mutual physical attractions exerted by minute particles circulating through an all-pervasive ether—the boldest hypothesis of all. In the end, however, the corpuscular conception of light failed Newton's own high standards of rigorous analysis and was reduced to a mere query in the *Opticks* of 1706. The related mechanical hypothesis of ether suffered a like fate not long after. In the last page of the *scholium generale* of the *Principia's* second edition, he wrote: "It is enough that gravitation actually exists and acts according to the laws we have exposed."

Sufficient publicly, perhaps, but certainly not in private. As with Kepler and his grandiose geometrical model of the universe, Newton had learned much from pursuing the one great theory he could never prove. How painful this printed disavowal of it must have been for him. Yet some three centuries after his death, we still lack a wholly satisfactory explanation of light, and the mystery of what gravity is haunts contemporary physicists even more than it did the virtuosi of seventeenth-century Europe. Newton may have fallen far short of what he believed himself capable of ultimately accomplishing, but then it was a far vaster and infinitely more complex universe than anyone of his time or place in history could possibly have imagined.

Chapter Five

A Kinde of Nothinge

Hope springs eternal in the human
breast:
Man never is, but always to be blest.

Pope

I

Isaac Newton was born a commoner and yearned to raise himself above
that position. His father, an illiterate yeoman, had died without any chance
to affect his son's character by direct example, but from the Ayscoughs the
boy had caught something of the grand air of privilege, if only in faded
form. His grandfather was a gentleman, as were his uncles, James and
William, who took a share, however small, in the rearing of their enigmatic
nephew. It was during the plague year of 1666 that Newton decided to
assert publicly his claim to a higher station. He attended the Herald's visitation at Grantham, inscribing himself "Isaack Newton of Wolstropp.
Gentleman, age 23."[1] Since his father had never enjoyed this distinction, it
might be thought that the son had overreached himself, but such was not
the case. Isaac's close family ties to the Ayscoughs, his university education,
and the ownership of a small manor and some additional land all combined
to support his claim. He was now a gentleman and a scholar, and if the latter
accomplishment meant little to the unschooled Hannah, the former meant
a great deal to her. Isaac's new status must have done much to still the
troubled waters that had disturbed their once close relationship.

The most recent outbreak of the plague having run its course, the
University reopened for the Easter Term on Lady Day, March 25, 1667.
With a bachelor's degree in hand and his status as a gentleman legally

recorded, Newton returned to Trinity to complete work on his Master of Arts. He had already risen above his origins, both social and intellectual, yet everything he cared for hung in precarious balance. He now faced the fellowship elections of October. If he was unsuccessful, his stay at Cambridge would soon be terminated. Should this come to pass, what would happen to him? Surely he must choose between the life of an obscure gentleman farmer or the rectory of some unknown village in the bowels of Lincolnshire. It was not the quest for a sinecure but the pursuit of knowledge that had borne him upward from the circumstances of his birth and from the rustic surroundings that produced him. Scholarship was his claim to dignity; he must continue with his intellectual labors or face the unthinkable—the terrible plunge backward into the abyss of anonymity. One thinks of Comenius' lines in the *Coriolanus*:

> *He was a kinde of nothinge, titleless,*
> *Till he had forged himselfe a name . . .*

Because of the plague, no elections to vacant fellowships had been held in 1665 or 1666. Still, there were only nine places to fill, and even that limited number would have been cut by at least a third had it not been for a series of bizarre occurrences. A Senior Fellow named Barton had recently been removed on grounds of "mental aberration"; what form his insanity may have taken is unrecorded. Not to be outdone, two other Trinity Fellows, both apparently disoriented by drink, failed to negotiate the staircases leading to their rooms, and in falling to the bottom sustained injuries serious enough to force their withdrawal from the University. It was probably one of those two fellowships that ultimately went to Newton.[2] If so, it is one of the ironies of his life that this inveterate moralist should have owed his permanent position at Cambridge to the very kind of coarse behavior he so despised. Finally, an additional vacancy occurred in July 1667 with the death of the poet and Fellow Abraham Cowley, who also fell victim to the tempations of alcohol. Cowley contracted a fatal fever as a result of lying in the fields after a drinking bout.

The timely intervention of Bacchus notwithstanding, Newton's candidacy for a fellowship, to be successful, required more direct support. In most instances, scholarly merit counted for very little at Restoration Cambridge. In certain respects he could thank his stars that it did not, for he obviously had paid precious little attention to the traditional curriculum after 1663. As a result, he was supposedly found wanting by Barrow during the scholarship examinations of 1664 and again, in 1665, by those who presided over the tedious medieval disputations leading to the B.A. Nevertheless, Newton had not been denied. Aside from the general state of academic torpor, his success owed much to the support of Benjamin Pulleyn and Humphrey Babington. For the fellowship Pulleyn, who had examined Newton during his candidacy for a Bachelor's degree, was not in a position to help,

but Babington most certainly was. The fellowship elections were con
ducted by the Master and eight Senior Fellows—the Seniority—whose
ranks the affable and protective Humphrey propitiously joined in 1667.

A brief account of the examination procedure, drafted by a Trinity student some twenty-five years after Newton's candidacy, survives. For three
days during the last week of September, the candidates assembled in the
Chapel to undergo oral questioning. On the fourth day each was given a
theme by the Master to be completed in writing within a period of six
hours. On October 1, "by ye tolling of ye little bell at 8 in ye morning y^e
seniours are called & the day after at one o'clock to swear them y^t are
chosen."[3] Newton answered the tolling bell and found himself among the
elect. He became a Minor Fellow on October 2, a step that guaranteed his
elevation to the rank of Major Fellow upon the completion of his Master of
Arts. Together with eight other students, he swore to embrace the "true
religion of Christ" and take holy orders after incepting M.A. At last, he had
been granted permanent citizenship in the academic community, which,
for all its faults, provided him with the solitude and economic wherewithal
to pursue his studies undisturbed. Never again would he have to face the
unsettling prospect of being wrenched from the labors he so loved or from
the cloistered haven that insulated him from the petty distractions and obligations of the outside world.

In addition to such perquisites as their board, an annual stipend of £2,
and a small livery allowance, new Fellows became entitled to free rooms
assigned, like everything else, on the basis of seniority. On October 5, 1667,
Trinity's Master, John Pearson, signed a document confirming the alloca-
tion of quarters. The last line on the page reads: "to Sr Newton—Spirituall
chamber."[4] The precise location of these quarters has never been determined, but Edleston conjectured that they were the ground-floor apart-
ment next to the Chapel, in the northeast corner of the Great Court. Even
if these became Newton's rooms, the chances were quite good that he did
not occupy them, at least not right away. Fellows had the option of sublet-
ting their quarters and pocketing the rent. Indeed, Newton's accounts for
1668 contain just such an entry: "Received for Chamber rent 1.11.0."[5] Perhaps this was simply money paid him by Wickins, who was younger and
soon to be made a Fellow himself. On the other hand, Stukeley was told (by
whom he does not say) that Newton's quarters "were those in the great
court, on the north side, between the Masters lodg and the chapel."[6] Whatever the case, he later settled permanently in rooms on the second floor just
to the north of the Great Gate, transforming these stone-walled chambers
into the living heart of the scientific revolution.

There were two periods in Newton's life when his natural tendency to
withdraw from contact with others gave way to what was for him an
unusual degree of sociability. They fell on either side of his greatest
discoveries. Just before and after being appointed Fellow, he temporarily

broke out of his shell by purchasing fine clothes, making occasional visits to local taverns, playing cards, and entertaining guests in his rooms. This, the briefer of the two periods, was cut short by the lure of intellectual puzzles yet unsolved. The second lasted far longer and came after he moved to London, where the crushing mental burden of the Cambridge years was significantly eased. Fortunately, Newton continued his earlier practice of recording his income and expenditures after returning to the University. These accounts and the supplemental information compiled by Edleston more than a century ago constitute almost the whole of our knowledge about his personal life during the immediate postgraduate years.

As a Fellow, Isaac had more money to spend than ever before. And having shed the despised title of sizar—if not the lingering stigma—he was determined to use that income for the dual purposes of living up to his new status as gentleman and Fellow, and of asserting, once and for all, his independence from the domineering Hannah.

With acquiescence and a degree of financial support from Wickins, Newton undertook a general refurbishing of their living quarters. A painter was engaged for the sum of 3s. 4d., and £1. 1. 8. more went for the services of a joiner. New floor coverings and furniture came next: carpets of "neats leather" costing 18s., a "Tickin for a featherbed 1. 10. 0.," and for "My part of A couch 0. 14. 1."[7] Far more impressive was the amount Newton expended on clothing during 1667 and 1668. His academic regalia aside, it has been calculated that more than twenty pounds went for new suits and shoes alone.[8] Added to this figure was the cost of $8\frac{1}{2}$ yards of "Woosted Prunella" for a Bachelor's gown, "18 yards of Tammy for my M^r of Arts Goune," cloth for lining, a hood, and tailor's fees. This was obviously a much better-dressed and more fastidious don than the down-at-the-heels professor known to Humphrey Newton some twenty years later or the crusty old Master of the Mint who spurned luxury in any form. In the inventory of Newton's possessions taken just after his death, only two suits were listed among the articles of apparel, a meager wardrobe indeed for a man of his wealth and prominence. This revelation prompted Richard de Villamil's wry remark: "This looks a little like one set on and the other at the wash."[9]

Ever since the Middle Ages it had been customary for the recipients of new degrees to celebrate with their friends at the tavern, and Newton proved no exception to the rule. He shelled out 17s. 6d. for food and drink on graduating B.A., and 15s. more when he was created Master of Arts on July 7, 1668. In between, he was "At the tavern several other times," which cost him an additional pound, not to mention 15s. lost playing cards. He also spent 12s. 6d. entertaining his cousin Ayscough and 10s. on an unidentified acquaintance.[10] Whether those last outings were occasioned by a natural desire to be in the company of others or the fear of being labeled a prude is difficult to say. Whiston, reflecting upon his experience at Clare, noted that it was not at all unusual for fellowship candidates to take the "side of the Drinkers" in the hope of gaining preferment.[11] While it is true that most of Newton's expenditures for extracurricular activities came at this critical juncture in his career, to attribute such a motive to him would be purely speculative. Whatever his intentions, this is one of the few instances where he attempted to live up to what was expected of him by his peers, to accept as his duty participation in the raucous and snobbish ways of young manhood. Still, the collective portrait drawn by those who knew him best always showed a solitary, taciturn figure, unconcerned with Falstaffian pleasures of the flesh. Not in adolescence, early manhood, or later life was he ever truly "one of the boys."

Distraught over the many academic abuses and moral failings he encountered at Restoration Trinity, Newton was later to draw up a plan titled "Of Educating Youth in the Universities." To his many proposals for changes in both the substance and the quality of instruction, he added the following: "All graduates wthout exception found by the Proctors in Taverns or other drinking houses, unless wth Travellers at their Inns, shall at least have their names given to ye Vice-chancellor who shall summon them to answer [for] it before the next Consistory."[12] Perhaps by this time the morally erect Lucasian Professor had forgotten that he, too, had taken some pleasure in hoisting a tankard as a newly minted Bachelor of Arts. In any case, his outline for reform was hardly the apology of the prodigal son who had only lately paused to repent the excesses of a misspent youth. If kept in perspective, the mild luxury Newton indulged in during his student days must be counted as nothing more than a minor detour from a normally undeviating puritanical path. In choosing such a course, he missed much of life by contemplating it only from a distance.

On December 4, 1667, Newton left Cambridge for Woolsthorpe, where he remained until February 12. He spent five shillings "for keeping Christmas" but collected the tidy sum of thirty pounds from Hannah, who managed the income from her son's property during his long absences.[13] Newton's fortunes continued to improve after he returned to the University. The anticipated promotion to Major Fellow came on March 16, 1668, and in July of the same year he was created M.A., the twenty-third entry on a list of 148 candidates signed by Thomas Burnet, the Senior Proctor.[14] Yet even though his name would grace Trinity's fellowship lists for thirty-four consecutive years, so entrenched were those above him that he never joined the select company of the eight who formed the Seniority.

Newton was now entitled to modest increases in the economic benefits he had been receiving since achieving the rank of Minor Fellow as well as some added perquisites denied those of lower degree. Besides a yearly stipend of £2. 13. 4. and a livery allowance of £1. 13. 4., all Major Fellows received a share of the surplus annual dividends accruing from the College endowments. The figure for those who, like Newton, had the least amount of seniority was £25. He also shared in the pandoxator's dividend, which derived from the profits of the College bakehouse and brewery. Calculated at 3s. 4d. weekly—enough to purchase ten penny loaves of bread, ten quarts of small beer at 1d. a quart, and ten quarts of ale at 2d. a quart—it gave each resident Fellow an additional £5 of income.[15] Add to this the value of his free lodging and meals, now taken at the prestigious high table, and Newton must have garnered in the neighborhood of £60 per year.[16] Nor, unlike some of his colleagues, did this constitute his only important source of income. The properties at Woolsthorpe and Sewstern produced an annual combined revenue of at least £80, and, while he might not have always collected the full amount from his mother, the money, as his accounts for 1667 and 1668 clearly indicate, was there for the asking. Hannah turned over at least £50 to him during this period, and there could have been even more that went unrecorded. A part of these funds was probably used to underwrite his ongoing activities as a moneylender. Only once while an undergraduate did he loan anyone as much as a pound, but now he was willing to risk up to twice that amount on a single borrower. Whatever the rate of interest, his thriving enterprise suffered no dearth of clients. Among the names listed by Newton are those of "Dr Wickins, Boucheret, Perkins," and "Wadsley," the last a Fellow from his native Lincolnshire. As before, an X was used to indicate the repayment of a debt in full.[17]

However much pleasure Newton may have derived from indulging in certain creature comforts previously beyond his financial reach, he never once thought seriously of abandoning his private studies. As early as the summer of 1668 he was preoccupied with a "Small Prospective" or experimental model of the famous reflecting telescope sent to the Royal Society in 1672.[18] When John Conduitt later asked him what craftsman he employed to carry out his design, Newton replied that he had made it himself. Conduitt then inquired about the special tools needed to fashion the speculum: "He said he made them himself & laughing added if he had staid for other people to make my tools & things for me I had never made anything."[19] Indeed, at the very time he and Wickins were refurbishing their rooms, Newton was busy transforming part of the space into a practical workshop. Among the implements he purchased were a "Lathe & Table, Drills, Gravers, a Hone & Hammer & a Mandrill." To these were added a magnet, compasses, glass bubbles, and three prisms. And this was only the beginning: In the autumn of 1669 Newton visited London for only the second time and while there acquired two furnaces and the supplies needed to conduct the first of untold scores of alchemical experiments. Given the hectic pace of his labors, it is little wonder that the length of Wickins's absences from the University increased in direct proportion to his roommate's devotion to the experimental principles of the new philosophy. Further evidence of Newton's intellectual interests is contained in his entries of book purchases: "Bacons Micelanys," Thomas Sprat's "Ye Hystory of y^e

Royall Society," and Lazarus Zetzner's "Theatrum chemicum," a multivolume anthology of alchemical treatises. He also bought a subscription to the
Philosophical Transactions, little knowing that his own papers would find
their way into the famous publication a few years later.[20] While he may not
have admitted it to others, Newton now considered himself worthy of being
counted among the virtuosi.

II

As Lucasian Professor, Newton became known for certain patterns of idio-
syncratic behavior, which Stukeley quaintly but quite correctly labeled
"pieces of inadvertancy"—a continuing manifestation of the youthful
habit of detaching himself from the outside world.[21] It also appears likely
that he possessed certain sensory powers considerably beyond the ordinary,
an offhand demonstration of which only added to his developing reputation as a gifted eccentric.

Public clamor and political pressure exerted by trading interests
plunged England into the Second Dutch War (1665–67). In June 1667 a
Dutch raiding force attacked Sheerness, severed the boom across the Med-
way, and sank three first-rates at their moorings. Two others were towed
back to Holland, one of which was the flagship, *Royal Charles.* According to
tradition, Newton came into the hall of Trinity College and told the Fellows
that there had been a naval action between the Dutch and English and that
the English had been bested. Asked how he knew this to be so, he replied,
"that, being in the observatory, he heard the report of a great firing of
canon . . . and that as the noise grew louder and louder, he concluded that
they drew near to our coasts; and consequently that we had the worst of it,
which the event verified."[22]

Newton's acute sense of hearing may have been complemented by even
more extraordinary powers of visual perception. As a child of nine he could
see perfectly well from Skillington, where he attended dame school, the
church spire at Grantham, which stood out "like a stick wch was six miles
off."[23] During his middle years, however, he suffered from myopia, an affliction not uncommon among those who devote long hours to intensive
study. As sometimes occurs in such cases, the defect disappeared completely in later life. Stukeley, a physician, noted that Newton's "full and
protuberant eyes" became better "by growing flatter. I saw him cast up the
treasurers accounts of the royal society, being a whole sheet of paper full,
without spectacles."[24] The myopic individual lives in a closed world and
can sometimes discern detail undetectable to the person with more normal
vision. Historians of science have long puzzled over Newton's ability to
count and measure the contiguous rings of colored light produced when a
lens is compressed against a flat piece of glass. In addition to his immense

powers of concentration, at least part of the answer may lie in an abnormal visual acuity at close range. It seems more than coincidental that the following passage made its way into the essay "Of Colours" about the time Newton anticipated an English defeat from the muffled roar of cannon—and initiated his experiments with thin films: "Mr Boyle mentions one that by sickeness became so undersighted as in ye dark night to see & distinguish plainly ye colours of ribbans . . . and of another y^t by a feaver became so tender hearing as to hear plainly soft whispers at a distance wch others could not at all perceive."[25]

The myopia afflicting Newton throughout much of his adult life was not only physical but cultural. He who could contemplate a universe of immeasurable dimensions lived in a small world whose boundaries formed a line connecting Woolsthorpe, Cambridge, and, later, London. Only rarely did he venture beyond its narrow confines, and never was he sufficiently tempted by curiosity to visit the Continent. To be sure, his library contained a number of the travel books so popular during the period, but most of these were purchased in later years and, with few exceptions, seem not to have been systematically consulted.

Not once did Newton visit London while an undergraduate, nor would he travel the 50 miles to the capital from Cambridge until his Master's degree was in hand. He wrote in the Fitzwilliam notebook that "I went to London on Wednesday Aug 5^t & returned to Cambridge on Munday Sept 28, 1668." Where he stayed, whom he visited, and how he occupied his time remain minor mysteries. Considering the calamitous events of 1666, he must have long regretted the decision to postpone the journey. Four-fifths of the city lay in ruin, a blackened monument to the massive devastation wrought by the Great Fire. Though reconstruction was well under way, London now seemed more a town than a great city. The rational street plans proposed by Christopher Wren, John Evelyn, and Robert Hooke had already been rejected by its suspicious citizens, who opposed any scheme involving the massive transfer of property to the government. The best that could be accomplished was a widening of the old traces and a prohibition of further half-timber construction. By the time of Newton's visit only eight hundred new brick and mortar houses stood where more than thirteen thousand dwellings had been just two years before. And not until 1672 would the rebuilt streets become substantially continuous.[26] Perhaps depressed by what he saw, Newton returned to Cambridge via Woolsthorpe. From Hannah he received £4. 5s., which substantially offset the £5. 10. 0. spent during his journey.[27] At least from a financial point of view the holiday was not a total loss.

Among the more curious but revealing documents left by Newton is a letter to Francis Aston, a Fellow of Trinity, who in the spring of 1669 was about to embark upon the Grand Tour. Aston wrote to Newton asking him to pass on whatever advice he thought would be of benefit to an English-

man abroad—this despite the fact that the sum total of Newton's experience as a traveler then consisted of a single visit to London. Aston's letter to
Newton has been lost, and since the latter's reply of May 18, 1669, was
never mailed, for it contains no postmark, it has been suggested that this
was simply a writing exercise incident to Newton's formal studies. Yet even
if it was the product of an academic assignment, there is little reason to
think that his frequently naive advice is less than honestly given. After all, it
is still Newton's letter:

> When you come into any fresh company 1, observe their humours; 2, suit your
> own carriage thereto by wch insinuation you will make their converse more free
> & open: 3 let your discourse bee more in Quaerys & doubtings yn premptory
> assertions or disputings, it being ye designe of Travellers to learne not teach. . . . .
> You will find little or noe advantage in seeming wiser or much more ignorant yn
> your company. 4, seldom discommend any thing though never so bad, or doe it
> but moderately, lest you be unexpectedly forced to an unhansom retraction. . . .
> 5 If you bee affronted, tis better in a forrain Country to pass it by in silence or
> wth jest though wth some dishonour than to endeavour revenge; For in the first
> case your credit's ne're the wors when you return into England or come into
> other company yt have not heard of the quarrell, but in the second case you may
> beare ye marks of ye quarrell while you live, if you out live it att all. But if you
> find your self unavoydably engaged tis best, I think, if you can command your
> passion & language, to keep them pretty eavenly at some certain moderate
> pitch, not much heightning them to exasperate ye adversary or provoke his
> freinds.[28]

Clearly, Newton's conception of face-to-face contact with other human beings is that it should be as dispassionate and as harmless as possible. To this
end he is willing to sacrifice temporarily the natural demands of curiosity
and, if need be, a considerable degree of personal honor as well. He wishes
neither to disturb nor to be disturbed. The most desirable form of conduct
is that which maintains a shield between himself and his fellow man, requiring not the slightest risk of emotional capital. What is more, Newton knew
well his own quickness to anger, and he decided early on never to let his
temper get the best of him in public, which it rarely did. This explains the
seemingly irreconcilable discrepancy in testimony between those who
knew him socially as an unemotional, even retiring presence and others
who became the target of his self-righteous wrath via private correspondence and third parties. Reason and passion have rarely been so at variance
in the same individual.

At the same time Newton, true to his character, urged Aston to utilize
his powers of observation to the full. In an incisive paragraph reminiscent
of Montesquieu's *Spirit of the Laws* (1748) he recommended the following:

> As 1 to observe ye policys wealth & state affaires of nations so far as a solitary
> Traveller may conveniently doe. 2 Their impositions upon all sorts of People
> Trades or commoditys yt are remarkeable. 3 Their Laws & Customes how far

they differ from ours. 4 Their Trades & Arts wherein they excell or come short of us in England. 5. Such fortifications as you shall meet wth, their fashion strength & advantages for defence; & other such military affaires as are considerable. 6 The power and respect belonging to their degrees of nobility or Magistracy. 7 It will not bee time mispent to make a Catalogue of the names & excellencys of those men that are most wise learned or esteemed in any nation. 8 Observe ye Mechanisme & manner of guiding ships. 9 Observe the products of nature in severall places especially in mines with ye circumstances of mining & of extracting metalls or mineralls out of their oare and refining them and if you meet wth any transmutations out of one species into another (as out of Iron into Copper, out of any metall into quicksilver, out of one salt into another or into an insipid body & c) those above all others will be worth your noting being ye most luciferous & many times lucriferous experiments too in Philosophy.[*] 10 The prizes of diet & other things. 11 and the staple commoditys of Places.[29]

Newton, whose deepening interest in alchemy had recently led him to establish a laboratory of his own, closed by advising Aston to gather all of the information he could concerning mining and the supposed transmutation of metals. He especially encouraged him to seek out one "Bory" (Giuseppe Francesco Borri) who "some yeares since was imprisoned by the Pope to have extorted from him some secrets (as I am told) of great worth both as to medicine and profit." He was thought to have escaped into Holland "where they granted him a guard." Aston could identify this wizard by his green clothing, but how he was to go about locating him, much less make personal contact, Newton did not say. About this time he gave further evidence of an interest in alchemy by labeling a section of his notebook "Philosophers Stone." Across the tops of the pages he inscribed the following headings: "Gross Ingredients, first preparation, Principles, Elements, Mercuries, Sulphurs, Salts, fires, Of y^e work wth common $\odot$ [gold], Times, Proportions, Hieroglyphicks, Progress of y^e Decoction, use of y^e stone, I Missellamies."[30] It was as if he had roughed out the answer to the ancient riddle of transmutation in his mind, looking upon its achievement as a foregone conclusion. But it was not to be. These pages (in contrast to hundreds of others devoted to alchemy) were destined to remain forever blank, a lasting reminder to future generations of what became of that fine dream.

III

Except for his friendships with Wickins, the rather mysterious chamberfellow, and Aston, a man of influential connections but mediocre intellect,

---

*A common phrase among alchemists used to distinguish between those experiments which advanced knowledge and those which enhanced income. Here Newton suggests the two need not be mutually exclusive.

Newton entered into few significant relationships with individuals his own age. As an adolescent and young adult, he was drawn to older, educated men such as Clark, Stokes, and Babington. Then, as he passed through middle life and old age, he seemed most content when in the company of younger admirers, including Edmond Halley, Nicolas Fatio de Duillier, David Gregory, and Stukeley. Those at the extremes were not perceived as a threat, whereas coevals—especially if gifted like himself—were almost inevitably viewed with a jaundiced eye. Hooke, for example, was less than seven years Newton's senior, while Leibniz and Flamsteed, both born in 1646, were only four years his junior. They became the supreme targets of his searing animosity. On the other hand, Isaac Barrow, Lucasian Professor of Mathematics, was a dozen years older than Newton, a difference in age magnified by Barrow's worldly background and prominent position in Trinity's rigid academic hierarchy.

According to his biographer Barrow was a man of only three faults.[31] First, he preached at excessive length, and this long-windedness deprived him of any hope of becoming a popular divine. (Once he was so moved by his subject—charity—that he spoke for three and a half hours, and once, while Royal Chaplain to Charles II, he detained his audience at Westminster Abbey so long that it got the organist to play "till they had blowed him down.")[32] Second, Barrow smoked to excess. Third, he was slovenly in dress and personal appearance. In his *Brief Lives,* the Wiltshire antiquarian John Aubrey related the following incident: "As [Barrow] was walking one day in St. James's parke, his hatt up, his cloake halfe on and halfe off, a gent. came behind him and clapt him on the shoulder and sayd, Well goe thy wayes for the veriest scholar that I ever mett with."[33]

However, if the following titles from the collection of Barrow's sermons are any indication of the man, he experienced the normal range of human weaknesses: "Against foolish Talking and Jesting," "Against Rash and Vain Swearing," "Of Evil-Speaking in general," "Of Quietness and doing our own Business," "Of a peaceabel Temper and Carriage."[34] For the last-named sermon Barrow took as his text Romans 12:18: "If it be possible, as much as lieth in you, live peaceably with all men." The operative word in this case was "if," for in addition to a penchant for sloppy dress, which he shared with Newton, Barrow, like the other Isaac, was capable of great anger, particularly when a young man. Other than a mutual love of scholarship, these were about the only qualities they had in common. In temperament, behavior, and social outlook Newton and Barrow were as different as two men passionately committed to the same ideas could be.

Isaac Barrow was born in London in 1630. His father, Thomas, was a prosperous merchant and linen draper to the ill-fated Charles I. Ann Barrow, Isaac's mother, died when the child was four, and he was subsequently sent to the newly opened Charterhouse for his first schooling. But books proved of little interest to the belligerent youth, taking a poor second to his love of sport and physical combat. Abraham Hill, Barrow's executor and first biographer, wrote that his father became so distraught over the boy's antisocial behavior "he often solemnly wisht, that if it pleased God to take away any of his Children, it might be his son Isaac."[35] The despairing draper finally transferred his wayward son to Felstead in Essex, where he eventually blossomed under the stern but deft guidance of its rigidly Puritan headmaster, Martin Holbeach.

Having mastered the classical curriculum at Felstead, Barrow matriculated as a pensioner at Trinity College in early 1647 and was elected to a scholarship later the same year. Although Puritan both in spirit and caste of mind, he refused to yield to the considerable pressures and temptations of subscribing to the Covenant, a measure of his respect for a father with Royalist connections. In the meantime Thomas Hill, the staunch Parliamentarian and Westminster Assemblyman, had been chosen to replace Thomas Comber as Trinity's Master. Well aware of Barrow's political sentiments, Hill nonetheless admired the young man's moral resolve. The new Master is said to have placed his hand on Isaac's head one day and told him that "thou art a good Lad, 'tis pity thou art a Cavalier." Somewhat later Barrow delivered an impassioned oration in which he condemned the treason committed by those involved in the Gunpowder Plot of 1605, which, considering the circumstances, was neither a timely nor a discreet topic of discussion. The address predictably aroused considerable animosity, and several Fellows agitated for Barrow's expulsion from the College. Once again Hill showed his admiration for the young scholar, silencing his critics with the comment, "Barrow is a better man than any of us."[36] Thus in the growing heap of academic deadwood choking the University, Barrow was recognized as one of those increasingly rare individuals entitled to a position on the basis of merit alone. Nothing else can account for the grudging acquiescence of his politically hostile peers when he was chosen Fellow in 1649.

Drawn to the writings of Gassendi, Gilbert, Bacon, Kepler, and Galileo, Barrow took as one of his M.A. theses the mechanical hypothesis of Descartes—"he who seems, not undeservedly, to shine with special pre-eminence among these philosophers, and so deserves the deepest applause of the whole world of letters." Notwithstanding his lavish praise for the French mechanist, Barrow balked at the unqualified acceptance of the Cartesian hypothesis for its lack of an empirical foundation: Descartes "proceeded in such a way that, first, he collected and set up metaphysical truths which he considered suitable to his theory from notions implanted in his own mind." Next, "he descended to general principles which, forsooth, he had framed without consulting Nature."[37] Newton was soon to harbor this same reservation, rejecting such *a priori* concepts as the vortex theory with his famous aphorism, *hypotheses non fingo*—"I make no hypotheses." But what most troubled Barrow about Cartesian mechanics also proved most disturbing to Newton: the elimination of God as an active agent in the natural order. Neither virtuoso could bring himself to accept Descartes's limited vision of God as Creator—the First Cause—and nothing more. Descartes, Barrow wrote, "thinks unworthily of the Supreme Master of things who supposed that he created just one homogeneous Matter, and extended it blockish and inanimate through the countless acres of immense space." God cannot be likened to "some carpenter or mechanic repeating and displaying *ad nauseam* his one marionnettish feat." He ended by asking Descartes to "pardon us if, while following your enquiring spirit, we forsake your conclusions." Though the "us" to whom Barrow referred is merely a rhetorical substitute for the first person singular, he could just as well have been speaking for Isaac Newton.

Having walked on eggshells ever since his arrival at Trinity, Barrow knew better than to tempt the hand of fate further. He applied for and was awarded a travel grant in the amount of sixteen pounds for each of the next three years. He raised additional funds by selling off his precious books and then set sail for France in June 1655.

One suspects that, aside from an ingrained aversion to facing new experiences and unfamiliar landscapes, Newton chose never to venture abroad for fear of being fatally exposed to that dread disease of the spirit which seventeenth-century Englishmen called Popery. As a contemporary, John Fuller, so gravely warned the prospective traveler: "They that go over maids for their religion will be ravished at the sight of the first Popish church they enter into."[38] Barrow's friends no doubt shared a similar concern for the soul of this innocent abroad, but his lively dispatches to the College, one of the conditions of his travel grant, quickly turned their fears into unqualified admiration. Barrow met and held his own with the Jesuits of Paris and Florence; charmed his hosts on the Greek isles, in Smyrna, and at Constantinople with his knowledge of the classics; and manfully defended his honor (and that of England!) by besting a loose-tongued Turk in hand-to-hand combat. The fighting skills acquired at the Charterhouse and honed in battles with the butcher's boys of St. Nicholas's shambles had come in handy after all, as they were about to again. In a fitting climax to an odyssey that might well have been spawned in the imagination of Gilbert and Sullivan, Barrow fought bravely when his ship was unsuccessfully attacked by Maltese pirates. Such stirring accounts of derring-do proved heady stuff to the sedentary, lackadiasical scholars of Trinity, who greeted Barrow on his return in 1659 as a conquering hero.

The ill political winds that drove Barrow into exile underwent a radical shift in direction. Charles II ascended the throne in 1660, and those influential Royalists who survived the Civil War and its aftermath without seriously compromising their religious and political convictions could expect to be handsomely rewarded. Barrow took holy orders and was promptly appointed Regius Professor of Greek, the chair denied him four years before. In 1662 he was concurrently appointed Gresham Professor of Geometry in London, where his name appeared on the Original Fellows list of the newly chartered Royal Society, although he attended few meetings and never took a very active part in its formal affairs. Scarcely a year later Barrow's stock in the academy took another jump when he became the first Lucasian Professor of Mathematics at Cambridge.

Not since Henry VIII established the five Regius professorships in 1540 had anyone endowed a chair at Cambridge. Aware of the pressing need for a scientific lectureship to match the Savilian chairs founded at Oxford in 1619, Henry Lucas, a Fellow of St. John's, directed his executors, Robert Raworth, a prosperous London lawyer, and Thomas Buck, a licensed printer to the University, to draft the necessary statutes. This they did on December 19, 1663, and the product of their labors, with certain revisions, was confirmed a month later by the royal warrant of Charles II.[39] Compared to Barrow's previous professorships, the new appointment was munificent. Enough land had been purchased in the Bedfordshire villages of Riseley and Thurleigh to provide the chair's occupant with an annual stipend of at least one hundred pounds, more than sufficient income to make up for the statutory prohibition against holding any other official position within the University, except for a fellowship. The income was not subject to taxation. A few years after succeeding Barrow to the chair, Newton became embroiled in a dispute with local tax commissioners, who approached "my tenants" concerning revenues believed due on the Bedfordshire property. Fearing they were about to tax his legally exempt stipend, Newton drafted a sharp response: "My lands are not in the nature of College lands but of a salary or stipend belonging to my Lectureship . . . & by consequence excuse me expressly from paying for any of the profits of my Professorship."[40] As was usually the case, Newton prevailed.

Not only was the income excellent, especially for an academic position, but the statutes governing the Lucasian professorship relieved its occupant of anything approaching an excessive teaching load. At a prearranged time once a week during the academic year of three terms, which then lasted a total of seven months, he was sworn to "lecture and expound" to a senior audience on "some part of Geometry, Arithmetic, Astronomy, Geography, Optics, Statics, or some other Mathematical Discipline" of his own choosing. As a safeguard against the sluggard, the statutes further required that the professor deposit in the University Library copies of no fewer than ten of the lectures he had given the previous year. Should he miss a lecture without sufficient cause, a fine of forty shillings was to be levied. He was subject to the further requirement of making himself available in his lodgings at stated times for two hours each week, during which students could ask questions relating to his lectures. In addition, he was to make globes and appropriate mathematical instruments available for the curious. (At Trinity College one may still examine compasses, prisms, and other assorted devices traditionally thought to have been Newton's.) As with the other Cambridge professorships, however, there was a glaring discrepancy between rule and deed. Barrow lectured no more than one term out of the required three, and Newton seems to have followed that unsanctioned precedent without a pang of conscience. Barrow also abused the requirement that he annually deposit the texts of ten of his previous year's lectures in the library, yet no sanctions were brought to bear. Newton also showed a flagrant disregard for the statutory timetable: Only twice did his lectures find their way into the library during the first fifteen years of his professorship.[41] Again, there is no evidence to indicate that anyone kept count or, for that matter, even cared.

To what extent can Isaac Barrow be considered Isaac Newton's intellectual father? Is their attraction to many of the same ideas, both religious and scientific, merely coincidential, or is it as least partially the result of Newton's perusal of the Lucasian Professor's speeches and writings—perhaps followed by stimulating conversations that lasted well into the night? Was it Barrow who awakened Newton's early interest in alchemy and started him down the tortuous path in a lifelong search for a more accurate chronology of ancient civilization? No one knows exactly when these two remarkable Isaacs first met, nor can one begin to gauge satisfactorily the nature of their intellectual association much before 1669.

Newton recalled having been in the audience when Barrow inaugurated the Lucasian lectures in 1664. He allowed that Barrow might have fueled his interest in the nascent calculus, but nothing more. There is no evidence to indicate that his vagueness on this point was selective, but with Newton one can never be certain. In contrast to Barrow, who was highly complimentary of those whose work he admired, something within Newton prevented him from giving credit where it was due.

In any case, Newton could hardly have been just another face in the crowd for the simple reason that there was no crowd. True, the letter patent of Charles II that confirmed the Lucasian statutes required that all undergraduates beyond their second year and all Bachelors of Arts until their third year be in attendance, enough potential auditors to fill a large lecture hall. Equally significant, however, is the fact that by the 1660s attendance requirements counted for almost nothing, because the college tutorial system had eclipsed university instruction. Students who ignored the lectures were no more subject to a statutory fine or other forms of discipline than was the uncaring professor who himself refused to put in an appearance. Even Barrow, a man of wit, erudition, and personal charm, could not muster an intellectual following upon being appointed Regius Professor of Greek. On the first anniversary of his inaugural oration he launched a new series of lectures with a trenchant observation: "Since you bade me that long farewell a year ago I have sat on my Chair incessantly alone—and I am sure none of you will, as an eye-witness, challenge the accuracy of that statement even if I should be lying." He referred to the pleasures he derived from inhabiting "this large and commodious domicile alone and undisturbed by a jostling and contentious crowd. I have breathed an atmosphere uncontaminated by the foul breath of yawners or by the fetid stench of perspiration."[42] Edmund Castell, who became the first Adams Professor of Arabic in 1666, did not show Barrow's perseverance. When his initial lectures stirred no student interest, he posted a caustic sign on the lecture hall door: "Tomorrow the Professor of Arabic goes into the wilderness."[43] Castell did not resign but taking the path of least resistance, turned his chair into a comfortable sinecure. The "wilderness" into which he retreated was nothing more than a petrified forest of academic deadwood. Since the new mathematics was far more alien to the student body than classical Greek (or for that matter Arabic), Barrow must have enjoyed even less success as Lucasian Professor. Under the circumstances, he may have marked Newton's stern countenance well.

As Barrow's successor, Newton was unable to improve upon his predecessor's record. Though he was appointed to the chair in 1669, no one is known to have been among his auditors before John Flamsteed attended a lecture in 1674.[44] And who can forget Humphrey Newton's remark, so chilling to the heart of the academic, that Newton oftentimes, "for want of Hearers, read to ye Walls"? In such a fashion did the most intellectually gifted man of his age fulfill his statutory obligations to the University. Yet even Newton was not immune to an occasional lapse of conscience, if the testimony of his amanuensis can be taken at face value. Sometimes, "when he had no Auditors," Newton returned to his rooms within a quarter of an hour or less of having departed, hardly a sufficient passage of time to have permitted the delivery of even the most cursory of lectures on any topic in natural philosophy.[45]

If Barrow did indeed become aware of Newton when delivering his early Lucasian lectures, as seems probable, then it is also possible that the younger Isaac, already deeply committed to the new science, was sufficiently motivated to overcome his ingrained shyness and approach his unusual professor. That Pulleyn was Newton's tutor there is no doubt, and even had he wanted to switch to Barrow, the statutes governing Barrow's professorship prohibited it, for Newton did not hold the required rank of fellow-commoner. On the other hand, he may have availed himself of the opportunity to visit Barrow in his quarters during the weekly hours set aside for consultation on matters raised in the lectures. There is more than a touch of the suspicious in Conduitt's story about Newton's tutor telling him "he was going to read Kepler's Opticks to some gentlemen commoners & told him he [Newton] might come to those lectures."[46] Rather than Pulleyn, could it not have been Barrow, deeply involved in optical studies at the time, who invited Newton to a session on the book with the fellow-commoners he tutored? Barrow had frequently bucked convention in the past, and he was certainly enough of an iconoclast to overlook Newton's inferior position. If there was contact between the two at this time, it would better explain why Pulleyn sent Newton to be examined in mathematics by Barrow. As close-mouthed as Newton was, it is also possible that he had not previously discussed Euclid with the Lucasian Professor and that the latter, having taken Newton's fundamental grasp of geometry for granted, was as surprised as Pulleyn to learn of his deficiency in that area. One can further imagine that Newton, embarrassed and usually submissive in the presence of superiors, would have offered no excuses nor otherwise defended himself.

Aside from the absence of documentation, the problem with such an interpretation is the narrow time frame in which these supposed events would have had to unfold. Barrow only began lecturing on March 14, 1664, and presumably examined Newton a month later. Whatever the truth of the matter, it does seem that the dismissal of Barrow as a formative influence on Newton is too sharp a reversal of tradition.[47] While a student at Christ's College, Cambridge, Darwin became familiar as "the man who walks with Henslow," his teacher of natural history. It seems not unlikely that Newton, for whatever period of time, was recognized by his peers as the man who walks with Barrow.

IV

In contrast to a voracious appetite for questions involving natural philosophy, Newton required a periodic goading to sustain his interest in mathematics. His blossoming attraction to alchemy and an abiding passion for optical studies caused him to set aside temporarily the fluxional method and the work on infinite series, which occupied so much of his time between 1665 and 1666. However, in September 1668 something happened that soon forced him to return to the subject, whether he wanted to or not. That was the publication of Nicholas Mercator's treatise, *Logarithmotechnia*. The book, as suggested by its title, contained a simplified method of calculating logarithms. This, of course, was hardly new ground to Newton, who had already ventured much beyond the mathematical horizons envisioned by the author. And where Mercator demonstrated the ability to construct a square (quadrature) equal in area to a hyperbola by an infinite series, Newton had extended the process to all types of curves.

In the spring or early summer of 1669 John Collins, a member of the Royal Society, sent a copy of Mercator's book to Barrow at Trinity. At the time Collins was an employee of the Excise Office, but he had previously taught mathematics in London and had published three treatises on various aspects of the subject between 1652 and 1659. He was especially interested in the solution of equations and interpolation. However, his lasting contribution to the discipline had less to do with original work of his own than with sustaining a vigorous correspondence among those committed to the new mathematics. In a manner somewhat reminiscent of the Socratics, Collins saw himself as an intellectual midwife, paving the way for the birth and dissemination of new ideas. Before he died in 1683, he established a thriving informational network binding England's most prominent mathematicians to their counterparts on the Continent.

Barrow, who was pleased to receive a copy of the new treatise, drafted an enthusiastic reply to Collins on July 20, 1669, the contents of which strike today's reader like a bolt from the blue. "A friend of mine here, that hath a very excellent genius to these things, brought me the other day some paper, wherein he hath sett downe methods of calculating the dimensions of magnitudes like that of Mr Mercator concerning the hyperbola, but very generall."[48] The papers referred to by Barrow were Newton's *De Analysi per Aequationes Infinitas* (On Analysis by Infinite Series). Barrow further whetted Collins's appetite by promising to send the material by the next letter. Impatient as he was to learn of new developments in the field, Collins must have wondered why Barrow chose not to enclose the papers with his communication announcing their existence. Nor had the Lucasian Professor given any clue to the identity of the friend possessed of "a very excellent genius." Such understanding, or perhaps one should say resignation, would only come later, after Collins became acquainted with Newton. Try as he might, his persistent attempts to persuade the young virtuoso to commit his work to print ultimately proved fruitless. The publication of *De Analysi* was delayed until 1711, when Newton was in his sixty-ninth year! By then its value as a current piece of pioneering research had greatly depreciated, though it remains ever significant as an early example of Newton's method of fluxions.

Barrow now had a selling job of his own to do. It was with much difficulty that he talked Newton into revealing his work to a wider world. This is all the more strange because Newton's very reason for hastily drafting *De Analysi* was a concern for establishing his rightful priority to the method announced by Mercator. A classical scholar, Barrow may well have built his case on humanistic grounds in the form of an appeal to the interests of posterity. If that proved ineffectual, it is not inconceivable that he reminded Newton of his plans to resign his endowed chair, pending, of course, the discovery of a worthy successor. Whatever his tactics, Barrow succeeded in bringing Newton around, but his victory was something less than total. On July 31 he again wrote to Collins. This letter is rife with caution and anxiety, which only Newton could have communicated to the normally self-confident savant. Indeed, it is almost as if Newton stood over Barrow's shoulder as the letter was being written: "I send you the papers of my friend I promised, which I presume will give you much satisfaction; I pray having perused them so much as you thinke good, remand them to me; according to his desire, when I asked him the liberty to impart them to you. and I pray give me notice of your receiving them with your soonest convenience . . . because I am afraid of them; venturing them by post."[49]

One would expect Newton's identity to have been revealed at this point, but Barrow was still operating under a peculiar form of interdict. No additional information could be imparted until Collins reacted one way or another. A reply reached Cambridge within a matter of days, and it was most enthusiastic. On August 20, in his third letter to Collins on the matter, Barrow introduced Newton's name to the world of scholarship. "I am glad my friends paper giveth you so much satisfaction. his name is Mr Newton; a fellow of our College, & very young . . . but of an extraordinary genius & proficiency in these things." Barrow also gave Collins permission to "impart the papers if you please to my Ld Brounker," President of the Royal Society.[50] This Collins apparently did, and he subsequently circulated copies of the manuscript among his associates. Years later, when Newton gained access to Collins's papers, he was surprised to learn just how widely his unpublished work had been read.

Thus, Newton and Barrow were well acquainted by 1669 and clearly had been for some time. Barrow's outgoing nature aside, a man in his position would hardly have referred to an obscure Fellow like Newton as a "friend" and "genius" without knowing him and his work rather well. Indeed, for what reason other than a familiarity with Newton's studies would Barrow have made the *Logarithmotechnia* available to him? Finally, Collins himself suggested that the Newton–Barrow relationship went back at least as far as 1667. He wrote to James Gregory in January 1671 that "Mr Newton informed me himselfe, that he invented and contrived this generall Method [of infinite series] above two years before Mercator published any thing, and communicated the same to Dr. Barrow, who accordingly hath attested the same."[51] *De Analysi*, then, was almost certainly the product of mathematical research to which Barrow had been privy for some time.

It is difficult to believe that Barrow was not satisfied in his own mind by this time as to who his replacement should be. Since he was keenly aware of Newton's mathematical prowess, there is reason to think that the elder Isaac was actively grooming the younger one to become the next Lucasian Professor. In all the time he spent at Cambridge, Newton took only three students as his tutees. The first was a young fellow-commoner with the wonderfully Dickensian name of St. Leger Scroope. Interestingly, Scroope became Newton's pupil on April 2, 1669, a few months prior to Barrow's resignation.[52] Had Scroope (of whom nothing else is known) matriculated as a sizar, there would be no reason to question this relationship. But because of the handsome fees collected by their tutors, fellow-commoners were rarely assigned to uninfluential first year Fellows. Hence it seems likely that Barrow, who had his eye on Newton, was already pulling strings for him in the background. Scroope did not graduate from Cambridge, but his name is linked with Newton's in the surviving record. Just as Newton would leave the name of a debtor on his student account until he received payment in full, carrying over a few names from year to year, so would Trinity's financial officer, the Junior Bursar. Scroope, in defiance of tradition, left no plate to the College on his departure. So for nearly thirty years, until Newton resigned his fellowship, the account book carried the following entry: "Plate not Received: From Mr Newton, Mr Scroope's."

More revealing of Barrow's intent was his request that Newton edit the manuscript of his optical lectures, due to be published in 1669. Historians have rightly puzzled over the result of this collaboration ever since. Barrow's book, drawing as it did upon the works of Kepler, Descartes, Scheiner, and Alhazen, was certainly no worse than most contemporary optical treatises and better than many. Among its original contributions was a method for finding the point of refraction at a plane interface. Far more critical, however, is the fact that significant parts of its contents had been rendered obsolete by Newton during the experiments conducted three years earlier. Yet the published work was devoid of even the slightest reference to the latter's revolutionary discoveries concerning the nature and origin of colors.

How is one to explain what appears to be a *prima facie* case of deceit on Newton's part, a hypocritical laughing up his sleeve at the work of a man who was about to advance his career? At least one scholar has speculated that Newton did in fact inform Barrow of his findings, but that Barrow was unwilling to embark upon a lengthy revision of his work at the very time he chose to quit the field of science.[53] Conversely, it has been argued, "It is quite inconceivable that Barrow would have permitted his book to be published if he had known about Newton's work. He was too able a scientist not to have recognized its importance and at least to have alluded to it."[54]

Newton probably did not reveal the full details of his optical studies to Barrow, but it does not necessarily follow that his reasons for keeping silent were perverse. Perhaps he thought that to tell Barrow everything would jeopardize the entire project, because no revision of the Lucasian Professor's work, however extensive, could do justice to his own. Realizing this, Barrow, a man of generous spirit, might have opted to withdraw his manuscript, an act that Newton hardly wanted on his conscience, especially in light of the kindness Barrow had shown him. Furthermore, Newton plainly entertained no desire to publish his own work at the time. The shy, unworldly loner was simply not ready to face up to the many tribulations associated with being thrust into the frightful vortex of celebrity. Self-doubt enveloped him like a shroud. He continually spurned Collins's pleas to print *De Analysi*, and the optics, which were unknown to the world of the virtuosi, could certainly wait until he had steeled himself mentally and the possibility of bringing embarrassment upon Barrow had passed. (As it turned out, Newton published in this field much sooner than he had anticipated, but under circumstances that could not at the moment be foreseen.)

At the same time we need not assume that his silence on such an important matter was dictated by wholly altruistic considerations reinforced by an irrational fear of scholarly rebuke. Like all complex human beings, Newton was a man of mixed motives. At what point Barrow told Newton that he favored him as his successor is difficult to say. Quite possibly they spoke of the matter several months in advance of Barrow's resignation. Even if they did not, Newton must have known fairly well where he stood. Thus he was not about to jeopardize his appointment to a position about which he had long dreamed by seriously criticizing his patron's manuscript. With perhaps one exception—his opposition to the royal mandate that Cambridge confer the Master of Arts degree on the Benedictine monk Alban Francis—Newton never bit either the hand that fed him or the one that promised to advance his career. Thus Barrow's optical lectures went to press in 1669. In the prefatory "Letter to the Reader" he generously thanked Newton, "a Man of great Learning and Sagacity, who revised my Copy and noted such things as wanted correction."[55] Shortly thereafter Barrow again turned to him for assistance, this time with the publication of his *Lectiones geometricae* (1670).

On October 29, 1669, Isaac Newton became Cambridge's second Lucasian Professor of Mathematics. As with the story of the apple, legend has again obtruded fact. Until quite recently it was generally believed that Barrow made way for Newton because of his mathematical genius and greater fitness for the post.[56] There is no reason to suppose that such considerations did not enter into Barrow's thinking. This knowledge may even have persuaded him to vacate the Lucasian chair sooner than he had anticipated. More important, however, is the fact that Barrow, then at the height of his powers, was a man of considerable ambition. His eye was cocked in the direction of an even more lucrative and influential preferment, one that he hoped would eventually lead to the Mastership of Trinity College itself. Raworth and Buck, the executors of Lucas's estate, who were responsible for drafting the Lucasian statutes, had relied heavily on Barrow for the completion of this cumbersome task in 1663.[57] Barrow naturally patterned the statutes of the position soon to be his after those governing the Regius professorships, with one notable exception: He did not prohibit the incumbent from holding an office in the college of which he was a Fellow. There is no way of knowing whether Barrow sincerely believed the King's advisers would grant such a boon, but it did not take long to find out. The letter patent confirming the statutes expressly forbade the Lucasian Professor to hold any administrative position. Thus after five years of working with only a handful of students and with no prospect of actively shaping academic policy, Barrow grew restless to move on. Besides, he had always considered himself more of a theologian than a natural philosopher, and when the opportunity arose to become chaplain to Charles II, he jumped at it. Less than four years later he was back in Cambridge, having been named Master of Trinity College by the King. Newton, in a rarely recorded moment of exuberance, expressed his feelings in a letter to Collins: "We are here very glad that we shall enjoy Dr Barrow again especially in the circumstances of Master, nor doth any rejoyce at it more than Sr Your obliged humble Servant Newton."[58] Alas, Barrow's tenure proved all too brief. The good Doctor fell ill during a visit to London in the spring of 1677. He sought to remedy what may have been a heart condition by taking opium, whose powers he observed firsthand while traveling in the East, but to no avail. "As he laye unravelling in the agonie of death, the standersby could hear him say softly, *I have seen the Glories of the World.*"[59] Barrow, only forty-seven, died on May 4 of an apparent drug overdose.

If he had not actually resigned his professorship in favor of Newton, there is little doubt that Barrow guaranteed his succession to it. According to the statutes, Buck and Raworth were invested with this power as long as they survived, but both had deferred to Barrow's judgment in the past and were now highly flattered to learn that their names graced the dedication of his new book on optics. They turned to him again, and Barrow turned to a twenty-six-year-old friend of "very excellent genius." Though not yet famous or for that matter even well known, Isaac Newton had finally won himself both a respected place and a name—destined to be "a kinde of nothinge" never more.

Chapter Six

Some Strangeness
in the Proportion

A man of genius has been seldom ruined but by himself.
Samuel Johnson

I

On November 26, 1669, a month after his dizzying elevation to the ranks of the professoriat, Newton left Cambridge for only his second visit to London.[1] Perhaps he thought he deserved a fortnight's diversion to celebrate his new dignity, but it seems more likely that the journey was undertaken at the urging of Barrow, who wanted to show off his young successor to certain influential members of London's scholarly community. Newton, as usual, left no record of the personal contacts he made, but John Collins, who for obvious reasons was informed of the impending visit, later described two meetings he had had with Newton in a letter to the Scots mathematician James Gregory:

> I never saw Mr Isaac Newton (who is younger than yourselfe) but twice viz somewhat late upon a Saturday night at his Inne, I then proposed to him the adding of a Musicall Progression, the which he promised to consider and send up. I told him I had done something in it, and would send him what Considerations I had about it, but his came up (before I sent him mine) without any Indication of his method. And againe I saw him the next day having invited him to Dinner: in that little discourse we had about Mathematicks, I asked him what he would make the Subject of his first lectures, he said Opticks proceeding where Mr Barrow left, and that himselfe was a practicall grinder of glasses, and had ground glasses

for a pocket tube, but 6 Inches long, that magnified the Object 150 times whereby he did frequently observe the Satellites of Jupiter, and that such a glasse was naught for a short distance.[2]

Collins longed to know more of the specifics regarding the young virtuoso's discoveries, but his thoughts must have returned to the strangely guarded correspondence with Barrow on the subject of analysis by infinite series. Realizing that all might be lost if he pressed his quarry too hard too fast, the circumspect Collins curbed his natural enthusiasm—"having no more acquaintance with him I did not thinke it becoming to urge him to communicate any thing."[3] Nevertheless, Collins was hopeful that the passage of time coupled with the natural human desire for recognition would loosen Newton's pen.

There is no way of knowing whether Newton met other prominent individuals during this visit. His accounts indicate that he had more on his mind than social discourse, however. Cambridge, as one would expect of a university town, had a large number of booksellers, but volumes on natural philosophy and related subjects, because of their limited appeal, were sometimes difficult to come by. Newton seems to have spent considerable time in London shops catering to the interests of the virtuosi, acting, perhaps, on the advice of Collins, who knew the decimated city's remaining booksellers well. He came away with Zetzner's six-volume *Theatrum chemicum*, for which he paid £1. 8. 0., and as proof that his interest in alchemy was more than theoretical, he made a number of other important purchases:[4]

	£	s.	d.
For Glasses at London	0.	15.	0
For Aqua Fortis, sublimate oyle perle, fine Silver, Antimony, vinegar, Spirit of Wine, White lead, Allome Nitre, Salt of Tartar	2.	0.	0
A Furnace	0.	8.	0
A tin Furnace	0.	7.	0

Newton returned to Cambridge on December 8, while back in London Collins waited anxiously to see if his new acquaintance would bother to answer his skillfully planted question about harmonic progression. As a subtle reminder to Newton of his promise to do so, Collins, via Barrow, sent him a copy of John Wallis's recently published *Mechanica*. Despite the fact that he was at the point of inaugurating his Lucasian lectures, Newton spent much of the next few weeks working out alternative solutions to the difficult problem. In a letter postmarked January 19, 1670, the first of their spasmodic correspondence, Collins found Newton to be a man of his word. He thanked Collins for his thoughtful gift and then turned to the more important business at hand:

The Problemes you proposed to mee I have considered & sent you here ye best solutions of one of them that I can contrive; Namely how to find ye aggregate of a series of fractions, whose numerators are the same & their denominators in arithmeticall progression. To doe this I shall propound two ways, The first my reduction to one common denominator. . . . The other way of resolving this Probleme is by approximation.[5]

The complex mathematics that followed appears to have been beyond Collins's understanding, but that mattered little to one whose primary aim was to discover and disseminate the innovative methods of gifted men. He soon presented Newton with a less knotty problem: "To know what rate (N per cent) an Annuity of B is purchased for 31 yeares at price A" (In other words, Collins wanted a formula for computing the rate of simple interest on an annuity continuing for a given number of years). It took Newton but a short while to formulate a satisfactory equation, which he forwarded to Collins on February 6.[6]

If by this time Collins permitted himself to become lost in the dream of persuading Newton to reveal all, he was quickly brought to his senses by the cold rub of reality. Upon receiving Newton's letter containing the solution to the annuity problem, Collins sent him a copy of Michael Dary's *Miscellanies*. Dary was a gauger in the Farthing Office, and his book dealt with certain technical questions associated with his profession. Dary wanted to know if Newton could devise a series for computing the zone of a circle. Newton complied once again, enclosing with his answer to Dary a curt note to Collins dated February 18. Collins had also asked Newton for permission to publish his formula for annuities, and there is no mistaking the undertone of incipient hostility in Newton's grudging acquiescence:

[I]f it will bee of any use you have my leave to insert it into the *Philosophical Transactions* soe it bee wthout my name to it. For I see not what there is desirable in publick esteeme, were I also to acquire & maintaine it. It would perhaps increase my acquaintance, ye thing wch I cheifly study to decline. Of that Problem I could give exacter solutions but that I have noe leisure at present for computations.[7]

Security, not fame, is what Newton valued most at this time, and now that he had achieved it he sought to wall himself off from the world. Collins recognized and heeded the storm warning. If only temporarily, the requests for solutions to mathematical problems immediately ceased.

During the upheaval following the Great Plague and the Fire of London, Collins assumed responsibility for seeing through the press a number of important books, both domestic and foreign, that might otherwise have gone unpublished. If disheartened over Newton's indirect but unmistakable rejection of his advances, the mathematical impresario was not about to abandon the effort to coax the find of a lifetime into the intellectual limelight. One of the books that Collins hoped to publish was a Latin translation of the *Algebra* by the Dutch mathematician Gerard Kinckhuysen. Shortly after their meeting in London Collins sent Newton the manuscript and with Barrow's support persuaded him to undertake a revision of it. Just how extensive that revision was to be became a source of some misunderstanding. In January 1670 Newton informed Collins that he had made some notes, but "I suppose you are not much in hast of it; wch makes me doe yt onely at my leisure." Collins apparently answered that he was expecting something considerably more detailed than what Newton contemplated; on February 6 Newton wrote, "You seeme to apprehend as if I was writing elaborate Notes upon Kinck-huyson: I understood from Mr Barrow yt your desire was only to have ye booke reviewed." He went on to add that "though the booke bee a good introduction I think it not worth the paines of a formall comment, There being nothing new or notable in it wch is not to bee found in other Authors of better esteeme."[8] Newton, no doubt at Barrow's urging, later relented and decided to undertake the extended revisions Collins originally had in mind.

From mid-February no word was forthcoming from Newton until July 11, when Collins finally heard from him. The tone of the letter is one of diffidence:

> I have here sent your Kink-Huysons *Algebra* wth those notes which I have intermixed wth the Authors discourse. I know not whither I have hit your meaning or noe but I have added & altered those things wch I thought convenient . . . and I guesse that was your desire I should doe. All & every part of what I have written I leave wholly to your choyse whither it shall bee printed together wth your translation or not.[9]

Newton had so extensively modified Kinckhuysen's work that he felt constrained to add, "if you print these alterations wch I have made . . . it may bee esteemed unhandsom & injurious to Kinck huysen to father a booke wholly upon him wch is soe much alter'd from what hee had made it." He therefore requested that the title page carry the inscription: "*ab alio Authore locupletata* [enriched by another Author] or some other such note."[10] For the second time in five months Newton declined the opportunity of having his identity revealed in print, and it would be a mistake to view his demurrer as an example of the false modesty that routinely made its way into the correspondence of seventeenth-century intellectuals. He was every bit as serious now as when he had previously informed Collins that public esteem "would perhaps increase my acquaintance, ye thing wch I cheifly study to decline." He closed with an apology for neglecting to write, trusting that Collins would be good enough to pardon him.

With characteristic forbearance, Collins chose to praise Newton rather than reproach him. He surely realized by now that a misstep on his part could result in a permanent rupture of their tenuous relationship. "I received yours with Kinckhuysens Introduction, and perceive you have taken great paines which god willing shall be inserted into ye Translation and printed with it, hereby you have much obliged the young Students of Algebra." And if Collins could not fathom Newton's obsession with protecting his anonymity, he nonetheless agreed without complaint—"why you should desire to have your Name unmentioned I see not, but if it be your will and command so to have it, it shall be observed."[11]

Collins remained undaunted in his endeavor to exploit Newton's gifts to the full. In the same letter he raised a more substantive matter: the marginal notes in which Newton criticized Kinckhuysen's inadequate treatment of surds. He sent Newton the works of Johann Scheubelius, Ludolph van Ceulen, and James Hume, requesting that he incorporate "what you think necessary" from them. It was a request Collins soon came to regret, for in Newton's reluctant assent his growing impatience with the overexuberent impresario again glowered forth: "I sometimes thought to have altered & enlarged Kinkhuysen his discourse upon surds but judging those examples I added would in some measure supply his defects I contented my self wth doing that onely. But since you would have it more fully done . . . send it back wth those notes I have made . . . & I will doe something more to it." Collins, sensing his mistake, moved quickly to smooth the choppy waters, but he chose an inept way of doing so—one that flew in the very face of Newton's dread of forfeiting his privacy:

> Perceiving by your last that you are willing to take some more paines at present with Kinckhuysen I remand the same [Newton's notes], but doe not presse your selfe in time, your paines herein will be acceptable to some very eminent Grandees of the R[oyal] Societie who must be made acquainted therewith, . . . as Algebra may receive a further Advancement for your future endeavours . . . than any man I know.[12]

If Collins thought that the reclusive young don would yield to the temptation of achieving recognition beyond anything previously suggested, he woefully underestimated Newton's resolve. Nine weeks passed before another letter from him crossed Collins's desk in late September. Newton attributed the hiatus to waiting for Barrow to return from London so that he might consult his library concerning solutions to cubic equations, and indeed a somewhat detailed discussion of the subject followed. But Newton had a more important purpose in writing, which he disclosed near the end:

> Upon the receipt of your last letter I sometimes thought to have set upon writing a compleate introduction to Algebra, being cheifely moved to it by this that some things I had inserted into Kinck-Huysen were not so congruous as I could have wished to this manner of writing. . . . But considering that by reason of severall divertisements I should bee so long in doing it as to tire you patience with expectation, & also that there being severall Introductions to Algebra already published I might thereby gain ye esteeme of one ambitious among ye croud to have my scribbles printed, I have chosen rather to let it passe wthout much altering what I sent you before.[13]

Newton then informed Collins·that he was returning the works by Scheubelius, van Ceulen, and Hume, but Kinckhuysen's *Algebra* "I presume to keepe by me till you have occasion for it." For the moment, this ritual sparring of opposites was over, and Collins had clearly lost. He was never again to lay eyes on the ill-fated manuscript, which would still have been in his hands had he not pressed Newton too hard. Ten months passed before Newton resumed the correspondence, and during the bleak winter of 1670 Collins's despair became manifest in a letter to Gregory. On December 24 he wrote of Newton: "observing a warinesse in him to impart, or at least an unwillingness to be at the paines of so doing, I desist, and doe not trouble him any more." [14] Collins could hardly have set a more proper tone for the austerity of a Puritan Christmas.

II

The statutes that governed the Lucasian chair required its occupant to dress in scarlet robes to distinguish him from the lesser teaching Fellows. Newton, who had just turned twenty-seven, would have hardly been inconspicuous as he self-consciously made his way across Trinity's Great Court to deliver the first of eight lectures during the Lenten term of 1670. He was free to select from a broad range of topics including mechanics, astronomy, optics, and mathematics. His creation of a systematic calculus, as set forth in *De Analysi,* would seem to have made it a likely subject of discourse, but he tended to look upon modern analysis as arid and tedious; something not to be valued in its own right but to be employed as a tool for expressing fundamental principles of natural law. This consideration, added to the numbing anxiety that almost always overcame him when he faced the prospect of publication, explains why he so rarely shared his findings in the field with others. He turned instead to optics, his first and greatest love among the experimental sciences:

> The late Invention of Telescopes has so exercised most of the Geometers, that they seem to have left nothing unattempted in Opticks, no room for farthur Improvements. . . . [I]t may seem a vain Endeavour and a useless Labour, if I shall again undertake the handling this Science. But since I observe the Geometers hitherto mistaken in a particular Property of Light, that belongs to its Refractions, tacitly finding their Demonstrations on a certain Physical Hypothesis not well established; I judge it will not be unacceptable if I bring the Principles of this Science to a more strict Examination, and subjoin, what to be true by manifold experience, to what my reverent Predecessor [Isaac Barrow] has last delivered from this Place. [15]

There is no record of how many were present to hear these bold opening remarks of what amounted to scores of lectures, all in Latin, delivered by Newton over the next twenty-five years. It is quite possible that, owing to the special occasion, he faced the largest audience of his undistinguished teaching career. If he did, no one was sufficiently impressed to have remembered being there. The problem lay not in what Newton had to say but in an esoteric style made even less intelligible by the advanced academic rigor mortis affecting the University. It mattered not what subject he chose to expound on—optics, analytic geometry, mechanics, or algebra—his students, with scarcely any exceptions, were too poorly prepared to benefit from his wisdom. Belatedly coming to grips with this problem about 1690, Newton proposed a number of basic reforms in his scheme "Of Educating Youth in the Universities." Rather than continue to rely solely on the tutorial system and a small group of university professors who lectured infrequently, each college, he felt, should appoint its own lecturers in Humanities, Greek, Philosophy, and Mathematics. "The Mathematick Lecturer to read first some easy & useful practical things, then Euclid, Sphericks, the projections of the Sphere, the construction of Mapps, Trigonometry, Astronomy, Optics, Musick, Algebra, &c. Also to examin & (if ye Tutor be deficient) to instruct in the principles of Chronology & Geography." Doubtless thinking back on his own unfulfilling relationship with the pupil monger Pulleyn, Newton favored a requirement that each tutor "accompany his Pupills to the Philosophy & Mathematick Lectures & ... examin them the next morning both in those Lectures & in his own & make them understand where they hesitate." And every student admitted to "Lectures in naturall Philosophy to learn first Geometry & Mechanicks. By mechanicks I meane here the demonstrative doctrine of forces & motions including Hydrostaticks. For w^{th}out a judgment in these things a man can have none in philosophy." [16]

A more rigorous program of instruction in mathematics would have placed greater demands on professors like Newton, because of both increased attendance at their lectures and a heavier tutorial load. Evidently, he delayed so long in committing his thoughts on reform to paper at least partly because he disliked teaching and feared having to take precious hours away from private study and research. After the *Principia* was published, he seemed more willing to suffer what the highly gifted often perceive as the indignity of stooping to the level of the newly initiated. Still, his motives were less than altruistic; if his plan had been adopted (and it was not), he singled out those who were to teach the new natural philosophy and mathematics as worthy of special remuneration, realizing from personal experience that they would carry a heavier burden than scholars trained in the more traditional disciplines. "Because the Philosophy & Mathematic Lecturer's office is laborious, for encouraging them to diligence ... all that will be auditors shall offer each of them a quarterly gratuity, suppose of 10^s ye sizar, 12 or 15^s ye Pensioner & 20 or 25^s ye Fellow-commoner." [17]

While genius is the main stuff of the Newtonian legend, the deft mythographer could weave an equally enthralling tale using what is known of his mental tenacity. Once in the grips of a complex problem, Newton returned to it time and again until the mists dissolved, revealing to him yet another level of scientific reality. So it was that he resumed the study of optical phenomena in the late 1660s, revising and amplifying his earlier insights and experimental data to formulate the masterly account delivered during the first Lucasian lectures. It was said of his predecessor "that when he was at study, [Barrow] was so intent [that] when the bed was made, he heeded it not nor perceived it, was so *totus in hoc* [absorbed]; and would sometimes be goeing out without his hatt on."[18] Dr. George Cheyne, a disciple of Newton during the early 1700s, reported that the latter's resolve was no less pronounced. While conducting his investigations into the theory of light and colors, "to quicken his faculties and fix his attention, [he] confined himself to a small quantity of bread, during all the time, with a little sack and water, of which, without any regulation he took as he found a craving or failure of spirits."[19]

It will be recalled that in the essay "Of Colours," completed in 1666, Newton told of placing a prism near the hole in a shutter of his darkened chambers. He then placed a second prism a few yards from the first and carefully observed the result. Blue light, on passing through the second prism, was refracted more than red light, just as it had been by the first prism. Nor were either of the two colors changed in any way: "ye purely Red rays refracted by ye second Prisme made noe other colours but Red & ye purely blew ones noe other colours but blew ones."[20] Newton had successfully challenged the accepted doctrine that spectral colors are produced when "pure" sunlight undergoes modification on passing through a dark medium, or when refracted at the boundary between two mediums: in this instance air and glass.

To confirm the results of this initial test, Newton spent much of the next three years designing and performing dozens of other experiments in which he employed two prisms. He appears to have been searching for nothing less than a single conclusive demonstration that would dispel the last of his own lingering doubts and at the same time lift the scales from the eyes of all but the most rabid of traditionalists. It is now generally recognized that no single experiment is capable of providing all the proof needed to verify a theory as sweeping as Newton's, but it is also agreed that he came as close to accomplishing this feat as anyone before or since. On a loose slip of paper in the manuscript of his first optical lectures Newton drew a diagram of his celebrated *experimentum crucis*. (Figure 6–1).[21]

While the phrase itself is nowhere employed in the *Lectiones opticae*, Newton left no doubt regarding its importance to him. In the famous letter to Oldenburg printed in the *Philosophical Transactions* for February 19, 1671–72, he presented this cogent analysis of his achievement, referring to it by the name coined in the *Micrographia* by Hooke to denote "a Guide or

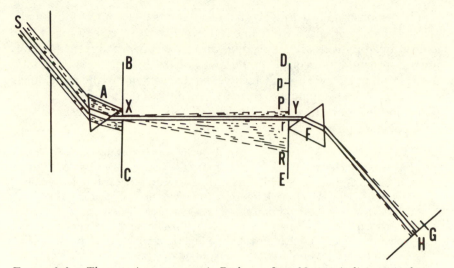

FIGURE 6–1.   *The experimentum crucis.* Redrawn from Newton's diagram in the optical lectures (*Lectiones opticae*).

Land-mark, by which to direct our course in the search after the true cause":

> The gradual removal of these suspitions at length led me to the *Experimentum Crucis,* which was this: I took two boards [BC,DE], and placed one of them close behind the Prisme at the window, so that the light might pass through a small hole, made in it for that purpose, and fall on the other board, which I placed at about 12 foot distance, having first made a small hole in it also, for some of that Incident light to pass through. Then I placed another Prisme behind this second board, so that the light, trajected through both the boards, might pass through that also, and be again refracted before it arrived at the wall. This done, I took the first Prisme in my hand, and turned it to and from slowly about its *Axis,* so much as to make the several parts of the Image, cast on the second board, successively pass through the hole in it, that I might observe to what places on the wall the second Prisme would refract them. And I saw by the variation of those places, that the light, tending to that end of the Image, towards which the refraction of the first Prisme was made, did in the second Prisme suffer a Refraction considerably greater then the light tending to the other end. And so the true cause of the length of that Image was detected to be no other, then that *Light* consists of *Rays differently refrangible,* which, without any respect to a difference in their incidence, were, according to their degrees of refrangibility, transmitted toward divers parts of the wall.[22]

What Newton had previously stated in promising theoretical terms could now be advanced as scientific fact. The *experimentum crucis* enabled him to demonstrate conclusively that all attempts to account for prismatic colors as modifications of white light must fail because of their inability to explain quantitatively the elongation of a spectrum when refracted by a prism onto a distant screen. The sine law advanced by Descartes applies not to the incident beam of light as a whole, but to every one of the individually colored rays which combine to form it. Sunlight is a mixture of all the rainbow colors, each of which obeys the laws of optics, for each is refracted at a different angle on passing through a prism:

> To bring forth my opinion more distinctly: In the first instance I find, that rays which are refracted more than others of the same incidence, exhibit purple and violet colours, while those exhibit red which are least refracted, and those blue, green, and yellow, which have intermediate refractions.
>
> Secondly & conversely, I find those rays which produce purple colours, are most refracted of all rays having equal incidence, and those least refracted of all which produce red, and that those have middle refractions which produce blue, green, and yellow colours. That is to say, I find that rays of equal incidence are gradually refracted more and more after their disposition to exhibit colours in this order: red, yellow, green, blue, & violet, with all their intermediate colours.[23]

Having effected the analysis of white light into its constitutent colors, Newton, as he had done in 1666, proceeded to reverse this process in order to demonstrate how the spectral colors could be recompounded into white light. In the first instance he had placed three prisms side by side so that their spectra overlapped. At the point of convergence the rainbows of colored light combined to form white. He repeated this experiment in 1669, but, as with his first demonstrations on refraction, he hoped to design an even more conclusive test, one that would serve as the counterpart of the *experimentum crucis*. This he did by drilling a $\frac{1}{3}$-inch hole in a shutter and intercepting the sun's light with a prism. He then placed a large lens some 4 or 5 feet from the prism so that the refracted colors would pass through it. At a further distance of about 10 feet the various rays were brought to the same focus and, as he anticipated, produced a patch of white light. Beyond this point the differently colored rays reappeared to form a spectrum in inverted order.

In a subsequent drawing, Newton introduced yet another lens to show that the process of alternately separating and recombining the prismatic colors could be continued *ad infinitum*.[24] And with the experimental apparatus shown in Figure 6–2 he ingeniously incorporated another device, a cogwheel made of wood. He inserted the wheel beyond the lens in such a manner that the series of teeth on its rim intercepted the rays of colored light just as they were about to converge. He rotated the wheel slowly at first, and the many colors appeared in succession at the focus. As the speed of the wheel was increased the colors merged into a brightly colored band, which on faster rotation yielded to gray and finally to white. There was no

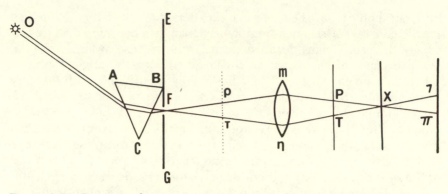

Figure 6–2. Reversing the *experimentum crucis*. Redrawn from Newton's diagram in the optical lectures (University Library Cambridge, Additional Manuscript 4002, p. 39).

longer any doubt in Newton's mind of the heterogeneity of light, a phenomenon no longer scientifically explicable in terms of modification, but of analysis alone.

III

By July 1671 Collins's patience was exhausted. For ten long months he had waited in vain for Newton to break his grating silence, but nothing was forthcoming. Feigning nonchalance, Collins decided to test the waters one more time. He found them to be almost as frigid as before.

As the gambit of a strategy designed to pique Newton's interest in publication, Collins reminded him of a letter he had sent to Barrow a year earlier, the contents of which Barrow had shared with Newton. In the letter Collins told of how James Gregory "by his owne Ingenuity [had] falne into your methods of infinite Series." Collins assured Newton, however, that "the said Mr Gregory being since informed by me that you had taken much paines in that harvest, and invented the method some yeares before Mercators *Logarithmotechnia* was printed, hath laid aside his Intentions of publishing anything."[25]

Collins chose not to belabor the point, but his implication was unmistakable: If Newton expected to receive credit for his mathematical discoveries, it was imperative that he commit his work to print. Gregory, the only one of Newton's British contemporaries who could hold a candle to him mathematically, had gracefully deferred, even though the two seem never to have met. Such restraint could hardly be expected of others, however, nor should it have been. To make matters worse, now that Collins had circulated *De Analysi* in manuscript, plagiariam had become a distinct possibility. Indeed, Collins declined to make the treatise available to John Wallis because of his known proclivity for treating the original work of others as his own. About 1677 Collins angrily faced Wallis with this very charge and, with a refreshing dose of candor, condemned his gifted colleague's intolerable behavior: "If I had been so minded I could ... at the beginning of 1669 have imparted to you a full treatise of his [Newton's] but did not, in regard you lye under a censure from diverse for printing discourses that come to you in private Letters without permission or consent as is said of the parties concerned." [26] If nothing else, this episode shows that Collins possessed a temper, which at times he must have labored mightily to control in his frustration over Newton's unresponsiveness.

Having delicately reopened the subject of publication, Collins next raised the much-vexed issue of Newton's introduction to Kinckhuysen's *Algebra*. This paper, as we have noted, had slipped through Collins's fingers several months earlier, and he worried—with good reason—that Newton might withdraw from the project altogether. Collins broached the matter on tiptoe, wanting to give his strange acquaintance no hint of his deep concern: "[T]he Bookseller [Moses] Pitts is not desirous as yet to put the Introduction to Algebra to the Presse, and I conceive you have made so many usefull additions thereto, that when it comes to the Presse it may very well beare the Title of your Introduction, and thereby find the better entertainment, and more Speedy Sale." [27] Collins also noted that he was forwarding a copy of Giovanni Borelli's new book, *De motionibus,* to Newton.

All this was briefly presented in half a page. The rest of the letter pertained to matters wholly unrelated to mathematics or natural philosophy. Collins told Newton of his recent move from a post in the Excise Office to one of Secretary to the Council of Plantations, while his wife had the good fortune to assume her sister's former position as Laundress of the Table Linen to the Queen. One senses that Newton cared very little about such matters, and Collins undoubtedly knew as much. It was simply his well-intentioned, if awkward, way of attempting to breathe new life into a virtually moribund intellectual relationship.

Newton's reply to Collins, dated July 20, began on a somewhat encouraging note. George Villiers, the Duke of Buckingham, had recently been appointed Chancellor of the University, and Newton had hoped to use the occasion of Villiers's induction to pay Collins a visit in London. He was prevented from doing so, however, by "ye suddain surprisall of a fit of sicknesse, wch not long after (God be thanked) I again recovered of." (The nature of the illness was undiagnosed. From all indications, Newton was again working at a fearsome pitch, and it is not unreasonable to suppose that he collapsed under the mounting burden of physical exhaustion compounded by unceasing mental strain.) Then, without warning, the young lion lashed out at his well-meaning correspondent:

> I am still become more your debtor both for the care you take about my concernes, & for Borrelius *de motionibus*. But for Borrellius I beg that I may bee accountable to you at our next meeting, & that you would not for ye future put your self to ye like trouble in sending any more bookes. I shall take it for a great favour if in your letters you will onely inform mee of ye names of ye best of those bookes wch newly come forth.[28]

As so often happens, genius was establishing its own peculiar rules of conduct, and those who sought to draw upon it had little choice but to live by them. The only debt Newton could tolerate was the one owed him by another.

Harsh though his words were, Newton's rebuff of Collins's advance was less than total. Suffering, perhaps, from a mild attack of conscience, he tossed Collins a welcome bone: "The last winter I reviewed the [Kinckhuysen] Introduction & made some few additions to it: And partly upon Dr Barrows instigation, I began to new methodiz ye discourse of infinite series, designing to illustrate it wth such problems as may ... be more acceptable than ye invention it selfe of working by such series."[29] Lest Collins mistake this for a sign that publication was imminent, Newton hastened to add that he had recently been diverted from his scholarly pursuits by business in the country and did not expect to return to the subject before winter. Presumably the business had to do with family matters in Lincolnshire, for he left Cambridge on April 17 and returned three weeks later, on May 11.[30] Since this was his only absence from the University in 1671, one must seriously question whether such a brief sojourn was as distracting as he led Collins to believe.

Be that as it may, Barrow had persuaded a reluctant Newton to expand upon his work with infinite series, a prospect of more than passing interest to John Collins. Unbeknownst to Newton, Collins had conversed with Barrow the previous autumn on this very topic.[31] He prudently kept silent about the mathematics in the cautious note to Newton of July 5, but Collins must have felt that Newton's voluntary mention of it in his reply more than made up for the stinging rebuke he had suffered for having made Newton a gift of Borelli's most recent book.

The new treatise was entitled *De Methodis Serierum et Fluxionum* (Of the Method of Series and Fluxions). Like the optical lectures, it rested on a solid foundation of earlier work, including the October tract on fluxions completed in the plague year of 1666 and the unpublished *De Analysi*, composed in haste in 1669 to establish Newton's claim to priority over that of Nicolas Mercator. In *De Methodis* Newton primarily concerned himself with the concept of uniformly flowing time, a subject to which he would return in Book II of the *Principia*. By employing his advanced method of fluxions, Newton was able to reduce all problems of local motion to two fundamental questions:

1. Given the length of the space continuously (that is, at every [instant of] time), to find the speed of motion of any time proposed.
2. Given the speed of motion continuously, to find the length of the space described at any time proposed.[32]

Many years later, after he became embroiled in a long and acrimonious struggle with Leibniz's defenders over priority to the discovery of the calculus, Newton wrote an unusually eloquent account of the fluxional method in a draft review of the *Commercium epistolicum*:

> I consider time as flowing or increasing by continual flux & other qualities as increasing continually in time & from ye fluxion of time I give the name of fluxions to the velocity, wth wch all other quantities increase. Also from the moments of time I give the name of moments to the parts of any other quantities generated in moments of time. I expose time by any quantity flowing uniformly & represent its fluxion by an unit, & the fluxions of other quantities I represent by any other fit symbols and the fluxions of their fluxions by other fit symbols & the fluxions of those fluxions I represent by the symbols of the fluxions drawn into the letter o & its powers o^2, o^3, &c: vizt their first moments by their first fluxions drawn into the letter o, their second moments by their second fluxions into o^2, & so on. And when I am investigating a truth or the solution to a Probleme I use all sorts of approximations & neglect to write down the letter o, but when I am demonstrating a Proposition I always write down the letter o & proceed exactly by the rules of Geometry without admitting any approximations. And I found the method not upon summs & differences, but upon the solution of this probleme: *By Knowing the Quantities generated in time to find their fluxions.* And this is done by finding not prima momenta but primas momentorum nascentium rationes.[33]

Newton had gone on to apply his powerful method to questions concerned with the maxima and minima of quantities, the drawing of tangents to curves, the determination of the curvature of curves, nascent ratios, and the calculation of areas under curves. The better to accomplish this last function he composed a masterful addendum to the *De Methodis* in the 1690s, which bears the title *De Quadratura Curvarum* (The Quadrature of Curves). Although *De Quadratura* circulated in manuscript, the first printed version of the tract in its original Latin did not appear until 1779.

IV

When, in searching for clues to the source of Newton's profound creativity, one pauses to consider the voluminous body of scientific and mathematical writings composed in the startlingly brief period between 1665 and 1671, one is reminded of something Newton confided to Henry Oldenburg in October 1676. Recalling his reaction to the publication of Mercator's *Logarithmotechnia* eight years earlier, Newton wrote: "I began to pay less attention to these things, suspecting that he either knew the extraction of roots as well as division of fractions, or at least that others upon the discovery of division would find out the rest before I could reach the ripe age for writing."[34] Exactly what did Newton mean when he employed this curious phrase, "the ripe age for writing?"

One explanation is that he had no plans to make his discoveries available to a wider audience. Having established his claim to priority with Barrow and Collins, Newton was unwilling to suffer the psychic trauma associated in his mind with publication. Unlike Galileo, he abhorred intellectual jousting and was disdainful of the cut and thrust of debate for its own sake. Hence the ripe age for writing might never come. Yet it is difficult to escape the feeling that some other force, equally compelling, was at work beneath the surface. Among Newton's strongest traits was his power of will, an unwavering tenacity of purpose, an instinctive faith that nothing, save serious illness or death, could prevent him from carrying out a project once it had matured in his mind. Time to him was the arch-enemy of man, something never to be squandered but devoured, before it devoured you. Exactly how many days or evenings he took off from his intellectual pursuits during his productive period one cannot say, but such occasions were about as rare as the men he called friend. There were no aimless walks along the Cam's green banks, no visits to the tavern, no late afternoon sessions of lawn bowling to put things back in proper balance at the end of a long and difficult day. Indeed, for Newton there was no natural end of the day, only work to the point of exhaustion, then merciful sleep.

Everything he had accomplished by 1671 (and all that followed) was the product of a patterned, almost inexorable mental process. First he would set forth his initial thoughts on a problem in one of his many notebooks. Then, by keeping "the subject constantly before me," he waited until "the first dawnings open slowly by little and little into the full and clear light." Satisfied, if only for the moment, that he could accomplish nothing further, he would focus his full attention on some other problem, which as often as not was unrelated to the one he had just left. It was only a matter of time, however, until fresh insights welled up from some inscrutable but immensely fecund depth: Newton then returned to his unfinished work, carefully refamiliarized himself with what had gone before, and, inventing anew, attained a higher level of synthesis.

He was not only quicker than anyone else to see the essentials of a problem but capable of developing its consequences with unequaled mastery. Uniting these extremes was a capacity for meditation without parallel, in both intensity and duration, in the history of science. Others have experienced like degrees of insight, but only in rare moments, followed by the inevitable and almost instantaneous closing of the curtain. Newton, by contrast, found it comparatively easy to get his mind fully under way. Still more important, he could sustain this trancelike state almost indefinitely. The passage of weeks mattered hardly at all, days even less, and hours counted for nothing. He was being deceptively candid when, asked how he made his discoveries, he said, "by always thinking unto them."

It is just this aspect of his genius that is so difficult to put into words. It can be appreciated fully, perhaps, only by reading the countless manuscript pages in the original. One can understand why so many Newton scholars have taken a dim view of most attempts to reduce the essence of his achievement to layman's terms, and why so relatively few have even bothered to try. It also might explain why one can spend an entire afternoon wandering about Woolsthorpe alone and undisturbed, whereas, at Stratford-on-Avon, a leisurely morning's drive away, the daunting lines of those waiting to be rushed without pause through Shakespeare's birthplace often number in the hundreds.

When Newton wrote to Oldenburg of a "ripe age" for writing, he was surely being intentionally vague, but far from devious. He simply could not predict at what point his creative powers would exhaust themselves, and to publish before that elusive plateau was reached must have seemed, to one of his gifts, an exercise in futility, as well as an open invitation to unending controversy. No matter what the field of endeavor—mathematics, optics, mechanics, alchemy, or religion—Newton thought of publication only in terms of definitive masterworks. The poet William Butler Yeats could well have had him in mind when he penned the lines:

> The intellect of man is forced to choose,
> Perfection of the life, or of the work.

Though there is much about Newton that remains a mystery, and forever will, his choice became clear almost from the moment he first set foot on the graveled walks of Restoration Trinity.

The unique value of any great intellectual endeavor lies not always in what it makes of the world, but in what it makes of the knower. To be truly revolutionary is to be alone, and this unshakable burden is sometimes so overwhelming that any attempt to communicate with those who do not share it seems doomed from the outset. And in those rare instances when a special vision can be reduced to rational principles, which even in the field of science is almost never, the acceptance of that vision by others cannot be willed; nor, if accepted, can it ever be appreciated in quite the same manner as it was by the visionary for whom the mists first parted. In the history of science, Newton is not the only seminal intellect to have kept silent about what he had discovered: What is true of him is also true of Copernicus and Darwin.

If part of Newton's reticence is explicable in terms of not wishing to declare himself prematurely, this can hardly account for his withholding of the introduction to a textbook on algebra, much less his refusal to have his identity revealed as the author of a simple formula for computing annuities. Neither did these instances present any peril of becoming embroiled in the kind of controversy that was bound to accompany the publication of his revolutionary optical studies. The key to the mystery behind Newton's mask lies in his almost total distaste for contact, let alone intimacy, with the outside world. History shows us Newton the man as through a glass darkly: We see his general outline but precious few details of a human face. At times Nature violates its own laws by bestowing great gifts upon an individual, then cruelly overcompensates by depriving him of even a modest share of others. "There is no Excellent Beauty that hath not some Strangeness in the Proportion," wrote Bacon. So it was with Newton, for whom some of the human parts were missing. He seems not to have been so much flesh and bone as a thinking machine, all wheels and gears, like the clockwork universe he helped to create.

Yet beneath his dour and impenetrable exterior lingered vestiges of the wounded child. Newton found himself a minority of one when his mother remarried and departed Woolsthorpe, again while an introspective, alienated scholar at Grantham and Cambridge, and still again as the "odd" professor of mathematics with scarcely a student to his name. Soon he would seemingly stand alone against some of the better minds of the Royal Society, as they took turns assaulting his disconcerting theory of light. Each successive wound penetrated more deeply, leaving permanent scars, which collectively constitute Newton's psychic stigmata. In an attempt to compensate, he became aloof, unwilling and ultimately unable to empathize with his fellow man. An emotional eunuch, Newton mixed in the world on occasion, but he was never of the world. His personality, much like his extraordinary mental powers, assumed dimensions beyond articulation. The whole of his pleasure centered on the isolation he guarded so tenaciously, something within the darkest night of himself that called him home.

V

Among his more curious habits, Newton developed a unique method of marking those passages in his books which had some special significance for him. Rather than simply "dogear" the corner of a page he planned to reexamine later, he went a step further by turning back (either up or down) the near corner, so that it pointed exactly to the line or individual word that most interested him. In several instances both the top and bottom corners of the same page were used.[35] Among the books he acquired as a young man in which dogearing was liberally employed is a copy of James Gregory's *Optica promota*, published in 1663. It was in this work that Gregory described his design for the catadioptrical or reflecting telescope, a concept that dated back at least a century.

As yet no one, including Gregory, had succeeded in building a working reflector, but not for want of trying. Gregory had visited London in 1663, where Collins put him in touch with Richard Reive, the capital's leading instrument maker. Reive attempted but failed to fashion the requisite object-mirror, causing Gregory to abandon the project. It was not long thereafter that Newton took up the challenge. "For when I first applyed my selfe to try the effects of reflexions," he wrote in May 1672, "Mr. Gregory's *Optica Promota* being faln into my hands . . . I had thence an occasion of considering that sort of constructions."[36] Why Newton should have been so strongly attracted to the idea of fashioning a working model of the new telescope is not difficult to understand. Unlike the refracting telescope, which forms an image of a distinct object with lenses, the reflector contains a parabolic mirror from which all light is reflected at the same angle. Hence the observer is not troubled by the phenomenon of chromatic aberration, which occurs because rays of light of differing wavelengths are not brought to the same focus on passing through a lens. Though no one but Newton realized it at the time, the reflecting telescope, by producing an image free from chromatic distortion, offered further conclusive proof of his new theory of light and colors:

> When I understood this, I left off my aforesaid Glass-works; for I saw, that the perfection of Telescopes was hitherto limited, not so much for want of glasses truly figured according to the prescriptions of Optick Authors, (which all men have hitherto imagined,) as because that Light it self is a *Heterogeneous mixture of differently refrangible Rays*. So that, were glass so exactly figured, as to collect any one sort of rays into one point, it could not collect those also into the same point, which having the same Incidence upon the same Medium are apt to suffer a different refraction. Nay, I wondered, that seeing the difference of refrangibility was so great, as I found it, Telescopes should arrive to that perfection they are now at.[37]

The reflecting telescope he was about to construct as a result of this radical insight proved nothing less than the technological vindication of the seminal *experimentum crucis*.

Newton was never one to advance promising but untested ideas as *faits accomplis*. His first extant reference to the reflecting telescope is contained in a letter of February 23, 1669, to an unidentified friend, which tells of a completed instrument capable of magnifying "about 40 times in Diameter which is more than any 6 foote Tube can doe, I believe with distinctnesse. . . . I have seen with it Jupiter distinctly round and his Satellites, and Venus horned." Newton had no doubt that a properly constructed 6-foot reflector would perform as well as any "60. or 100. foote Tube made after the Common way." He also realized how puzzling this claim might appear to his friend: "it may seeme a Paradoxicall assertion, yet it is the Necessary consequence of some Experiments which I have made concerning the Nature of Light."[38] And so it was. Eight months later, during his first meeting with Collins in London, Newton had spoken of himself as "a practical grinder of lenses" and discussed the same telescope described in his correspondence, the difference being that by presumably crafting a better speculum, he had elevated its powers of magnification to the dubious range of "150 times."[39]

Word of important technical innovations spread quickly in the close-knit world of the virtuosi. It did not take long for news of the reflecting telescope to reach the members of the Royal Society, who erroneously hailed Newton as its inventor, a circumstance reminiscent of 1609, when Galileo constructed the first refracting telescope in Italy from knowledge imparted to him by Jacob Badovere, a correspondent in France. But no matter; Newton's was a technical achievement worthy of celebration in its own right, especially at a time when the Royal Society was avid for answers to its critics, who characterized the activities of natural philosophers as much ado about nothing. This time Newton was unable to resist the pressure exerted by his peers and entrusted the product of his labors to Barrow, who proudly carried the little instrument to London at the end of 1671.

The telescope's arrival created something of a sensation. Robert Moray, Sir Paul Neile, Christopher Wren, and Robert Hooke carried it to Whitehall, where the newly restored Stuart King, Charles II, was favored with a special demonstration of its powers. "They had so good opinion of it, as they concluded, that a description and scheme of it should be sent by the secretary [of the Royal Society] in a letter to Mons. Huygens then at Paris."[40] In the meantime an excited John Flamsteed, soon to be named the first Astronomer Royal of England, wrote to Collins concerning "this prodigie of arte; pray if you have [seen it] let me understand its dimensions & effects from you." Three months later Flamsteed again wrote to Collins, having learned that the Royal Society had commissioned the London optician Christopher Cock to construct two reflectors, one 4 feet the other 6 feet in length. "I must beg of you further to informe mee by ye next at which rates they are valued & with what effect they performe. I meane how may I buy one of about $2\frac{1}{2}$ feet & whether it will doe as well as a long one [meaning a refractor] of 30 feet."[41] Huygens, on receiving word of the reflector, was to write of it as nothing less than *du merveilleux telescope de Mr. Newton.*[42]

Newton himself first became aware of the stir he had created via a flattering letter he received in January 1672. It came from an unexpected quarter and began: "Your Ingenuity is the occasion of this address by a hand unknowne to you. You have been so generous, as to impart to the Philosophers here, your Invention of contracting Telescopes." Its author was none other than Henry Oldenburg, the porcine but intellectually acute Secretary of the Royal Society.

Oldenburg, a German by birth, came to England during the 1650s as consul to the Commonwealth from his native city of Bremen. He later became the tutor of Richard Jones, nephew of Robert Boyle, a position which took him to Oxford in 1656. It was at Oxford that Oldenburg met John Wilkins and other members of the scientific circle who were instrumental in founding the Royal Society following the Restoration. Oldenburg was named one of the Society's two secretaries in 1662, a position he held until a few months prior to his death in September 1677. While his responsibilities were many, Oldenburg's greatest service was as a correspondent, which involved receiving and replying to an average of a half-dozen letters a week, many of them long and taxing documents. Beginning in 1666 he instructed his foreign contacts to write by post to "Grubendol, London." Under this anagram the letters were received and paid for by the Secretary of State, who milked them of important intelligence from the Continent before passing them on to Oldenburg. Even though the privilege of corresponding with foreigners was granted to the Royal Society in its charter, the practice was not without its considerable pitfalls, a hard lesson learned at its Secretary's expense. During the summer of 1667 Oldenburg, who did not become an English citizen until the last year of his life, received an all too vivid taste of the Tower's dankness for some ill-advised criticism of the government contained in a letter he sent abroad. Thereafter he was careful to limit his statements to scientific matters alone. It was Oldenburg who inaugurated the *Philosophical Transactions* in 1665 and who supervised their publication until June 1677. The first of what are now classified as purely scientific journals, the *Transactions* became the principle medium of intercourse between English and Continental scientists. Oldenburg also played an important role in persuading the Royal Society to undertake the publication of separate scientific monographs. Finally, he gave much encouragement to young natural philosophers, as is evidenced by his unsolicited letter to Newton.

The Secretary informed Newton that he wished to send a detailed explanation of the telescope to Huygens "to secure this Invention from ye Unsurpation of forreiners." He enclosed a drawing and an account of the instrument, requesting Newton to modify it in any manner he might choose so that it could be forwarded to the Dutch savant, who was then living in Paris. What most pleased Newton about the letter, however, was word that he had been nominated by Seth Ward for membership in the Royal Society.[43] Ward made the nomination on December 21, 1671 and at the next meeting, held on January 11, 1672, Newton was duly elected.[44] A description of the telescope was also read and formally entered in the register book. Coincidentally, a letter from Francis Vernon, Secretary to the English Ambassador to Paris, was read at this same meeting. It contained a description of Picard's accurate method for measuring the earth, which eventually enabled Newton to correct his 1666 calculations on planetary attraction. There is no indication that he took serious notice of Picard's work at the time, however.

Despite Newton's ingrained aloofness, even he could not help but warm to Oldenburg's generous praise. He remained a Fellow of the Royal Society the rest of his life and would come to know most of the important scientists of the early seventeenth century, but Newton never came closer to a feeling of security and trust in his professional dealings with other men than at this rare moment. It was not without a touch of false modesty that he began his reply:

> At the reading of your letter I was surprised to see so much care taken about securing an invention to mee, of wch I have hitherto had so little value. And therefore since the R. Society is pleased to think it worth the patronizing, I must acknowleg it deserves much more of them for that, then of mee, who, had not the communication of it been desired, might have let it still remained in private as it hath already done some years.[45]

Except for a few technical details, Newton strongly approved of Oldenburg's description of the telescope. He urged the Secretary to be certain that Huygens would understand "it represents things distinct & free from colours." Nor, in contrast to his earlier correspondence with Collins, was there any outward hesitancy to have his name publicly linked with his work. Still awaiting word on his candidacy for the Royal Society, Newton concluded with a promise that, if elected, "I shall endeavour to testify my gratitude by communicating what my poore & solitary endeavours can effect towards ye promoting your Philosophicall designes."[46] So innocuous was his phrasing that no one, including Oldenburg, could have guessed what, if anything, he meant by "my poore & solitary endeavours." If the Secretary gave any serious thought to the matter, which seems unlikely, he did not have long to wait for further developments. Newton's next letter, which arrived in London on January 19, 1672, gratefully acknowledged his election to the Royal Society a week earlier. What is more, it made clear his feelings that the development of a reflecting telescope was of far less importance than the discovery of the scientific principle that made it possible. The years of feverish study and isolation were about to culminate at last in a wider justification, or so it seemed. What Newton sought from the Royal Society was unqualified, universal approbation. He would become, like his ideal, Robert Boyle, one of the intellectual inner circle, not merely a member of the supporting cast, not a maker of ingenious gadgets, but a creater of scientific law:

> I desire that in your next letter you would inform mee for what time the Society continue their weekly meetings, because if they continue them for any time I am purposing them, to be considered of & examined, an accompt of a Philosophicall discovery wch induced mee to the making of the said Telescope, & wch I doubt not but will prove much more gratefull then the communication of

that instrument, being in my Judgment the oddest if not the most considerable detection wch hath hitherto been made in the operations of Nature.[47]

Obviously in an expansive mood rarely associated with his suspicious and secretive character, Newton was referring to nothing other than his theory of light and colors confirmed by the *experimentum crucis* a few years earlier. It is as if the ghost of Shakespeare's Cleopatra stood whispering in his ear:

> *Give me my robe, put on my*
> *crown. I have*
> *Immortal longings in me.*

Newton was indeed destined to achieve immortality; but the crown he was about to don contained many more thorns than the laurels of which he now dreamed.

Chapter Seven

The Killing Ground

*Ride on! Rough-shod if need be. Smooth-shod if that will do. But ride on!
Ride on over all obstacles, and win the race.*

Dickens

I

Judging by the contents of Isaac Newton's voluminous papers, he was one
of those unusual individuals who do not always know the exact course of
their mental processes until they put pen to page. Indeed, there are times
when he appears to have been two separate personalities: on the one hand,
the familiar abstracted thinker, who for weeks on end remained oblivious to
his surroundings and his fellow human beings; on the other, often but less
clearly seen, an incredulous bystander who emerged from the shadows,
following some specific act of creation, to pay homage to the genius from
whom he was irrevocably estranged. Newton would make draft after draft
of a given manuscript, changing only a word here or a phrase there, until he
reassimilated his original ideas as if they had been put forth by another.
Five or six drafts of the same paper are commonplace; a dozen or more are
not unusual. The compulsion of the exacting artist applying the never quite
final touches of pigment to his canvas pales by comparison.

Now that he had so boldly committed himself to divulging his discovery
that white light is composed of the primary colors, Newton immediately
suffered another attack of acute anxiety. With nothing left to be done save
condense his findings for publication, he affected an indifferent tone when,
on January 29, 1672, he informed Oldenburg that "I hope I shall get some
spare howers to send you also suddenly that accompt [of light] wch I prom-

ised in my last letter."[1] The Royal Society Secretary may well have puzzled how "the most considerable detection wch hath hitherto been made in the operations of Nature," as it was described to him by Newton only ten days earlier, could be so quickly downgraded that its discoverer was now uncertain whether he could find the time required to put it on paper. Newton, it seems, had come to repent his boastfulness and was readying his psychological defenses in advance against possible attack.

He set about composing the lengthy letter to Oldenburg which, if he had published nothing else, would be sufficient of itself to guarantee him a place among the immortals of modern science. The original, dated February 6, is lost, but a transcript in Wickins's hand survives. (The obscure chamberfellow soon came to understand that there was no living on equal terms with so dominant a personality as Newton and had long since accepted the subservient role of amanuensis and scientific assistant.) Newton begins at the turning point of his optical research, 1666: "I procured me a Triangular glass-Prisme, to try therewith the celebrated *Phaenomena* of *Colours.*" The major steps in his prolonged quest are then traced in roughly chronological order, culminating in "the gradual removal of these suspitions [that] at length led me to the *Experimentum Crucis.*" In order to avoid making "a discourse too tedious & confused," Newton presented the heart of his findings in a series of masterfully drafted propositions which have taken unassailable form:

1. As the rays of light differ in degrees of Refrangibility, so they also differ in their disposition to exhibit this or that particular colour. Colours are not *Qualifications of Light,* derived from Refractions, or Reflections of natural Bodies (as 'tis generally believed,) but *Original* and *connate properties,* which in divers Rays are divers.

2. To the same degree of Refrangibility ever belongs the same colour, and to the same colour ever belongs the same degree of refrangibility. The *least Refrangible* Rays are all disposed to exhibit a *Red* colour . . . the most *refrangible* Rays . . . to exhibit a deep *Violet* colour. . . . And so to all the intermediate colours in a continued series belong intermediate degrees of refrangibility.

3. The species of colour, and degree of Refrangibility proper to any particular sort of Rays, is not mutable by Refraction, nor by Reflection from natural bodies, nor by any other cause, that I could yet observe.

4. Yet seeming transmutations of Colours may be made where there is any mixture of divers sorts of Rays. For in such mixtures, the component colours appear not, but, by their mutual allaying each other, constitute a midling colour.

5. There are therefore two sorts of colours. The one original and simple, the other compounded of these. The Original or primary colours are, *Red, Yellow, Green, Blew,* and a *Violet-purple,* together with Orange, Indico, and an indefinite variety of Intermediate gradations.

6. The same colours in *Specie* with these Primary ones may be also produced by composition: For, a mixture of *Yellow* and *Blew* makes *Green*; of *Red* and *Yellow* makes *Orange*; *of Orange* and *Yellowish green* makes *yellow*.

7. But the most surprising and wonderful composition was that of *Whiteness*. There is no one sort of Rays which alone can exhibit this. 'Tis ever compounded, and to its composition are requisite all the aforesaid primary Colours, mixed in due proportion.

8. Hence therefore it comes to pass, that *Whiteness* is the usual colour of *Light*; for, Light is a confused aggregate of Rays indued with all sorts of Colours, as they are promiscuously darted from the various luminous bodies.

9. These things considered, the *manner*, how colours are produced by the Prisme, is evident. For, of the Rays, constituting the incident light, since those which differ in Colour proportionally differ in refrangibility, *they* by their unequal refractions must be severed and dispersed into an oblong form in orderly succession from the least refracted Scarlet to the most refracted Violet.

10. Why the Colours of the *Rainbow* appear in falling drops of *Rain*, is also from hence evident. For, those drops, which refract the Rays, disposed to appear purple, in the greatest quantity to the Spectators eye, refract the Rays of other sorts so much less, as to make them pass beside it; and such are the drops in the inside of the *Primary Bow*, and on the outside of the *Second* or Exteriour one. So those drops, which refract in greatest plenty the Rays, apt to appear red, toward the Spectators eye, refract those of other sorts so much more, as to make them pass beside it; and such are the drops on the exterior part of the *Primary*, and interior part of the *Secondary* Bow.[2]

The letter concluded with the request that if any of the Royal Society Fellows were interested enough to repeat the experiments, "I should be very glad to be informed with what success." Yet his final sentence leaves little doubt about his own feelings on what he had accomplished: "That, if any thing seem to be defective, or to thwart this relation, I may have an opportunity of giving further direction about it, or of acknowledging my errors, if I have committed any."[3]

The matter was now out of Newton's hands, and there was nothing further to do but anxiously await a response from London. During the interim his thoughts surely turned back to 1669 when Barrow, acting on his behalf, sent a copy of *De Analysi* to Collins. The reception of the little treatise had been most favorable, and Newton may have taken this as the good omen it proved to be.

As luck would have it, the paper reached Oldenburg on February 8 as he was preparing to attend a meeting of the Royal Society that very afternoon. It was placed third on the agenda of items for consideration, after correspondence from Dr. John Wallis concerning the possibility that the moon's perigree and apogee might influence the rising and falling of the mercury in a barometer, and a letter from one Signor Cornelio, a Neapolitan who, with the assurance of a savant among the heathens, had concluded that folk stories of the "odd effects" of the tarantula's sting are fictitious.[4] After the meeting adjourned, Oldenburg wrote Newton that the "reading of your discourse concerning Light and Colours was almost thei only entertainment for that time. I can assure your, Sir, that it there meti with a singular attention and uncommon applause."[5] Oldenburg requested that Newton allow him to publish the paper in the *Philosophical Transactions* later in the month.

No natural philosopher could have asked for more from this, the most distinguished body of his peers, and Newton's spirits soared when he received the news of his triumph. In none of his extant papers is there any greater approximation of outright joy than in the letter of acknowledgment he sent Oldenburg on February 10:

> Twas an esteem of ye R. Society for most candid & able Judges in philo- sophicall matters wch encouraged mee to present them with that discourse of light & colours, wch since they have so favorably accepted of, I doe earnestly desire you to returne them my cordiall thanks. I before thought it a great favour to have been made a member of that honorable body; but I am now more sensi- ble of the advantage. For beleive me Sr I doe not onely esteem it a duty to con- curre with them in ye promotion of reall knowledge, but a great privelege that in- stead of exposing discourses to a prejudic't & censorious multitude (by wch many truths have been bafled & lost) I may wth freedom apply my self to so judicious & impartiall an Assembly.[6]

As to the printing of the letter, "I leave it to their pleasure." But again Newton thought it important to emphasize that what he had submitted was only a brief synthesis of a far larger and more complex body of work; if published by itself, it could leave him open to attack on the grounds of superficiality. He thus wisely reserved the right "to supply the afforesaid defects" by making available additional supporting experiments for inclu- sion in subsequent issues of the *Transactions*—a right he would soon invoke.

The efforts of John Collins, so barren of positive results, to persuade Newton to publish his mathematical discoveries were no secret to Henry Oldenburg. Having gained Newton's consent to print the letter of Febru- ary 6, he moved with a swiftness rarely exceeded in the history of scientific publication. "A Letter of Mr. Isaac Newton . . . containing his new Theory about Light and Colors" appeared in the *Philosophical Transactions* for February 19, 1671–72. It was followed five weeks later by "An Accompt of a New Catadioptrical Telescope invented by Mr. Newton." From this point onward, no matter how hard he tried, Cambridge's eccentric Lucasian Pro- fessor would never again be able completely to resume the spectral ex- istence that up to then had protected his anonymity.

II

The founding, during the 1660s, of the *Philosophical Transactions* by the Royal Society and the *Journal des Sçavans* by the *Académie Royal des Sciences* are axial events in the history of modern science. Previously books were the most common vehicle for conveying new scientific ideas, but their availability and readership were usually limited, and their authors' original contributions, as in the case of Kepler's *Astronomia Nova* and *Harmonice Mundi,* were often lost within a convoluted web of hypothesis. A regular journal, on the other hand, carries from one researcher to another the concise observations that are of common interest, affording to many the opportunity to contribute individual pieces to the larger puzzle. It also provides institutional support and sanction to the Baconian idea of science as a collective undertaking with utilitarian aims. No longer would it be necessary for each natural philosopher to erect his own formal system of nature; he could now concentrate on the narrower problems of perfecting his experimental techniques and methodology. Although seminal treatises of great length were yet to be written, including Newton's own *Principia* and *Opticks,* the journal article was destined to replace the book as the primary medium of communication among members of the scientific community.

No better proof is to be found that the world of nature was opening up as never before than the titles of articles appearing in the *Transactions* during Oldenburg's editorship. Few areas of inquiry in the natural sciences and related fields were left untouched. The physician would surely have found interesting "An Observation made upon the Motion of the Hearts of two Animals, after their being cut out," or "A letter of Doctor John Wallis . . . concerning the said Doctors Essay of Teaching a person Dumb and Deaf to Speak, and to understand the Language." The astronomer could take delight in "New Observations of Spots in the Sun," while the budding entomologist was doubtless much enlightened by the "Extracts of three Letters; one with a Table of 33 sorts of Spiders to be found in England." The naturalist and the alchemist were not overlooked: The former could immerse himself in ". . . Inquiries, Directions and Experiments, concerning the Motion of Sap in Trees," even as the latter pondered the implications for his crucible of a "Letter . . . concerning the Principles and Causes of the Volatilisation of Salt of Tartar and Other Fixed Salts." Perhaps somewhere in England a restless shipwright decided to exchange the Old World for the New after perusing "An Account of the Advantages of Virginia, for building ships." If so, he would have been well advised to prepare for his testing colonial experience by mastering certain other practical skills. These, too, he could have acquired in part from the pages of the *Transactions* in such articles as "Brief Directions [on] how to prepare and Tan Leather" and "The way of making Vinegar in France."

Among the journal's many gifted foreign contributors, Anton van Leeuwenhoek, the Dutch naturalist and pioneer in microscopy, was perhaps the most prolific. He submitted his first article in 1673 and made regular appearances in the *Transactions* over the next several decades. Typical of Leeuwenhoek's work with the microscope is his "Observations ... concerning the Optic Nerve." Robert Hooke, no stranger to the microscope himself, and the Royal Society's indefatigable Curator of Experiments, was the most frequent contributor among the English virtuosi, with Boyle a respectable second.

It is hardly surprising that during this incubative period at least as many men of mediocre talent and intellect, including a fair number of crackpots, published in the journal. Oldenburg, like the age in which he lived, evidenced an unusual and at times morbid concern for teratology and all manner of medical curiosities. Early issues of the *Transactions* abound with accounts of two-headed calves, werewolves, animated horsehairs, hermaphrodites, gargantuan tumors, and the birth of animals of one species to parents of another. One physician published "An account of a Foetus that continued 46 Years in the Mother's Body" while a colleague reported on opening "the big-belly'd Woman near Hamon in Stropshire, who was Supposed to have continued many Years with Child." What he discovered was a large mass of benign tissue rather than the anticipated remains of a calcified infant.

To further their biological understanding the virtuosi sacrificed every kind of creature on the altar of the new science with dispassionate regularity. In May 1661, amid remarks in his diary on dining with the King and a comment on Francis Walsall's sermon "concerning Love & Charity," Evelyn wrote of "Discourses at our Society about poysons againe. We gave *Nun Vom:* to birds that killed them out-right." A few days later "we tried more *Vipers* & poysond arrows to dogs &c: but they succeeded not."[7] Another time Evelyn and his associates watched with keen interest as a model of Boyle's vacuum pump made quick work of a chicken by asphxiation but proved unsuccessful in dispatching a snake. According to Samuel Pepys, poisons were still very much in vogue four years later, in 1665. After a "very pleasant" dinner he and two friends visited Gresham College, the home of the Royal Society "where we saw some experiments up[on] a hen, a dog, and a cat of the Florence poyson," a concoction derived from tobacco. "The first it made for a time drunk, but it came to itself quickly. The second it made vomitt mightily, but no other hurt." Pepys did not stay to see the effect of it on the third, because he was called away on business. Two weeks later the dyspeptic civil servant was back at Gresham to see a repeat of the experiment. This time success: "Saw a cat killed with the Duke of Florence's poison."[8]

Doubtless the most cruel experiments of all were the hundreds of operations performed on live dogs, the results of which were frequently written up for the *Transactions*. All but the most hardened vivisectionists must have paled on reading Hooke's account of his investigations into animal respiration in 1664: "The other Experiment . . . was with a dog, which, by means of a pair of bellows, wherewith I filled his lungs, and suffered them to empty again, I was able to preserve alive as long as I could desire, after I had wholly opened the thorax, and cut off all the ribs, and opened the belly." Hooke himself suffered intense pangs of guilt over this drastic form of mutilation and decided that "I shall hardly be induced to make any further trials of this kind, because of the torture of the creature." Nevertheless, the talented experimentalist wistfully speculated, "the enquiry would be very noble, if we could . . . find a way to stupefy the creature, as that it might not be sensible."[9] Hooke, asked to repeat his experiment before the Royal Society in 1667, sought and obtained a postponement. Two doctors then tried their hand but failed. Hooke was later ordered to perform the operation and, according to the *Philosophical Transactions* for October 24, "succeeded well."

Thomas Sprat drew up an extensive list of medical and anatomical experiments performed by Royal Society Fellows, which included the following: cutting out the spleen of a dog; preserving animals in spirits of wine, oil of turpentine, and other liquids; injecting various substances into the veins of different creatures; destroying mites with fumes; making insects with cheese (the belief in spontaneous generation being very much in vogue); killing newts and toads with salts; and testing whether a spider could be enchanted by placing it in a circle of unicorn's horn or Irish earth. A spur was grafted onto the head of a cock to determine if it would grow (which it did), strangled animals of various species were revived by blowing air into their lungs, and the blood of one animal was transfused into another.[10] This last experiment was a favorite of Hooke's, as Pepys noted in his diary entry of November 16, 1666: "This noon I met with Mr. Hooke, and he tells me the Dogg which was filled with another dog's blood at the College the other day, is very well, and like to be so as ever. And doubts not its being found of great use to men; and so doth Dr. Whistler, who dined with us at the tavern."[11] This experiment had actually been performed on humans in France, with the inevitable tragic result. Only in the twentieth century, with the development of biochemistry and immunology, would Hooke's prediction come true.

The work of the vivisectionists appears to have held little interest for Newton, who his niece claimed paled at all forms of animal suffering. His was a world of mathematical symbols and of matter in motion, of primary colors and of gyrating celestial orbs. Yet when it came to a dispute over ideas, especially if it involved his own, he showed a kind of animal ruthlessness, granting no mercy and asking none. Ever reluctant to enter the public arena, once aroused he proved an awesome opponent. The lion of Woolsthorpe was a born virtuoso of the intellectual killing ground.

III

Notwithstanding the troubled times, Henry Oldenburg's extraordinary commitment to the dissemination of scientific ideas had guaranteed the regular publication of the *Philosophical Transactions* since its inception in 1665. Only twice during the twelve years of his editorship did brief interruptions occur. The first resulted from what Sprat, in a chilling metaphor, described as "the Arrow that flies in the dark"—the Great Plague—the second when Oldenburg was briefly incarcerated in the Tower. In all, some seventy-nine issues of the journal had been printed before Newton's paper on light made its appearance. Yet in those hundreds of pages no one had advanced an experimentally based proposal for the radical revision of a scientific theory, nor had anyone initiated a major international discussion and debate within the confines of this or any other journal.[12] It was Newton's destiny to do both.

The letter to Oldenburg was published only eleven days after it reached London, but the debate that Newton so morbidly anticipated had begun to take place even before its printing. When presented with data on recent discoveries, the Royal Society customarily conducted its own independent investigation of the findings. In addition to instructing Oldenburg to thank Newton on behalf of the Society for his "ingenious discourse," the group further ordered that "Mr. Boyle, and Mr. Hooke be desired to peruse and consider it, and bring in a report of it to the Society."[13] Boyle, who seems to have been preoccupied with other matters, was content to leave the task to Hooke, his onetime assistant at Oxford. Notwithstanding the complexity of the issues at hand, Hooke, with characteristic self-assurance, presented his analysis on February 15, only one week after the formal reading of Newton's paper. A few months later, after Newton had launched a bitter and humiliating attack on his report, Hooke admitted to Lord Brouncker that he "had not above three or 4 hours time for the perusall of Mr. Newton's paper and the writing of my answer." Hooke lamely sought to justify his dereliction with the disclaimer, "I never intended it for Mr. Newton's perusall."[14] Even if this were so, and had Oldenburg not sent Newton a copy of Hooke's critique, the Curator of Experiments had done little justice to a theory that had taken untold grueling hours in the proving. Thus did the following lines, quoted from the first paragraph of Hooke's superficial report, initiate a bitter conflict that lasted until their author's death in 1703:

> I have perused the Excellent Discourse of Mr. Newton . . . and I was not a little pleased with the niceness and curiosity of his observations. But though I wholy agree with him as to the truth of those he hath alleged, as having by many hundreds of tryalls found them soe, yet as to his Hypothesis of salving the phaenomena of Colours thereby I confess I cannot yet see any undeniable argument to convince me of the certainty thereof. For all the expts [experi-

ments] & obss [observations]: I have hitherto made, nay and even those very
expts which he alledged, doe seem to me to prove that light is nothing but a pulse
or motion propagated through an homogeneous, uniform and transparent
medium.[15]

Like Achilles and Hector, Newton and Hooke seemed foreordained by
the gods to do battle. Hooke's biographer astutely chose to begin her portrait not with his birth but with an analysis of the personal factors that gave
rise to one of the most celebrated quarrels in the history of modern science.
She could not have chosen a more fitting quote to open her work than the
line penned by their contemporary John Dryden: "Two such as each
seem'd worthiest when alone."[16]

Hooke, seven years Newton's senior, possessed intellectual gifts that
bordered on genius and in temperament was cast in a mold reminiscent of
Leonardo. His thought process, like that of the great Florentine, was a
strange mixture of sudden uprisings and agonizing downfalls, of crescendo
and diminuendo. Such restless, peripatetic brilliance quickly burns itself
out, then rises brightly from the ashes to attack a new problem, regardless of
whether the old one has been satisfactorily resolved. Like a meteor skipping
across the outer atmosphere of a planet, talent of this variety is both explosive and spectacular, but its product often disintegrates before reaching
terra firma. Richard Waller, Hooke's first biographer and successor as Secretary of the Royal Society, observed: "The fertility of his Invention . . .
hurry'd him on, in the quest of new Entertainments, neglecting the former
Discoveries."[17] Indeed, it is on men like Hooke that fate plays one of its
cruelest jokes, blessing them with great prescience and creative intuition,
yet denying them the powers of abstraction and the analytical capacity
needed to transform their most promising visions into concrete principle.
Always painfully conscious of more than he could prove, Hooke, who never
plucked the heart out of a profound scientific mystery, watched in agonized
distress while his contemporaries reaped the harvest for which he had frequently prepared the ground. Had he lived at another time or place, Hooke
might have shone like a star of the first magnitude. Instead, his historical
image has been blurred by that projected by Newton, the rising sun of
modern science.

Whereas Newton outgrew his childhood infirmities, Robert Hooke was
marked in both body and physiognomy for life. No portrait of him survives,
but Pepys noted of the young Hooke that he "is the most and promises the
least of any man in the world that I ever saw." John Aubrey, one of Hooke's
few lifelong friends, was even more graphic: "He is but of midling stature,
something crooked, pale faced but little belowe, but his head is lardge; his
eie full and popping, and not quick; a grey eie." As if to soften his frank
description of one he deeply admired, Aubrey hasted to add, "He haz a
delicate head of haire, browne, and of an excellent moist curle."[18] By the
time Hooke reached middle age he had become permanently stooped, con-

veying the aspect of a man far beyond his years. At a time when major physical deformities were far more commonplace than today, Waller's characterization of Hooke's person as "dispicable, being very crooked" is all the more revealing. It is little wonder that "His Temper was Melancholy, Mistrustful and Jealous, which more increas'd upon him with his Years." [19]

Whether or not Robert Hooke inherited his melancholy countenance from his father we do not know, but something so unbearable came to dominate the existence of John Hooke that he was moved to take his own life by hanging in 1648, an especially disgraceful end for a man of the cloth. Robert, then age thirteen, received a modest inheritance of £100 and was promptly packed off to London with the intention that his artistic talents be nurtured under the mastership of the painter Sir Peter Lely. Shortly thereafter Hooke was befriended by Richard Busby, the legendary "flogging master" of Westminister School. In this instance the rod was spared. Busby recognized Hooke's special talents for what they were and became so eager to see them developed that he took the youth into his home. Having mastered the classical curriculum taught at Westminster, Hooke moved on to Oxford's Christ Church, enduring the indignities incidental to making his way as a virtual servant, an experience familiar to Newton. Coincidentally, Hooke "lay in the chamber... that was Mr. Burton's, of whom 'tis whispered that, *non obstante* all his Astrologie and his booke of *Melancholie*, he ended his dayes in that chamber by hanging him selfe." [20]

There is no record of Hooke's ever having taken a bachelor's degree (he was nominated for the M.A. in 1663), yet his arrival at that lofty center of learning could not have been better timed. At Oxford he joined the group around which the Royal Society crystallized after the Restoration, a far greater assemblage of scientific figures than greeted Newton at Cambridge, where Barrow walked the lecture halls in seldom broken solitude. Thomas Willis, a member of the Oxford Club for whom Hooke first worked as a scientific assistant, introduced him to the illustrious Robert Boyle shortly after the latter's migration to Oxford. A bond formed between the two men, and Hooke accepted Boyle's offer to become the chemist's paid assistant, a position he held until late 1662 when, with Boyle's influence, he was named Curator of Experiments for the Royal Society.

It is difficult to imagine anyone better suited for this position than Hooke, although most men would have found the demands of the office highly excessive in relation to its meager financial rewards. The Society, by its charter, required Hooke to furnish each of its weekly meetings with three or four "considerable Experiments" in addition to performing any others proposed by the Fellows. The pace at which Hooke labored was nothing short of frenetic, yet he thrived on the intense pressure, for it nourished his inner compulsion to skip from idea to idea without respite. Furthermore, the innumerable responsibilities associated with his position brought a kind of order into the life of the natural philosopher that otherwise would have been lacking. As long as he worked at fever pitch, the brilliant ideas poured forth in endless variety: astronomy, blood transfusion, mechanics, cartography, skin-grafting, optics, botany, geology, springs, horology, engines, telescopes, microscopes, and so on. Hooke, nearly as fascinated with flight as Leonardo had been, claimed to have invented some thirty ways of flying, which, he told a friend, could not be revealed lest another steal his ideas. It was Hooke's great weakness always to claim more than he could prove, but there is no denying that his contributions to the science and technology of his day were multifarious. Only when his health began to deteriorate rapidly and the demands placed on him by his peers gradually slackened did Hooke's creative powers wane: His mind then became flaccid, his ability to innovate permanently impaired.

The dark quirks of temperament that repelled Hooke and Newton from each other, as with the like poles of magnets in juxtaposition, were the misshapen progeny of similar youthful experiences—delicate health, lost fathers, extraordinary creativity, the need to work their way through Oxford and Cambridge as outsiders seeking their due. Here, however, the similarity seems to have ended. While Newton preferred the isolation and intellectual freedom of a university town bordering the desolate and windswept fens, the more culturally attuned and gregarious Hooke reveled in the renaissance of a great metropolis rising from its fertile ashes, a rebirth which, as a gifted architect and one of the capital's official surveyors, he helped engineer. (The income from this lucrative employment was discovered shortly after his death in an old iron chest that had evidently lain unopened for many years.) The swirl of ideas and gossip that enveloped London's intellectuals was as invigorating to Hooke as it was to his illustrious acquaintances, Aubrey, Wren, and Pepys. Hooke became such a prominent figure in his own right that Thomas Shadwell drew most of the ammunition for *The Virtuoso*, his popular farce, from the pages of the *Micrographia*. (Against his better judgment, Hooke was persuaded to attend a performance of the play in June 1676, where his worst fears were realized: "Damned Doggs. *Vindica me Deus.* [God, avenge me.] People almost pointed.")[21] An incurable lover of sweets and a lifelong insomniac, Hooke relished the conversation of the coffeehouse, where he imbibed prodigious quantities of "chocolatt," and took his evening meals in any of several taverns, often lingering with a friend over a glass of claret until closing time at one or two in the morning. Newton, if he dined at all, usually did so alone in his rooms and almost never visited a tavern after becoming Lucasian Professor. Hooke took several mistresses, including his brother's coquettish and unfaithful daughter Jane, and recorded his orgasms in his *Diary* with the astrological symbol Pisces.[22] Newton never knew a woman and seems scarcely to have wanted to, except, perhaps, in what he obliquely referred to as the "uncleane" thoughts and dreams of his early manhood.

A comparison of Newton's and Hooke's life-styles and attitudes toward their fellow human beings might be considered less relevant were it not for the influence these divergent attitudes exerted on their respective philosophies of science. Hooke's approach to the new learning was classically Baconian in that he believed the advancement of scientific knowledge demanded the joint labors of many men. His outlook was consonant with that held by the founders of the Royal Society, who drew their inspiration from the model of cooperation supplied by Sir Francis in the *New Atlantis*. Hooke, like Bacon, was resolved to apply the experimental method to the whole range of nature's wonders. In the plentitude of discovery that must inevitably follow, the interests of utility could not but be served. Hooke's was thus an expansive and pragmatic view of the world. The fall of man from grace was not irreversible after all. Through careful coordination and application of his knowledge, it was only a matter of time until man regained his previous dominion over nature. The retreat from darkness was at hand.

Hooke has been pictured as a Don Juan of science, who made quick and easy conquests as myriad insights crowded in upon one another.[23] As the person who almost singlehandedly fashioned the scientific outlook of eighteenth-century England, Newton eschewed Hooke's facile tendency to spread his talents too thin. He did not share Hooke's Baconian faith that natural philosophy is most productive when pursued collectively, nor did he optimistically look forward to a future when the pleasures of life would be fundamentally enhanced by what was learned in the laboratory. Newton became the very symbol of what a genius might accomplish on his own. Once he became fixed upon an idea, he pursued it without regard for time or physical discomfort. He performed his own experiments and fashioned the required instruments with his own hands. Whereas Hooke spent many happy hours in collaboration and social discourse with laborers and skilled craftsmen such as the clockmaker Thomas Tompion, Newton may never have spoken intimately with a workman during his creative period, save as master of his manor at Woolsthorpe.

Of equal importance was Newton's obsession with preserving the inductive purity of his rigorous scientific method. Few things rankled him more than public speculation of the type so freely indulged in by Hooke, a persistent violator of Occam's injunction against the needless multiplication of conceptual entities. Very early on Newton became determined to publish only those hypotheses which yielded to airtight demonstration. *Hypotheses non fingo* became the anthem of his methodological approach, and woe to the person who attacked his work on the basis of conjecture, no matter how promising. (Privately, of course, he speculated on a grand scale.) Finally, by focusing his attention on a narrower but more profound range of scientific questions than that treated by Hooke, Newton guaranteed the dominance of the abstract studies of mathematics, physics, and astronomy through the eighteenth century and beyond. Of the various fields into which he delved, only alchemy resisted his attempts to formulate a major unifying principle.

IV

It took Newton only a day to draft a brief and, for him, rather subdued reply to Hooke's critique. He promised Oldenburg a more detailed answer at a later date. For the moment he was content to assert that Hooke had offered no hard evidence to refute his essential premise. "I received your Feb 19th. And having considered Mr. Hooks observations on my discourse, am glad that so acute an objecter hath said nothing that can enervate any part of it." Hooke implied that he had repeated certain of Newton's experiments when he wrote that the *experimentum crucis* "is not that, which he soe calls." Newton, of course, knew better, for Hooke had only a week in which to test the complex data, where months were required. Though he stopped short of calling Hooke a liar, the implication is certainly there: "I . . . doubt not but that upon severer examinations it will be found as certain a truth as I have asserted it." [24] How suddenly Newton's admiration for the author of the *Micrographia*, a work from which he drew so heavily, had been reduced to contempt.

Newton must have taken some solace from learning that Hooke's paper was not, as had been suggested at the Royal Society meeting of February 15, to be published in the issue of the *Philosophical Transactions* that contained his own: "It not being thought fit to print them together, lest Mr. Newton should look upon it as a disrespect, in printing so sudden a refutation of a discourse of his, which had met with so much applause at the Society but a few days before." [25] As it turned out, Hooke's paper never made its way into the journal, whereas, much to his embarrassment, Newton's brilliant but needlessly abusive reply was published in November. One senses that the intellectual force and explosiveness of temper revealed in Newton's correspondence with Collins and Oldenburg had already combined to cast something of an inhibiting spell over certain of the more influential Fellows, who gained access to the letters.

Hooke possessed a considerable instinct for the jugular, and Newton carelessly left him an opening through which to strike. While the numbered propositions at the heart of Newton's findings proved to be unassailable, their author, by also choosing to argue for the corpuscular conception of light, violated his own tenet against mixing hypothesis with scientific fact. Just before his assertion to Oldenburg that "I shall not mingle conjectures with certainties" by speculating on "what Light is . . . and by what modes or actions it produceth in our minds the Phantasms of Colours," Newton observed that "it can no longer be disputed . . . whether Light be a Body." [26] Hooke, who leaned toward the wave theory, rightfully took Newton to task for making this unprovable claim. It is doubtless partly for this reason that Newton delayed his reply. Angry with himself for not being more circumspect, and thus giving his critic the opportunity to turn his attention away from the central doctrine of heterogeneity, Newton resolved never to make the same mistake again.

Hooke's challenge went beyond the mistaken assumption that the doctrine of corpuscularity constituted the key argument of the paper on light. On the very day Newton's reflecting telescope was examined and applauded at a meeting of the Royal Society, "Mr. Hooke, seeing this Tellescope to obtaine esteeme," rejected its necessity:

> Mr. Hooke moreover affirmed *coram multis* that in the year 1664 he made a little Tube of about an Inch long, to put in his fobb, which performs more than any Tellescope of 50 foot long made after the common manner; but the Plague happening, wch caused his absence, and the fire, whence redounded profitable employmts about the Citty, he neglected to prosecute the same, being unwilling the Glass grinders should know any thing of the Secret.

Hooke now sought to secure his priority "with a Cipher containing the Mysterie the which he disclosed to the Lord Brouncker, and Dr. Wren, who report plausibly of it, and what is done in this way is performed by Glasse Refraction."[27]

The acerb and ill-timed manner of its expression aside, Hooke's claim seems more than the product of sour grapes, though we may seriously question whether he truly fashioned an acceptable miniature version of a refractor small enough to fit on the fob of his watch. More important was Hooke's assumption that Newton's construction of the reflector sprang from the belief that it promised the best way of attacking the problem of chromatic aberration, as opposed to offering a technical demonstration of a profound new scientific principle. Newton himself lent a degree of credence to this view in his paper on light: "I left off my aforesaid Glass-works; for I saw, that the perfection of Telescopes was hitherto limited, not so much for want of glasses truly figured according to the prescriptions of Optick Authors . . . as because that Light it self is a *Heterogeneous mixture of differently refrangible Rays*." Having done extensive research on lenses, Hooke reacted by challenging what to him was a serious error of judgment in the first part of Newton's statement, while disregarding the latter. "I am a little troubled that this assumption should make Mr. Newton wholy lay aside the thoughts of improving telescopes and microscopes by Refractions, since it is not improbable, but that he . . . would, if he had prosecuted it, have done more by Refraction."[28]

The very first volume of the *Philosophical Transactions* contained Hooke's ingenious idea on fashioning achromatic lenses for refracting telescopes. He proposed to do so by means of fastening a plano-convex lens to a plane glass slab with a metal ring, then filling the space between them with a liquid such as water, oil of turpentine, spirit of wine, or a saline solution.[29] Hooke argued that the refractive indices of differing liquids would improve the focus of the lens, sharply reducing the negative effects of chromatic aberration. The fact is, Newton had also realized the possibilities of constructing achromatic lenses, but the February paper contained not the slightest hint of his thoughts on the matter.[30]

For all the stir it created, Newton's telescope proved more a novelty than a triumph of superior craftsmanship. Its speculum soon became so tarnished as to render the instrument useless. The larger versions commissioned by the Royal Society, about which Flamsteed waxed enthusiastic, proved even less effectual. Not until 1722, when James Hadley succeeded in grinding a parabolic mirror, could the reflector compete successfully with the refractor. At that time Newton's design was discarded in favor of those developed by James Gregory and Guillaume Cassegrain. Yet the truly important point, which Hooke did not come to appreciate until it was too late, is that Newton's reflector was primarily intended as a test piece for his thoughts on light and color. Hooke had chosen to reject this central argument out of hand, a foolish act which Newton, above all men, would not suffer lightly.

Had Newton been allowed to choose the two natural philosophers he most wanted to accept his "Paradoxicall assertion," Robert Hooke almost certainly would have been one of them. For the other, he would have turned to Paris—"the City of Light"—where Huygens had taken up residence in 1666 on the promise of a handsome pension from Jean-Baptiste Colbert, Louis XIV's brilliant Minister of Finance. Disappointed and angered over Hooke's patronizing and superficial treatment of his theory, Newton took heart from Huygens's warm comments on the reflecting telescope, which Oldenburg had forwarded to Cambridge along with a copy of Hooke's critique. In the brief communication to Oldenburg denouncing Hooke's objections, Newton, in pointed contrast, referred to the "severall hansome & ingenious remarques" in "Monsieur Hugenius's letter." Meanwhile Oldenburg sent a copy of the *Philosophical Transactions* containing the paper on colors to Paris, reminding Huygens that its author and *"l'Inventour du telescope Cata-dioptrique"* were one and the same. Huygens's response reached Oldenburg within less than a month, the key passage from which the Secretary dispatched to Newton on April 9: "[T]he new Theory of Mr. Newton concerning light and colors appears highly ingenious to me."[31] Coming from the greatest of the European virtuosi and on the heels of Hooke's hardbitten analysis, this was welcome and heady praise indeed.

Less elevating were the comments of Sir Robert Moray and Ignance Gaston Pardies. Moray had played a key role in persuading Charles II to grant the original charter incorporating the Royal Society, and he had the honor of serving as its first President from March 1661 until July 1662. He proposed four adolescent and essentially groundless experiments by which to test the theory of light, including "mov[ing] the Prism so, as the End may turn about the middle being steady." Still, considering who the author was, Newton felt compelled to draft a polite and dignified reply. Pardies, a French Jesuit and professor of Rhetoric at the *Collège de Louis-le-Grand* in Paris, had raised more weighty though equally manageable concerns. Oldenburg, as if anticipating the worst, rather indelicately cautioned Newton to "expedite an answer for him and yt in the same language, wherein he writeth." Somewhat surprisingly, Newton took no umbrage. Indeed, he responded in an almost apologetic manner, which may have caused Oldenburg to wonder whether Collins had not misjudged his character after all: "I herewith send you an answer to the Jesuite Pardies considerations, in the conclusion of wch you may possibly apprehend me a little too positive, but I speake onely for my selfe. I am highly sensible of your good will in communicating to me . . . observations . . . & I desire you to continue that favor to me." [32]

Pardies recognized that Newton's "very extraordinary hypothesis" must, as its originator argued, stand or fall "on that experiment with the glass prism, in which the rays entering a dark chamber through a hole in the window [shutter] and then received on the wall or a sheet of paper did not form a round shape . . . following the received rules of optics, but appeared stretched out into an oblong shape." None of the various modification theories were able to account quantitatively for this elongation of the spectrum, a problem brought to a brilliant resolution by Newton when he applied the existing laws of refraction not to the incident beam as a whole, but to each of the colored beams contained in the compounded white one. Nevertheless, Pardies fell back on the traditional argument employed by Grimaldi and others who had also observed this phenomenon, namely, that because the sun is circular in form, rays emanating from different parts of its surface strike the prism at different angles, producing the oblong spectrum. "Since therefore a manifest cause may be brought forward to show why the shape of the rays is oblong, that cause springing from the very nature of refraction, it seems unnecessary to run after another hypothesis or to admit a divers refrangibility in those rays." [33] Pardies also employed this argument to refute the *experimentum crucis*.

Newton could do nothing but reiterate the experimental results detailed in his published paper, findings that both anticipated and more than adequately dealt with the argument raised by Pardies. As in his reply to Moray, Newton's manner was restrained for the most part, if not indulgent. But while he might excuse a respected member of the international scientific community for exercising insufficient rigor in evaluating certain aspects of his experimental work, Newton was justly irritated by the Jesuit's incautious assumption that he, like some rank amateur, was chasing after a chimera in the guise of a hypothesis. The final paragraph of his reply, which he had allowed might be "too positive" for Oldenburg's taste, gave vent to his sense of indignity:

> I am content that the Reverend Father calls my theory an hypothesis if it has not yet been proved to his satisfaction. But my design was quite different, and it seems to contain nothing else than certain properties of light which, now discovered, I think are not difficult to prove, and which if I did not know to be true, I should prefer to reject as vain and empty speculation, than acknowledge them as my hypothesis.[34]

Pardies was shaken by both the sharpness of tone and the intellectual force of Newton's response. When he next wrote to Oldenburg, enclosing another comment for Newton, he beseeched the Royal Society Secretary to examine its contents carefully before sending it on to Cambridge. "And if there are any words which might offend Mr. Newton even a little please remove and change them." Pardies sought to assure Newton that he had not employed the word "hypothesis" out of contempt: "When I called the theory an hypothesis I did so of no set purpose, but used the first term that came to mind."[35] He also informed Newton that, having more thoroughly studied the arguments against the theory of unequal incidence, he was withdrawing his protest "on that account." However, Pardies now suggested that the diffusion occurring after light passes through an aperture might serve as an alternative explanation of the disputed phenomenon.

This objection, too, had been met and overcome by Newton in his paper. Thus in responding to it in a second letter, he could not refrain from turning Pardies's own admonition concerning hypotheses against the well-meaning but outmanned Reverend Father:

> In answer to this, it is to be observed that the doctrine which I explained concerning refraction and colours, consists only in certain properties of light, without regarding any hypothesis, by which those properties might be explained. For the best and safest method of philosophizing seems to be, first to inquire diligently into the properties of things, and establishing those properties by experiments and then to proceed more slowly to hypotheses for explanations of them. For hypotheses should be subservient only in explaining the properties of things, but not answered in determining them; unless so far as they may furnish experiments. For if the possibility of hypotheses is to be the test of the truth and reality of things, I see not how certainty can be obtained in any science.[36]

Pardies now knew something of how Job felt when answered from out of the whirlwind by an intelligence far greater than his own. And like the ancient Hebrew, he had no choice but to submit: "I am quite satisfied with Mr. Newton's new answer to me. The last scruple which I had, about the Experimentum Crucis, is fully removed. And I now clearly perceive by his figure what I did not before understand.... I have nothing further to desire."[37]

Would that others had been equally compliant.

V

Accounts with Pardies had no sooner been settled when another disturbing communication reached Newton via Oldenburg in early May. Jean-Baptiste Denis, a doctor and medical adviser to the Sun King, had published an article on a reflecting telescope designed by Guillaume Cassegrain, Professor of Physics at the Collège de Chartres. Oldenburg copied parts of the article for Newton's consideration, then added this goad, as if one were needed: "I am of [the] opinion you will find cause to controle the confident assertions of this Author; as I am apt to believe you will make good your Theory of Light."[38] Cassegrain claimed priority over Newton, but the Frenchman could produce no proof of his assertion, for there was none. On the other hand, the Cassegrain reflector ultimately proved superior to the instruments designed by Gregory and Newton, because its mirrors tended to correct rather than to reinforce the effects of spherical aberration, providing a better definition of the image, and because it allowed a greater proportion of light to enter through the aperture of its convex mirror.

As his March correspondence with Oldenburg clearly illustrates, Newton was much troubled by the relatively poor performance of his own telescopes.[39] Yet he affected an arrogant, all-knowing tone when drafting his analysis of Cassegrain's design. After listing seven particular "disadvantages"—some of which later proved wholly erroneous—Newton concluded: "I could wish therefore M Cassegraine had tryed his designe before he divulged it; But if for further satisfaction he please hereafter to try it, I believe the successe will informe him that such projects are of little moment till they be put in practice."[40] Newton may have derived some measure of satisfaction from attacking the work of a fellow virtuoso even as his own discovery was under fire, but it is the Cassegrainian reflector, not the Newtonian, that has found favor among modern astronomers.

Nearly ten months had passed since Newton received his last letter from John Collins, whose long and patient efforts to convince the intractable scholar that he must publish his mathematics had sadly come to naught. Collins still entertained hopes of prying loose the treatise on infinite series, but he realized that a direct approach on the matter, at least for the present, was out of the question. He wrote a chatty missive to Newton on April 30, disarmingly suggesting that the latter's decision not to publish De Analysi may have been for the best after all, "because our Latin Booksellers here are avers to ye Printing of Mathematicall Books" for want of buyers. "[I]n stead of rewarding the Author they rather expect a Dowry with ye Treatise."[41] Collins had learned from Barrow that Newton was preparing his optical lectures for the press at Cambridge. If by any chance the plans to publish there should change, he wrote, "I shall most willingly affoard my endeavour to have it well done here, and if so, what you have written might be sent up the sooner in order to the Preparing of Schemes." Collins also wanted to assure Newton that he was in total sympathy with his struggle against those who rejected the new theory of light: "I could not but observe with much content that you were pleased to become a Member of the R. Society, though withall sorry that it should redound to your Charge, especially seeing it was your Designe thereby to enrich Learning with your excellent Contemplations about ye same."[42]

Newton, who in the months since his last letter seems to have forgotten the harsh scolding he gave Collins for sending him unsolicited books, responded warmly to the impresario's praise. Yet he could not agree to the proposed undertaking, "finding already by that little use I have made of the Presse, that I shall not enjoy my former serene liberty till I have done with it." After a paragraph outlining the current state of his mathematical studies, Newton petulantly returned to the subject that now vexed him more than any other:

> I take much satisfaction in being a Member of that honourable body the R. Society; & could be glad of doing any thing wch might deserve it: Which makes me a little troubled to find my selfe cut short of that fredome of communication wch I hoped to enjoy, but cannot any longer without giving offence to some persons whome I have ever respected. But tis no matter, since it was not for my own sake or advantage yt I should have used that fredome.[43]

As if the atmosphere was not already sufficiently charged, Oldenburg was pressing Newton for a publishable reply to Hooke's offensive critique. He had requested something for the March *Transactions*, but Newton wrote that he would "not be ready for them, because I intend to annex to that answer some further explications of the Theory." Oldenburg broached the matter a second time in April and was again put off.[44] By May, when he reminded Newton of his promise for yet a third time, Oldenburg had become almost as much concerned about the wording of the reply as about its content. One need not look far for an explanation.

Only after Newton's letter to Pardies had been read before a meeting of the Royal Society was it sent to France. Though there is nothing in the minutes to suggest that its pointed conclusion had raised any eyebrows, Oldenburg's letter of May 2 strongly implies that at least some of the Fellows, perhaps including Hooke, had taken offense. In a manner less diplomatic than before, Oldenburg repeated his previous warning against the mixing of personalities with scientific discourse:

> I shall suggest, yt when your answer to Mr. Hooks and ye Jesuit Pardies objections shall be thought fit to be printed, ye names of the objectors, especially if they desire it may be so, be omitted, and their objections only urged: since those of the R. Society ought to aime at nothing, but the discovery of truth, and ye improvemt of knowledge, and not at the prostituting of persons for their misapprehensions or mistakes.[45]

Oldenburg was surely relieved when Newton, not without a touch of pique, seemed to acquiesce: "I am not at all concerned whether Objections be printed wth or wthout ye Objectors names. And those of ye Jesuite Pardies may be conveniently so printed if he desire it." It seemed to him useless to apply the same rule to Hooke, however, "since the Contents will evidently discover ym to be his. And besides, it is publiquely known that he hath writ objections & my answer is expected."[46]

Indeed, the more Newton thought about Oldenburg's cautionary advice the more he smoldered. A fortnight later the Royal Society Secretary received a rather ominous note: "I told you that it was indifferent to me whether they [Newton's objections] were printed wth or without the Authors names.... But yet I understand not your desire of leaving out Mr. Hooks name, because the contents would discover their Author unlesse the greatest part of them should be omitted." This, in turn, would raise "new objections & require another Answer then what I have written. And I know not whether I should dissatisfy them that expect my answer to these that are already sent to me."[47]

Newton was not about to attack Oldenburg directly, for he and Collins represented the savant's only important link to the wider scientific community. But Newton could punish in other ways. By all indications, the Lucasian Professor had planned to make his reply to Hooke the occasion for unveiling the full range of his optical studies, including the phenomena of colors in thin plates—but not any longer. He now retreated, placing the blame on those who by their criticism would establish limits to his freedom of communication: "[U]pon the receipt of your letter I deferred the sending of those things which I intended, and have determined to send you alone a part of what I prepared."[48] What Newton so vengefully chose to deny the world for more than three decades was nothing less than the *Opticks* itself, the consummate product of his experimental genius. Looking back on this decision four years later, even as the controversy on light continued unabated, he wrote: "Then frequent interruptions that immediately arose deterred me from [publication] and caused me to accuse myself of imprudence, because, in hunting for a shadow hitherto, I had sacrificed my peace, a matter of real substance."[49] Not only was his anonymity forfeit, but the offspring of years of contemplative ecstasy had been defiled at the hands of lesser men.

Newton sincerely believed that the travail of the past few months would be ended if only he ceased to publish, and to a degree he was right. In the classical sense, the passion for creation that was a major part of his unique excellence (*arete*) was proving itself a tragic flaw (*hamartia*) when employed in defense of his discoveries. What had raised him to the very heights of scientific understanding at one moment plunged him into an emotional abyss at another. No clearer illustration of this is to be found than his reply to Hooke of June 11, 1672, where brilliance and hatred intertwined to form a double helix of incongruous proportions.

The period preceding publication was always a time of terror in which Newton experienced the breakdown of his self-image, a time when the limits of his sanity were severely tested. To bolster his fragile ego and to maintain a sense of equilibrium, he drafted multiple copies of the reply, gradually whipping his private sea of anger into a white-hot foam. In the cover letter that accompanied the paper he informed Oldenburg "that I have industriously avoyded ye intermixing of oblique & glancing expressions in my discourse." Instead, Newton opted for the foot soldier's bludgeon, preferring to dispatch the enemy with a series of crushing body blows rather than lightly wound him with the gentleman's foil. To this end he chose to reject Oldenburg's warning against using Hooke's name, hoping "it will be needlesse to trouble the R. Society to adjust matters." While he did promise to "readily give way to ye mitigation of whatsoever ye Heads of ye R. Society shall esteem personal,"[50] a betrayal perhaps of a bad conscience, one must question his sincerity on this point, since there are something like thirty direct references to Hooke, nearly one for each year the *Opticks* was to be delayed. Hence Newton was not only challenging a critic, he was daring the London virtuosi to back up their implied threat of censorship, knowing full well that if they refused, Hooke's humiliation would be all the greater. Should the tide flow in the opposite direction, he would have yet another excuse to withdraw from the public eye.

Newton began in a manner no less condescending than the one assumed by Hooke in his critique of the optical studies: "I must confesse at ye first receipt of these Considerations I was a little troubled to find a person so much concerned for an *Hypothesis*, from whome in particular I most expected an unconcerned & indifferent examination of what I propounded." Then came the real attack:

> The first thing that offers it selfe is less agreable to me, & I begin with it because it is so. Mr. Hook thinks himself concerned to reprehend me for laying aside the thoughts of improving Optiques by *Refractions*. But he knows well yt it is not for one man to prescribe Rules to ye studies of another, especially not without understanding the grounds on wch he proceeds. Had he obliged me by a private letter on this occasion, I would have acquainted him with my successes in the tryalls that I have made of that kind, wch I shall now say have been less then I sometimes expected, & perhaps lesse then he at present hopes for.[51]

He had not, as Hooke charged, given up on the refracting telescope, only those "of ye ordinary construction, signifying that their improvement is not to be expected from ye *well figuring* of Glasses as *Opticians* have imagined." To this end Newton "examined what may be done not onely by *Glasses alone,* but more especially by a *complication of divers successive*

Mediums," including glasses in combination and crystals with water and other fluids inserted between them.

He turned next to Hooke's criticisms of the theory of light: "And those consist in ascribing an Hypothesis to me wch is not mine; in asserting an Hypothesis wch as to ye principall parts of it is not against me; in granting the greatest part of my discourse if explicated by that Hypothesis; & in denying some things the truth of wch would have appeared by an experimentall examination." Newton gritted his teeth and admitted that he had argued for "the corporeity of light, but I doe it without any absolute positivenesse, as the word *perhaps* intimates, & make it at most but a very plausible consequence of the Doctrine, and not a fundamentall supposition." He went on to show that indeed it did not matter which mechanical hypothesis one might subscribe to, whether corpuscles, waves, or some other, the doctrine of colors remained unaffected. He next called Hooke's bluff on the critical matter of experimentation: "I see no notice taken . . . that if any colour at the Lens be intercepted, the whitenesse will be changed into other colours." Thus, "if there be yet any doubting, tis better to put the event on further *circumstances* of the *experiment,* then to acquiesce in the possibility of any Hypotheticall Explication."[52]

And so it went, the vicious but brilliantly struck blows raining down from page after merciless page; the keen, masterful mind, its force and subtlety of reason flawed by an incurable cynicism and ingrained distrust of all mankind. Throughout Newton treated Hooke as *agent provocateur,* putting the lowest interpretation on his every assertion, refusing to believe that he was moved by anything but the basest of motives.

VI

That bad blood existed between Robert Hooke and Henry Oldenburg has long been known. Its origins are thought to date from 1675, when Huygens applied for a patent on a spring-balance watch. Hooke, an expert in the field of horology, launched a vehement protest, claiming that the balance spring had been invented by none other than himself in 1658. Whether this was true or simply another case of Hooke's creative intuition having outstripped his powers of application is impossible to say, but Oldenburg took Huygens's part in the dispute, something for which Hooke never forgave him. He attacked Oldenburg in the postscript to *Helioscopes,* and Oldenburg retaliated in kind in the pages of the *Philosophical Transactions.* The Royal Society itself was divided, but a majority of the Council supported its Secretary, causing Hooke seriously to contemplate resigning. On November 8, 1675, he made the following entry in his *Diary:* "Saw the Lying Dog Oldenburg's *Transactions.* Resolve to quit all employments and seek my health."[53]

Oldenburg placed considerable, if not excessive, faith in the exercise of critique and rebuttal to elicit scientific truth. Unless someone made a special request for secrecy, he regarded virtually everything received and discussed at the regular meetings of the Royal Society as public information. And he took special pains to communicate the substance of the meetings to absent Fellows, whether native, like Newton, or foreign, like Huygens and Hevelius. Oldenburg attended a Council meeting on the morning of June 12 when, thanks to the diligence of the Cambridge carrier John Stiles, Newton's reply to Hooke, dated the previous day, was already in his hands. Because the minutes contain only three terse lines, all pertaining to the printing of a new book by the Italian anatomist Marcello Malpighi, there is no way of knowing if the strident attack on Hooke was brought to the attention of the Council members, but that very afternoon when the Society convened, Oldenburg "produced Mr. Newton's answer to Mr. Hooke's considerations upon his discourse on light and colours; which answer was read in part."[54]

Hooke, who was present at the meeting, must have been thunderstruck, and so must the other assembled Fellows. Picture the downturned eyes, the uneasy shifting in chairs, but most telling of all the strained silence that enveloped the room at the presentation's end. Perhaps because the lengthy paper was read only in part, some care was exercised to eliminate certain of its more repugnant passages. When Oldenburg published the text in the November *Transactions,* he frequently substituted the term "considerer" for Hooke's name and attempted to soften its harsher tones, albeit with little success. All the while, Hooke's arrogant but less abusive critique remained private, ignored by Oldenburg, thus sowing the seeds of future friction between the two men.[55] Hooke was allowed to present a brief rebuttal to Newton's paper at the meeting of June 19, after which he "was desired to make more experiments . . . for a farther examination of Mr. Newton's doctrine."[56] But his longer reply, addressed to Lord Brouncker, President of the Society, received no hearing and never appeared in the *Transactions.*[57] He who had been chosen the Royal Society's first Curator of Experiments must have believed himself forsaken by those who sought the favor of an upstart and rarely seen Cambridge professor who was yet to set foot inside Gresham's learned halls.

While the drama in London was being played out, Newton, as if attempting to flee farther from the stage, left Cambridge for the first time in more than a year. He paid a visit to Bedfordshire for the purpose of examining the lands supporting his professorship and then, after briefly returning to Trinity, went to Woolsthorpe on June 18. It proved impossible for him to concentrate on other matters, however, and he wrote Oldenburg a short note, requesting that he print nothing more concerning the theory of light "before it hath been more fully weighed." By July 6 he had moved on to a Mrs. Arundell's house at Stoke Park in Northamptonshire, from where he asked Oldenburg to suspend the printing of Pardies's second letter.[58] Oldenburg politely responded that he was unable to comply, for the paper was "already in ye presse."

In this same letter Newton sought to preempt further arguments of the kind raised by Hooke and Pardies and to focus scientific attention on his experimental work alone:

> I cannot think it effectuall for determining the truth to examin the severall ways by wch Phaenomena may be explained, unlesse where there can be a perfect enumeration of all those ways. You know the proper Method for inquiring after the properties of things is to deduce them from Experiments. And I told you that the Theory wch I propounded was evinced to me, *not by inferring this thus because not otherwise,* that is not by deducing it onely from a confutation of contrary suppositions, but *by deriving it from Experiments concluding positively & directly.* The way therefore to examin it is by considering whether the experiments wch I propound do prove those parts of the Theory to wch they are applied, or by prosecuting other experiments wch the Theory may suggest for its examination.[59]

There is a persistent precept in science that if a complex world is to be reduced to human dimensions one must not multiply hypotheses unduly and without reason. This principle of parsimony was first enunciated by the thirteenth-century English scholastic William of Occam. "What can be done with fewer [assumptions] is done in vain and with more," he wrote. No scientist of his or any other time has wielded "Occam's razor" with greater flair than Newton. Yet right as he was, by having to pit a theory with which he had lived on intimate terms for a few years against centuries of tradition, Newton placed himself in a position similar to Galileo's during the 1620s. What had become common sense to them embodied the very antithesis of reason to almost everyone else. Neither the telescope nor the *experimentum crucis,* powerful tools of modern science that they are, could begin to make a clean sweep of the field against such odds. Thus did both men learn a most bitter lesson: to demonstrate a great truth and still not have it accepted is the heaviest burden an original thinker can bear.

Chapter Eight

"I Desire to Withdraw"

I have not loved the world, nor the world me.
Byron

I

Not until 1689, when Newton was in his forty-seventh year, did Godfrey Kneller paint the scholar's first and most famous portrait. Almost all that is known of his physical appearance as a young man is that he was of about medium stature and had turned prematurely gray by the age of thirty. Wickins half teasingly told him it was "ye Effect of his deep attention of Mind," to which Newton jestingly replied that he experimented so frequently with quicksilver "as from thence he took so soon the Colour." (The Wren Library at Trinity College still possesses what is purported to be a lock of Newton's silver hair.) Wickins was also a frequent witness to Newton's famed "forgetfullness of his food, when intent upon His Studies; And of his Rising in a pleasant manner wth ye Satisfaction of having found out some Proposition without any concern for or seeming want of his Nights Sleep wch he was sensible he has lost thereby." Nor did the recurrent bouts of hypochondria escape Wickins's notice: "He sometime[s] suspected Himself to be inclining to a Consumption, & ye Medicine He made use of was ye Lacatellus Balsam w^{ch}, when he had composed Himself, He would now & then melt in Quantity ab^t a Q^r of a Pint & so drink it" in order to bring on a heavy sweat.[1] Newton's formula for this repugnant concoction has survived: turpentine, rose water, beeswax, olive oil, sack, red sandalwood, and oil of St. John's wort. To make matters worse, the sack—which even by the standards of a confirmed teetotaler must be considered among the most

palatable of the ingredients—was poured off before the preparation was imbibed. No wonder Newton thought the brew equally useful for external application to "green wounds" and the bite of a mad dog.[2]

Yet it is difficult to fault the individual who chose to play physician to
himself during the seventeenth century. There was not a single disease that
could be cured by the doctors of Newton's day, something very much on
Dryden's mind when he wrote of "the malice of their art." Every disorder,
serious or slight, was attended to by bleeding, cupping, purging, sweating,
or a combination thereof. It is remarkable that the medical profession survived and got away with so much without raising a hue and cry from a
dangerously victimized public. Though Newton's pharmacopeia is enough
to make the stout of stomach queasy, it was hardly atypical considering the
fact that every known plant, extracts of every known metal, and every conceivable diet were prescribed on no sounder basis than trial and error, usually resulting in precisely that sequence. Nor could Newton hold a candle to
Robert Hooke, a hypochondriac who enjoyed every day of ill health that
came his way. Hooke's *Diary*, filled with references to headaches, insomnia,
noises in the head, catarrh, and dizziness, begins with the words, "Drank
iron and mercury," and, so far as his health is concerned, continues
downhill from there. The Lacatellus Balsam favored by Newton seems
almost ambrosial compared to Hooke's insatiable appetite for all the
potable metals, in addition to "senna, aloes, rhubarb, wormwood,
laudanum, flowers of sulphur, resin of jalop, Aldergate cordial, Dulwich
water, Epsom water, North Hall water, lignum vitae, sal ammoniac, hagiox,
and others."[3] Newton's exemption from serious illness is largely attributable to an iron constitution wisely protected against the rampant
phlebotomy of seventeenth-century medical practice and reckless dosing à
la Robert Hooke.

To anyone meeting the Cambridge don for the first time, especially a
student, he must have appeared much older than he was. Newton was only
in his thirtieth year in 1672, but the long gray hair and absorbed
countenance gave the impression of someone with a great weight of experience behind him. Stiff and reserved as he was, one imagines him as
reluctant to offer his hand as was Washington, afraid of contracting some
disease but fearing even more the human touch. If not understood by his
contemporaries at Trinity, Newton was nevertheless held in awe by them,
as if, as Emerson remembered of Thoreau, he spoke when silent and was
present after he had departed. He took an occasional turn in the Fellows'
gardens, now and then sketching out diagrams in the freshly raked gravel
with a stick. These, out of respect, the other Fellows spared by walking
round them, so that they sometimes remained for a good while.[4] (In later
life he wistfully pictured himself as a child playing on a vast seashore of undiscovered truth, a beach with *his* footprints only.) And yet he brought no
sense of high purpose or passion to his teaching as he did to his research and

writing, with their prospect of worldly immortality. It was likely a raspy voice, easily strained from lack of use, that on occasion drifted upward from a bent, red-robed figure behind the lectern. Half-consciously he would read his notes in Latin while a glazed look came over the blue, somewhat protuberant eyes. But what could have been mistaken for boredom was in reality the abstracted look of a man engaged in an intense inner quest. For beneath this preoccupied exterior dwelled the troubled, insecure figure who constantly labored beyond all reason to prove himself. Where there was immense fulfillment and ecstasy we have also discovered inadequacy and self-scourging, a considerable but skillfully masked appetite for competition (if conducted on his own terms), and a shocking lack of tolerance toward his brother scientists—for he must ever decree himself the best. If Newton was the Moses of seventeenth-century science, he was also its Cain, the perpetrator of intellectual fratricide.

II

During 1672 John Ivory, a citizen of Cambridge, published what was the seventeenth-century equivalent of the modern university catalog. The entry for Newton's college was worded as follows:

> TRINITY COLLEGE—A Master, sixty Fellows, sixty-seven Scholars, three publick Professors, four Conducts, thirteen poor Scholars, a Master of the Choristers, six secular Clerks, twenty Beads-men, besides Exhibitioners, Officers and Servants of the Foundation, with many other Students being in all 400.

Of the other fifteen foundations, only St. John's, with a population of 372, came close to Trinity in size. Next largest were Christ's (206) and Emmanuel (170). The entire University, including students, officers, and servants of the foundations, numbered 2,522.[5]

Not only was Trinity the largest and perhaps the most visually appealing of the several colleges; it was also first in the King's eyes, just as it had been for each of his predecessors dating back to its founding by Henry VIII, a century and a quarter before. Charles II had paid his first official visit to Cambridge on October 4, 1671. The next day's *London Gazette* reported that, after dining, the King abandoned his coach and marched through the streets to the cheers of the entire student body, arranged according to their orders and degrees. As he passed by the marketplace, the fountain "ran with Claret wine." He was received at the schools by the Duke of Buckingham, Chancellor of the University, and presented with a "fair Bible." Following visits to the public library and its counterparts at King's, St. John's, and Trinity, Charles was entertained at a great dinner given by Trinity's Masters of Arts. Afterward he conferred the honor of knighthood upon

Charles Caesar and watched a comedy, with which his Majesty expressed himself to be well pleased. For the first time in his life Newton had the opportunity to lay eyes on a royal personage, and even his legendary sang-froid must have given way to an elevated heartbeat beneath his scarlet gown. Kings, too, were flesh and blood, not sainted martyrs of the kind he remembered from an adolescent fascination with *Eikon Basilike*. How gratifying it was to this former country boy to learn, only four months later, that his little telescope had been touched by the appreciative hands of royalty, hands which he himself had seen gesture, even as his name was on royal lips, which he had heard speak.

Newton's next contact with an important figure of the Restoration monarchy was as a participant rather than a mere spectator. On July 11, 1674, the King addressed a letter to the University removing the scheming George Villiers, Duke of Buckingham, from the office of Chancellor and "recommending" as his successor James, Duke of Monmouth, Charles's favorite, but illegitimate, son by his mistress while in exile, the beautiful Lucy Walter. At the election held three days later, Monmouth received "197 sufferages, no man giving a vote for any other person."[6] The installation was scheduled to take place at London's Worcester House on September 3.

As one of many Masters of Arts chosen to take part, Newton left for London on August 28 and remained in the capital until September 5.[7] The entourage, numbering more than 480 persons, gathered at Darby House on the appointed day, awaiting Monmouth's pleasure. The Prince's summons came at four in the afternoon. The procession formed behind the King's mounted Life Guard, which made way through the large crowd straining to catch one last glimpse of dying postmedieval splendor. The gates of Worcester House, destination of the dignitaries, were guarded against the assembled throng by musketeers, and once inside the Masters of Arts formed a double line through which the Vice-Chancellor and Doctors passed. Monmouth, handsome, almost effeminate in appearance, met them at the door and listened with pleasure as the Vice-Chancellor expressed to those assembled "the singular content & satisfaction of the University in their Relation to a Personage whose virtues were as eminent as his place & Fortune." After the investiture—and another round of speeches bordering on the obsequious—came the reception and dinner, complete with "excellent Musick both vocall & Instrumentall."[8]

Even had he wanted to, which at this juncture seems most unlikely, Newton could not have attended a meeting of the Royal Society, for it had gone into adjournment June 18 and did not reconvene until November 12. And if he took advantage of the opportunity to meet individually with any of the London virtuosi, no record survives. Oldenburg would have been his most likely contact, but Hooke, who saw the Royal Society Secretary on September 1 and 3, makes no mention of their having discussed Newton by

William Stukeley's early eighteenth-century drawing of the manor house at Woolsthorpe. *(By permission of the Royal Society.)*

The manor house at Woolsthorpe as it appears today with a reputed descendant of the famous apple tree in the foreground. *(Photograph by the author.)*

Isaac Barrow (1676) by David Loggan. (*By permission of the National Portrait Gallery, London.*)

Trinity College in the late seventeenth century with Newton's chambers in the lower right-hand corner between the Great Gate and Trinity Chapel. *(From David Loggan, Cantabrigia Illustrata, 1690.)*

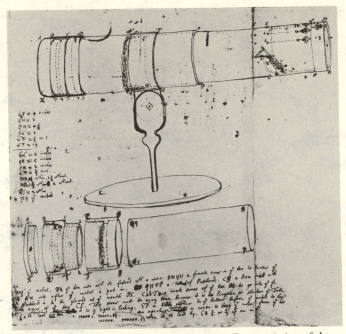

Newton's drawing of his reflecting telescope. *(By permission of the Syndics of Cambridge University Library.)*

The reflecting telescope given to the Royal Society by Newton in 1671. *(By permission of the Royal Society.)*

PHILOSOPHIÆ
NATURALIS
PRINCIPIA
MATHEMATICA.

Autore *JS. NEWTON*, *Trin. Coll. Cantab. Soc.* Matheseos
Professore *Lucasiano*, & Societatis Regalis Sodali.

IMPRIMATUR.
S. PEPYS, *Reg. Soc.* PRÆSES.
Julii 5. 1686.

LONDINI,

Jussu *Societatis Regiæ* ac Typis *Josephi Streater.* Prostat apud
plures Bibliopolas. *Anno* MDCLXXXVII.

The title page from Newton's *Principia.*

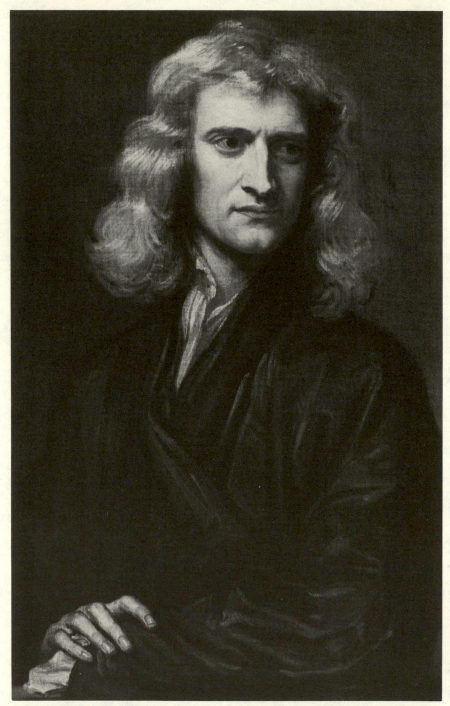

Newton's first portrait (1689) at age forty-six by Sir Godfrey Kneller. *(By permission of Lord Portsmouth and the Trustees of the Portsmouth Estates.)*

Above left, Edmond Halley as a young man by Thomas Murray. *(By permission of the Royal Society.) Above right*, Charles Montague, the first Earl of Halifax (c. 1703-1710) by Sir Godfrey Kneller. *(By permission of the National Portrait Gallery, London.) Left*, Nicolas Fatio de Duillier by an unknown artist. *(By permission of the Bibliothèque Publique et Universitaire, Geneva.)*

A coining press in operation. (*From Denis Diderot*, Encyclopédie.)

The edging machine invented by Peter Blondeau. (*From Denis Diderot*, Encyclopédie.)

name. Still, there is reason to believe that Hooke was aware of his nemesis's presence. Hooke rarely noted events in which he was not somehow personally involved, yet the last line of his diary entry for September 3 reads: "Duke of Monmouth treated Cantabrigians at Worcester house."[9]

Newton found his outlook and style of living at Cambridge altered to some extent by the events of the previous year. The King, who certainly had his cynical side, nonetheless proved capable of recognizing and rewarding the talent in his realm. After three years in London, where Charles had heard him preach as one of his royal chaplains, Isaac Barrow—mathematician, classical scholar, virtuoso, theologian, and good friend of Isaac Newton—was returned to Trinity, where he occupied the prestigious Master's Lodge until his untimely death in 1677. Had he lived longer, this man, whom the King called the best scholar in England, might have made significant headway in combating the ennui and sloth that enervated not only his own college but the University as a whole.

Nevertheless, Barrow's tenure proved far from futile, for his vision and zeal were the catalysts that produced the most magnificent of Trinity's buildings, the Wren Library. As the story goes, Barrow was unable to convince an assembly of University officials that the risk of constructing a much needed theater should be undertaken on the basis of subscriptions alone. "Piqued at this pusillanimity," he stormed out of the meeting declaring that he would lay out the foundations with his own hands, together with the ground plan of a "stately library." He proved as good as his word: That very afternoon, in what must have been a curious sight, the Master of Trinity College, assisted by his gardeners and servants, staked out the ground on which the library now stands.[10] His pique aside, he had a considerable ace up his ill-fitting sleeve in the figure of Christopher Wren, the scientific and architectural genius. Barrow persuaded his friend of London days to render his much sought-after services gratis. The finished result, which Barrow did not live to see, is a true masterpiece of Restoration architecture. It consists of two levels and two classical orders: an arcade below with a flat ceiling, supported by Doric columns, and the principal story, decorated with three-quarter Ionic columns of elegant proportions. Along the white plastered walls of the exquisite interior stand the tall, hardwood shelves, also of Wren's design. Grinling Gibbons, the master carver, was commissioned to fashion the delicate ornamental woodwork on the shelf ends, comprising limewood coats of arms of the principal subscribers. Newton, who warmly applauded Barrow's return as Master, contributed the sum of £40 toward the project, not enough to earn him a place of honor within its precincts even had he then possessed the requisite heraldry, which he did not. Still, it is Barrow's name that has since faded into the margin of history, while Newton's papers and personal library are today counted among the Wren's greatest treasures.

Newton was in Woolsthorpe for three weeks in the spring of 1673. He would not return for two and one-half years, further evidence, perhaps, that things were somewhat less than idyllic on the home front.[11] With the inscrutable Wickins in tow, he appears to have occupied the rooms in E–4 Great Court later in the year, and there he continued to live until his departure for London in 1696.[12] These quarters, as previously noted, are located on the second floor (called the first in England) between the Great Gate and the Chapel, on the side adjoining the Gate. The rooms below were later occupied by Macaulay and Thackeray. The advantages of this arrangement were considerable: The garden (which still exists and contains a commemorative apple tree) was at the disposal of the occupant; the "observatory," such as it was, was then housed atop the Great Gate;[13] and, most important, there was sufficient space to pursue the alchemical experiments so critical to Newton during the 1670s and 80s.[14] Though he entertained infrequently, Newton acquired new furnishings befitting the dignity of a university professor: sixteen chairs, ten cushions, a writing desk, a chest, and two Spanish tables with neat's leather carpets.[15] In 1747, half a century after Newton departed Cambridge and twenty years after he died, Richard Cumberland, the grandson of Richard Bentley, the gifted but tyrannical Master of Trinity from 1700 to 1742, came up as a freshman. Cumberland was shown Newton's old rooms by their current occupant, Richard "Our Hat" Walker of *Dunciad* fame. Previous to Walker, Daniel Hopkins and Nathaniel Hanbury had also lived there, yet the young visitor testified that every relic of Newton's "studies and experiments were respectfully preserved to the minutest particular, and pointed out to me by the good old Vice-Master with the most circumstantial precision."[16]

III

Having long since despaired of winning Robert Hooke's approval for his theory of light, Newton, with Oldenburg as his agent, had succeeded in recruiting only one important ally from among the ranks of the international scientific community. Now, in early 1673, even this minimal support was showing every evidence of imminent collapse. After first reading Newton's paper in the spring of 1672, Christiaan Huygens had characterized its contents as "highly ingenious." A few months later Huygens waxed almost as enthusiastic: "I confess that up to now," he wrote Oldenburg, "it has seemed very probable to me, and the *experimentum crucis* (if I understand it aright, for he has written a little obscurely) confirms it very well." But by September of that year Huygens's confidence in the doctrine of colors had begun to come unraveled. Communicating with Newton, as always, through Oldenburg, his protective filter, Huygens wrote, "What you have put in your last Journals from Mr. Newton confirms still further his doctrine of colors. Nevertheless the thing could very well be otherwise, and it seems to me that he ought to content himself if what he has advanced is accepted as a very likely hypothesis."[17]

It was bad enough that Hooke had criticized the *experimentum crucis* without even deigning to repeat it for himself. Now Huygens, who had so recently testified to its validity, refused to accept it for what it was, choosing instead to relegate it to the realm of hypothesis. A priori principle had again impinged upon scientific fact—and won. As was the case with Newton's other critics, Huygens could not entirely separate himself from the scholastic attitude against which he was rebelling. In vain, Newton had repeatedly explained that he was not writing about hypotheses but of experimentally established law, which could be rendered invalid only by repeating his experiments and proving them incorrect. His was such a new conception of the function of science that no one among his contemporaries completely grasped it; any lingering illusions he might have entertained on this score were totally dispelled by the contents of Huygens's fourth and most detailed commentary on the doctrine of colors. It was forwarded to Newton in a letter from Oldenburg dated January 18, 1673, and appeared in the July *Philosophical Transactions* as a scarcely disguised extract of a letter "by an ingenious person from Paris:"

> I have seen, how Mr. *Newton* endeavours to maintain his new Theory concerning *Colours*. Me thinks, that the most important Objection, which is made against him by way of *Quaere*, is that, Whether there be more than two sorts of Colours. For my part, I believe, that an *Hypothesis*, that should explain mechanically and by the nature of motion of Colors *Yellow* and *Blew*, would be sufficient for all the rest, in regard that those others, being only more deeply charged (as appears by the Prisms of Mr. *Hook*) do produce the dark or deep-Red and Blew; and that of these four all the other colors may be compounded. Neither do I see, why Mr. *Newton* doth not content himself with the two Colors, Yellow and Blew; for it will be much more easy to find an *Hypothesis* by Motion, that may explicate these two differences, than for so many diversities as there are of other Colors. And till he hath found this Hypothesis, he hath not taught us, what it is wherein consists the nature and difference of Colours, but only this accident (which certainly is very considerable,) of their *different Refrangibility*.[18]

Huygens's third letter had reached Cambridge during the autumn of 1672, a time when Newton chose to curtail his correspondence on the theory of light. Even though his frustration must have been total, he had somehow checked his temper and remained silent. It appeared that he might take the same tack on this occasion, for two months went by during which Oldenburg heard nothing. Then, in early March 1673, the Royal Society Secretary received one of the briefest and surely the most startling letter of their tortured correspondence. "As for Monsieur Huygens Observations I conceive that they are but the abstract of a private letter sent to you, & therefore concern not me to take notice of them." Then, this: "Sr I desire that you will procure that I may be put out from being any longer fellow of ye R. Society. For though I honour that body, yet since I see I shall neither profit them, nor (by reason of this distance) can partake of the advantage of their Assemblies, I desire to withdraw."[19]

Oldenburg's reply is lost, but we nonetheless have a good idea of what it contained. On the back of Newton's letter of withdrawal the Secretary penned the following synopsis: "... represented to him my being surprised at his resigning for no other cause, than his distance, wch he knew as well at the time of his election, offering wthall my endeavor to take from him ye trouble of sending hither his qterly paymts & without reflection."[20] Oldenburg, of course, was well aware that Newton's decision to resign had nothing to do with his inability to join regularly in the meetings of the Royal Society. His reluctance to correspond with anyone but Collins and himself, let alone visit London more frequently, was ample proof of that. Recognizing the resignation for what it was—a plea for emotional support—Oldenburg subtly shifted psychological grounds. What better way, without directly confronting the troubled genius with the obvious, than an offer to cancel his quarterly dues?

That Oldenburg acted wisely becomes apparent in a letter Newton wrote to Collins on May 20. The real reason for his decision to withdraw is here revealed: "I suppose there hath been done me no unkindness, for I met wth nothing in yt kind besides my expectations. But I could wish I had met with no rudeness in some other things. And therefore I hope you will not think it strange if to prevent accidents of that nature for ye future I decline that conversation wch hath occasioned what is past."[21] The letter was shown to Oldenburg, who, now that Newton had permitted his true feelings to surface, sought to speak to them more directly through expressions of high praise and deep affection:

> I could heartily wish, you would passe by the incongruities, yt may have been committed by one or other of our Body towards you, and consider, that hardly any company will be found in the world, in wch there is not some or other yt wants discretion. You may be satisfied, that the Body in general esteems and loves you, wch I can assure you of, *fide viri boni*.[22]

Newton's reply came two weeks later, and he was not wholly placated:

> The incongruities you speak of, I pass by. But I must, as formerly, signify to you, yt I intend to be no further sollicitous about matters of Philosophy. And therefore I hope you will not take it ill if you find me ever refusing doing any thing more in yt kind, or rather yt you will favour me in my determination by preventing so far as you can conveniently any objections or other philosophical letters that may concern me.[23]

Despite his frustration, Newton chose not to press the issue of his resignation, and Oldenburg, doubtless having breathed a sigh of relief, discreetly let the matter drop without further comment. This was not the first time Newton threatened to sever his slender ties to the outside world, nor would it be the last.

As if the tension of the moment were not sufficient, Newton had begun to entertain second thoughts about permitting Huygens's latest commentary to go unanswered. Finally in April, nearly three months after receiving the critique, he reversed himself and drafted a reply in the familiar form of a letter to Oldenburg, which duly found its way into the pages of the *Philosophical Transactions.*[24] In his cover letter to Huygens, a somewhat embarrassed Royal Society Secretary thought it necessary to apologize for Newton's behavior on two counts. First, he attributed the considerable delay to Newton's having been absent from Cambridge "for several weeks," when in fact he had been away for only twenty-one days. Second and more important, Newton, in what was becoming an all too familiar pattern, again resorted to the blunt and aggressive tone of one who had become prey to a siege mentality. "I can assure you," Oldenburg wrote, "that Mr. Newton is a man of great candor, as also one who does not lightly put forward the things he has to say." He trusted Huygens would look beyond a newcomer's indiscretion, "that you will read it not without pleasure, and that it will give you the opportunity of meditating further on this fine and important matter."[25]

While less abrasive and self-righteous, Newton's paper is reminiscent in style of the June 1672 attack on Hooke:

> It seems to me that M. Hugens takes an improper way of examining the nature of colours whilst he proceeds upon compounding those that are already compounded. . . . Perhaps he would sooner satisfy himself by resolving light into colours as far as may be done by Art, and then by examining the properties of those colours apart, and afterwards by trying the effects of reconjoyning two or more or all of those, & lastly by separating them again to examine wt changes that reconjunction had wrought in them. This will prove a tedious & difficult task to do it as it ought to be done but I could not be satisfied till I had gone through it. However I onely propound it, and leave every man to his own method.

"Nor," he continued, "is it easier to frame an Hypothesis by assuming onely two originall colours rather then an indefinite variety," unless one assumes that there are only two types of corpuscles or waves and two fixed speeds at which they move. Since no one questions "the indefinite variety of waves of the sea or sands of the Shore," why should the properties and movement of light be any less varied?

> But to examin how colours may be thus explained Hypothetically is besides my purpose. I never intended to show wherein consists the nature and difference of colours, but onely to show that *de facto* they are originall & immutable qualities of the rays wch exhibit them, & to leave it to others to explicate by Mechanicall Hypotheses the nature & difference of those qualities; wch I take to be no very difficult matter.[26]

Oldenburg was not wrong in his assumption that Newton's reply would provide Huygens with ample grist for further meditation. However, if he truly hoped that Huygens would look upon its contents "not without pleasure," he was bound to be disappointed. It seems doubtful that the Dutch savant had ever before been lectured in so blunt a manner by one of his peers, and he was not about to countenance it now. Replying to Oldenburg on May 31, Huygens wrote: "Touching the Solutions given by M. Newton to the scruples by me propos'd about his Theory of Colors, there were matter to answer them, and to form new difficulties; but seeing that he maintains his opinion with so much concern,[*] I list not to dispute."[27]

At some earlier point, perhaps before he received Newton's criticism of his fourth and final letter, Huygens had dispatched presentation copies of his newly published *Horologium Oscillatorium* to a dozen men of science in England and Scotland, including Newton. The books' shipment, as was common, had been delayed, and their arrival chanced to coincide with Oldenburg's receipt of Huygens's letter of refusal to debate any more. Newton received his volume early in June, and his subsequent actions suggest that he experienced some measure of guilt for having been so hard on its well-intentioned author. His next letter to Oldenburg contained not one but two requests that Huygens be thanked for his generosity. Assuming a more moderate tone than before, Newton once again informed Huygens of the best way to test the validity of his theory.[28] With that, Huygens heard no more from him, which hardly seems surprising, for Newton's final comments were contained in the very letter requesting that Oldenburg relieve him of further philosophical considerations. Unless some of his letters from this period are missing, Oldenburg, who by now had every reason to be suffering from acute dyspepsia, received nothing from his Cambridge correspondent during the next year and a half.

IV

Apart from the visit to London to participate in Monmouth's installation as Chancellor of Cambridge University and a change of living quarters, the record of Newton's activities between mid-1673 and early 1675 is a virtual blank. Nor does the pitifully meager correspondence shed much light on what he was about. So far as is known, he wrote only a single letter during the last six months of 1673, a brief note to Collins indicating that he was expecting a visit from James Gregory, who was returning to Scotland via Cambridge after an extended stay in London.[29] If, in fact, the anticipated meeting did take place, which is by no means certain, it brought together

---

*Actually "*chaleur,*" more frequently translated as "heat" or "warmth."

Great Britain's two most gifted theoretical mathematicians, marking the first time Newton had come face to face with a virtuoso of the first rank.

The curtain of silence promptly descended for eight more months, to be broken only in June of 1674, after Collins sent Newton a copy of Robert Anderson's *Genuine Use and Effects of the Gunne demonstrated.* Newton called the work "ingenious" but seriously questioned Anderson's claim that a bullet "moves in a Parabola." This, he pointed out, would occur only if its horizontal velocity were uniform. Employing his analytical method of extracting the roots of an equation, a technique first exemplified in *De Analysi,* Newton was able to show that a projectile's speed is subject to constant change. However, he neither wished to have his current labors interrupted nor wanted to risk still another controversy: "If you speak of this to ye Author, I desire you would not mention me becaus I have no mind to concern my self further about it."[30] Try as he might, Collins was unable to engage him in a further exchange of letters. Five years earlier, after a fruitless attempt to convince Newton that he should publish the calculus, Collins had informed Gregory that "I desist and doe not trouble him any more." When, on September 30, 1675, Collins wrote the following to Gregory, history virtually repeated itself: "I have nott writt to or seene [Newton] this 11 or 12 months, not troubling him as being intent upon Chimicall Studies and practices, and both he and Dr. Barrow & c beginning to thinke mathcall Speculations to grow at least nice and dry, if not somewhat barren."[31] Indeed, the letter on gunnery brought to a virtual conclusion Newton's fitful and often turbulent mathematical correspondence with the persistent but, in the end, deeply disappointed John Collins.

In Greek mythology, Hydra, the monstrous offspring of Typhon and Echidna, was endowed with the extraordinary capacity to regenerate two heads for each one severed from its body. Newton, steeped from his youth in classical studies, must have known that Hercules, in carrying out one of the twelve labors of immortality demanded by Hera, finally dispatched the horrible beast by cauterizing each neck after severing its head. As if in emulation of the Greek champion, he had attempted to eradicate the multiple sources of criticism directed at his doctrine of light with the heat of impassioned rejoinders—and to seal his own psychic wounds in the process. Hooke had been humiliated, Pardies muted, and Huygens had chosen to retire from the field. Yet the monster lived—if only in Newton's mind—and thereby hangs the tale.

On September 26, 1674, one Francis Hall (Linus or Line as he called himself), Professor of Hebrew and Mathematics at the College of English Jesuits at Liège, drafted a letter to Oldenburg that proved sharply critical of Newton's experimental work. Whatever his intellectual merits as a young man, Linus, who was by now almost eighty, had crossed the boundary separating a dignified old age from the facile musings of senility. He too had experimented with prisms and, like Newton, had produced an elongated spectrum on the wall of his darkened chamber; "but [I] never found it so when the sky was cleare and free from clouds, neere the sunne." From this "I conclude that the spectrum, this learned Author saw much longer than broad, was not affected by true sunne beames, but by rays proceeding from some bright cloud.... And by consequence, that the Theory of light grounded upon that experiment cannot subsist."[32]

Oldenburg could not have had a very high opinion of Linus's analysis, but by letting the matter pass he would have both compromised his neutrality as a scientific intermediary and undermined his credibility with the Liègeois. Besides, he had heard nothing from Newton since June of the previous year, and difficult though the Lucasian Professor was to deal with, it seemed a price worth paying in order to reestablish contact. The Secretary dutifully transcribed the letter from Linus and dispatched a copy to Cambridge in October. Several weeks passed before Newton bothered to pen a brief reply, which perfectly expressed his isolationist mood. "I am sorry you put your self to ye trouble of transcribing Mr. Linus's conjecture, since (besides yt it needs no answer) I have long since determined to concern my self no further about ye promotion of Philosophy." However, if the Jesuit wished to avoid the embarrassment of "slurring himself in print wth his wide conjecture," Oldenburg should refer him to Newton's second answer to Pardies in the *Philosophical Transactions,* and also "signify (but not from me) that ye experiment as it is represented was tryed in clear days."[33]

So much for the bothersome but innocuous Linus. Of more pressing concern to Newton at the moment was the realization that his fellowship at Trinity would expire within the next several months. He could, like Wickins and so many others, simply pursue the prescribed path by taking holy orders and thereby retain his sinecure at little more than the price of a gentleman's oath. But Newton nourished a healthy Puritan aversion to the taking of any pledge, especially one requiring that he uphold the doctrine of the Trinity, a religious principle with which he profoundly disagreed. According to tradition, he had already made an unsuccessful attempt to retain both is fellowship and lay status the previous year. Dr. Robert Crane, the holder of Trinity's only law fellowship, died in February 1673. Newton, along with Robert Uvedale, who was two years his senior, became a candidate for the vacancy. Barrow, who had just been appointed Master, had little choice in the matter. He decided in favor of Uvedale, "saying that Mr. U. and Mr. N. being (at that time) equal in literary attainments, he must give the fellowship to Mr. U. as senior."[34] Because this information comes from Uvedale's great-grandson, there is reason to question its accuracy. On the other hand, it is interesting to note that Uvedale had bested Newton not on merit, as one would expect in a story concocted to glorify a family ancestor, but on the credible grounds of seniority, far and away the most sacred cow of the contemporary academy.

At any rate, Newton's resolve to pursue all reasonable alternatives before submitting himself to holy orders is well documented. He left Cambridge for one of his rare visits to London on February 9, 1675. In his luggage was a petition of indulgence asking Charles II to allow him to retain his fellowship without becoming a cleric so long as he held the Lucasion Professorship. His confidence could hardly have been buoyed by the fact that Francis Aston, one of his well-connected friends at Trinity, had failed in a similar attempt just weeks earlier, despite the backing of Sir Joseph Williamson, Principal Secretary of State. However, Barrow, who took Aston for the opportunist and mediocre intellect that he was, had opposed the application. That the new Master's mordant wit had lost none of its bite is evident from the following extract from his letter of remonstrance: "Indeed a Fellowship with us is now so poor, that I cannot think it worth holding by an ingenuous person upon terms liable to so much scruple."[35]

Newton stayed in London for well over a month, awaiting action by the King on his petition. Since he had no reasonable excuse to further postpone personal contact with the resident virtuosi, he attended his first meeting of the Royal Society at Gresham College on February 18. Newton signed his name in the register and was thereby formally admitted to membership. He even made small talk with Robert Hooke, explaining his method of polishing metal with pitch.[36] Hooke kept a curious and wary eye on Newton's movements throughout his visit and noted that he also attended the following week's meeting on February 25. Indeed, Newton, who was finding that the Royal Society had a human face after all, seems to have missed none of its gatherings during his more than five weeks in the capital.

By then he had probably forgotten the ill-founded objections raised by Francis Linus to his theory of light. The tenacious old Jesuit had not taken kindly to his peremptory dismissal by Newton, however, and an even more adamant letter arrived from Liège during the latter's stay in London. The semicircular ends of the elongated spectrum, as they had been described by Newton, "are never seene in a cleare day," Linus foolishly reiterated.[37] A better opportunity to conduct a definitive test of the theory could not have presented itself, and Hooke was ordered to have the necessary apparatus in readiness for the Royal Society meeting of March 18. Unfortunately the weather proved unfavorable, and Newton, who had delayed his departure to be on hand for the demonstration, returned to Cambridge the following day. He considered Linus's second letter no more deserving of a reply than the first, and even the normally conscientious Oldenburg chose to disregard it, which upset Linus and his colleagues all the more. Aside from this mildly unpleasant occurrence, the visit proved a resounding success. Newton had beem warmly received by the Royal Society and, even more important at the moment, assured by the King's advisers that his petition to hold a fellowship without taking holy orders would soon be approved. Five weeks later, on April 27, 1675, Charles II made good their word.

V

Even during the seventeenth century seven months seemed an inordinately long time to await an answer to one's correspondence. This was especially true for the senescent Linus, who had recently become an octogenarian. Having received no reply to his February letter from either Oldenburg or Newton, the Jesuit decided that a reminder was in order. Accordingly, he wrote Oldenburg on September 11 to ask why his second letter had not been published in the *Philosophical Transactions*, as had the first: "[It] is rather to my disadvantage, and praejudice of the Truth, being apt to make the Reader conceave that Mr. Newton's Theory of light is still good; and that I sayd more agaynst the same then I could mayntayne."[38] Oldenburg must have been sorely tempted to return the letter with "Amen!" boldly scrawled beneath Linus's unconscious self-indictment, but he forwarded it to Newton instead.

Linus had admitted in his first letter that his criticism of Newton's theory was based in part on experiments conducted "neere 30 yeares ago." After reviewing what his detractor had written, Newton could only guess that this must be the continuing source of his error: "I suspect he has not tryed ye expt since he acquainted himself wth my Theory, but depends upon his old notions taken up before he had any hint given to observe ye figure of ye coloured image." Though he would have been well within his rights not to comment further, Newton chose to draft detailed instructions on how the experiment should be conducted. He added that he would be happy to perform the same before the Royal Society when next he visited London, although "if Mr. Line persist in his denyal of it, I could wish it might be tryed sooner there."[39]

As a natural philosopher, Linus had accomplished nothing worthy of Newton's respect, but his most recent communication was not without a certain transitory value. Newton, ever a prisoner of his own complexities, often revealed his primary purpose in writing a letter only near the end, and this occasion proved no exception. Unable to admit freely that he nurtured a considerable need for recognition, he used his reply to indicate a willingness to reopen the channels of scientific communication. His manner, as befitted the established pattern, was deceptively insouciant. First, the characteristic feint: "I had some thought of writing a further discourse about colors to be read at one of your Assemblies, but find it yet against ye grain to put pen to paper any more on yt subject." Then, the deft thrust: "But however I have one discourse by me of yt subject written when I sent my first letter to you about colours.... This you may command wn you think it will be convenient if ye custome of reading weekly discourses still continue."[40] As had so often happened to Collins, Oldenburg was being cleverly manipulated into assuming the role of a supplicant whose only recourse is total obeisance. Of course he wanted whatever Newton had to offer and hastened to tell him so in a letter since lost. Oldenburg's expression of gratitude for work he was yet to lay eyes on must have bordered on the effusive, however, for Newton, in mock humility, later wrote that his efforts were "not worth ye ample thanks you sent."[41] For his part, Newton savored every moment of his low drama, prolonging the tension to the last by resorting to his familiar tactic of delay. His next letter was written on November 30, two and a half weeks after the reply to Linus. Imagine Oldenburg's disappointment when he opened it, only to find a drawing and description of an ear trumpet that Newton had promised to "an ancient Gentleman I met at your Assemblies (whose name I cannot recollect)." He assured Oldenburg that he had intended to send "ye papers this week but upon reviewing them it came into my mind to write another scribble to accompany them."[42] Finally, on December 7 the Cambridge carrier John Stiles was summoned to Newton's chambers, where an unusally thick packet was handed to him for delivery to Oldenburg's residence in London.

VI

The truly great scientific mind distinguishes itself in two ways: first, in demonstrating a capacity for original thought; second, in forging a principle of unity by which both new and previously existing knowledge is synthesized and made whole. Newton gave evidence of possessing both these attributes when he began drafting the *Quaestiones quaedam Philosophicae* while an undergraduate of twenty-one. What is more, he was already being torn between the self-imposed rule that he proceed by methodically rational analysis and the compulsion to effect a grand synthesis—the scientist's supreme paean to nature. And yet the face of Newton the visionary had always been carefully concealed from public view. Only once before, in his published paper on light, had he let his guard down, and Hooke promptly took him to task for violating his own tenet against mixing hypothesis with scientific fact. So chilling was the effect that Newton refused to divulge the rest of his research, much less the sweeping philosophical principle on which it rested in his mind.

The latest visit to London had wrought a gradual change in Newton's thinking, for the high praise lavished on his work had come as a pleasant surprise. Even Hooke, misshappen and painfully stooped, seemed much less of a threat now that Newton had viewed him in the flesh. Indeed, he thought (quite mistakenly as it turned out) that the Curator of Experiments "had changed his former notion of all colours being compounded of only two Originall ones . . . & accorded his Hypothesis to this my suggestion of colours."[43] After a withdrawal spanning some two and a half years, Newton seemed prepared to risk further controversy by forwarding his most daring and complex paper to date: "An Hypothesis explaining the Properties of Light."

Newton, who had categorically rejected all hypothetical speculation on the nature of light by others, now found himself in the awkward, if not embarrassing, position of having to defend a sweeping hypothesis of his own. That he was painfully aware of this glaring contradiction is clear from his cover letter to Oldenburg: "Sr. I had formerly proposed never to write any Hypothesis of light and colours, fearing it might be a means to ingage me in vain disputes." Nor would he allow himself to be dragged into dubious battle in this instance: "I hope a declar'd resolution to answer nothing that looks like a controversy . . . may defend me from yt fear." He was making his hypothesis known only because it "would much illustrate ye papers I . . . send you."[44] In recognition, no doubt, that this looked more like a declaration of principle than an explanation of his actions, Newton went a step farther in the introduction to the paper itself. But his tortured attempt to rationalization through blame-shifting and pretended indifference only served to compound a dilemma admitting of no satisfactory solution. He finally retreated by donning the mantle of one too dignified to enter the fray, vainly hoping thus to elude the censure that would almost surely follow.

> And therefore because I have observed the heads of some great virtuoso's to run much upon Hypotheses, as if my discourses wanted an Hypothesis to explain them by, & found, that some when I could not make them take my meaning, when I spake of the nature of light & colours abstractly, have readily apprehended it when I illustrated my Discourse by an Hypothesis; for this reason I have here thought fitt to send you a description of the circumstances of this Hypothesis as much tending to the illustration of the papers I herewith send you. And though I shall not assume either this or any other Hypothesis, nor thinking it necessary to concerne my selfe whether the properties of Light, discovered by me, be explained by this or Mr. Hook's or any other Hypothesis capable of explaining them; yet while I am describing this, I shall sometimes to avoyde Circumlocution & so represent it more conveniently speak of it as if I assumed it & propounded it to be believed. This I thought fitt to Expresse, that no man may confound this with my other discourses, or measure the certainty of one by the other, or think me oblig'd to answer objections against this script. For I desire to decline being involved in such troublesome & insignificant Disputes.[45]

Its inherent philosophical contradictions notwithstanding, the importance of the "Hypothesis of Light" is attested to by the fact that it monopolized the Royal Society's calendar during the next several weeks. The meetings of December 9 and 16 were almost entirely devoted to its reading. Then, following a Christmas recess, intensive deliberations over its contents were undertaken during the meetings of December 30 and January 13.

The first and in many ways the most fascinating part of the paper presented a sweeping system of nature, a resounding cosmographical trumpet blast from one who had made and cultivated a reputation as a positive scientist. Newton propounded the existence of a universal ether as the agent by which the various forces acting on matter throughout the universe are generated. This subtle matter is like air but far rarer and much more elastic; and though it cannot be seen or felt, the reduction of a pendulum's motion in a "glasse exhausted of Air almost as quickly as in the open Air, is no inconsiderable argument" for its presence. While all-pervasive, the ether is rarer within the dense bodies of the sun, stars, and planets than in the great interstellar spaces between them. In his mind's eye, Newton pictured the earth and the other celestial bodies as giant sponges, steadily soaking up a stream of ethereal matter, which constantly presses down upon their surfaces. This same stream of ether, by virtue of its sustained impact, "may beare down with it the bodyes it pervades with force proportionall to the superfices of all their parts it acts upon," the first known written musings on the hypothesis of gravitation. Once the ether has penetrated into the "bowells" of a planet or star it is somehow transformed and ascends into space where the process begins anew. (An incensed Newton, in his letter to Halley of June 20, 1686, made reference to this very passage as proof that his thoughts, not Hooke's, had first turned toward universal gravitation.) Here, too, was the seventeenth-century equivalent of the law of the conservation of matter, according to which material is neither created nor destroyed but simply takes on different forms in an eternally cyclical pattern:

> For nature is a perpetuall circulatory worker, generating fluids out of solids, and solids out of fluids, fixed things out of volatile, & volatile out of fixed, subtile out of gross, & gross out of subtile, Some things to ascend & make the upper terestriall juices, Rivers and the Atmosphere; & by consequence others to descend for a Requitall to the former. And as the Earth, so perhaps may the Sun imbibe this Spirit copiously to conserve his Shineing, & keep the Planets from recedeing further from him. And they that will, may also suppose, that this Spirit affords or carryes with it thither the solary fewell & materiall Principle of Light; And that the vast aethereall Spaces between us, & the stars are for a sufficient repository for this food of the Sunn & Planets.[46]

Newton sought to explain not only gravity in terms of the ethereal medium, but many of the other phenomena that so puzzled the scientific community of his time: cohesion, electricity, animal sensation, reflection, and refraction. Content with a theory only so long as it yielded empirical data, he filled the paper with detailed accounts of experiments designed to reinforce his overarching hypothesis. Most fascinating to the Royal Society was an experiment with electricity. Newton had placed a piece of glass, 2 inches wide and set in a brass ring, on a wooden support about one-sixth of an inch above a table. He then rubbed the glass briskly with some "ruff and rakeing stuffe," until small fragments of very thin paper scattered on the table under the glass began to be attracted, moving to and fro. He watched as the papers became increasingly agitated, sometimes "leaping up to the

Glasse & resting there a while, then leaping downe & resting there." He could think of only one satisfactory explanation of this phenomenon:

> Now whence all these irregular motions should spring I cannot imagine, unless from some kind of subtill matter lyeing condens'd in the glasse, & rarefied by rubbing as water is rarified into Vapour by heat, & in that rarefaction diffused through the Space round the glasse to a great distance, & made to move & circulate variously & accordingly to actuate the papers, till it returne into the glasse againe & be recondensed there. And as this condensed matter by rarefaction into an aethereall wind (for by its easy penetrating & circulating through Glass I esteme it aethereall) may cause the odd motions, . . . so may the gravitating attraction of the Earth be caused by the continuall condensation of some other such like aethereall Spirit.[47]

The first attempts to duplicate this relatively simple experiment before the Royal Society failed, which seems puzzling considering Hooke's acumen with scientific apparatus. Not until Newton drafted two more letters of explanation to Oldenburg was Hooke able to make the following entry in his *Diary*: "Thursday, January 13th [1675-76]. I tryd Mr. Newtons experiment of the electricity of glasse by Rubbing, and it succeeded with a rubbing brush but especially with the whalebone haft of a knife."[48]

The "Hypothesis of Light" provides indisputable proof that Newton, like Pythagoras and Kepler, was fully capable of sustained flights of scientific imagination. Here was a mind every bit as subtle and elastic as the universal ether it propounded. And here, too, were the order, harmony, and aesthetic sensitivity found wanting in his often bitter relationships with men. System, order, connection—these he was compelled to seek; an unanalyzed whole was an impossible conception for him. In 1738, the year Voltaire's popular *Elémens de la Philosophie de Newton* was published, its author wrote to his former teacher, "I do not see why the study of physics should crush the flowers of poetry. Is truth such a poor thing that it is unable to tolerate beauty?" So far as Newton the physicist is concerned, the answer to the *philosophe's* question is plainly discernible in the "Hypothesis of Light." Despite the paper's contradictions and ambiguities, it is above all an unqualified declaration that the universe is a coherent and rationally ordered entity, and that infinite variation in the absence of law is a synonym for chaos. Preoccupied with this perceived tendency toward pattern and symmetry in nature, Newton quested for nothing less than a unifying cosmological principle with which to bind the microcosm to the macrocosm, the invisible particle to the flaming star. Who among the immortal poets has attempted more?

As indicated by its title, "An Hypothesis explaining the Properties of Light" contained a detailed elaboration of Newton's mechanical theory as it applied to optical phenomena. He speculated that when corpuscles of light pass through the ether, the medium's varying density alters the speed and direction in which the corpuscles move, thereby producing the familiar effects of reflection, refraction, and diffusion. But colors themselves do not arise from the modification of the heterogeneous corpuscles. Instead, the spectral band results when these particles are separated from each other. This theory was bolstered and elaborated upon in a second paper, "Discourse of Observations," sent to London with the first. Substantially identical to Book II of the *Opticks*, it was read and discussed at the three Royal Society meetings held between January 20 and February 10.

Newton based the paper on a demanding series of experiments begun in 1666, after reading Hooke's *Micrographia* account of the colors of thin films. Hooke, as related in Chapter Five, had observed "differing consecutions or orders of Colours" in the "lamina" or film of air created when a piece of curved glass is placed on top of a flat one. Intrigued by this phenomenon, Newton sought more than a hypothetical explanation of it, hoping to relate future findings to his wider optical studies. He realized that any adequate theory of light must also explain the permanent colors of solid bodies, something still to be achieved. Hooke's casual work with thin films seemed a potentially productive avenue of approach; but where Hooke had relied exclusively upon observation, Newton was to employ meticulous measurements and quantitative analysis, ultimately enabling him to expropriate yet another piece of Hooke's promising intellectual ground.

As Newton informed the Fellows who assembled to hear the reading of his paper, he began by compressing two slightly convex prisms "hard together" at their sides. The place at which they touched became "absolutely transparent," as if they formed a single piece of glass. At the same time, "a black or dark spot" appeared in the area where the air remained trapped between them, indicating that the light was being reflected. He also found that the spot became much broader when additional pressure was put on the glass. Rotating the prisms about their common axis produced "many slender arcs of colours [which] increased and blended more and more about the said spot, till they were compleated into circles or rings incompassing it"—"Newton's rings," as they were soon to be called.

"To observe more nicely the order of the colours," Newton exchanged his small prisms for two large convex lenses designed to fit reflecting telescopes of 14 feet and 50 feet. By lashing one lens to the other, he succeeded in increasing the diameter of the colored rings by nearly five times. "I pressed them slowly together, to make the colours successively emerge in the middle of the circles, and then slowly lifted the upper glass from the lower, to make them successively vanish again in the same place." He varied the experiment by employing other lenses and introducing water rather than air into the intervening space: the result was always the same. Though much remained to be done, he was now reasonably convinced of a direct correlation between particular colors of rings and the thickness of the layer of entrapped air or water. If this was so, then solid bodies, like thin films, are composed of transparent particles whose thickness also determines the colors they reflect.[49]

Armed only with a compass and his naked eye, Newton launched a scarcely believable assault on the phenomenon that bears his name. Each measurement of a ring's diameter, he decided, must be accurate to one-hundreth of an inch—or less! (As so often with Newton, one gropes for an inhuman simile, but in this instance none comes to mind. Suffice it to say that when Thomas Young devised an explanation of the rings based on the revitalized wave theory of light and the revolutionary principle of interference, he employed Newton's very measurements to calculate the wavelengths of the spectral colors, attaining results consistent with those accepted today.) Newton, by means of a simple arithmetic progression, was able to show that the colors return periodically with a measured increment of thickness. He first measured the dark circles between the colored circles and then squared their diameters. If the thickness of the first dark circle is 2 units, for example, successive circles will appear at thicknesses of 4, 6, 8, 10, and so forth, while the colored circles between them will appear at thicknesses of 1, 3, 5, 7, and 9 units. By virtue of Newton's magnificent experimental acumen, the concept of periodicity had been advanced from the realm of hypothesis to that of scientific truth.

VII

The "Hypothesis of Light" and "Discourse of Observations" marked the end of an extraordinary decade of scientific creativity for Newton. He had passed almost the whole of the previous ten years in a state of intellectual and physical quarantine, whipsawed between the brightest hope and the blackest despair, like a man in an endless dream. He had experienced both the bitter self-doubt born of failure and the tranquil ecstasy of the illusive "moment of truth"—that oceanic sense of wellbeing which defies rational analysis. Newton knew too much for others to grasp yet not enough to satisfy himself. He could, in the best Cartesian tradition, acquaint the Royal Society with his grandiose conception of a mechanistic universe, even as he suffered deep personal disappointment at not being able to corroborate this metaphysical hypothesis by mathematical demonstration. To add to this considerable strain, he believed himself the object of constant surveillance by jealous members of the scientific brotherhood, "little smatterers" he later called them, who hoped to uncover a major flaw with which to undermine his credibility. So be it, but they could expect even less cooperation from him than in the past. When Oldenburg asked permission to publish the "Hypothesis" and "Discourse" Newton refused, effectively limiting their readership to the London Fellows who had direct access to the Royal Society's register.

There was just enough of the conspirator in Robert Hooke to fan the flames of Newton's paranoia. On December 11, two days after the first part of the "Hypothesis" was read before the Royal Society, Hooke, accompanied by Sir John Hoskins, Master of Chancery, went to "Joes coffee house where we began a New Clubb. Mr. Hill, Hoskins, Lodowick and I, at last Mr. Aubrey discoursd about Mr. Newtons new hypothesis." The next meeting of the group, which had expanded to include Christopher Wren, took place on January 1, 1676: "Resolved upon Ingaging ourselves not to speake of any thing that was then revealed *sub sigillo* to any one nor to declare that we had such a meeting at all." Once again Newton's paper was the main topic of conversation. "I showd," Hooke wrote in his *Diary*, "that Mr. Newton had taken my hypothesis of the puls or wave."[50]

What John Aubrey had described as not a quick but a gray eye in Hooke had suddenly sprung to life, flashing at the mere mention of Newton's name. Nor did Hooke confine his expression of pique to this ostensibly secret forum. After having listened to a second reading from the "Hypothesis" on December 16, he told the assembled Fellows "that the main part of it was contained in his *Micrographia*, which Mr. Newton had only carried farther in some particulars."[51]

Newton, as a member of scholars have pointed out, would probably have won Hooke's support or, failing that, his lasting respect, if only he had proved capable of acknowledging his indebtedness to another. But no one was more susceptible to the myth that Isaac Newton operated as a wholly independent intellectual agent than Isaac Newton himself, a belief aptly characterized as "the magic circle of his infallibility."[52] This is not to say Newton is undeserving of the many laurels heaped upon him, for the insights of his fellow virtuosi, including Hooke's, usually bottomed out where his began. It is equally true, however, that the *Micrographia* had provided him with some fruitful ideas—the periodicity of colors in thin films, for example—even as it served as the catalyst for others. Newton had also profited from a paper on diffraction presented by Hooke during his most recent visit to London. Still, he preferred to make an enemy by damning with faint praise rather than to cultivate an ally by the simple act of giving credit when it was deserved. Hooke, as was his habit, reacted by laying claim to things he had not discovered, which in turn angered Newton the more. As in 1672, matters of substance were obscured by a resounding clash of giant egos, the echos of which continue to reverberate across the centuries.

Even though the participants were sworn to secrecy, the chances are quite good that Newton got wind of the philosophical discussions then taking place in the coffeehouses of London. Regular gatherings of such prominent figures, all Fellows of the Royal Society, were bound to arouse suspicions, and neither Hooke nor Aubrey was particularly noted for keeping his own council. Moreover, Oldenburg, Newton's primary source of information, had an additional reason for keeping a close watch on Hooke's behavior. The two were now at loggerheads over Hooke's claim that he, not Christiaan Huygens, whose part Oldenburg had taken, was the true inventor of the spring-balance watch. It has long been argued that the animosity

Oldenburg felt toward Hooke over this issue caused him deliberately to provoke Newton into striking out at his old nemesis. Hooke was keeping equally close track of Oldenburg. On Friday, October 15, he wrote of "A Grubendolian [anagram for Oldenburgian] Caball at Arundell house."[53]

For reasons that suggest something more than chance, Oldenburg's letters to Newton from this emotionally charged period have disappeared without a trace. Yet, as is obvious by now, the slightest goad, let alone a suggestion of plagiarism, was sufficient to provoke an aggressive counteraction on Newton's part. "As for Mr. Hook's insinuation yt ye sume of ye Hypothesis I sent you had been delivered by him in his Micrography," Newton wrote Oldenburg on December 21, "I need not be much concerned at the liberty he takes in yt kind." Indeed, "I may avoyd ye savour of having done any thing unjustifiable or unhansome towards Mr. Hooke." On the other hand, could Hooke in good conscience say the same about his relationship to Descartes, from whom he borrowed most of the essential elements of his hypothesis of light? Newton contended that the only debt he owed Hooke was for introducing him to the phenomenon of thin films, whose full explication was beyond Hooke's limited powers: "He having given no further insight into it then this yt ye colour depended on some certain thicknes of ye plate: though what that thicknes was at every colour he confesses in his *Micrographia* he had attempted in vain to learn & therefore seing I was left to measure it my self I suppose he will allow me to make use of what I tooke ye pains to find out."[54] Like it or not, Newton's hauteur was solidly backed by the knowledge that the state of genius is no republic. To Hooke, Newton was a fearsome rival; to Newton, Hooke was nothing more than an intolerable nuisance, a skulking jackal unfit to feed among the lions.

Newton wrote a second letter on January 10, thanking Oldenburg for "your candor in acquainting me wth Mr. Hook's [most recent] insinuations. It's but a reasonable piece of justice I should have an opportunity to vindicate my self for what may be undeservedly cast on me." He then took back the niggardly credit bestowed on Hooke in his previous letter: "I desire Mr Hook to shew me therefore, I say not ye summ of ye Hypothesis I wrote, wch is his insinuation, but any part of it taken out of his *Micrographia*: but then I expect too that he instance in what's his own."[55]

The best evidence in support of a charge that Oldenburg deliberately baited Newton in order to get at Hooke derives from the Secretary's handling of Newton's first letter. The explanatory material on how to conduct the electrical experiments that had so far baffled Hooke was read at the Royal Society meeting of December 30. Nothing was said of Newton's heated remarks regarding Hooke's "insinuations," however. Three weeks later, on January 20, Oldenburg read the openly critical passages of the letter, apparently without warning Hooke in advance. Hooke was naturally furious, the more so because he had been similarly humiliated before his peers in 1672. In his diary entry that evening Hooke accused Oldenburg of instigating a quarrel between Newton and himself by "fals suggestions." What is more, Hooke decided that he must take matters into his own hands: "Wrot letter to Mr. Newton about Oldenburg kindle Cole."[56]

Difficult though it is to believe, every known letter Newton had received to that date from another natural philosopher had first passed through the guiding hands of either Oldenburg or Collins. Thus, Newton could tell at a glance that this letter was written in an unfamiliar hand. One can almost see the graying, normally tight-knit eyebrows involuntarily transformed into half-moons of wonder as he read the heading: "Robert Hooke—These to my much esteemed friend, Mr. Isaack Newton." Hooke did not mention Oldenburg by name, but he made it clear that he considered himself the victim of "sinister practices" designed to arouse Newton's suspicions concerning his intentions. "I do noeways approve of contention or feuding and proving in print, and shall be very unwillingly drawn to such kind of warr." Hooke went on to praise Newton's accomplishments as presented in the "Hypothesis of Light." His tone was nevertheless rather patronizing; he, not Newton, had initiated these important studies, and only the press of other business had prevented him from bringing them to a successful conclusion:

> I doe justly value your excellent Disquisitions and am extremely well pleased to see those notions promoted and improved which I long since began, but had not time to compleat. That I judge you have gone farther in that affair much than I did, and that as I judge you cannot meet with any subject with a fitter and more able person to inquire into it than yourself, who are every way accomplished to compleat, rectify and reform what were the sentiments of my younger studies, which I designed to have done somewhat at myself, if my other troublesome employments would have permitted, though I am sufficiently sensible it would have been with abilities much inferior to yours.

Hooke concluded with a proposal that in the future they correspond directly in order to prevent any reasonable differences of opinion from erupting into open hostility: "[T]wo hard-to-yield contenders may produce light yet if they be put together by the ears of other's hands and incentive it will produce rather ill concomitant heat which serves for no other use but . . . kindle cole."[57]

If anything, Newton's reply was even more conciliatory in tone and lavish in its praise than Hooke's letter had been. He addressed Hooke as "his honoured Friend" whose "generous freedom" in writing "becomes a true Philosophical spirit." Newton professed to loathe intellectual warfare as much as Hooke: "There is nothing wch I desire to avoyde in matters of Philosophy more than contention, nor any kind of contention more then one in print." He gladly embraced Hooke's proposal for inaugurating a private correspondence because "what's done before witnesses is seldome wthout some further concern then that for truth: but what passes between friends in private usually deserves ye name of consultation rather than conquest, & so I hope it will prove between you & me." Newton was laying on the cement of goodwill with a trowel, which causes one to wonder if he was taking himself seriously. Aside from the comments on controversy, almost everything else he wrote was contradictory to his temperament and character. Did the Newton we have come to know really mean it when he assured Hooke that there was "no man" better able to furnish him with "objections" to his scientific work than Hooke himself? And are we to believe that "equity and friendship" were more important to him than "philosophical productions"? Equally misleading is Newton's oft-quoted allusion to genius: "What Des-Cartes did was a good step. You have added much several ways, & especially in taking ye colours of thin plates into philosophical consideration. If I have seen further it is by standing on ye sholders of Giants."[58] Though a gifted and clever man, Hooke was no intellectual giant, especially not to Newton. It appears that for once politeness—or most probably an affectation of it—had simply gotten the best of Newton as he competed with Hooke for the honor of seeing who could deprecate his own work the most. One scholar has pictured them as onetime country boys who, having made good, were now affecting the manners of Restoration courtiers—"flattering each other, overpraising, scraping and bowing, doffing their scientific plumed hats in grand gestures."[59] The disingenuous quality of this exchange is best illustrated by the fact that it resulted in only the uneasiest of truces and never produced the enlightened correspondence both antagonists professed to desire.

The agitation stemming from Newton's latest scientific offering was heightened by the arrival of yet another letter from Liège. Linus had died during an epidemic the previous November, but the Hydra immediately sprouted another head in the form of John Gascoines, the deceased Jesuit's fanatically loyal pupil. "Wonder not to see yours of November 18, 1675, to Mr. Francis Line answered by another hand," Gascoines wrote Oldenburg on December 15, "nor that hereafter the same happen as often as new occasions of writing present themselves."[60] At the very least Newton must have winced on reading these lines; possibly he even sinned against his Christian conscience by breaking the Third Commandment. Gascoines requested that Linus's second letter be published without further delay and bristled at Newton's suggestion that his mentor had been less than diligent in subjecting the theory of colors to rigorous experimental analysis. Indeed, so anxious was Gascoines to impress others with the depth of Linus's commitment to experimentation that he unintentionally caricatured the old man, but with a hilarity dampened by a wholly unsubstantiated claim: "For as Mr. Line was always at home, and in his chamber, and ordinarily kept his Prisme just ready before the hole, so we think it probable he hath tried his experiment thrice for Mr. Newton's once." It was the contemptuously self-assured tone of Gascoines's closing remarks that proved most galling to

Newton: "And, therefore, you may assure Mr. Newton, that when the Sunne and Season shall serve for the tryall [before the Royal Society] nothing shall be certify'd to him."[61]

Newton's reply was appended to the very letter to Oldenburg (January 10, 1676) in which he had rejected, for the second time in three weeks, Hooke's assertion that the *Micrographia* was the true mother of his invention. His counterarguments were almost equally divided, as if by conscious choice, between brilliant empirical analysis and withering invective. Offering no hint of the impending emotional outburst, Newton painstakingly retraced each step in the experiment, suggesting any number of possibilities as to where Linus might have gone wrong. Should Gascoines care to oblige, he would be pleased to receive a description of Linus's experimental procedures "to consider what there is in that wch makes against me." Then, in mid-paragraph, the tone suddenly shifts. It was Linus, not he, who employed experimental results reaching back nearly thirty years. Neither had he done anything to prevent the publication of Linus's second paper, as Gascoines seemed to suggest. "All I think that they [the Jesuits] can object to," he fumed, "is that you [Oldenburg] were of a stand becaus you could not ingage me in ye controversy, & me yt I had no mind to be ingag'd: a liberty every body has a right to & may gladly make use of ... especially if he want leisure or meet wth prejudice or groundless insinuations."[62] Gascoines, who was obviously no match for Newton, soon passed from the scene, but not before enlisting the services of a fellow Liègeois to uphold the honor of the fraternity. The Hydra, Newton was learning, was the most fecund of beasts.

VIII

Four years and more had gone by since Newton first apprised the Royal Society of the prismatic experiment so persistently challenged by the Jesuits of Liège. Yet the Society's only attempt to confirm it had been aborted because of cloudy skies in March 1675.[63] Finally on Thursday, April 27, 1676, "a very clear sun-shine day," the "experiment of Mr. Newton, which had been contested by Mr. Linus and his fellows at Liège, was tried before the Society, according to Mr. Newton's directions, and succeded, as he all along had asserted it would do."[64] Hooke, who in his *Diary* assiduously recorded everything else he did and witnessed that day, including Tompion's repair of Lady Wilkins's watch and the directing of a drain through cellars, could not bring himself to acknowledge yet another triumph by his rival—which he, as Curator of Experiments, had helped engineer.[65] Newton could scarcely hide his pleasure on receiving the news from Oldenburg. Surely now "Mr Linus's Friends will acquiesce."[66]

They did not. Professor Anthony Lucas, the new Jesuit standard-bearer, drafted a long letter to Oldenburg in May 1676, arguing "against Mr Newtons new Theory of light and colors." A more accomplished experimenter than Linus, his aged predecessor (though by no means gifted), Lucas verified the oblong form of the spectrum, but he took exception to Newton's claim that the length of the colored image was five times its breadth. Lucas also included an account of nine other experiments that seemed to cast additional doubt on Newton's beleaguered theory.[67]

Newton received a copy of Lucas's letter in early June, but he chose not to reply until August 18, perhaps allowing his temper to cool during the interim. When he did respond, he attributed the alleged discrepancies to a lack of technical acumen on the Jesuit's part.[68] Lucas had also attempted to duplicate the *experimentum crucis,* but with even less success. He found red rays among the purple and so dismissed the theory of colors out of hand. Newton urged his detractor to alter his methodological approach, "presum[ing] he really desires to know what truth there is." "Instead of a multitude of things try only the *Experimentum Crucis.* For it is not [the] number of Expts, but weight to be regarded; & where one will do, what need of many?"[69] This is not, as some have claimed, a contradiction of Newton's philosophy of science. Rather, it was his way of encouraging a mediocre intellect to focus its full attention on one scientific question at a time. Indeed, Newton found Lucas's eclectic approach to experimentation only slightly less disturbing than the groundless multiplication of hypotheses, for both must inevitably produce bad science.[70]

Lucas acceded to Newton's request and attempted the *experimentum crucis* once more, with much the same result. Lapsing into the bankrupt thought pattern of his neoscholastic training, the Jesuit concluded that the phenomenon must be attributable to "some extrinsecall and accidentall cause" which violates "the received lawes of Logick."[71] It was clear by now that so far as their standards of accuracy were concerned, Newton and Lucas inhabited different planets, while Linus had been of another universe.

Isaac Newton was bound to anger as the sparks flew upward. His first letter to Lucas, while polite, offers every evidence of intense strain, a pen barely able to contain a mounting reservoir of passion. He admitted as much to Oldenburg a few days later: "I have not sent you an answer so full as I intended at first but perhaps more to ye purpose considering who I have to deale with, whose business it is to cavill."[72] Now, as he brooded over his reply to the "plotting" papist's second letter, Newton again made Oldenburg privy to his thoughts. "I see I have made my self a slave to Philosophy," he lamented, "but if I get free of Mr Linus's business I will resolutely bid adew to it eternally, excepting what I do for my privat satisfaction or leave to come out after me. For I see a man must either resolve to put out nothing new or become a slave to defend it."[73] Amid the rancor and despair was a brief but revealing allusion to what Newton hoped his place in history would be: What he did not publish in his lifetime, others someday must.

Newton's reply of November 28 to Lucas's latest letter was nothing less than incendiary. He accused the Jesuit of fomenting a dispute by running "from one thing to another before we come to a conclusion." Overlooking the fact that it was he who had exhorted Lucas to investigate the theory of colors further, Newton characterized the renewed criticism of the *experimentum crucis* as an "attempted digression" designed to muddy the waters and inpugn his integrity as a natural philosopher:

> If Mr. Lucas hath not yet procured a Prism with an angle about ye bigness of mine there used, & with sides not at all concave, but plain or onely a very little convex, let him onely upon ye receipt of this signify without any insinuation of suspicion yt he sees no reason to oppose or question me further upon ye experiments they have hitherto made and proceeded on, & I shall acquiesce and begin with his Objections.[74]

François La Rochefoucauld, a French moralist and contemporary of Newton, observed, "No man deserves to be praised for his goodness unless he has the strength of character to be wicked. All other goodness is generally nothing but indolence or impotence of will." Anything but indolent or lacking in resolve, Newton had long believed his uncontrollable temper sinful and feared that it could one day lead to his undoing. Though it would seem otherwise, he had struggled mightily, if unsuccessfully, since adolescence to dampen the inner fires of resentment, suspicion, and intolerance. In the confession of his youthful sins we recognize the progenitor of the wrath directed against Hooke, Pardies, Linus, Gascoines, and Lucas: "Threatning my father and mother Smith to burne them and the house over them; Wishing death and hoping it to some; Punching my sister; Peevishness with my mother; With my sister; Falling out with servants; Beating Arthur Storer." Nor, as it might also seem, was Newton guilty of doubletalk when he expressed his abhorrence of controversy. Yet the fact remains that he was born to controvert. Every one of his major discoveries, from the calculus to the law of universal gravitation, was attended by an acrimonious and long-drawn-out dispute. Unable to control his temper through the normal exercise of will, Newton took the more radical path of periodic withdrawal, hoping thereby to salve his conscience and to gain a respite from the maddening ghost in the machine.

Such was the state of affairs when fate took a hand. However much confidence Newton may have had in his work, he was more covetous of recognition, and less willing to risk all on the unaided judgment of posterity, than he had recently let on to Oldenburg. Besides, his critics would give him no rest. Goaded by still another disparaging letter from Lucas in February 1677, Newton resolved to publish the full range of his optical studies, including the correspondence arising from them. The work was apparently well under way when fire, that most feared enemy of scholars through the ages, broke out in his quarters, consuming the notes and nearly all of the unfinished manuscript, along with other invaluable papers. Wallis, in the Preface to his *Algebra* dated November 20, 1684, wrote: "I have only given some specimen of what we hope Mr Newton will himself publish in due time.... But most of those papers have since (by a mischance) been unhappily burned." The most vivid account of Newton's personal tragedy is contained in the diary of Abraham de la Pryme, an undergraduate at St. John's in the early 1690s. De la Pryme seems to have fashioned his version from the anecdotes of fellow students:

> There is one Mr Newton (whom I have very oft seen) Fellow of Trinity College, that is mighty famous for his learning, being a most excellent Mathematician, Philosopher, Divine, &c. He has been fellow of the Royal Society these many years . . . but of all the Books that he ever wrote there is one of colours & light established upon thousands of Experiments which he had been 20 years of making, & which cost him many a hundred of pounds. This Book which he valued so much & which was so much talked of had the ill luck to perish, & be utterly lost just when the learned Author was almost at putting a conclusion to the same, after this manner: In a winter's morning leaving it amongst his other Papers, on his Study table whilst he went to Chapel, the Candle which he had unfortunately left burning there too, catched hold by some means of other papers, & they fired the aforesaid Book, & utterly consumed it, & several other valuable writings, & which is most wonderful did no further mischief. But when Mr Newton came from Chapel and had seen what was done, every one thought he would have run mad, he was so troubled thereat that he was not himself for a Month after. A long account of his system of light & colours you may find in the Transactions of the Royal Society which he had sent up to them long before this sad mischance happened unto him.[75]

In contrast to de la Pryme's dramatic, richly embroidered account, Newton's recollection of the incident is, like almost all his references to himself, disappointingly laconic. Queried by Conduitt about the fire the year before he died, Newton recalled being "in the midst of his discoveries" when "he left a candle on his table amongst his papers & went down to the Bowling green & meeting somebody that diverted him from returning as he intended the candle set fire to his papers & he could never recover them." Asked whether they related to his optics or to the method of fluxions, he replied, "there were some of both."[76] Humphrey Newton also told Stukeley that before he went up to Cambridge several sheets of his future employer's optics were burnt by a candle left in his room. If Stukeley is correct the fire destroyed far more than the papers on optics and mathematics: "He wrote likewise a piece of chymistry, explaining the principles of that mysterious art upon experimental and mathematical proofs, & he valued it much, but it was unluckily burnt in his laboratory which casually took fire, he would never undertake that work again."[77] Whether, in the face of such adversity, Newton became temporarily unhinged, as de la Pryme had heard, is impossible to say—one more instance of the tyranny of the anecdote.

For another year Newton thought seriously of reviving the project; he even requested that Lucas supply him with copies of his first three letters, since those made for him by Oldenburg had gone up in flames. Sympathy briefly overcame rancor, and a commiserating Lucas wrote, "I am glad I am in a capacity to repair in greatest part your losse."[78] He sincerely regretted that he could not also provide Newton with a copy of Gascoines's letter, but the onetime pupil of Linus had left Liège for parts unknown. Newton's problems were compounded by the sudden death of Henry Oldenburg in September 1677 and his replacement by none other than Robert Hooke. When Newton wrote to the new Secretary for specific information about the crucial prismatic experiment that had proved the Jesuits wrong, he received a polite but less than encouraging response. "[T]he Particulars of them I doe not at all remember nor can I tell whether any account bee taken of them in our Journalls." Hooke did promise to conduct a search, "though I fear nothing is entred. Oldenburg of late omitting all things done by me."[79] Faced with these daunting obstacles, Newton lost what remained of his ebbing enthusiasm. He would publish nothing more on his theory of light and colors until the *Opticks* went to press in 1704, the year after Hooke was laid to rest.

Oldenburg's death, coupled with that of his friend Isaac Barrow the previous May, strengthened Newton's resolve to retreat even more deeply into the academic cloister. Only one item of unfinished business remained, his quarrel with the Jesuits of Liège, Anthony Lucus in particular. So anxious was Newton to divest himself of this last entanglement that he wrote not one but two letters to Lucas on March 5, 1678. The first began and ended on a heated note: "The stirr you make about your Objections draws this from me to let you see how easy it was to answer your Letters had not other considerations (particularly your contemming my profer to answer one or two which you should recommend for ye best) made me averse from meddling further wth your matters." All Lucas's objections, he wrote, "run upon two general mistakes . . . ye question of different refrangibility & ye notion of primary colours." Yet all the experiments "are on my side." In Newton's eyes Lucas had compounded his felony by taking part in a conspiracy to protect Linus: "[I]nstead of acknowledging his mistake ye business was represented as if none had been committed by him." By the time the Cambridge professor was ready to draft a sequel he had whipped himself into a fine fury, exposing the fissures of his sanity:

> Pray trouble your selfe no further to reconcile me wth truth but let us know your own mistakes. . . . Do men use to press one another into disputes? Or am I bound to satisfy you? It seems you thought it not enough to propound Objections unless you might insult over me for my inability to answer them all, or durst not trust your own judgement in choosing ye best. But how know you yt I did not

think them too weak to require an answer & only to gratify your importunity complied to answer one or two of ye best? How know you but yt other prudential reasons might make me averse from contending wth you? But I forbeare to explain these things further for I do not think this a fit Subject to dispute about, & therefore have given these hints only in a private Letter. . . . I hope you will consider how little I desire to explain your proceedings in public & make this use of it to deal candidly wth me for ye future.[80]

He could have said the same thing in a single sentence: "I am without rival and acknowledge no judge but myself."

Captain Robert Pugh, Lucas's courier, delivered the Jesuit's reply (since lost) via John Aubrey in late May. Hooke, who had been permitted to read it, suggested in a letter informing Newton of its arrival that the contents were anything but conciliatory. In what smacks of gleeful insincerity veiled by mock sympathy, Hooke wrote, "I must admire your patience that you will trouble your self wth such an extravagantly impertinent—, who never will yeald be the matter never soe plain."[81] Both distraught by what he had seen of himself and weary of a debate that held out no promise of a favorable resolution, Newton hadn't the kidney to continue.

Surprising as it may seem, the denouement came not with a resounding thunderclap but with a sigh of resignation, as though the mere thought of solitude were sufficient to restore his equilibrium. "I understand you have a letter from Mr Lucas for me," he wrote Aubrey in June. "Pray forbear to send me anything more of that nature."[82] Serene days where trouble rarely penetrated were to be his at last—days given over to the alchemist's crucible and an exhaustive examination of the patristic fathers. Not until 1684, when Edmond Halley traveled to Cambridge seeking his thoughts on planetary attraction, did Newton voluntarily turn outward once again.

Chapter Nine

The Treasures of Darkness

The Alchemical meditatio is an inner dialogue with some-
one who is invisible, as also with God, or with oneself,
or with one's good angel.

Ruland the Lexicographer

I

In May 1727, Dr. Thomas Pellet, a respected member of the Royal Society, was faced with what was undoubtedly one of the most daunting tasks of his professional life. Pellet had recently received a commission from John Conduitt to determine which of the papers in Isaac Newton's estate seemed worthy of publication. Newton's personal possessions were relatively few and worth surprisingly little for a man who had grown wealthy during his long tenure as Master of the Mint. The papers, voluminous to the point of incredulity, presented a wholly different and far more knotty problem. The writings on theology and chronology alone totaled at least 1,400,000 words; the alchemical notes and manuscripts have been conservatively estimated at 1,000,000.[1] It is doubtful that Pellet himself realized that the endless pages through which he sifted contained upwards of 4,000,000 words. Even this prodigious figure cannot be taken as the full measure of Newton's lifework, for fire, both accidental and intentional, had claimed untold additional pages. An unattended candle resulted in the destruction of at least one major manuscript and possibly many more miscellaneous papers in the winter of 1677–78, and Conduitt noted that not long before he died Newton burned many "foul draughts of his Mathematical treatises & Chronology & abstracts of the history." Conduitt also wrote of having personally assisted Newton in burning box upon box of written matter pertaining to his three decades of service at the Mint.[2]

Precisely what criteria the hard-pressed Pellet established for publication is impossible to say, but whatever his standard, none of the alchemical papers measured up. Time and again Pellet scrawled his unequivocal judgment across the wrappers in the boldest of letters: "Not fit to be printed." History long concurred with Pellet's harsh assessment. Not for upwards of two and a half centuries was the darkness pierced by even the faintest rays of scattered light.

The Earls of Portsmouth fell heir to Newton's papers via his niece Catherine Barton Conduitt, and they permitted certain scholars not only to examine the vast collection but to publish from it as well. During the eighteenth century Bishop Samuel Horsley brought to press his five-volume *Isaaci Newtoni Opera* (1777–85), but it contained not a whisper of the alchemy, which as an art had fallen into scandalous disrepute in the half-century after Newton's death. The good Bishop also chose to overlook a number of Newton's heterodox theological writings, particularly those that challenged the doctrine of the Trinity.

Sir David Brewster, a brilliant nineteenth-century scholar and Isaac Newton's first major biographer, harbored an even deeper horror of the alchemy than had Horsley. Newton was Brewster's untarnished hero—that is, until the alchemical papers surfaced in the course of his research. Even then Brewster's treatment of his subject took a decidedly hagiographic tone, although intellectual honesty compelled him to take note of the "damaging" evidence. "There is no problem of more difficult solution," Brewster wrote, "than that which relates to the belief in alchemy and to the practice of its arts by men of high character and lofty attainments." Surely Newton was not one of those who embraced "a process . . . commencing in fraud and terminating in mysticism." He must have been driven, like Boyle and Locke, by "a love of truth alone, a desire to make new discoveries in chemistry." Brewster's strong Presbyterian conscience could possibly excuse Newton's attempts at the transmutation and multiplication of metals as an aberration of the age in which he lived, but "we cannot understand how a mind of such power . . . could stoop to be even the copyist of the most contemptible alchemical poetry, and the annotator of a work, the obvious production of a fool and a knave."[3] Even in the nineteenth century hero worship had its limits.

In the 1880s the Portsmouths made a gift of the Newton papers to the University Library, Cambridge, with the stipulation that all materials deemed lacking in scientific value be returned to the family. The manuscripts on alchemy and theology were soon back in the hands of their donors. Finally, in 1936 the Portsmouths sold the remainder of their collection at an auction conducted in the London galleries of Sotheby and Company. It was at this sale that John Maynard Keynes acquired fifty-seven of the 121 lots on alchemy, which he left to his own King's College at his death in 1946.

An entranced Keynes spent many hours pouring over his boxes of unique treasure, and he later prepared a paper on his disturbing findings. What Keynes had read caused him to challenge the eighteenth-century conception of Newton as "the first and greatest of the modern age of scientists, a rationalist, one who taught us to think on the lines of cold and untinctured reason." But if Newton was not a rationalist, what was he? According to Keynes, "he was the last of the magicians, the last of the Babylonians and Sumerians, the last great mind which looked out on the visible and intellectual world with the same eyes as those who began to build our intellectual inheritance rather less than 10,000 years ago." And why did Keynes choose to call Newton a magician?

> Because he looked on the whole universe and all that is in it *as a riddle*, as a secret which could be read by applying pure thought to certain evidence, certain mystic clues which God had hid about the world to allow a sort of philosopher's treasure hunt to the esoteric brotherhood. He believed that these clues were to be found partly in the evidence of the heavens and in the constitution of elements (and that is what gives the false suggestion of his being an experimental natural philosopher), but also partly in certain papers and traditions handed down by the brethren in an unbroken chain back to the original cryptic revelation in Babylonia. He regarded the universe as a cryptogram set by the Almighty—just as he himself wrapt the discovery of the calculus in a cryptogram when he communicated with Leibnitz. By pure thought, by concentration of mind, the riddle, he believed, would be revealed to the initiate.[4]

Keynes, an extraordinarily creative twentieth-century thinker, had seemingly confirmed the eighteenth-century suspicions of Bishop Horsley and the nineteenth-centry conclusions of David Brewster. "Interesting, but not useful," was Keynes's assessment of the alchemical papers, "wholly magical and wholly devoid of scientific value."

II

Perhaps no one before or since has formulated a more succinct statement of the theory underlying the alchemist's quest than Arnold of Villanova, a Catalan, who was born near Valencia in the first half of the thirteenth century. A physician, astrologer, diplomat, social critic, and alchemist, Arnold spoke not only for himself but for his fellow adepts when he observed: "That there abides in Nature a certain pure matter, which, being discovered and brought by art to perfection, converts to itself proportionally all imperfect bodies that it touches." This "pure matter" is the legendary philosopher's stone, a substance so powerful that it could transform base metals—lead, copper, tin, iron, and mercury—into the most precious metals, namely, silver and gold. Moreover, the exoteric or outward alchemy of which Arnold wrote was paralleled by the rise of an esoteric or mystical inner alchemy credited by its practitioners with a power even more wonderful than that of physical transmutation. It was believed that the ingestion of the *Elixir Vitae* (Elixir of Life), one of many names given to the philosopher's stone, would bestow earthly immortality on its discoverer, provided, of course, he had deported himself in a manner "pleasing in the sight of God." Thus for the esoteric alchemist the transmutation of metals became a mere symbol of the far more profound transformation of sinful man into a creature worthy of Divine grace. In practice the two types of alchemy were frequently intermixed, often making it impossible for later generations to determine whether a given alchemist was motivated primarily by crass materialism, transcendent religious principles, or a combination of both.[5]

Despite its oriental roots, Western alchemy was largely shaped by the Greek intellectual tradition. Just as in the study of astronomy, medicine, and biology, Greek thinkers of the early Christian era made every attempt to wed alchemy to the scientific teachings of Aristotle of Stagirus. It was Aristotle's beliefs concerning the constitution and unity of matter that gave hope to the ancient and medieval alchemist alike, and those beliefs survived virtually intact until the time of Isaac Newton.

Aristotle had taught that all things in nature are composed of a combination of basic elements. Instead of the large and ever increasing number accepted today, the Stagirite settled on the four elements identified in the fifth century B.C. by Empedocles of Agrigentum: earth, water, air, and fire. These four elements were thought to follow certain inexorable patterns of movement. Because earth and water are heavy, they express their ideal nature by moving downward, while air and fire, because they lack weight, rise naturally toward the heavens. Aristotle's elements are not immutable, for all substances in nature, even those bearing the names of the elements themselves, are composed of at least two of the four elements, though one or another is always predominant. In solid, heavy bodies such as metals, for example, the earth content is very high, but water is also present. All organic objects, including the human body, are also made up of the basic elements. Thus Shakespeare consciously took the part of an Aristotelian when he gave a dying Cleopatra the line: "I am Fire and Ayre, my other elements I give to baser life."

Aristotle also held that the four elements express themselves in both the internal composition and the temperament of the individual. Each element corresponds to one of the four body fluids or humors, which in turn govern basic human emotions. Earth is present in the form of black bile, an excess of which produces a state of melancholy. Water manifests its presence in the form of phlegm, too much of which produces a sluggish or phlegmatic temperament. Fire is naturally equated with blood and a passionate disposition, while air, present in the form of yellow bile or choler, is associated with anger and illtemper. Aristotle's humoral conception of medicine, later expanded upon by the teachings of Galen, continued to hold sway long after his geocentric universe had been abandoned. As late as 1799 physicians continued to prescribe phlebotomy for certain physical ailments, as in the case of George Washington, who succumbed that year to pneumonia complicated by debilitation resulting from controlled bleeding.

Aristotle further held that the elements may be transformed into one another, a principle fundamental to the alchemist's art. To fire he attributed the complementary qualities of hot and dry, the opposite to water, which is cold and moist. Earth, on the other hand, is dry and cold, while its opposite, air, is moist and hot. The transformation of one element into another takes place most readily between two elements that have one quality in common, such as heat, cold, dryness, or moistness. In theory, however, the transmutation of complete opposites, of water into fire and of earth into air, is also possible. If this is so, then it must follow that any type of substance can be transformed into any other type by subjecting it to forms of treatment that alter the proportions of its elements to accord with the proportions of the elements to be created. Hence all matter must have a common "soul," which alone is permanent. The physical bodies in which matter manifests itself are transitory, ever changing into other forms. Such is the foundation on which the alchemical theory of transmutation rested, the supreme justification for a lifetime spent in stooped and bleary-eyed labor over furnaces that burned round the clock for months, even years, on end. "If lead and gold both consist of fire, air, water, and earth, why may not the dull and common metal have the proportions of its elements adjusted to those of the shining precious one?" Why not, indeed. Never mind the fact that such reasoning was based upon two *a priori* assumptions: first, Aristotle's belief in the unity of all matter; second, the existence of the all-powerful agent of transmutation, Arnold of Villanova's elusive philosopher's stone.

The Greek alchemists of the early Christian period, following the precedent set by scholars in other fields of study, sought to legitimate their formative doctrines by attributing them to the most respected thinkers of the distant past. Many of the works that inspired alchemists of later generations, including Newton, were thought to be of great antiquity, but they had actually been composed during the second and third centuries. Moses, reared in the kingdom of the ancient priests and pharaohs, was surely an adept, it was argued, as was Miriam, his sister. Solomon, son of David and King of Israel, had left ample, if veiled, proof of his mastery of the art in the Old Testament writings. And would Jason and the Argonauts have risked all by crossing the raging Pontic Sea to lay claim to the Golden Fleece of Colchis had it not contained the formula for making unlimited quantities of gold? Plato also knew the secret of the Stone, as had Hermes, Cleopatra, and Isis before him. Christ, who openly turned water into wine and multiplied bread and fishes, gave added encouragement to credulous adepts. Indeed it was believed that the Holy Grail itself, the chalice from which the Master drank at the Last Supper, would, if found, resolve forever the maddening riddle of transmutation. In the melting pot of the Alexandrian Empire all knowledge and religion were treated as one. The very fact that so many disparate schools of thought appeared to confirm the wildest of alchemical speculations lent full credence to the doctrine of transmutation. Persian Magi, Chaldean astrologers, Egyptian priests, Greek natural philosophers, Gnostics, Hebrews, Christians, and Neo-Platonists could not all be wrong.

What did the philosopher's stone look like, and how did one produce this supreme catalyst that could transform leprous metals into perfect gold and base man into a living saint? As one might imagine, there were nearly as many variations in the theory of how to accomplish the Great Work as there were practicing alchemists. Still, a number of general principles were thought to apply. Color was the alchemist's most important guide, and over the centuries a generally accepted rule evolved regarding the order in which color changes should occur, regardless of the ingredients with which the alchemist initiated his operation. He could be fairly certain that he was on the right track if his bubbling crucible first yielded a black substance— the color of impurity, of evil, and of death. Gradually, black must give way to white, and white to iridescent, followed in order by yellow, purple, and, finally, red, the color of blood, the very symbol of life and therefore the color of the philosopher's stone. Little wonder that transmutation was frequently called "tingeing" in the literature, while the Stone itself was dubbed the Tincture.

Fire was by far the alchemists' most important agent, and their accounts of its varieties and powers are legion. Then as now, fire or heat was graded according to the different temperatures it produced. The process began with a relatively low heat, which was gradually increased by increments of a few degrees, a remarkable accomplishment in itself when one considers the relatively primitive nature of their equipment. Too little or too much heat could wipe out months of unremitting labor: Woe to the hapless apprentice who disturbed his master's slumber to report a change in color brought about too rapidly or out of sequence by an ill-tended flame.

Accounts of operations or processes deemed essential to success are also voluminous. These include calcination, congelation, fixation, solution, digestion, distillation, sublimation, separation, ceration, fermentation, multiplication, and projection. Calcination, for example, involved the reduction of a solid to a powder, while sublimation consisted in heating a substance until vaporization occurred, then condensing the vapor back into a solid by means of rapid cooling. Though some of these processes have long since been discarded, others were adapted to modern chemistry with little if any substantive change.

The number, sequence, and duration of the subsidiary operations became matters of considerable disagreement. Paracelsus, the famous sixteenth-century Swiss physician and medical alchemist, believed that the Stone could be created by employing seven basic processes. But a century earlier George Ripley, Canon of Bridlington in Yorkshire, had insisted on no less than twelve operations much like those enumerated above. Moreover, each major process was related to a different sign of the zodiac, requiring the alchemist to command an intimate knowledge of astrology. The positions of the planets and stars were considered by many to be every bit as crucial to success as the controlled use of fire and the sequential appearance of colors. The ancient Babylonian idea that each of the planets was connected with a specific metal also appealed to students of alchemy: the Sun and Moon (planets to the ancients) stood for gold and silver, while Mars, Mercury, Venus, Jupiter, and Saturn signified iron, mercury (quicksilver), copper, tin, and lead, respectively.

The amount of time required to produce the stone was also a subject of much speculation. Everyone could agree that, above all else, alchemy was the art of the patient and the long-suffering. The Canon's Yeoman of the *Canterbury Tales* had spent so many wearing hours attending his master's furnaces that, like the contents of his crucible, he underwent a gradual physical transformation, and in the process became a stranger to himself:

> *I am so used in the fyre to blowe,*
> *That is hath chaunged al my colour I trowe ...*
> *Though I was wont to be right fresh and gay*
> *Of clothing, and of other good array,*
> *Now may I were an hose upon myn head;*
> *And where my colour was both fressh and red,*
> *Now it is wan, and of leden hewe,*
> *Why-so it useth, sore shal he rewe.*

Newton was obviously jesting when he told Wickins that his hair had turned silver from experimenting too frequently with mercury, but he knew better than anyone of his day the price the adept must pay for his devotion to this ancient art.

III

It seems almost certain that Newton's introduction to the fire and the crucible came during those formative years when he lived with the apothecary Clark while studying at King's School in Grantham. There is no evidence to suggest that Clark was a seeker of the Grand Elixir, but in the absence of patent remedies his livelihood dictated that he establish some type of laboratory in which to prepare the medicines required by the local populace. And like many another seventeenth-century calling associated with the rise of modern science, the apothecary's art was in a state of considerable intellectual ferment.

That state resulted substantially from the efforts of Phillipus Aureolus Paracelsus, the widely traveled, much-maligned Swiss physician and alchemist of the sixteenth century. Eschewing the humoral theory of disease championed by the entrenched Galenists, Paracelsus indelicately shouted the need for experimentation: "For every experiment is like a weapon which must be used according to its power, as a spear to thrust, a club to strike."[6] (One can almost picture Newton nodding his approval.) Paracelsus had little patience with alchemists and physicians who, having failed in their attempts to transform lead into gold, joined in the quest for a universal medicine, the long-dreamed-of panacea for all the ills of mankind. He became the advocate of a radical new medical doctrine called iatrochemistry, which involved the application of specific chemical remedies (laudanum, mercury, sulfur, iron, lead compounds, zinc, salts, and arsenic) to specific diseases. Paracelsus's work *On Diseases of Miners* was the first study solely devoted to occupational illnesses; he was also the first to discern a hereditary pattern in the transmission of syphilis, to associate cretinism with goiter, and to establish a causal relationship between severe head injuries and paralysis of the limbs.

Believing himself to be chosen by God for his special work, the brilliant but caustic physician made enemies of almost everyone with whom he came into professional contact, which contributed to the delay in acceptance of his iconoclastic theories. Paracelsian principles of treatment were at last coming into their own in England by the mid-seventeenth century, however, and apothecaries like Clark were increasingly attracted to them, as were such self-healers as Robert Hooke, who polished off potable chemicals by the tankard. If the new remedies could not cure disease, at least they gave the afflicted some measure of symptomatic relief. That the adolescent Newton was an attentive observer of Clark's professional activities is attested to by the recipes and nostrums written down in his early student notebooks, along with others copied from John Bate's *The Mysteries of Nature and Art* and Francis Gregory's *Nomenclatura*. Also worth recalling is the fact that Clark's brother, Joseph, a frequent visitor to the High Street apothecary shop, was a physician and an usher at the very school were Isaac studied. The deceptively quiet, unobtrusive youth may well have been privy to the brothers' animated discussions of medical chemistry in particular and of alchemy in general, adding to a mounting list of scientific mysteries already tugging like the waxing moon at the widening sea of his curiosity. And if, as seems likely, Clark was something of a father figure to Isaac, such an experience would have meant even more to an alienated boy who already loved ideas more than men. One also thinks of what Stukeley described as the "great parcel of books" stored in the Clark attic after Hannah called her reluctant son home to Woolsthorpe. They belonged to the recently deceased Dr. Clark and, like an irresistible magnet, drew the intellectually starved Isaac back to his old garret quarters whenever he was sent to Grantham on family business. Stukeley wrote that the library comprised works on "physic, botany, anatomy, philosophy, mathematics, astronomy, and the like."[7] Considering both the breadth and size of the collection and Dr. Clark's profession, it would have been odd had he not also owned a number of chemistry works.

As with so many facets of his intellectual development, the turning point in Newton's study of chemistry and alchemy came at Trinity College before his swift rise to the ranks of the professoriat. The most profound and original writing in the *Quaestiones quaedam Philosophicae* is that on the optics, but Newton's commonplace book also contains guidelines on how to make a crucible endure for half a year by coating it with a mixture of "tobaccopipe clay & salt of Tartar".[8] He further paused to consider the virtues of "Fier" and its product "Heate," and reflected upon the changes wrought when different substances are exposed to wide variations of temperature. Most important to the subsequent development of his thoughts on alchemy, however, were the many notes concerning the composition and behavior of particulate matter. Newton wrote five pages on "Attomes," seven on "Motion," and three on "Matter" in addition to describing "Attraction," both magnetic and electrical, "Flux & Reflux," "Sympathy and Antipathie," "Density," and "Filtracion." Here was the true seed of the alchemy, if not the flower.

By 1666 the notebook containing the *Quaestiones* was filled to overflowing, and Newton's knowledge of the new science, which burgeoned like a cell dividing, compelled him to begin another. The young savant extended his writings on many of the subjects previously addressed, but he also expounded upon a number of new topics, including chemistry. An autodidact, Newton had earlier drawn on Robert Boyle's works as one of his main sources for mastering the intriguing new mechanical philosophy. He now devoured Boyle's just published *The Origine of Formes and Qualities,* copious notes from which appear in the second or so-called chemical notebook under such headings as "Of formes and Transmutations wrought in them".[9] The depth of Newton's early interest in this subject is further revealed by a dictionary of chemical terms he composed about the time he began taking notes from Boyle. The writing, which dates from 1666 or 1667, is so cramped that it has been described as microscopic, an indication that Newton's eyesight at close range was nothing short of extraordinary. He again drew heavily on Boyle's *Of Formes*; indeed, Boyle's work is the only outside source cited. It must be obvious, however, to anyone who has squinted his way through these several quarto pages, that their twenty-four-year-old author was already something more than a raw initiate. Take, for example, Newton's detailed knowledge of furnaces, which could not be found anywhere in Boyle's works:

> ffurnace. As l ye Wind furnace (for calcination, fusion, cementation &c) wch blows it selfe by attracting ye aire through a narrow passage 2 ye distilling furnace by naked fire, for things y^t require a strong fire for distillation. & it differs not much from ye Wind furnace only ye glasse rests on a cross barr of iron under wch is a hole to put in the fire, wch in ye wind furnace is put in at ye top. 3 The Reverbetatory furnace whereby ye flame only circulating under an arched roof acts upon ye body. 4 ye Sand furnace when ye vessel is set in Sand or sifted ashes heated by a fire made underneath. 5 Balneum or Balneum Mariae when ye body is set to distill or digest in hot water. 6 Balneum Roris or Vaporosum ye glasse hanging in the steame of boyling water. Instead of this may bee used ye heat of hors dung (cald ventur Equinus) i:e: brewsters grains wheat bran, Saw dust, chopt hay or straw, a little moistened close pressed & covered. Or it may in an egg shell bee set under a hen. 7 Athanor Piger Henricus, or Furnace Acadiae for long digestions [ye vessel] being set in sand heated wth a Turret full of Charcoale wch is contrived to burn only at the [bottom] the upper coales continually sinking downe for a supply. Or the sand may be heated [by a] Lamp & it is called the Lamp Furnace. These are made of fire stones, or bricks.[10]

During a visit to London two years later, in 1669, Newton purchased two small furnaces, one made of tin, for the combined sum of fifteen shillings.[11] He constructed most of his own experimental apparatus, however, and Conduitt later reported that Newton's furnace at Cambridge, like his living quarters, was "preserved religiously & shewn to strangers."[12]

It has been observed of Newton's early notebooks that he did not "stumble" into alchemy, find it wanting, and abandon its pursuit in favor of "sober, rational, chemistry." Rather, he took the opposite path, beginning with "sober chemistry," which he soon forsook for a more profound study of alchemy.[13] In fact, it is not clear that such a distinction exists. There was no specific body of knowledge that could be called chemistry during the seventeenth century. The belief still prevailed—and Newton, like Boyle, adhered to it from the very outset—that every substance was formed by the rearrangement (that is, transmutation) of minute particles of one universal or catholic matter. Modern chemistry, on the other hand, derives from the principle that matter forms distinct chemical elements, each with its own unique type of particle. If anything, alchemy was the more clearly defined doctrine, while rational chemistry remained very much in its germinal state, a condition reflected in the broad diversity and widely varying quality of the new chemical literature. Not until that brilliant generation of eighteenth-century thinkers that included Joseph Priestley, Henry Cavendish, and, most important, Antoine Lavoisier did chemistry become a clearly distinguishable branch of the new science.

The contents of the early papers also belie the notion that Newton strictly differentiated between alchemy and rational chemistry in his own mind. There are indeed more references to what might be called purely chemical terms and processes than to alchemical ones, but the latter category is by no stretch of the imagination conspicuous by its absence. Furthermore, many of the terms now classified as chemical were subsumed under the more encompassing alchemy three centuries ago, just as astronomy and its nomenclature remained the handmaiden of astrology until well into the Renaissance. What we see is a young thinker striving to master the terminology and experimental techniques of an emerging science that was to become his obsession for the next thirty years. Success would require that he progress from the more simple and straightforward to the highly complex and abstruse, his strategy from the very outset, which he pursued with like fervor in mathematics, optics, and mechanics. The chemical dictionary contains such alchemical terms as "*Sanguis Draconis*," the blood of the dragon, which Newton defined under the heading "Vermillion" as "ye flowers of Mercury and sulfur sublimed together." The many entries on mercury and its various derivatives also shade off into alchemy, especially when he observes that *Mercurius Sublimatus* possesses the power to "open" or break down tin, copper, and silver, but not gold itself. Temporarily donning the alchemist's mantle with its lining of purest optimism, Newton conjectures that another sublimate may well be found that can reduce this most perfect of metals to its fundamental constituents.[14] Dobbs has pointed out that a number of other "alchemically sensitive" terms—"Alcahest," "Anima," "Elixar," "Minera Work," and "Projection"—are listed by Newton but left undefined.[15] Whether or not these omissions are philosophically significant is open to question, however, for he also left blank the spaces reserved for "Isinglasse," "Laudanum," "Senna," and "Soape." The very inclusion of alchemical terms, even if they are not always defined, would seem the more telling point. Most important of all, perhaps, is the entry containing directions for making the star regulus of Mars, soon to become one of the brightest orbs in the vast galaxy of Newton's alchemical thought.

IV

The exact point at which Newton made the critical transition from his initial role as a collector of alchemical data to that of an active experimentalist remains unknown. It could have occurred as early as 1667 and certainly not later than 1669, as indicated by his record of expenditures for that year. He visited London in November and while there purchased Zetzner's six-volume *Theatrum Chemicum*, two furnaces, some glass equipment, and a few pounds' worth of sundry chemicals. Equally revealing is his letter of

May 18 to Francis Aston, who was about to embark on the European Grand Tour. Newton, it will be recalled, had urged his friend to gather all the information he could on the transmutation of metals and to seek out the mysterious green-clad alchemist Giuseppe Borri, a possessor of "secrets of great worth." [16] According to Newton, these alleged secrets were mainly of a medical nature; we may surmise with some measure of assurance that he believed the archetypal "Bory" capable of greatly extending the normal human life span.

Newton's chemical notebook contains a section titled "Medical observations" in which he wrote out the recipe for a "*primum ens* of Baulm," said by Paracelsus to help restore lost youth. The concoction was to be "taken early every morning so much in good wine as will give it a tincture till ye nailes hair & teeth fall of[f] & lastly the skin be dryed & exchanged for a new one." Newton further noted that "Monsieur L. F. (Le Fever)" told Boyle of a friend who had tried the recipe on himself "for a fortnight till his nails came of[f] without any pain." What is more, the friend gave the *ens* to a woman of seventy "for 10 or 12 days till her Menstrual came copiously upon her so as to fright her." [17] The "Le Fever" Newton referred to is unquestionably Nicolas Le Fèbre, the famous Parisian chemist on whom Charles II in 1660 conferred the dual titles of Royal Professor of Chemistry and Apothecary in Ordinary to the Royal Household. John Evelyn was taken by the King to his laboratory one September afternoon in 1662, where "his *Chymist* (who had formerly ben my Master in *Paris*) compos[ed] of Sir *Walter Raleighs* rare *Cordial,* he making a learned discourse before his Majestie in French on each Ingredient." [18] Samuel Pepys, accompanied by Lord Brouncker and Sir Robert Moray, later gained access to Charles's sanctum sanctorum, but with a disappointing result: "into the King's little elaboratory under his closet, a pretty place, and there saw a great many Chymicall glasses and things, but understood none of them." [19] The year of Pepys's visit was 1669, the very time when Newton was undergoing his alchemical baptism by fire. One wonders what Newton would have given had he been able to switch places, however briefly, with the indifferent future Secretary to the Admiralty?

Appearing at various points in the margins of Newton's notebooks and manuscripts is the carefully drawn figure of a pointing hand, a device for singling out passages deemed most critical to the subject under study, whatever it might be. The hand's appearance in the early alchemical papers is symbolic of something other than the purpose for which Newton intended it, however, for it again raises the difficult question of whether he acted alone or directly benefited from the guiding hand of some so far unidentified master.

There is some evidence that the tiny circle of natural philosophers who inhabited Trinity College during the 1640s and 50s maintained a communal laboratory for the study of "chemistry." Isaac Barrow was among them, as were John Ray, the brilliant naturalist who had matriculated the same year as Barrow, and John Nidd, an older Fellow in whose rooms their informal meetings were probably held. Nidd appears to have taken charge of their equipment, which included, among other things, a large iron retort. Dobbs speculates that when Nibb died in 1659, Ray, who remained at Trinity until 1662, took responsibility for the apparatus and that it eventually passed to Newton in the late 1660s.[20] If Newton did fall heir to the equipment, he probably restored it to good working order after a considerable period of disuse, for by then only Barrow was left from the original group.

Owing to his absence from Cambridge from 1655 to 1659, Barrow had probably given up active experimentation before he made Newton's acquaintance. Still, there is little question that he approved of the study of alchemy and encouraged Newton in his pursuit of it. Perhaps also influential in this regard, if only indirectly, was Henry More, the renowned Cambridge Platonist and Grantham native. The son of "a gentleman of fair estate and fortune," More was educated at Eton and elected Fellow of Christ's College in 1639, a position he retained throughout the Restoration. He declined all ecclesiastical and academic preferments, including the mastership of Christ's, which passed to his fellow Platonist, Ralph Cudworth, in 1654. More had been Dr. Joseph Clark's tutor at Cambridge, the same physician whose library drew young Newton back to the Grantham apothecary shop whenever he could get away from Woolsthorpe. When and under what circumstances Newton was introduced to More is not known. While it seems unlikely that a lowly Trinity sizar would have kept company with an august scholar from another college, Newton was obviously familiar with More's *The Immortality of the Soul* (1659) well before he took his B.A. On the third page of the *Quaestiones quaedam Philosophicae* under the heading "Of Attomes" he wrote, "The excellent Dr Moore in his booke of ye immortality hath proved beyond all controversie" that matter "cannot be divisible in infinitum."[21]

As a young scholar Henry More had been profoundly attracted to Cartesian mechanics, perceiving it as a means of creating a rational theology by reconciling Scripture with the new natural philosophy. More was nevertheless deeply troubled by Descartes's absolute severance of matter and spirit, for it must inevitably lead to the unthinkable: that God takes no direct part in the day-to-day operations of the universe. Unable, through personal correspondence, to convince Descartes of the need to qualify his stand, More devoted much of his intellectual energy to a refutation of what he viewed as incipient atheism. The mechanical universe must not be separated from its Creator. With this Newton readily agreed.

More's interest in alchemy was a product of his attempt to prove that certain phenomena—transmutation, magnetism, and electricity, for example—result from more than the mere impact of atom upon atom. Behind these operations More perceived what he termed the "Spirit of Nature," which mediates between an ever active God and the purely mechanical universe. Although Newton's Platonism was not entirely at one with the Platonism of More, both men sought to extend their understanding of the Creator and His universe by elucidating the behavior of insensibly small particles. Indeed, Newton's ultimately unsuccessful attempt to ally the world of mechanical forces with the world of active principles lay at the very heart of his philosophy of nature—and of his monumental alchemical quest as well.

Because Isaac Barrow left Trinity College for London in 1669 and the solitary More seems not to have done any actual experimentation until after 1670,[22] we may be fairly certain that Newton had no true alchemical master but himself, just as he had no one to teach him advanced mathematics, optics, or mechanics. Besides, he was ever a loner; one senses that he kept even Barrow at arm's length, picking the brain of his older friend while telling him only what he wanted him to know. Leon Edel's metaphor for Thoreau's reclusive and suspicious manner applies equally well to Newton: "[He] struggled to keep the parcel of himself from becoming unwrapped and scattered."[23] To have worked closely with another would have required not only a sharing of himself and of the credit for any discoveries but a tacit admission that he, like other men, was in some sense ordinary.

Whatever the subject under study, Newton would not rest until the tracks of his erudition were spread across an endless succession of pages. He sought a depth of understanding rarely dreamed of by others, aiming, in Humphrey Newton's words, "at something beyond ye Reach of humane Art and Industry." To attain that understanding in alchemy Newton probed the vast corpus of extant literature with a thoroughness never equaled before or since.[24] In an effort to organize his knowledge to make it available whenever needed, he compiled a remarkable document during the 1680s, the *Index chemicus*. It exceeds 100 pages and contains 879 separate headings. No one has as yet counted each of the individual page references to the scores of catalogued works, but a figure of 5,000 has been suggested by Richard S. Westfall, who spent a week conducting his preliminary quantitative analysis: "It is hard for me to imagine that anyone could have composed it in less than a thousand times that week," he wrote, "although I hasten to add I cannot find room in Newton's career for any period approaching a thousand weeks."[25] In the face of such massive evidence Westfall could not but concur with Dobbs: "[A]lchemy has never had a student more widely and deeply versed in its sources."[26] This development is all the more extraordinary in light of the fact that Newton contemporaneously emerged as the greatest mathematician, physicist, experimentalist, and authority on optics the world had yet known.

Newton literally read everything alchemical that he could buy or borrow, fearing, it seems, that the smallest unturned stone might conceal the key to the universal matter that he believed must explain the very structure of the world. Whereas books dealing with most scientific subjects were purchased in significant numbers by the University and Trinity College libraries, their alchemical holdings were not very substantial. Nor was Barrow's personal library, to which Newton had ready access, of much help to him in this pursuit. The inventory taken at Barrow's death contained a total of only six titles on alchemy and chemistry.[27] Newton, therefore, had little choice but to build his own alchemical library which, as with so much else he chose to do, was accomplished in masterful fashion.

Among the many hundreds of volumes from Newton's personal collection now in the possession of Trinity College are 169 titles on alchemy and chemistry, or about 10 percent of the total. Since 156 of these volumes were published before 1696—the year he moved to London—we may be quite certain that the ratio of alchemical works to the whole was considerably greater at one time.[28] Neither does this figure take into account the "3 dozen of small chymical books," nor the "hundred weight of wast[e] books and Pamphlets" since lost or dispersed at auction among unknown buyers.[29]

Newton purchased his first alchemical treatises in the late 1660s, showing no partiality as between authors old and new. Using Zetzner's *Theatrum Chemicum* and Elias Ashmole's *Theatrum Chemicum Britannicum* as his base, he soon began accumulating the works of individual alchemists: Johann Becher, Kenelm Digby, Geber, Samuel Hartlib, Helvetius, Raymund Lull, Michael Maier (a favorite), Samuel Norton, Paracelsus, George Ripley, Michael Sendivogius, George Starkey (another favorite), and Basil Valentine, to name but a few. Who his favorite booksellers were we do not know, but the dearth of interest in natural philosophy at Cambridge suggests that London served as his main source of supply. Many of his alchemical books were purchased secondhand and still bear the names and annotations of their previous owners, an indication that at least some volumes were bought in bundles at book sales, perhaps by an agent acting on his behalf. The Great Fire had wreaked havoc on the London book trade, which, for reasons of supply, needed more time to recover than most other enterprises. William Cooper, with a shop at the sign of the Pelican in Little Britain, was the leader among the few venders of alchemical literature in the capital. Cooper also translated and published the works of many of the medieval and postmedieval alchemists and was in daily contact with numerous practitioners of the art. It would be difficult to believe that Newton long remained a stranger to Cooper's bustling shop. A much-used copy of the bookman's *A catalogue of Chymicall books* (1675) was a welcome addition to his private collection.[30]

To Isaac Newton, a pathological hoarder of manuscripts, the printed word, however important, was never to be taken as the last word on any subject, alchemy included. Though the facts are relatively few and scattered, there can be no doubt that he had direct access to a considerable range of unpublished alchemical materials, some of which came into his possession either at or very near the outset of his foray into the field. Among the papers purchased by Keynes is a substantial bundle of alchemical essays, hardly any of which have been published. These are written in several different hands, none of which match the writing of Newton or his amanuensis-chamberfellow Wickins.[31] It does not appear that Newton purchased the manuscripts from a bookseller or through an agent. If he had, he would not have bothered to copy several of the treatises, which he did, or take extensive notes on others. Indications are that he borrowed the manuscripts and for some unexplained reason did not return them to their owners. Other copies of unpublished alchemical tracts in Newton's hand also survive, lending additional support to the argument that he was somehow connected with "the largely clandestine society of English alchemists."[32]

Exactly how and with whom are both matters of conjecture. One again thinks of Barrow, especially during the late 1660s, for as yet Newton had no known association with a broader circle of natural philosophers. On the other hand the elder Isaac, a member of the Royal Society since 1662, had wide-ranging intellectual contacts. Though he may not have been actively involved in experimentation at this time, Barrow's interest in alchemy remained strong. Secret manuscripts undoubtedly would have been placed at his disposal for the asking. While Newton and More were probably not well acquainted during the 1660s, More and Barrow almost certainly were, and this link, as we shall see, was quite possibly of considerable benefit to Newton. In addition to his and More's mutual interest in the new science, Barrow was deeply sympathetic to the philosophical position espoused by More and his fellow Platonists, most particularly their rejection of rigid Cartesian mechanism.

Henry More led a reclusive existence similar to Newton's, but with at least one important difference. He occasionally departed Christ's College for Ragley in Warwickshire, the country estate of his "heroine pupil" Anne Finch, Viscountess Conway. More made Anne's acquaintance through her brother John Finch, one of his favorite students at Cambridge. At Ragley and in a long correspondence, the celebate Platonist, who became known as the "Angel of Christ's," carried on a Platonic love affair with the frail but sensitive and gifted woman he called his "dearest dear." Lady Conway reciprocated by making Ragley a center for More's intellectual circle, which included Samuel Hartlib, the renowned London alchemist and self-appointed disseminator of knowledge of all sorts. It was Hartlib's dream to institutionalize his far-reaching and voluminous correspondence by gaining government funding for his brainchild, the Office of Address. In this he failed, but Henry Oldenburg, in the tradition of Hartlib, was later to provide a public forum for the free exchange of scientific ideas by launching

Philosophical Transactions. The alchemist George Starkey, the pseudony-
mous Eirenaeus Philalethes, became a close associate of Hartlib, as did
other prominent London adepts: Robert Boyle, Frederick Clodius (Hartlib's son-in-law), and Sir Kenelm Digby, in whose Covent Garden house
they often met to conduct experiments and exchange ideas. In the late
1660s Newton copied from an unpublished version of Philalethes's (Starkey's) *Exposition upon Sir George Ripley's Epistle to King Edward IV* and
also took detailed notes from this same author's manuscript of *Ripley
Reviv'd* a decade before its publication. Equally intriguing is the fact that
the publication of *Ripley Reviv'd* was undertaken by none other than the
bookseller William Cooper, who printed at least two other manuscripts
copied by Newton.[33] Hartlib died in 1662, but members of his circle were
frequent visitors to Ragley for many years thereafter. More would have had
little difficulty obtaining unpublished literature from them, which, on being taken to Cambridge, could have easily passed from his hands to
Barrow's and from Barrow's to Newton's. Later, after Newton was appointed Lucasian Professor, there would have been no further need for a
go-between. This, in addition to the demand for secrecy, would also explain
the absence of any private correspondence from this period on the subject
of alchemical manuscripts.

Additional evidence of a tie to the Hartlibians survives in the form of a
letter Newton received from Oldenburg in September 1673. "I herewth
send you Mr. Boyle's new Book of Effluviums, wch he desired me to present to you in his name, wth his very affectionat service, and assurance of ye
esteem he hath of your verture and knowledge." In the same packet were
two more copies of the book, "one for Dr. Barrow, and ye other for Dr.
More, wch he intreats you to send to ym from him."[34] Though it is not clear
whether the two men had as yet met face to face in London, Boyle was obviously aware that Newton's scientific interests extended well beyond those
thus far communicated to the Royal Society. The son of the Earl of Cork
treated the yeoman's son as if he were the senior rather than the junior
member of the intellectual triumvirate to which he dispatched copies of his
book. It was an age that still respected the single mind.

While Barrow and More were quite likely Newton's earliest sources of
unpublished material, he clearly developed other contacts during the
1670s. On a manuscript bearing the title "Manna," which is written in an
unknown hand, Newton penned this sentence: "Here follows several notes
& different readings collected out of a M.S. communicated to M[r] F. by
W. S. 1670, & by M[r] F. to me 1675."[35] "M[r] F." has been tentatively identified as Ezekiel Foxcroft, a Fellow of King's College from 1652 to 1675, the
year he died.[36] Foxcroft's uncle was Benjamin Whichcote, a member of the
little band of Platonists that included Henry More. Both Ezekiel and his
mother, Elizabeth Whichcote Foxcroft, were repeatedly referred to by
More in his extensive correspondence. In the succeeding years Newton

made a few equally oblique references to the alchemists from whom he obtained manuscripts, but nothing beyond this. However, his reading of Philalethes-Starkey and the use of initials rather than names to identify his contacts are reminders that he also devised an alchemical pseudonym for himself: "*Jeova sanctus unus*," an anagram derived from the Latinized *Isaacus Neuutonus*. Whether the anagram was significant to Newton for reasons other than secrecy is a fascinating question. It could also have been a private declaration that Jehovah is the one and perfect God, a scarcely veiled expression of his well-documented antitrinitarianism.

V

For Victorian biographers, candor was among the least prized of virtues. They often conspired with their subjects—both living and dead—to protect them from posthumous harm, making themselves allies of reputation and of legend. They placed their heroes on Jovian pedestals and would not dream of regarding what the Victorian writer John Morley contemptuously called "the hinder parts of their divinities." Sir David Brewster's discovery that Newton had indiscriminately immersed himself in the writings of alchemical adepts therefore came as a bitter blow. He was haunted by the specter of Newton as a cloaked and bearded magus personified in literature and limelight by Marlowe's Doctor Faustus and Shakespeare's Prospero. More appalling still were the images of such real-life historical figures as Dr. John Dee, an original Fellow of Newton's own Trinity College. "Hee had a very faire rosie complexion," wrote Aubrey, who remembered Dee from his childhood, and "a long beard as white as milk." A handsome man, tall and slender, "he wore a Gowne like an Artist's gowne, with hanging sleeves, and a slitt," the very embodiment of how the mind's eye would picture a disciple of the occult.[37] Dee was arrested in 1553, shortly after Mary I's accession to the throne. Accused of trying to murder his Queen by means of magic, he was brought before the Star Chamber, England's version of the Inquisition. Acquitted of treason, no mean act of legerdemain in itself, Dee was back at court in 1558, where he employed astrology to determine the most propitious day for Elizabeth's coronation. Dee was both a scryer and an alchemist. His globe of polished smoky quartz is now in the collection of the British Museum, along with waxen tablets containing the crypitc figures he employed in his magical rites. Dee also traveled to Prague, then the center of alchemical studies in Europe, and there "he had more than once seen the Philosopher's Stone."[38] Back home in England his furnaces blazed forth day and night, and children dreaded the thought of walking past this shadowy figure's home, because they thought him a sorcerer. So did their parents. In 1576 a mob assembled around Mortlake, Dee's country residence, and, attacking the house, broke his instruments and demolished his large and costly library. The magus and his family barely escaped with their lives. A few years later Dee, acting in apparent good faith, joined forces with the notorious but ingratiating mountebank Edward Kelly (alias Talbot), a onetime apothecary's apprentice and convicted forger. Together they conducted seances and alchemical experiments. Kelly, who was never without the black skullcap that hid the wounds inflicted when the authorities cropped his ears in the pillory, succeeded in convincing the learned but gullible Dee that he had unearthed a supply of the philosopher's stone amid the ruins of Glastonbury Abbey. What is more, he argued that it was prepared by none other than Saint Dunstan himself! Kelly even had the gall to send Queen Elizabeth a piece of gold which he claimed had once been nothing more than an iron cutting from a warming pan. It was against this historical background of fraud and mysticism that Brewster came to grips with Newton's alchemy. One can understand why he blanched and drew back in horror.

The questions that confronted Brewster a century and a quarter ago and that persist today can be put rather simply: Was Newton a mystic and an adept in the mold of a John Dee? If not, then what did he expect to learn by becoming the world's most knowledgeable expert on alchemical writings, present and past? Finally, but most important of all to the biographer, what deep obsessions and strange tensions were asserting themselves beneath that ever taciturn exterior?

After cutting his alchemical eyeteeth on the published works of Robert Boyle, Newton found his appetite for the writings of other adepts growing to omnivorous proportions. Especially tantalizing to him were the esoteric treatises of Michael Maier (1568–1622), a German physician and aristocrat. Newton had read Maier's *Symbola aureae mensae duodecium natorium* (Frankfurt, 1617) by early 1669, for he drew upon his extensive notes from it when composing the May letter to Francis Aston.[39] (It was from this book that Newton learned of Giuseppe Borri some fifty-two years after Maier had written about him. Yet if his correspondence is to be our guide, Newton thought it likely that Borri was still among the living.) The *Symbola*, like the other eight works by Maier that Newton came to own, makes little clear chemical sense. Rather, it shows great scholarly erudition in mythology and ancient history. Nothing, according to Maier, is so praiseworthy or sublime as the pursuit of *Chemia*, the ancient art of arts. Its true devotees must be rescued from the calumny heaped upon them by their implacable adversaries: "May the Father of lights shower down all good and perfect gifts on those who vindicate the chastity of this virginal science."[40] It was for these same enlightening gifts that Newton so desperately hungered.

Maier was obsessed by the grandiose notion that the whole of Egyptian and classical mythology was an allegorical expression of the alchemical process.[41] He embraced the centuries-old argument that the "Queen of Arts" was fathered by Hermes Trismegistus, or Hermes the Trice-Great, the Greek counterpart of the Egyptian god Thoth. In legend Thoth is credited with the invention of writing in the form of hieroglyphics, and the earliest books are supposed to have been written either in his hand or under the direct guidance of the god himself. Hence the word hermetic, taken in its broadest sense, meant inspired by Thoth (Hermes). Varying estimates exist as to the number and content of these works. Clement of Alexandria mentioned forty-two hermetic books, which supposedly contained the sum of all knowledge, human and divine. That number had been reduced to fifteen in the West by 1460, when a monk in the employ of Cosimo dé Medici brought a Greek manuscript of the *Corpus Hermeticum* to Florence.[42]

To Maier, however, Hermes was far more than the mere product of a highly creative ancient mythology. He was in fact the first in a line of illustrious Egyptian philosopher-kings, a probable contemporary of the Hebrew Patriarch Abraham, to whom he was almost certainly related. Hermes' all-encompassing wisdom derived from Seth, the son of Adam, in pure and uncorrupted form. Among his many invaluable treatises was the *Tabula Smaragdina,* or *Emerald Table of Alchemy,* allegedly rescued from the ancient king's tomb by none other than the conveniently ubiquitous Alexander the Great. The principle tenets of this and other hermetic books can be briefly summarized: God made the cosmos by his word out of fluid matter; the human soul is a union of light and life and derives from the cosmic soul; life and death are merely changes in form, for nothing is destructible; the soul transmigrates; passion and suffering are the result of motion. By tradition Hermes was even thought to have possessed knowledge of the doctrine of the Holy Trinity. So far as Maier was concerned, all of history's true adepts—Plato, Democritus, Avicenna, Albertus Magnus, Arnold of Villanova, Raymond Lull, Roger Bacon, and many others—had merely feasted on generous servings from Hermes' inexhaustible banquet.

Newton's preoccupation with the great alchemical masters from Hermes Trismegistus to Michael Maier was an outgrowth of the Renaissance belief in the *prisca sapientia,* or wisdom of the ancients. Most elaborately developed by Marsilio Ficino and Pico della Mirandola, the leading members of the Platonic Academy at Florence in the late fifteenth century, the doctrine proclaimed that the great truths of nature had been recorded at some time by the most brilliant and morally upright thinkers of the distant past. Not surprisingly, this anciently given wisdom had been preserved in a veiled and enigmatic form, the better to protect it from the avaricious hands of the vulgar. It was the researcher's sacred duty to decipher the language and rationally interpret the myths and allegories in which the *prisca* was cast. Had not the prophet Isaiah promised enlightenment to the true believer?

> And I will give thee the treasures of darkness, and hidden riches of secret places, that thou mayest know that I, the Lord, which call thee by they name, *am* the God of Israel.[43]

As an alchemist and devoted Biblical scholar, Newton lived by this promise. He appended the following to the "Manna" manuscript obtained from Foxcroft in 1675:

> For Alchemy tradeth not wth metalls as ignorant vulgars think, wch error hath made them distress that noble science; but she hath also material veins of whose nature God created Handmaids to conceive & bring forth his creatures.... This Philosophy is not of that kind wch tendeth to vanity & deceipt but rather to profit & to edification inducing first ye knowledge of God & secondly ye way to find out true medicines in ye creatures ... ye scope is to glorify God in his wonderful works, to teach a man how to live well.... This Philosophy both speculative & active is not only to be found in ye volume of nature but also in ye sacred scriptures, as in Genesis, Job, Psalms, Isaiah & others. In ye knowledge of this Philosophy God made Solomon ye greatest philosopher in ye world.[44]

What he had written as a young man was scarcely erased by old age. Newton was well into his eighties when he informed Conduitt, "They who search after the Philosopher's Stone by their own rules [are] obliged to a strict & religious life."[45]

Newton's self-imposed regimen was previously likened to that of a medieval monastic, and Humphrey Newton's description of his employer's approach to alchemy during the 1680s serves to strengthen the analogy:

> So intent, so serious upon his studies y^t he eat very sparingly, nay ofttimes he has forgot to eat at all.... He very rarely went to Bed, till 2 or 3 of ye Clock, sometimes not till 5 or 6, lying about 4 or 5 hours, especially at spring & fall of ye Leaf, at wch Times he used to imploy about 6 weeks in his Elaboratory, the fire scarcely going out either Night or Day, he sitting up one Night, as I did another till he had finished his Chymicall Experiments, in ye Performance of wch he was ye most accurate, strict, exact. What his Aim might be, I was not able to penetrate into, but his Pains, his Diliginc at those sett Times made me think, he aimed at something beyond ye Reach of humane Art & Industry. I cannot say, I ever saw him drink, either Wine, Ale or Bear, excepting Meals, & then but very sparingly.[46]

Though he was no Boswell, Humphrey well knew that he was in the presence, if not the company, of abstracted genius impelled by an intense moral force, a man who could never accept John Donne's somber pronouncement on the emergent modern world: "'Tis all in peeces, all cohaerence gone."

Alchemy has always served as a metaphor for character transformation, and with Newton something more than chemicals and metals was being tested in the crucible. In his most private moments he dreamed of highborn kinsmen who never were and of a nonexistent entitlement based on earthly wealth and power. But in reality he always knew that his true entitlement was an irresistible power of mind, God-given and therefore to be employed in the Creator's service. His Puritan sense of mission was nothing if not awesome, and never was it more evident than when he immersed himself in the alchemist's art. The "little smatterers" who demanded replies to their facile critiques of his optical discoveries had trodden on hallowed ground. They now compounded their blasphemy by interrupting his intellectual communion with the ancient ones. He rewarded them by striking out in anger or else, with a glacial silence, refusing to acknowledge their existence. To step aside, even for a moment, would require that he expose himself to the fearsome risk of losing his place forever.

Dwarfs did indeed abound, but God in His infinite wisdom had not permitted the race of giants to become extinct. Newton's feeling of rejection by this world was further proof to him of his favor in the next. It was still a time when the pursuit of perfection was a possible calling. His life dramatized a high idealism with all its attendant pitfalls: self-righteousness, a lack of sensitivity toward others, moral arrogance, and egotistical narrowness. Standing atop the peak of a great intellectual divide, he beheld beautiful valleys on either horizon—ancient and modern—and both called him home. He chose to dwell in neither one and attempted instead to forge both into an even grander habitable creation—a *science universelle*.[47] However chimerical, the *prisca sapientia* provided Newton with a personal mythology that bore him from a limited present to the scientist he was to become.

Lest there be any misunderstanding, let it be stated unequivocally that, unlike certain of his deeply religious contemporaries, Newton did not hear voices or consider himself a passive vessel waiting to receive holy text. Nor was he a mystical lawgiver in the tradition of Pythagoras and Kepler, as some have suggested.[48] Rather, he shared the belief, common in the seventeenth century, that natural and divine knowledge could be harmonized and shown to be one. Only when the two were joined did their objects and events acquire full significance and existential import. Although mankind could not possibly hope to attain final truth, neither was the human species condemned to live in the benumbing dark. If properly investigated, "the mechanics manifested in natural phenomena served as windows through which man could catch fleeting and imperfect glimpses of the higher reality and ultimate purpose of creation."[49] But Newton waited for no instantaneous flashes of insight of the kind that impelled Descartes to undertake a pilgrimage of thanksgiving to the Lady of Loreto's shrine in Italy. If he was to illuminate the treasures of darkness, everything hinged on heart-rending, soul-ravishing research underpinned by reason—what the Cambridge Platonists metaphorically spoke of as "the candle of the Lord." The rational faculties must be concentrated on the ancient and esoteric in order to derive the commonsensical, and logic dictated that the most obscure literature guarded the choicest secrets. Newton believed that by carefully removing the patina of mysticism accumulated over the centuries, he could lay bare the truth—not quite what Keynes had in mind when he characterized the savant as "the last of the magicians."

To argue otherwise requires Newton to have left his special gifts at alchemy's door, something he would not and, indeed, could not do. "He brought with him standards of intellectual rigor born of mathematics, which he applied to his own experimentation." He also brought with him "the mechanical philosopher's sense that nature is quantitative."[50] Hence his mastery of the alchemical literature proceeded hand in glove with experimentation in the laboratory, the early results of which we are about to explore. If he was not a chemist in the current meaning of the word, Newton nevertheless understood that chemical processes, not mystical flights into the unknown, lay at the vital center of the hermetic art.

Still, it is a foolhardy student of Newton and of the seventeenth-century history of science who does not weigh most carefully the dangers of speaking in absolutes. In assessing Newton's alchemical endeavors I have come down on the side of what one writer has termed "the bowdlerizing rationalists."[51] This is because experimental verification proved no less important to Newton the alchemist than to Newton the physicist. On the other hand, Keynes was perhaps less wide of the mark than recent scholarship suggests. Driven, like all first-rate scientific thinkers, by the challenge of solving great puzzles, Newton did tend to look on the universe as something of a "cryptogram set by the Almighty." And there is no denying his belief in and strong attraction to an original wisdom "laid about the world" by God for the benefit of an "esoteric brotherhood." This, more than any other consideration, gives the rationalist pause. For Newton's reasoned mechanistic approach to alchemy would seem to have collided head-on with his belief in a transcendental reality.

Support for Newton's Calvinist belief in a special grace and mission was present in his posthumous birth on Christmas day and in memories of his mother's account of a miraculous survival in the face of overwhelming physical adversity. Evidence that Newton thought himself among the elect abounds in his actions and in his work. He was nothing if not compulsive; asked how his discoveries were made, he responded with a Puritan's candor: "By always thinking unto them." Like a prophet in the wilderness, he chose to cut all ties with the outside world and allowed the great outflow of his mind to replace the need for personal love felt by those who have no such profound resource. A virtual recluse, he was at home only within the narrow but sanctified confines of his silent study, where his ideas doubled and tripled in volume, like so much mental yeast. As one historian succinctly observed, "Newton was in Cambridge but not of Cambridge."[52] He spurned all but the most superficial collaboration and could rarely force himself to bestow credit upon another lest such recognition jeopardize the uniqueness of his own creations. Oh, how he fumed, fretted, and raged on learning of Leibniz's independent discovery of the calculus! The cry of plagiarism went forth, later replaced by the ringing indictment "Second inventors count for nothing!" Then, too, there was the *annus mirabilis* itself: the calculus, the optics, and the inverse square law were at one time known to but a single human being and to God. How often, one wonders, did

Newton ask himself why he, above all men, had been blessed with the vision and insight to make such profound discoveries? Except for certain private papers, including many relating to the Mint, Newton preserved almost every variety of written material imaginable: random jottings, adolescent notebooks, and book-length manuscripts were all treated like Holy Scripture. The constant recopying of his papers may be taken as evidence of self-conscious awe and of a struggle for perfection. And now hidden alchemical truth appealed to him not because of any materialistic value associated with its discovery but because it would serve as his calling card to posterity, providing additional proof of inner grace and favor in the sight of the Lord while guaranteeing him a place among the great minds of history, that life beyond life of which Milton so eloquently wrote. In Newton's way of thinking there was no place for warfare between science and religion as there was in Voltaire's—and in our own. His attempt to seize the key to the pristine truth lost in classical times was a wonderous scientific enterprise and at the same time a religious pilgrimage of the holiest sort, an effort to burst the chafing bonds of his unrequited self.

VI

Although he was no more attracted to Robert Hooke the man than smoke to a magnet, Isaac Newton could not but agree with a comprehensive statement of the Royal Society's objectives that Hooke had drafted in 1663, shortly after the scholarly body received its second charter. "The business and design of the Royal Society," Hooke wrote, is "to attempt the recovery of such allowable arts and inventions as are lost," and "to examine all systems, theories, principles, hypotheses, elements, histories, and experiments of things naturall, mathematical and mechanicall, invented, recorded, or practiced by any considerable authors ancient or modern." Nor will the Society "own any hypothesis" until "by mature debate and clear arguments, chiefly such as are deduced from legitimate experiments, the truth of such experiments be demonstrated invincibly."[53] Newton had achieved an unmatched level of experimental invincibility in the field of optics, and he now brought his finely tuned skills to bear on alchemy with a passion of even greater intensity and duration. Humphrey Newton had not yet arrived on the scene, but the much inconvenienced Wickins was there to assist him. Newton later told Conduitt that his chamberfellow, who was stonger than he, had helped him move his heavy kettle about. What is more, Wickins tolerated the installation of several furnaces in his own chambers to facilitate Newton's "chymical experiments."[54] And, as before, Newton lost track of time and place. The days and dates he was to record for several experiments did not correspond to those of the calendar, Old Style or New.[55]

The extant manuscripts indicate that Newton's alchemical methodology comprised three distinct stages. He first selected the material to be studied, usually concentrating on the most esoteric works because they promised to reveal the choicest secrets known to the ancients. The second step was rational analysis, whereby he compared the writings of several adepts on the same subject; mythology, mystical insights, chemical knowledge, and everything else that seemed even marginally relevant received the closest scrutiny. (True knowledge was one, but it had been stated in varying forms waiting to be reconciled.) The second step corresponded to the hypothesis–forming stage of the scientific method—the effort to derive the general from the specific—and resulted in statements of relationships that required testing. Finally came the actual experimentation itself. When faced with results of dubious value or the prospect of outright failure, Newton retraced his steps by returning to stages one and two.[56] Evidence of his capacity to discriminate between the promising and the lackluster surfaced almost at once. After striking out a passage from his early notes on one of many alchemists represented in the *Theatrum Chemicum,* he set down this opinion in typical Newtonian terms: "*Credo hic nihil adeptus*" ("I do not believe this author is an adept").[57]

The primary thrust of Newton's earliest alchemical experiments was in the direction of the structure and composition of metals, an interest spawned while he fashioned speculums for the first reflecting telescopes. Deeply influenced by the work of Boyle, Newton sought a method of "opening" metals, that is, making them permanently change their physical characteristics. Beginning at the outmost layer of a body, he attempted to penetrate to the innermost, invisible, seedlike core. He believed that at this core rested the "philosopher's mercury," the first matter of all metals and the source of all activity in the universe. Elias Ashmole, a contemporary, described the "philosophic mercury" as "that Universal and All-piercing Spirit, the One operative Vertue and immortall Seede of Worldly things, that God in the beginning infused into the Chaos, which is every where Active and still flows through the world in all kindes of things by Universall extension."[58] To liberate this special mercury from its fixed form in metals would not only open wide the doors to transmutation and the conquest of disease but would make it humanly possible to understand the very process by which God actively sustains and governs all creation, to identify that supreme interface where matter and spirit meet.

Of the seven metals with which the alchemists traditionally experimented—gold, silver, iron, mercury, copper, tin, and lead—only mercury or quicksilver took the form of a liquid. Yet when melted, the six solid metals displayed many of the qualities associated with mercury. Hence all seven were characterized by a "mercurial principle." Though the philosopher's mercury and common quicksilver were not believed to be one and the same, it was thought that quicksilver contained the highest proportion of the philosopher's mercury, thus making it a major object of experimental interest. Newton's initial method of extracting the mercury of metals derived from a technique published in the 1669 edition of Boyle's *Physiological Essays.* He first dissolved a quantity of quicksilver in *aqua fortis* or nitric acid, a highly reactive oxidizing agent. To this Newton gradually added an ounce of lead filings, which "will bee corroded dissolving by degrees." A "white praecipitate like a limus" appeared on the sides of the crucible, and he took this for the ordinary mercury dissolved in the *aqua fortis.* But in the bottom of the vessel appeared a liquid or "running mercury," which Newton believed to be the mercury of lead. Encouraged by this result, he went a step beyond Boyle. He now substituted tin for lead, and, as before, a white precipitate appeared, while the original solution at the bottom remained colorless. So far so good. When he next added copper to the mercury and *aqua fortis,* however, it turned blue, an indication that some of the metal itself was likely to be penetrating the solution.* "I know not whither y^t [mercury] came out of ye liquor or of [copper] for ye liquor dissolves [copper]."[59] Newton was correct in his suspicion; he had actually obtained only ordinary mercury from that added in the beginning, a far cry from the philosophical mercury of his dreams. An experiment that seemed to offer great promise had abruptly ended in failure, thwarting his initial attempt to unravel alchemy's tangled web.

The second of Newton's recorded experimental endeavors also focused on the extraction of mercury from metals. This time he attempted to open the bodies of metals with a combination of heat and even more powerful reactive agents, including sal ammoniac (ammonium chloride) and corrosive sublimate (mercuric chloride). His approach was no less systematic than before: he used equal quantities of sublimates and subjected them to the same level of heat, varying only the metals to be opened. Wickins's willingness to assist him with a heavy kettle notwithstanding, the majority of Newton's recorded experiments involved no more than a few ounces of chemicals, and often far less. In such instances the furnace could be comfortably forsaken for the candle. So fine were his measurements that he sometimes mixed his ingredients on a "looking-glass that none of them might be lost," and counted out not only individual grains but fractions of grains "wth ye pointe of a knife."[60] It would behoove anyone who might question this assertion to recall Newton's precision in the field of optics, where he employed fractions of a hundredth of an inch in his measurements of colored rings. "I am unable," wrote Westfall, "to conceive of that deadly serious countenance deliberately deluding itself" in this regard.[61] Why should it have? Newton had nothing to gain by making false claims in his private notes. Where he erred, if it can be labeled error, was in using

---

*Newton was a master of contemporary alchemical symbolism. To facilitate the reader's understanding, I have deleted the symbols, inserted the specific names of the substances for which they stand, and enclosed the names in brackets.

equal weights for the metals in his experiments rather than the chemically equivalent weights, which were not yet known.[62]

The laboratory notes contain no clear-cut statement of his conclusion, but indications are that Newton thought these experiments a success. Negative signs like the change of color, which had caused him to abandon his initial method extracting the mercury of metals, were apparently absent from this process. In point of fact, however, he had only repeated his previous mistake. Mercury, a chemical element, is always recoverable in its original form when part of a compound, something he could not have known, although Boyle seems to have suspected it. Newton's use of mercuric chloride thus resulted in the recovery of the mercury from the very sublimate he had introduced to achieve the desired reaction. Still, he was too cautious to believe that he had succeeded so easily where scores of other gifted adepts had failed, that the true philosopher's mercury now graced the bottom of his crucible. Additional experiments, either lost or not written up, were to convince him that contaminants from the metals made his mercury unfit for the Great Work: "[T]his [mercury] hath as many cold superfluities as common [mercury] hath . . . wch makes it more remote from ye philosophik [mercury] than ye common [mercury] is."[63]

Neither bent nor bowed, Newton continued his search for a method of extracting the philosopher's mercury from metals, which moved him ever deeper into alchemy's bewitching realm. By 1670 his attention had focused on the regulus of antimony, a substance which was to remain near the center of his thoughts for as long as he pursued the hermetic art. We know antimony as a metallic element, a hard, extremely brittle, glistening, silver-white, crystalline material used in a wide variety of alloys. To the alchemists, however, antimony was not the metal itself but stibnite, the lead-gray ore from which it was extracted by heating it with charcoal or some other mild reducing agent. The metallic antimony sinks to the bottom, and this (our element) is what the alchemists called the regulus of antimony. The name probably derived from the Latin *regulus*, meaning petty king. Because the regulus of antimony combines readily with gold (the king of metals) it became important to the process of refining the precious metal and an object of considerable experimental interest to seventeenth-century adepts. The regulus was also separated from stibnite by the introduction of various metallic reducing agents, in which case it became the regulus of Venus (copper), the regulus of Jupiter (tin), the regulus of Saturn (lead), or most importantly, the regulus of Mars (iron). It was thought, quite erroneously, that the "seed" of the metal used to reduce the regulus from the ore remained embedded in the regulus itself, thus raising all sorts of tantalizing possibilities in Newton's mind.

The earliest evidence of his interest in the different reguli had surfaced in the form of notes copied into the chemical dictionary between 1666 and 1667.[64] Now, some two years later, the young adept felt sufficiently emboldened to compose his own essay on their preparation. As usual, he wrote with the confidence born of firsthand experience:

> These rules in general should be observed. 1st y^t ye fire bee quick. 2dly y^t ye crucible bee throughly heated before any thing bee put in; 3dly y^t metalls be put in successively according to their degree of fusibility [iron], copper, antimony [stibnite], tin, [lead]. 4^{thly} That they stand some time after fusion before they be poured off accordingly to ye quantity of regulus they yield [iron], copper, tin, [lead]. 5^{thly} That noe salt bee thrown on, unlesse upon [iron] to keep it from hardeing.... 6 That if you would have ye saltpeter flow wth out two great a heat, you may quicken it by throwing in a little more saltpeter mixed wth 1/8 or 1/16 of charcoal finely powdered.

Newton went on to enumerate the many telltale "signes" of failure that, in their turn, had disturbed the rapt tranquility of his laboratory. But with the perfection of his experimental technique success was soon assured: "Thus wth a good quick & smart fire 4 of [iron] to 9 of [stibnite] gave a most black & filthy scoria & ye Reg after a purgation or two starred very well."[65]

The term "starred" was here employed by Newton in its most literal sense. For if the antimony has been properly purified as in this instance, it forms long and slender crystals. During cooling the crystals in turn form triangular branches around a central point, taking on the aspect of a silver star. Masters of the symbolic, the alchemists named this heart of antimony ore after Regulus, the bright double star near the heart of the constellation Leo. When the star regulus of antimony was achieved with the aid of a metallic reducing agent rather than a nonmetallic one, it was given the planetary name associated with the given metal. Having selected iron as his primary reducing agent in the above experiment, Newton had produced the star regulus of Mars. Further confirmation of his success is contained in a letter to Oldenburg of January 1672: "What the stellate Regulus of Mars (which I have sometimes used)... will doe" as a reflecting mirror in a telescope "deserves particular examination."[66] Yet it was for profounder reasons than the fashioning of better telescope mirrors that Newton long remained concerned.

At exactly what point and under what particular circumstances Newton began to contemplate seriously the principle of attraction between physical bodies is impossible to say. The general idea of gravity, though far from well developed, is certainly hinted at in the "Hypothesis of Light," the controversial paper he sent to the Royal Society in December 1675. It has been observed that the lines of crystals that appeared to radiate out from the center of the star regulus "might just as well be considered as radiating *in* to the center, which gives them the character of attraction rather than the character of emission." If, indeed, Newton viewed the star regulus in this light, then the very concept of gravitation "in which the lines of attraction run in to and converge in a center point" may have suggested itself to him.[67]

Present in this diminutive terrestrial orb was the invisible cosmic glue that binds the planets to the stars and the solar systems to the galaxies of the macrocosm.

Most probably, however, the idea of gravitation had not taken such definite form in Newton's mind by the early 1670s, though there is no question that its roots eventually found ready nourishment in the fertile field of his alchemical thought. Nor, if his notes on Basil Valentine are accepted at face value, did he mistake the star regulus for the philosopher's stone, as had more than a few bedazzled adepts.[68] Instead, he looked upon the star as a most promising step in the creation of the philosophical mercury, the *materia prima* or first matter from which all substances are formed.

In the mid-1670s he composed a paper of some 1,200 words entitled *"Clavis"* ("The Key"). This intriguing document, so concise and polished, gives evidence of being the last in a succession of drafts, the compilation of which had by then become one of Newton's distinctive intellectual trademarks. The contents represent the distillate of years already spent in the meticulous study of the star reguli in the hope of extracting philosophical mercury from common metals. It was clearly Newton's belief that he had succeeded in doing just that.

He began with the star regulus of Mars, which was fused with a small quantity of pure silver, the "doves of Diana." To this he added common mercury, amalgamating the mixture in a sealed vessel over a "slow fire." The amalgam was then ground for "1/8 of an hour in a mortar ... until it spits out its blackness." Repeated flushings, grindings, and washings left an alloy "like shining and cuppellated silver." A series of seven to nine more distillations and washings produced a mercury seemingly capable of dissolving all metals, especially intractable gold. The *cauda pavonis*, the multicolored tail of the peacock described by ancient alchemists, unfolded before Newton's very eyes:

> I know whereof I write, for I have in the fire manifold glasses with gold and this mercury. They grow in these glasses in the form of a tree, and by a continued circulation the trees are dissolved again with the work into a new mercury. I have such a vessel in the fire with gold thus dissolved, but extrinsically and intrinsically into a mercury as living and mobile as any mercury found in the world. For it makes gold begin to swell, to be swollen, and to putrefy, and to spring forth into sprouts and branches, changing colors daily, the appearances of which fascinate me every day. I reckon this a great secret in Alchemy.[69]

For a brief while the prolonged and painful tension that precedes the luminous moment of discovery was alleviated. Newton partook at once of exhilaration and tranquility. The fiercest of all competitors experienced the fulfillment of a transcendent passion put to creative use, and he was deeply humbled in the process. Mere pleasure gave way to ecstasy, a Puritan thinker's most trustworthy proof that he has indeed achieved membership in the regenerate, esoteric brotherhood.

VII

The dissolution of gold, not its multiplication, is what most interested Isaac Newton. He measured the magnitude of his supposed achievement against Boyle's oft-repeated alchemical dictum: "It is easier to make gold than to destroy it." In other words, once someone has solved the knotty riddle of what a substance is made of, producing that substance should be comparatively easy, a familiar enough notion to the student of modern chemistry.[70]

Newton's pursuit of the true philosophical mercury had caused him to draw heavily upon the works of George Starkey, who, as previously noted, published under the pseudonym Eirenaeus Philalethes. Nine of Starkey's books graced the shelves of Newton's library when he died, a number matched only by the indispensable treatises of Count Michael Maier.[71] The last of the great philosophical adepts, Starkey's views on the alchemical mediation of special mercuries were set forth in a manner strikingly similar to those expressed by Newton in the "*Clavis*" manuscript.[72] Moreover, Starkey sought to put this knowledge to practical use by effecting the process of transmutation.

The son of a New World Puritan divine, Starkey had emigrated to England in 1650, and associated himself with Samuel Hartlib's alchemical circle in London. Starkey brashly informed his new colleagues of his contact with a New England adept (probably John Winthrop, Jr.), from whom he claimed to have learned the secret of transmutation. A joint commercial operation—aided by a £1,200 advance from an anonymous Dutch investor—was formed for the express purpose of manufacturing gold. Starkey and Boyle took the lead during the early stages of experimentation, but Boyle gradually lost interest when the anticipated breakthrough was not forthcoming. By 1663 it had become obvious, even to the hard-bitten Starkey, that far more had been promised than could be delivered. In a letter to Boyle, who was back at his estate in Ireland, Hartlib described "Dr. Stirk" as "altogether degenerated," a man who "hath undone himself and his family." Twice imprisoned for debt and twice bailed out by his friends, Starkey abandoned his house in London to live obscurely. Hartlib's curt but trenchant assessment of this depressing turn of events echoes the fate of many a failed alchemist through the centuries: "[H]e stands more in need of us, than we of him."[73]

Newton, it appears, became privy to this sad tale.[74] Yet his esteem for the hapless adept seems not to have diminished, at least judging from the many scholarly references he made to Starkey's writings. Nor were the main participants in the gold venture seriously discouraged by the obvious failure of their friend. Hartlib immediately established a new laboratory in his back kitchen for his gifted son-in-law, Frederick Clodius, while Boyle retained his faith in the possibility of transmutation, as is evident in his rather curious paper published in the *Philosophical Transactions* of 1675–76.

Robert Boyle's friends described him as a contemplative and seemingly dispirited presence, the perfect example of what contemporary physicians diagnosed as melanchlolia, the mental depression brought on by an excess of black bile. In 1680 Henry More wrote to his friend Dr. John Sharp that Newton's countenance was "ordinarily melancholy and thoughtful."[75] It was probably during Newton's extended visit to London in early 1675 that melancholy countenance came face to face with melancholy countenance for the first time. Not surprisingly, each liked what he saw. The intellectual bond already forged from the perusal of each other's scholarly writings was strengthened during an engrossing exchange of scientific ideas.

Several months after this meeting, Oldenburg published a paper entitled "Of the Incalescence of *Quicksilver* with *Gold*, generously imparted by B. R." Newton, who did not get around to reading his issue of the *Transactions* until April 1676, had no trouble identifying the author as Robert Boyle. Boyle wrote of having discovered a special mercury that grows hot (incalescent) when mixed with gold. He considered it a breakthrough in the preparation of medicines, but he was also chary of the great harm its disclosure might do. For if Boyle had refined a true philosophical mercury, a discovery Newton privately claimed as his own, it could be used by "ill hands" to multiply gold, thus lifting the lid from a Pandora's box of endless "political inconveniences." Boyle sought advice from the "wise and skilful" as to whether he should make known to the world the specific ingredients of his recipe for the mercury.[76]

Newton, it seems, was the only adept who chose to reply, at least in writing. He cautioned Oldenburg to "keep this letter private to your self." His usual desire for secrecy was underscored by the knowledge that the attempted transmutation of metals was legally punishable by hanging. As an alchemist, Newton could not but question Boyle's optimistic conclusion regarding the mercury. He had explored methods similar to Boyle's, only to abandon them for more encouraging prospects. Still, Newton counseled caution, partly perhaps to avoid alienating a respected colleague, and partly because Boyle might know more than he had let on:

[I]t may possibly be an inlet to something more noble, not to be communicated wthout immense dammage to ye world if there should be any verity in ye Hermetic writers, therefore I question not but that ye great wisdom of ye noble Author will sway him to high silence till he shall be resolved of what consequence ye thing may be either by his own experience, or ye judgmt of some other ... that is of a true Hermetic Philosopher ... there being other things beside ye transmutation of metalls (if those great pretenders bragg not) wch none but they understand.[77]

While Newton doubtless shared Boyle's concern for the dire economic and social consequences that must follow from easy transmutation, one senses that this cautionary advice was rooted in other than altruistic grounds. No common "goldmaker," Newton's personal anxiety surfaced when he employed the self-revealing phrase, "there being other things beside ye transmutation of metalls." If Boyle were to disclose this great secret of the ancients, Newton's belief in his special relationship with the Almighty must suffer irreparable harm. The gates of the *prisca sapientia* would have been breached, and to the vulgar materialists would belong the desecrated spoils.

As a member of the inner circle that directed the general course of Royal Society activities, Boyle was surely aware of Newton's reticence in scientific correspondence. Yet it seems doubtful that Boyle was taken aback when Oldenburg informed him of the Lucasian Professor's response to his recent paper. Indeed, he had good reason to think that his newfound friend might have written even more. Boyle, after all, was the seventeenth century's most astute practitioner of "chymistry," and he had been present some months earlier during the reading of Newton's much-debated "Hypothesis of Light." Interpreted by most as the treatise on mechanical philosophy Newton meant it to be, the paper's equally profound if veiled alchemical implications could hardly have escaped Boyle, especially considering that Newton accepted and elaborated on a number of his ideas.

In the earlier discussion of the "Hypothesis" it was noted that Newton conceived of nature as "a perpetuall circulatory worker, generating fluids out of solids, and solids out of fluids, fixed things out of volatile, & volatile out of fixed." He further conjectured that a subtle and elastic ether or spirit was the causative agent of such diverse phenomena as gravity and magnetism. We can discern the mind of a master alchemist behind these thoughts. For to Newton the universe had become nothing other than the mighty crucible of God in which the first matter of creation is being constantly transformed:

> Perhaps the whole frame of Nature may . . . be nothing but various Contextures of some certaine aethereall Spirits or vapours condens'd as it were by praecipitation, much after the manner that vapours are condensed into water or exhalations into grosser Substances . . . at first by the hand of the Creator, and ever since by the power of Nature, wch by vertue of the command Increase & Multiply, became a complete Imitator of the copies sett her by the Protoplast.[78]

Elsewhere, in an essay "On the Gravity and Equilibrium of Fluids," Newton attempted to "show that the analogy between the Divine faculties and our own is greater than has formerly been perceived by Philosophers. That we were created in God's image holy writ testifies. And his image," he continued somewhat wistfully, "would shine more clearly in us if only he simulated in the faculties granted to us the power of creation in the same degree as his other attributes." Yet this could not be, for "we ourselves are created beings and so a share of this attribute could not have been equally granted to us."[79] Man the microcosm must content himself with the reason and intellect bestowed upon Adam. The alchemist's *raison d'être*, no less than that of the mathematician and physicist, is to duplicate and understand insofar as is humanly possible the grandiose operations of the macrocosm.

Newton's embrace of the ancient metaphysics of the macrocosm and the microcosm was tied directly to an equally time-honored doctrine called the Chain of Being. Its adherents—and they were many in the seventeenth century—pictured nature as a giant flight of stairs beginning with primal matter and ascending through man to God. It was the avowed purpose of mechanical philosophers like Descartes to separate body from spirit in nature, to deny, as it were, that any "occult" forces, such as attraction and repulsion, are manifest in this vast chain. Everything, they argued, is explicable in terms of matter in motion.

Though a mechanist tried and true, Newton could never be persuaded that spirit was absent from the operations of nature. The universe must be understood as God had made it, not as Descartes perceived it. Thus Newton conceived of atoms as incessantly active particles imbued by God with what might be described as an unconscious sense of awareness imparted by the ether in which they move. Atoms are the immutable seeds from which the phenomena of nature arise: birth, growth, decay, magnetism, transmutation, and so forth. For Newton, as for other alchemists, there was no hard and fast distinction between the worlds of the organic and the inorganic. In a paper titled "The Vegetation of Metalls" he expressed his belief "that metalls vegetate [i.e. germinate and grow] after the same laws" as plants and animals. Moreover, "vegetation is the sole effect of a latent spirit, and . . . this spirit is the same in all things, only discriminated by its degrees of maturity and rude matter." Lastly, "nature can only nourish" protoplasts, "not form them, That's Gods mechanism."[80] It was for this generative hermaphrodite containing the seeds of both male and female that Newton quested, a fundamental matter that gives rise to all varieties of physical transformation, making nature "a perpetuall circulatory worker" dependent upon the will of God.

As always, Newton appealed to an overarching principle of cosmic order. An understanding of nature must be based on the immutability of the ultimate particles that consitute it and of the subtle alchemical spirit present in all bodies, which is the perpetrator of their activity. In one sense Newton was enthralled by the very questions that so enliven current scientific debates. Black holes, those Dantean pits in the firmament, are fascinating partly because their explanation promises to aid in the unification of the great and the small. The two critical theories of twentieth-century physics are relativity, which is played out across the numbing void of trackless space, and quantum mechanics, which seeks to open invisible worlds of micromatter. The great problem is to formulate a principle that combines the two, which has thus far eluded all aspirants, including a deeply disappointed Einstein. Yet in another sense Newton thirsted for even more. A man's view of the world most often reflects what he thinks of himself. As an alchemist and favorite of the Lord, Isaac Newton attempted nothing less than a "metascientific" synthesis, from the Greek word meaning beyond. That he was destined to fail from the outset cannot be doubted. There is no *Principia Chemica* to match the *Principia Mathematica*. We have here another example of a heuristic principle unprovable by empirical standards. But we have also a further demonstration of intuitive prescience that would soon lead Newton to the very threshold of universal gravitation and its attendant concept of force. Samuel Johnson, articulate master of the King's English that he was, could reject Bishop Berkeley's philosophical proof that matter does not exist only by impulsively "striking his foot with a mighty force against a large stone, till be rebounded from it," exclaiming in anger, "I refute it *thus!*"[81] Two centuries later the architect Ludwig Mies van der Rohe would observe that "God is present in the details." Transfixed by the incessant bubbling of his crucibles, Newton had long since found the most convincing evidence of God's presence in the details as well.

An anonymous French poem of the fourteenth century divides the life of a man into twelve ages corresponding to the months of the year. At eighteen, the youth trembles like the winds of March that herald the approach of spring. At twenty-four, he yields to April love. At thirty-six, he is at the summer solstice, when whatever talents he may possess reach full flower.[82] Consider what Newton had already accomplished by 1678, the year of his thirty-sixth birthday. But remember, too, that his greatest triumph still lay in the future. During the early stages of his alchemical studies Newton was drawn to the Vicar of Malden's poetic rendering of the alchemist's quest, *The Hunting of the Greene Lyon.* The poem also serves as a metaphor of things to come.

> But our Lyon *wanting maturity,*
> *Is called* greene *for unripeness trust me,*
> *And yet full quickly can he run*
> *And soone can overtake the* Sun.[83]

Chapter Ten

Heretic: *Sotto Voce*

A *little philosophy inclineth man's mind to atheism, but depth in philosophy bringeth men's mind's about to religion.*

Francis Bacon

I

In mid-May 1679 Newton visited London for the first time in two years, returning to Trinity after an absence of only nine days. In the meantime Hannah Smith was summoned to Stamford, where her son, Benjamin, had come down with what Conduitt described as a "malignant feaver." The young man eventually recovered, but Hannah subsequently contracted the malady. The family, fearing the worst, sent word of this ominous development to Newton, and the depth of his apprehension may be gauged by his failure to sign the College Exit and Redit book before hastening to her side.[1] Upon reaching Stamford, Newton learned that Hannah's condition was indeed grave. He decided to take personal charge of his mother's care and medical treatment. "[He] sate up whole nights with her," Conduitt wrote, "gave her all her Physick himself, dressed all her blisters with his own hands, & made use of that manual dexterity for wch he was so remarkable."[2]

The imminence of death is enough to give even the most private soul pause; Newton's heart surely went out to the frail, ulcerated patient who was now totally dependent upon his gentle ministrations. But the protracted, agonizing vigil may have also released some bitter memories long held in suppression—memories of a childhood spent in a chronic state of emotional dislocation because of a mother who absconded with a stranger at the time of her son's greatest need, only to return years later with equally strange progeny and an iron resolve to make of her firstborn a respectable tiller of the earth. Newton never looked upon his father's face; in his mother's he saw that of an alien. Still, he did not abandon Hannah in her hour of peril.

Newton was capable of much, but saving Hannah proved beyond even his exceptional powers. She died in late May or early June, and the body was borne to Colsterworth for burial on the church's north side, near the illiterate yeoman who was her first love. (Isaac père had died in his thirty-seventh year, the very age of their only child now.) The rector dutifully recorded the somber event in the parish register: "Mrs Hannah Smith Wid. was burried in woolen June ye 4th 1679."[3] Newton's natural parents were reunited at last, a thought which may have provided the son with some measure of comfort as he stood, in the warmth of late spring, with the other mourners over the freshly dug grave. If Hannah's children by Barnabas Smith objected to the arrangement, it was to no avail. Their mother's will appointed Isaac executor, giving him the right to bury her as he "shall think fit."[4]

The death of a parent—especially an only parent—is a crucial event in life. Everything is suddenly changed, and in a manner almost impossible to anticipate. It is a deliverance as well as a blow: There is no longer any need to explain or to apologize. The surviving son is "dreadfully, wonderfully free."[5] Next to himself, Hannah was Newton's harshest censor. The infrequency of his visits to not very distant Lincolnshire after the plague years may be partially attributed to his ever deepening attraction to scholarship. At the same time, mother and son had too little in common to draw him away from Cambridge; she, a semiliterate widow concerned with gossip, hearth, and home; he, an inscrutable university professor who had unaccountably turned away from his people and agrarian heritage. Their conversation probably wore thin in a matter of hours, if not sooner. Restless, repelled by Hannah's unspoken but lingering disappointment and by the ghosts of a troubled youth, Newton grew anxious to be on his way. Now that Hannah was dead he must have found Woolsthorpe much as it had been during her first absence so many years before, except that this time she would return only in the selective glow of memory.

The will was proved at Lincoln on June 11, 1679, one week after the funeral. Following the Christian example of Newton's father, Hannah left five pounds to the poor of Colsterworth and Woolsthorpe, a significantly greater amount than the forty shillings left them by her husband in 1642, a measure of her more comfortable circumstances. Each of the servants received forty shillings with the exception of one William Cottam, who "in consideration of his true and faithful services" merited five pounds. For the first time we learn that Hannah Smith had a sister, Sarah Cook, who inherited no more than the steadfast Cottam. Hannah's brothers, William and James Ayscough, are not mentioned in the will, for reasons not ex-plained. The most plausible guess would be that they were well provided for through inheritance from a landed father, or else they had predeceased their sister. Additional monies went to Newton's half-brother and half-sisters; these bequests were relatively modest, however, for Mary, Hannah, and Benjamin had inherited from their father Barnabas Smith. All of the lands Hannah purchased with the hundreds of pounds she inherited from the Reverend Smith and "all the rest of my goods and chattels ... I give unto my said sonn Isaac Newton."[6] The master of Woolsthorpe Manor sud-denly found himself a very well-to-do gentleman, the fulfillment of at least part of his mother's lifelong dream.

Except for the involuntary periods of exile from Cambridge during the plague years of 1665–67, Newton's absence from the University after Han-nah's death was the most lengthy of his academic career. He returned to Trinity on July 19 but stayed little more than a week. July 28 found him back on the Great North Road to Lincolnshire, where he chose to remain until November 27.[7] Newton's only direct reference to his dealings during this period is characteristically terse. He replied, on November 28, to a let-ter sent by Hooke during his absence: "I have been this last half year in Lin-colnshire cumbred wth concerns amongst my relations till yesterday when I returned hither; so yt I have had no time to entertein Philosophical medita-tions or so much as to study or mind any thing els but Countrey Affairs."[8] Such affairs were obviously attendant upon his dual responsibilities as ex-ecutor and heir.

The meager documentation from this period would suggest that Han-nah Smith's management of the family property on the eve of her death left something to be desired, at least from the perspective of her eldest son. Among her tenants was Eduard Storer, stepson of the apothecary Clark, whose cherry desserts Isaac confessed to stealing as a child. The tables had since turned, however, for Eduard and his sons were not living up to the terms of their lease, thus depriving the Newtons of their just deserts. But because Hannah was a close friend of Storer's mother and indebted to the woman for having provided Isaac with a home in Grantham during his stu-dent days, she may have been reluctant to press the matter for fear of creating hard feelings. Newton, on assuming the role of landlord, sought to collect his due. Storer's sons, it seems, cleverly evaded him while he was in Woolsthorpe. Nor did they respond to his written inquiry concerning their mounting debt. Frustrated by his tenants' dereliction, yet mindful of his long-standing ties with their family (not to mention the kindness shown him by the Storers' great-uncle Humphrey Babington), Newton, after returning to Cambridge, presented his case in the form of a plea rather than the usual harsh indictment that often greeted those who crossed him: "I wrote to you about 4 or 5 months since, but you were then from home, & not having heard from you since, I have sent this to beg a line or two from you to let me know how my affairs stand & when I may find you and your brother in mind to have his arrears ready against yt time you appoint, & I will wait upon you."[9] The approach did not work. A few months later Newton felt a need to broach the subject again, more explicitly: "You know You are now 3 years & almost a $\frac{1}{2}$ behind hand. Pray therefore let me know what money you have in readiness for me & I will order ye Kethrin [Kettering?] Carrier to call upon you for it at what time you shall appoint. Or if I know at what time I might find you & your Brother at home I would wait upon you."[10] Wait upon the Storers he did. Eight years later their rent was still chronically in arrears, and Newton finally had to face the laggardly clan with a lawsuit to force them to complete extensive repairs of his decaying fences and buildings. During this same period his relations with Arthur Storer, Eduard's more enterprising brother, remained distinctly cordial. The man Newton admitted to "beating" when they were boys left England for Maryland in 1678 but continued to keep in touch with the Cambridge professor on astronomical matters.[11]

Newton was far less inclined to exercise forbearance when it came to the collection of an even more substantial debt affecting his inheritance. A man identified in name only as "Mr Todd" owed Hannah £100 when she died. Every effort on Newton's part to obtain a settlement had come to nothing. His patience exhausted, Newton drafted a letter in which he both upbraided Todd for his shameful conduct and informed him of impending legal action:

> About your pretenses of ye money's being ready long since & of a jugment wch you would have me believe I had against you I do not think it material to expostulate. I shall only tell you in general that I understand your way & therefore sue you. And if you intend to be put to no further charges you must be quick in payment for I intend to loos no time. I desire you therefore to pay it to my Sister Mary Pilkington at Market Overton as soon as you can & take her acquittance for your discharge.[12]

Whether or not Todd paid up as ordered is not known, but it would seem unwise to bet against such a tenacious creditor as Newton. This letter is simply another of many examples that prove he did not have to go to war to wage war. His appeals to the legal system promised the very kind of clearcut justice he so desperately sought but rarely obtained when embroiled in long-running scientific disputes.

The unusual length of Newton's absence from Cambridge, a total of nearly six months, suggests that the yeoman's son was paying far closer attention to the seasonal round than he ever had before. His mother died shortly after the spring planting, yet he remained in the country to monitor the summer cultivation and autumn harvest, doubtless settling accounts with the tenantry before departing late in November. Perhaps he also took advantage of the opportunity to visit old haunts and renew acquaintances that had languished over the years. William Walker, master of his old grammar school, asked him to carry eleven pounds to Humphrey Babington when he returned to Trinity, evidence that he had at least found his way to Grantham.[13] The Cambridge professor whom in his youth Hannah's servants had pronounced "fit for nothing but the 'Versity" had proved just how fit he really was. The admiring glances and deferential posturing that greeted his passage among the local folk were proof of that. Because he no longer feared the prospect of becoming one of them, he could move about with self-assured ease. Newton had crossed his social and economic Rubicon, never to enter the waters of commonality again. Although papers containing scattered theological notes and a few mathematical equations indicate he was not being completely candid when he wrote Hooke that he minded nothing else but country affairs while in Lincolnshire, for the first time in his adult life he felt no compulsion to hurry. That there was considerable business to settle is self-evident. But Newton's decision to linger long after discharging his filial responsibilities leads one to question whether this unprecedented hiatus was not only a form of release, but of expiation and of tribute, a private memorial to a mother who had spent almost the whole of her life on the land.

II

Of the four people closest to Newton during the 1670s three were dead by the end of the decade, while the fourth was about to become permanently estranged from him. Henry Oldenburg's sudden demise—"being stricken speechless and senselesse," according to Hooke's *Diary*—not only dealt a telling blow to scientific activity in England but brought to a virtual halt Newton's productive, if fitful, written exchanges with the Royal Society. Barrow's death, if less disruptive, proved more painful to him personally, for Trinity's late Master had been many things to Newton: model thinker, fellow scientist, cautious publicist, patron, and not least, respected friend. With Hannah now dead, only John Wickins remained.

By 1680 the man with whom Newton had lived for nearly two decades was absent from Cambridge for all but one or two weeks a year. Wickins did not resign his fellowship until 1684, but it appears that he had already accepted the comfortable living as Rector of Stoke Edith, near Monmouth, an appointment procured for him by John North, Barrow's highly eccentric, if not periodically deranged, replacement as Master of Trinity. Exactly why the entrenched and sedentary Wickins, who was in his mid-thirties, chose to bestir himself at that particular time is not certain: financial considerations, the desire to wed and start a family, a sudden welling up of repressed religious zeal, and Newton's incessant scholarly demands and smothering possessiveness are all possibilities. We do know that Wickins last visited

Cambridge for three weeks in March 1683. He then left for good on the twenty-eighth, and Newton exited for parts unknown that same day.[14] Whether this was merely coincidence or the result of some profound emotional rupture is impossible to determine, but the two never set eyes on each other again. The dearth of correspondence between them at least raises the prospect that their parting left some deep scars. Aside from the story of their meeting as distraught undergraduates, which was related by Wickins's son Nicholas eight years after his father's death, almost everything else John Wickins could have recorded for an information-starved posterity accompanied him to the grave. Nicholas did note that Newton conscientiously sent the dividends and rents due his nonresident father before the senior Wickins resigned his fellowship. And Newton performed one additional service for which both father and son were deeply appreciative:

> But there is one thing upon acc^t of wch not only my Father did but my Self also shall always pay a peculiar Regard to His Memory; wch was a Charitable Benefaction wch has privately passed from Him through my Fathers & since his Death through my own hands. We have been ye Dispensers of many Dozens of Bibles sent by him for poor people ... w^{ch} ... bears ye great regard he had to Religion.[15]

Newton's loss of his erstwhile chamberfellow cum factotum moved him to seek a replacement in 1683. He was assisted in his search by Dr. Walker, Henry Stokes's successor as master of King's School in Grantham. Newton especially missed Wickins's services as a part-time amanuensis. Thus Walker arranged to have Humphrey Newton, apparently one of his better students, join the most illustrious alumnus of Grantham School at his lodgings in Cambridge. The young scribe lived with Newton for the better part of five years, the most creative period of the professor's life. "By his Order," Humphrey wrote Conduitt in 1728, "I copied out, before it went to ye Press," that "stupendous Work."[16] Humphrey was referring, of course, to nothing other than the *Principia Mathematica*.

While Mrs. Vincent supplied the image of the young Newton, Humphrey fashioned the stereotype of the absent-minded but tireless professor, for which posterity owes him a considerable debt. Yet for the most part the amanuensis regarded the great man and his creations with the same lack of penetration and understanding that is the very hallmark of the copyist's uninventive calling. Though Humphrey spent many sleepless nights tending the fires under Newton's gurgling crucibles, he cheerfully admitted to Conduitt that, "What his Aim might be, I was not able to penetrate into." This same ignorance extended to mathematics, for Newton blamed his scribe for several of the errors that found their way into the first edition of the *Principia*—"its being copied by an amanuensis who understood not what he copied."[17] In fairness to Humphrey, it must be said that few other men, no matter the level of their erudition, would have truly grasped what his hired pen was required to reproduce. And while he may be faulted for fumbling a rare Boswellian opportunity to detail genius in action, there is something affecting about Humphrey's decision to christen his only son "Isaac" in honor of the man he remembered to Conduitt as his "dear deceased Friend."[18]

The Master's Lodge at Trinity College, left vacant when Barrow died in 1677, was next occupied by John North, the fifth son of Lord North, Baron of Kirtling. Three years Newton's junior, North was admitted a fellow-commoner at Jesus College in 1661, the same year Isaac made his less conspicuous debut as a Trinity sizar. North was very much the scholar and student, a rather unusual circumstance for the son of a seventeenth-century peer. His brother described him as "bent with perpetual thinking and study which manifestly impaired his health." Like another young scholar living but a short walk away, North "could not be pleased with such insipid pastimes as bowls or less material discourse such as town tales, or punning, and the like." Later, after meeting Newton and observing his habits, North was in an excellent position to assess the depth of this trait as reflected in another. He believed Newton would have "killed himself with study" were it not for the timely diversion of conducting experiments "with his hands." North's one entertainment besides books "was keeping great house spiders in wide-mouthed glasses, such as men keep tobacco in." The arachnids were amply supplied with bread crumbs and freshly caught flies, while their keeper watched in childlike fascination as they plied their mysterious and delicate trade of spinning gossamer webs.[19]

Books and spiders were not the only things that captured North's overactive imagination; he lived in perpetual fear of the dark and of the terrible things it must conceal. Dr. Cook, his tutor at Jesus, came home one evening to find North in bed with only his crown poking out from beneath the blankets. Cook, "indiscretely enough," pulled the youth's hair, then quickly "slunk down." North burrowed deeper into the bedclothes, but the tension finally became more than he could bear: "with a great outcry the scholar sprang up, expecting to see an enorm[ous] spectre." Instead, he was greeted by Cook's much amused visage. To calm his fears (and perhaps for more intimate reasons) "he lay with his tutor."[20] Newton apparently gave the spiritual underworld considerably less credence. Abraham de la Pryme related that a house opposite St. John's College was believed haunted in May 1694. Several people were gathered outside, and others "had rushed in armed with pistols," when Newton chanced on the scene. He became openly contemptuous upon learning the cause of the commotion: "Oh ye fools! says he, will you never have any wit? Know you not that all such things are mere cheats and impostures? Fie! fie! go home for shame. And so he left them scorning to go in."[21]

This "slight and feeble machine," as John North was described by his brother Roger, later moved on to Trinity. Five years later he was Master of the College. North's tenure marked the onset of a tragicomic winter of discontent, both for himself and for the foundation he was chosen to serve. The position was one of increasing difficulty for which the sickly and neurotic scholar was by nature unfit. The ancient Statutes notwithstanding, the elections to fellowships by royal mandate had expanded to the point where favor at court was virtually the only way to preferment; discipline was at its lowest ebb in decades; and the Seniority had effectively blocked the Master's ability to institute significant reforms. One night, as North was walking the cloister, some merrymaking students were alleged to have spotted him from above and rushed down the stairs to harass their leader. On another occasion, while North was entertaining friends in his lodgings, a stone shattered a window, just missing the assembled company. Delusion set in as the pressure mounted, "and accordingly, his mind always lighted upon extremes." The physically and mentally besieged Master began asking his friends to scrutinize his every act "to discover his own failings." He, of course, felt free to repay their efforts with interest. By 1680 North had reached the end of his tether; he suffered a stroke while administering a severe tongue lashing to some students for an unspecified transgression. North recovered sufficiently to resume his administrative duties, but it had become apparent to all that the end was drawing near. He died in 1683, mourned by few but the members of his family. He was buried, as he requested, in the antechapel beneath a flat stone bearing only the initials J. N., so that the Fellows "might trample upon him dead as they had done living."[22]

While there is little to suggest that the two were fast friends, North held Newton in high regard, and we have some reason to think that Newton returned the feeling, at least until the early 1680s. North gave the Lucasian Professor a copy of *A Philosophical Essay on Musick*, authored by his elder brother Francis, hoping Newton would comment on it. Newton responded favorably, if somewhat cautiously, in a cordial letter dated April 21, 1677. (Barrow did not die until the following month, and North was then living in London.) Indeed, Newton had called at North's lodgings the previous week and expressed the hope of having "an opportunity to wait on you again. In ye mean time I rest, wth my thanks to you for your kind acceptance of my former Letters."[23] Those letters have long since disappeared, and with them any record of other dealings between these not dissimilar spirits.

North was succeeded by John Montague, fourth son of the first Earl of Sandwich, who retained the mastership for the remainder of the century. Montague quickly proved himself as unfit for the position as his predecessor. But where North had attempted to govern with an iron hand, Montague exercised hardly any discipline at all. His neglect and indifference were compounded by his being absent from Trinity more than he was in residence. James II, who came to the throne early in 1685, issued as many ill-advised mandates for elections to fellowships as his brother, yet he met no resistance from Montague, as Charles II had from Isaac Barrow. The single accomplishment of Montague's administration was the completion and fitting out of the Wren Library in 1695. For this the Fellows were eternally grateful, and not only for reasons of scholarship. During two decades of limping construction the building's final cost had mushroomed to well over £16,000, more than double the original estimate. Newton, it will be recalled, donated £40 during the initial attempt to raise the required funds. In 1680, shortly after coming into his inheritance, he loaned the College an additional £100, the very sum he was simultaneously attempting to recover from the uncooperative Todd through legal action. Not until 1689, when the College borrowed funds from another Fellow, did Newton receive his money, which was repaid without interest.[24] He was hardly alone. By 1695 nearly every Trinity Fellow knew from sobering firsthand experience the precise meaning of that now hackneyed saying, "robbing Peter to pay Paul."

Not once in the decade and more since his appointment as Lucasian Professor had Newton deigned to accept a tutee. This situation changed, however briefly, in June of 1680, when he agreed to take on as his pupil one George Markham, a fellow-commoner. Little more is known of their association than of the equally transitory relationship between Newton and his first student, St. Leger Scroope. All that remains is the partial draft of a letter Newton wrote to Sir Robert Markham announcing his son's arrival and financial situation:

> Your son is well arrived hither, by whom I received 5 guineys. An aquitance for it you will receive inclosed in his letter. He shall have money of me as he has occasion for it, & his accounts you shal receive quarterly as the custome is. I suppose it may be enough to keep his french if a French M^r [Master] come to him once a week. If you think more necessary be pleased to let me know your mind.[25]

Since Newton spoke no French and read it very poorly, we may assume that he was only too pleased to entrust someone else with a share of Markham's instruction. Perhaps his demonstrated indifference toward the calling of pedagogy played its part in hastening the youth's departure. In any event Markham, like Scroope, never matriculated in the University. But unlike his faceless counterpart of the late 1660s, Markham left a silver tankard as his contribution to the College plate, sparing his erstwhile tutor the further indignity of having his name permanently associated with another student debtor in the Junior Bursar's accounts.

Of no immediate interest but of ultimately far greater concern to Newton was the arrival, while he was away in Lincolnshire, of Charles Montague, an eighteen-year-old fellow-commoner and nephew of Trinity's Master. Montague was born in Northamptonshire on April 16, 1661, the fourth son of George Montague and his wife Elizabeth, a native of Boston in Lincolnshire. Before coming to Cambridge, Charles had distinguished himself at Westminster School by making up witty epigrams based on theses appointed for King's scholars. For whatever reason, Montague was drawn to Newton and Newton to the young man, though it is clear he never served as Montague's tutor. It is possible that Montague attended Newton's lectures, as Newton had attended Barrow's, and afterward sought him out during the hours when the Lucasian Professor was required by statute to assist students with questions. In 1685 Newton informed Aston of an aborted plan to form a philosophical society in Cambridge partly "pushed forward" by "Mr. Charles Montague."[26] By this time Montague had been created a Master of Arts and elected a Fellow of Trinity by royal mandate. His authorship of some "ingenious and fulsome" verses on the death of Charles II attracted the attention of the Earl of Dorsett, the great patron of the day, who invited him to London where he soon became known for his wit and satirical bent. It was at this time that John Dryden published *The Hind and the Panther,* announcing to the world his conversion to the Roman Catholicism of the new monarch, James II. The poem cast the Catholic Church in the less than convincing role of an innocent deer caught in the cruel feline jaws of a predatory Anglican Church. Montague seized the opportunity to enter the literary lists against the publicly unpopular Dryden. In collaboration with William Prior, he wrote a stinging parody on the poet laureate's work entitled *Country Mouse and the City Mouse.* It carried London by storm, securing Montague's reputation in the bargain. He left Cambridge for good in 1689, but his ties to Newton remained strong. They served together in the Convention Parliament following the accession of William and Mary. Montague eventually rose to become Chancellor of the Exchequer and took a leading part in forming the Bank of England in 1694. When his Recoinage Bill was passed by Parliament in 1695, he was able to secure Newton's appointment as Warden of the Mint. More will be heard about this talented, social-climbing rogue who became not only Newton's unlikely patron but the even more unlikely lover of his niece, the gay and witty Catherine Barton.

III

It was a tradition among Cambridge Fellows to donate books and rare manuscripts periodically to their respective college libraries. A somewhat reluctant benefactor, Newton gave but few volumes during his long tenure at Trinity. His first gift, made in 1675, was a just-published edition of St. Irenaeus's *Against Valentinus and similar Gnostic heresies.* Newton thought so much of the work he shelled out 14s. for a second copy of his own, which became dog-eared over the long course of his endless religious studies. Four years later he gave the library another theological work, Bishop Huet's

Demonstratio Evangelica. His last contribution of the period came in 1681 in the form of Nehemiah Grew's *Musaeum Regalis Societatis,* a catalog and description of the rarities belonging to the Royal Society. Like the volume by Irenaeus, it was duplicated in Newton's private collection.[27] He would later sell the library a Silesian version of the Lord's Prayer for 10s. and donate a copy of his own *Principia,* a work of science to be sure, but of science steeped in profound religious overtones whose roots reached backward into the Lincolnshire of its author's youth.

Amid the endless and confusing motion of every other point in the firmament of childhood stand the parents, unmoving and dependable, like twin polestars. They alone remain a steadfast guide while all other objects shift their bearings and orientation. How did Margery Ayscough explain the absence of that steady, reassuring parental light to her questioning grandson? Perhaps she took him to an upstairs bedroom window and pointed to the eastern horizon. On the left stood the Norman tower of Colsterworth Church in whose shadow Isaac's father lay; on the right, much farther in the distance, was the tiny spire of North Witham Church, where his mother prayed on Sunday mornings. It would have been only natural to tell the child that these happenings were God's will and that he must submit to it without further question. Under the circumstances, protest would have been futile on Isaac's part. If he wanted to know more about this mysterious power called God he must seek the answers for himself and in doing so experience his first encounter with the universe.

According to William Stukeley, Hannah made a gift to Isaac of some "2 or 300 books . . . chiefly of divinity and old editions of the fathers."[28] These had belonged to her late clergyman husband Barnabas Smith. The collection passed to Humphrey Newton many years later, after he returned to his native Grantham to pursue the practice of medicine. Humphrey in turn gave some of the books to Stukeley, who in 1726 also settled in Grantham, where he made it his business to collect reminiscences of the early life of Newton. Yet judging from the extent of Newton's patristic holdings now in Trinity College Library (61 individual titles), it is likely that a number of these books were once owned by his stepfather. The same can be said of some of the other 416 titles on theology, which together with the patristic writings account for 27.5 percent of the Newton collection, about three times the number of volumes on any other subject.[29]

The earliest clues to Newton's intimate thoughts on God are those contained in the confession penned in Shelton shorthand in 1662, when he was nineteen and a sizar at Trinity. "Not turning nearer to Thee for my affections. Not living according to my belief. Not loving Thee for Thy self. Not loving Thee for Thy goodness to us. Not desiring Thy ordinances. Not long[ing] for Thee. Not fearing Thee so as not to offend Thee. Fearing man above Thee."[30] Though hardly explicit in doctrinal terms, these eight sins, all camouflaged by a covering of venial adolescent transgressions, reveal much. Try as he might, Newton could no more forge a bond of deep affection with his Creator than with his fellow man. His future relationship with God was to be an intellectual rather than an emotional one, in which Christ, the loving and forgiving Redeemer, played a secondary role. It was Yahweh, the omnipotent Creator, harsh Taskmaster, and imperious Judge of the Old Testament, who commanded Newton's lifelong attention and obedience. When he next wrote about God it was in the philosophical notebook of 1663–1665. By then he had become steeped in the Platonism of Henry More and its compelling emphasis on the Book of Nature. If one would truly know the Creator one must study the natural scheme of things—the original ordering of matter and the laws that govern its composition and motion. "Were men & beasts & c made by fortuitous jumblings of attomes," he wrote under the heading "Of God," "there would be many parts useless in them, here a lumpe of flesh there a member too much. Some kinds of beasts might have had but one eye some more y^n two." Under the rubric "Of ye Creation" he interpreted Moses' statement in Genesis "y^t ye evening & ye morning were ye first day & c" as proof that God had also created time as understood by men.[31] Many more examples could be cited, but they all point in the same direction. Newton would spend the rest of his life trying to weld into a single philosophy of nature and religion two not entirely compatible principles. The first holds that nature is governed by an inviolable system of physical laws intelligible to rational men. The second holds that an omnipotent God continues actively to express His will within this law-bound system of matter and motion. The danger, of course, was that radical mechanical philosophers like Descartes and Hobbes would separate God from His works, thus eliminating the chief argument for His existence, or if not His existence then the need for His continued presence. What was required, from Newton's point of view, was a God who never abandoned His creation, but also one who would never arbitrarily change the rules by which it functions so as to make His handiwork inexplicable in scientific terms. Newton lived with the agonizing tension generated by these thoughts all his adult life, experiencing release only when he immersed himself in ancient Scripture and the commentaries of the Fathers or performed experiments that promised to open new vistas on the universe. Yet having found his God, he never once denied Him.

IV

When Henry VIII issued royal letters in 1546 calling for the creation of a new Cambridge college in honor of the "Holy and Undivided Trinity," the monarch never dreamed that its most gifted scholar would one day reject the very Christian doctrine for which the institution was named. Nor did that scholar's widowed mother foresee the irony of christening him Isaac, from the Hebrew Yishaq, "he who laughs." Sometime in the early 1670s, unbeknownst to anyone but himself, Newton became a heretic, hardly a laughing matter in a century marked by profound religious fervor and sectarian strife. He did so by embracing the teachings of Arius, an Alexandrian priest of the fourth century who steadfastly denied that Jesus was of the same substance as God.

If Newton experienced serious scruples before 1670, they were well hidden. By signing for his B.A. degree in 1665, he attested to his belief in the Thirty-nine Articles of the Anglican Church. Orthodoxy, as spelled out in the Act of Uniformity of 1662, was further served when a fellowship was conferred upon him in October 1667. Together with eight other candidates, he swore to embrace the "true religion of Christ." And on incepting M.A. the following year, he unhesitatingly signed the article requiring acceptance of the creed as taught by the Church. Finally, he agreed to do the same when he assumed the Lucasian Professorship in 1669.

But by 1673 conscience and ambition were on a collision course, and Newton became desperate to keep from sacrificing one at the expense of the other, another turn of the emotional screw already dangerously overtight from fending off repeated challenges to his paper on optics. To retain his fellowship, he would have to avow his orthodoxy publicly a fifth and last time by taking holy orders. Equating ordination with blasphemy, Newton first sought to evade the prospect of resigning his chair by obtaining the exempt law fellowship vacated on the death of Robert Crane. Barrow, however, for reasons of seniority, chose the other candidate, Robert Uvedale, for the vacancy. At that point Newton may have toyed with the idea of seeking to retain his chair without holding a fellowship, but the prospects were less than appealing. In the first place, there was no established precedent for this unusual action. Even more important, any refusal to accept ordination, especially when it entailed no additional responsibilities, was bound to invite embarrassing questions that could easily lead to exposure, something he was determined to avoid at all costs. What Newton would have done had he been forced to choose between his professorship and his religious beliefs remains a matter of conjecture. For in April 1675, most probably through Barrow's influence at court, his petition to hold a fellowship without taking holy orders was granted by Charles II. The soft-voiced heretic was now protected from the threat of exposure and revile.

Heresy in the form of dissenting opinions about the Trinity was almost as old as the Christian religion itself. Try as it might, the Church had never succeeded in stamping out this noxious weed; by Newton's day it had cropped up once more and soon became the most nettling problem faced by orthodox theologians. Of its several varieties, Socinianism and Arianism were among the most prevalent. Often regarded as virtually the same thing by conforming Anglicans, the differences between them were of no mean significance to their respective adherents. Socinianism, the product of a sixteenth-century religious movement organized by Italian humanist Faustus Socinus, interpreted the Scriptures in the light of the new rationalism. The sacraments were viewed as spiritual symbols, and Jesus (the Nicene and Athanasian creeds notwithstanding) was held to be only the *human* instrument of divine mercy, the Holy Spirit merely the activity of God. In contrast, the Arians did not deny Christ's divinity. Rather, they believed God created, before all things, a Son who was the first creature, but who was neither *equal* to nor *eternal* with the Father. Such distinctions were lost on the orthodox: To make of Christ anything but a creature consubstantial with God was to subvert the whole Christian theory of salvation.

The rationale for this impulse to heterodoxy is succinctly encompassed in a couplet by Dryden:

> *Faith is not built on disquisitions vain;*
> *The things we must believe are few and plain.*

Antitrinitarianism was just one element of a profound reaction by the Puritan mind against too much theology. "Get rid of the 'metaphysics'," this spirit demanded, "and reduce that which we must believe to the 'few and plain' basic truths."[32] And where are these truths to be found? Why, in Scripture, of course, carefully recorded by God's anointed spokesmen; all the rest is irrelevant. Thus to Scripture did Newton turn as the purest form of worship. He achieved a mastery of the Bible equaled by few theologians, enabling him to string out citations like a concordance. There, through the application of reason, he also encountered and embraced a Christ quite different from the Savior of orthodox teachings. Adam of St. Victor had anticipated Newton's sin well before its commission:

> *Of the Trinity to reason*
> *Leads to licence or to treason.*

The earliest of the surviving theological manuscripts, which total well in excess of a million words, were composed about 1672. Since few of them bear internal dates, it is impossible to determine the exact order of their composition. But fortunately for the researcher, Newton was nothing if not a creature of deeply ingrained habits; he approached every subject in virtually the same manner, theology being no exception. Beginning with the blank pages of a notebook, he would enter several headings as a way of establishing initial parameters of inquiry. He then summarized his findings under each heading, sometimes writing only a few sentences, but more often filling sheet upon sheet of white folio. At this stage he inevitably drew upon the most authoritative sources. Having grasped the basics, he went on to broaden his investigative boundaries and began recording original insights of his own. Rapidly and inexorably ignorance—if indeed one can call it that—yielded to complete mastery.

Newton recorded some of his earliest and most extensive writings from Scripture in a notebook that is now part of the Keynes collection at King's College. The Latin headings offer no hint that heresy, even in an incipient form, was in the air: "Idolatria," "Deus Pater," "Deus Filius," "Christi Incarnatio," "Christi Satisfactio & Redemptio," "Spiritus Sanctus Deus," "Angeli boni et mali," and "Predestinatio." Upon closer examination of the manuscript's contents, however, one senses a certain uneasiness on the part of its author. He had gleaned the New Testament for extended references on the nature of Christ and His relationship to the Father. These he collected under the heading "Deus Filius" or "Son of God." Copying verbatim from Colossians 1:16–17, he wrote: "For by him were all things created that are in heaven & y^t are in earth visible & invisible.... And he is before all things & by him all things *consist*" (Newton's emphasis). So there would be no confusion as to who the "him" meant, Newton inserted "God ye Father" after the pronoun. He went on to copy verse 18: "And he is the head of the body, the church, who is the beginning, ye first born from ye dead." This time Newton inserted "Christ" after "he," an indication that the Father and Son were already assuming separate and unequal status in his mind, a development consonant with Arian teachings. This impression is reinforced a bit farther down the page, where, borrowing from II Corinthians 4:4, he refers to Christ as "ye Image of God." Finally, Newton was sufficiently impressed with the first chapter of Paul's epistle to the Hebrews to quote from it at length. He thought the ninth verse especially enlightening: "Thou [Christ] has loved righteousness & hated iniquity, therefore God, even *thy God*, hath annoynted thee with ye oyle of gladness above thy fellows." In the margin across from the words he had underlined for emphasis, Newton wrote, "Therefore the Father is God of the Son."[33]

A reprise of Arianism's emergent theme plays even more lucidly across another manuscript page titled "Deus Pater," or "God the Father." The opening phrase derives from I Timothy 2:5, "There is one God & one Mediator between God & Man ye Man Christ Jesus"; the second from I Corinthians 11:3, "The head of every man is Christ, & ye head of ye woman is ye man, & the head of Christ is God"; the third from verses 23 and 24 of the same book and chapter, "All are yours & y^e are Christ's & Christ is God's"; the fourth from I Corinthians 15:28, "When all things shall be subdued unto him then shall ye son also himself be subject unto him y^t put all things under him y^t God may be all in all"; the fifth and last, with revealing emphasis, from Luke 1:32, "He shall be great & shall be called ye son of y^e *most high*."[34] The very reading undertaken by Newton to prepare himself for holy orders had suddenly become a major stumbling block to ordination itself.[35]

By asking the simple question, What does the Bible say? Newton implicitly put his finger on the two great weaknesses of the trinitarian credo.

Its more perceptive followers had never been able to prove—even to their own satisfaction—that their doctrine is unequivocally scriptural. What is more, trinitarianism defies a certain fundamental logic. Some things, of course, must remain forever mysterious and unintelligible to the Christian—the Virgin birth and the Resurrection, for example—but these things, albeit miraculous, are not, it was argued, necessarily contrary to reason, while the principle that three equals one and one equals three is no more applicable to rationalist theology than to the fields of physics and mathematics.

In an effort to confirm his deepening suspicions, Newton broadened the base of his theological studies by rapidly mastering the voluminous writings of the patristic fathers. At the opposite end of the Biblical notebook, under a new set of headings, he filled scores of pages with reading notes from Athanasius, Augustine, Jerome, Gregory, Cyprian, Tertullian, Basil, Origen, Clement, Beda, Rufinus, Eusebius, and several more. He would compare holy writ for himself with the commentaries of the earliest and most venerated theologians. If there was any significant evidence of misinterpretation or, worse, outright conspiracy and fraud, it must be traced to its origins, branded for what it was, and mercilessly expunged, root and branch. Yahweh, as the Hebrews had so painfully learned, was anything but a God of patient tolerance. Above all else, He had commanded the children of Israel, "Thou shalt have no other gods before me." Nor, despite the moderating influence of a loving Christ's compassionate entry into history, did the record show that the jealous Creator of the Old Testament had experienced a profound change of heart. Any and all corruptions of Scripture, especially one of the magnitude of trinitarianism, constituted a form of idolatry—the sin of sins—a thought much on Newton's mind and conscience as evidenced by the prominent place given to the subject in this the earliest of his theological writings. Here was a man truly created in the image of his God.

Whether or not Newton was aware of it in the beginning, he had been preparing himself for a spiritual voyage of no return. Now, having mastered the principles of exegetical navigation, he set sail for the fourth century of the Christian era, the age that gave birth to the doctrine of trinitarianism. No figure of that time was more fascinating (and, in the end, more repugnant) to him than St. Athanasius, Patriarch of Alexandria, Doctor of the Church, and indomitable champion of orthodoxy during the Arian crisis. As a young man of twenty-eight Athanasius took part in the Council of Nicaea convened by the Roman Emperor Constantine in 325. After a long and acrimonious debate, the Council condemned Arius's view that Christ is not identical in substance with God, accepting in its stead the formulary known as *homoousion,* which holds that Christ is of the same substance as the Father. A long and sometimes violent struggle ensued. After attaining the bishopric of Alexandria upon the death of his superior, Athanasius was faced with a conspiracy led by Eusebius of Nicomedia, who sought to return the exiled and condemned Arius to Egypt. When Athanasius demurred, a pro-Arian council was convened at Tyre, which convicted him of everything from dishonest grain dealings to sacrilege and murder. Thus began a series of five exiles and alternative reinstatements, the last occurring in 366. It was during a period of hiding with hermit monks in the Egyptian desert that Athanasius wrote *Discourses Against the Arians,* the most compelling exposition of his Christology.

Under the headings "Observations upon Athanasius's works" and "De Athanasio, & Antonio," Newton carefully recorded the anti-Arian principles that became Christian orthodoxy, paying special attention to the concept of *homoousion.* He was no less thoroughly steeped in the patristic literature of Athanasius's opponents, as the entries headed "De Trinitate" and "De Arrianis et Eunomianis et Macedonianis" attest.[36] The more deeply Newton delved, the more clear-cut the evidence became to him: The New Testament had been corrupted intentionally by scheming power brokers who would stop at nothing to achieve their temporal ends. Pope and emperor had together abetted Athanasius in his idolatrous blasphemy by pronouncing anathema any and all who opposed their heresy and cleverly limiting Scriptural access to all but a select and compliant few, a ruse ultimately penetrated by Luther in the sixteenth century. The horned beast described in the Revelation of St. John was to Newton nothing other than the Church of Rome itself. He knew well the consequences of being seduced by its power and authority: "If any man worship the beast and his image, and receive *his* mark in his forehead, or in his hand, The same shall drink of the wine of the wrath of God . . . and he shall be tormented with fire and brimstone."[37] Orthodox Anglicans could hardly rest easy, for as inheritors of the trinitarian falsehood their souls were also imperiled. Now that he possessed "the truth," Newton yielded to a full embrace of the Arian credo, hoping to save himself from the eternal darkness looming beyond the pale. Privately but with unwavering purpose, he drafted a list of twelve points (c. 1673), which summarize his lifelong view of the nature of Christ.[38]

Newton could not have been more explicit. He declared the Father alone to be supreme. The Son is a separate being, different from the Father both in substance and in nature. Christ is not truly God but is the so-called Word and Wisdom made flesh, divine to be sure, but only so far as divinity is communicated by the Father. It was almost as if the young professor were consciously renewing with God his Whitsunday covenant of 1662, in the tradition established by the Hebrew patriarchs of old. "Apage Satanas! I will bear the mark of the beast no more."

Idolatry is a breach of the first & greatest commandment. It is giving to Idols the love, honour & worship wch is due to the true God alone. It is forsaking the true God to commit whoredome with other lovers. It makes a Church guilty of Apostasy from God as an Adulteress forsakes her husband. It makes her guilty of spiritual whoredome with other lovers. It makes her become the Church of the Idols, fals Gods, or Daemons whom she worships, such a true Church is as in Scripture is called a Synagogue of Satan.[39]

The day was not distant when Newton would further sharpen his already trenchant metaphor by charging that the Church of the venerated Virgin was nothing other than a reincarnation of the Bible's supreme symbol of spiritual fornication—the blasphemous Whore of Babylon. He would have been wholly unmoved, if not confounded, by the claim of nineteenth-century historian Henry Adams that Constantine's acceptance of the Cross and the Virgin in the form of Catholicism had brought to Europe the highest level of creative energy previously known to humankind, a force embodied in the soaring spires and stained glass of Chartres cathedral and the stupendous Benedictine abbey erected atop the rocky face of tide-bound Mont-Saint-Michel.[40]

V

Newton kept his thoughts on the Trinity a closely guarded secret. Not one of the incriminating manuscripts—and there are many besides those discussed here—was entrusted to his chamberfellow for reproduction. Unless Wickins engaged in some surreptitious research of his own, there is every likelihood that he left Cambridge without even so much as sensing the heresy in his midst. Still, Newton could never rest easy. He thought long and hard about the consequences that must follow discovery and of how best to defend himself should it ever come to that. Perhaps the most enlightening clue to what he would have done is contained in a later ireni-cal manuscript that appears to date from the early eighteenth century. Under the heading "The fundamental requisites to communion in the Church of England," Newton listed the following: First, "To renounce the Devil & all his works . . . & the carnall desires of the flesh." Second, "To profess the faith conteined in the . . . Apostles Creed. And the profession of faith in the primitive Church the Apostle [Paul] calls faith towards God & the resuirection of the dead & eternall judgment." Third, "To keep Gods commandments; that is the ten commandments, as is explained in the Church catechism. These & baptism & laying on of hands are all the fun-damentalls requisite to communion in the Church of England." Therefore, "to excommunicate any man for any thing else is contrary to the fundamen-tal constitution of the Church. . . . It is to excommunicate a man who . . . became a member of Christ a child of God & an inheritor of the kingdom of heaven, & may he still go as much as he was at his admission into communion for any thing objected ag^t him."[41] Newton's defense would be the truth as he perceived it. The theological chips must fall where they may—or so he argued in the abstract to himself on paper. Whatever the earthly consequences of his actions, they would be as nothing compared to the wrath of an angry God when directed against the persecutors of the chosen through all eternity; and let there be no mistake about it, Newton counted himself among the chosen. And yet when William Whiston, his friend and successor at Cambridge, was drummed out of the Lucasian chair for professing to be an Arian in 1710, Newton uttered not a word in Whiston's defense. Evangelical passion seems to have been as foreign to him as most other forms of heartfelt emotion, especially if its indulgence threatened the social respectability and financial security he had always coveted. Job, after all, went to his grave a wealthy, revered citizen, and Newton saw no reason why he should not do the same, another example of his preference for the moral lessons taught by the Father in the Old Testament over those of the Son revealed in the New.

Such historical might-have-beens notwithstanding, Newton's succinct list of qualifications for church membership reflected an abiding Puritan preference for simple moral truths as opposed to complex and tortured theologies. He attacked orthodox trinitarianism, the papacy, the invocation of saints, and the veneration of the Virgin as illegitimate intrusions of metaphysics into the only truly revealed religion, primitive Christianity. Nor was it by chance that this same rationalist thread was deeply woven into his philosophy of science. The motto *"hypothesis non fingo"*—aimed at speculative reason, not at critical reason—was assigned important double duty by Isaac Newton.

We observed in some detail how Newton applied the Platonic doctrine of the *prisca sapientia* to alchemy. The theological manuscripts reveal that he no less ardently pursued a parallel religious wisdom or *prisca theologia*, from which corrupters of Scripture like Athanasius had so perilously strayed. Here too, the Old Testament became the primary focus of his studies. The scientists and philosophers of the most revered ancient civilizations—Babylon, Egypt, Israel, and Greece—were also their chief priests and religious leaders. While practicing two fundamental forms of science, chemistry and astronomy, certain of these venerable sages had become convinced that a first and only cause was responsible for the creation and the ongoing operation of nature. Eschewing the false gods of polytheism, they became the first monotheists. As vehicles of God's eternal truth, the old priest-scientists recorded their discoveries and revelations on precious manuscripts to be decoded by a select few from yet unborn generations. In his generation Newton believed himself to be just such a chosen interpreter of God's master plan, both natural and divine. He was *homo universalis* from whom nothing discernible to the human mind could long remain obscure. On the one hand, there are "the first principles and fundamentals of religion conteined in the doctrine of baptism and laying on of hands and in the Creed which all are to learn, and wch the Apostle therefore compares to milk for babes." Adherence to these precepts is sufficient for membership in the community of true Christians. On the other, "there are many truths of great importance but more difficult to be understood." Though not an absolute requirement of salvation, "these the Apostle compares to strong meat for men full of age who by use have their senses exercised to discern both good and evil. With these truths the mind is to be fed continually as the body is wth meats."[42] The man so unusually gifted will value his wisdom "above all other treasures in the world by reason of the assurance and vigour it will add to thy faith, and steddy satisfaction to thy mind wch he onely can know how to estimate who shall experience it."[43] That Newton placed a high estimate on his own powers cannot be doubted. The general contempt he harbored for lesser scientific minds was brought to bear with equal Puritan force in the world of religious thought, although in the latter sphere he sought to avoid controversy at all costs.

When formulating their doctrine of the *prisca sapientia* during the late sixteenth century, Renaissance thinkers were intrigued by ancient references to a certain Moschus, thought to have been a Phoenician who lived before the Trojan War and the first advocate of atomism. In 1598 the Friesian philologist Arcerius identified Moschus with Mochus, another Phoenician, whose followers Pythagoras had allegedly met while visiting Sidon. Arcerius speculated further by suggesting that Moschus or Mochus was none other than the chief patriarch of the Old Testament. This identification gained additional currency among such Protestant leaders as Gerard Vossius and John Selden after Isaac Casaubon (1559–1614) confirmed that Mochus was indeed the Tyrian name for Moses.[44]

Newton was well acquainted with the writings of Selden and Vossius. Even more crucial is the fact that Cambridge Platonists Ralph Cudworth and Henry More also leaned heavily toward this identification, and Newton had mastered their treatises chapter and verse. Moses became a pivotal figure not only in Newton's theology but in his conception of early science as well. The Hebrew leader had dissolved the golden calf by alchemical means and acquired all the wisdom known to the ancient Egyptians. The double revelation of God written about by Galileo—the one contained in His word, the other contained in His works—was known to Moses centuries and civilizations past. Newton was searching for this same consonance, this same unifying structure of matter and spirit, which he attributed to the great monotheist. He attempted to fit everything in the heavens and on earth into a grandiose, rational conceptual framework, while measuring himself against the cosmos. Though he periodically protested that no one must try to know the unknowable, he could not help but go right on thinking about it.

It was inevitable that this search for universal truth in sacred texts would lead Newton to Solomon, King of the ancient Hebrews, son and successor of David, and in Newton's words "ye greatest Philosopher in ye world."[45] Solomon constructed the Hebrew Temple at Jerusalem, ensuring the city the central position in Israel. Newton was aware that the astronomer-priests of Egypt, Chaldea, and Greece had built temples that served as models of the universe, thus preserving their arcane knowledge of the macrocosm in mortar and stone. But Solomon had been the first to accomplish this feat, and he had done so in the name of Yahweh, the one and only God.

> So then twas one designe of ye first institution of ye true religion to propose to mankind by ye frame of ye ancient Temples, the study of the frame of the world as the true Temple of ye great God they worshipped. . . . So then the first religion was the most rational of all others till the nations corrupted it. For there is no way (wth out revelation) to come to ye knowledge of a Diety but by ye frame of nature.[46]

Newton set himself the arduous task of reconstructing Solomon's Temple on paper. An avid student of the Apocalypse, he may have been further inspired by the angels' charge to St. John: "Rise, and measure the temple of God, and the altar, and them that worship therein." His best source of information was the prophet Ezekiel, who preached to the Jews of the Babylonian Captivity during the sixth century B.C. Solomon's Temple having been destroyed, Ezekiel infused his people with the promise of a brilliant restoration crowned by a new center of worship built to Solomon's plan. By probing the prophet's eloquent but frequently obscure pronouncements, Newton succeeded in drafting a detailed floor plan, which survives in the Babson College Library. He wrote of the structure that it "commends itself by the utmost simplicity and harmony of all its proportions."[47] This, however, was true of countless other buildings, sacred and secular, including the Wren Library under construction no more than 500 feet from Newton's lodgings. So far as can be determined from the manuscripts, the supposed cryptogrammic plan of Solomon's Temple yielded up none of its alleged secrets, although a scaled-down version of it was eventually published in Newton's *Chronology of Ancient Kingdoms Amended*. Its significance lies not in what it reveals of Solomon but of the lengths to which Newton was willing to go to increase his knowledge of God and man.

Given the depth of his interest in theology, it might seem somewhat surprising that Newton's attendance at church and chapel was irregular at best. This pattern was established shortly after he settled in Cambridge. He confessed to missing chapel in 1662 but made little if any effort to rectify his conduct. The sins committed following his Whitsunday confession include "Negligence at the chapel" and missing four Sunday sermons at St. Mary's Church. Humphrey Newton recalled how his employer seldom went to early morning chapel, "that being the time he chiefly took his repose." (These were the days when Newton "rarely went to bed till *two* or *three* of the clock, sometimes not until *five* or *six*.") As for the afternoons, "his earnest and indefatigable studies retained him, so that he scarcely knew the house of prayer."[48] Newton may well have reasoned that what took place in his study and laboratory was as fundamental a part of genuine religious experience as any prayers he might offer to the Creator from a hard-backed pew. To be incessantly probing nature's secrets and studying God's plan was the more demanding form of worship. Besides, he knew far more theology than the local clergy, whose "milk for babes" was no substitute for the "strong meat" demanded by a voracious intellectual lion.

In theory Trinity College, like its sister institutions, was a community made up of dedicated scholars pledged to the principle of celibacy and, by implication, sexual abstinence. In reality Newton was one of an ever dwindling minority who zealously embraced these ideals. He accepted Paul's advice to the early Christians that the celibate life, difficult though it may be, is preferable in the eyes of God to the state of matrimony. "To renounce . . . the carnal desires of the flesh" was part of Newton's first rule for admission to communion in the Church of England. In later life he told Conduitt, "They who search after the Philosopher's Stone by their own rules [are] obliged to a strick & religious life." Though he was not a member of a religious brotherhood, the similarities between his abstemious existence and that of the regular clergy did not escape Newton's notice. In a history of the Church begun in the late 1670s he devoted considerable attention to the practice and spread of monasticism, placing particular emphasis on the problem of sexual temptation in the cloister. Newton offered a shrewd analysis of the shortcomings of those unable to cope with the demands of celibacy:

> I find it was a general complaint among them y^t upon their entering into ye profession of a Monastick life they found themselves more tempted in ye flesh then before & those who became strickter professors thereof & on that account went by degree further into ye wildernesses then others did, complained most of all of temptations. The reason they gave of it was that ye devil tempted them most who were most his enemies & fought most against him: but ye true reason was partly that the desire was inflamed by prohibition of lawful marriage, & partly that ye profession of chastity & daily fasting on y^t account put them perpetually in mind of what they strove against, & their idle lives gave liberty to their thoughts to follow their inclinations.

Writing from deep personal experience, Newton then opened a rare psychological window through which we can survey his ongoing struggle to keep the Devil and the demons of lust at arm's length:

> The way to chastity is not to struggle directly with incontinent thoughts but to avert ye thoughts by some imployment, or by reading, or meditating on other things, or by convers. By immoderate fasting the body is also put out of its due temper & for want of sleep the fansy is invigorated about whatever it sets it self

upon & by degrees inclines towards a delirium in so much that those Monks who fasted most arrived to a state of seeing apparitions of weomen & their shapes & of hearing their voices in such a lively manner as made them often think the visions true apparitions of ye Devil. Thus while we pray that God would not lead us into temptation these men ran themselves headlong into it.[49]

Idleness is the Devil's playground. Monasticism failed as an institution because it demanded a vacuous seclusion and discouraged the investigation of nature, in whose folds the Creator's mark was most deeply embedded. Newton met the Prince of Darkness head on by laboring at fever pitch. Thus were the "uncleane thoughts words and actions and dreamese" of his youthful concupiscence denied; thus did he seemingly excel the daydreaming monks of medieval Christendom. But in 1693, during a period of sleepless mental derangement, Satan carried the day. Newton accused John Locke of a preposterous attempt to "embroil me with women," a projection of his unbearable personal guilt over lustful thoughts upon another. One may also ponder whether his exceptionally liberal use of the words "whore" and "whoredom" in the theological manuscripts, especially to condemn the "false" Church of Rome, served double duty as a form of psychosexual release. And was Hannah, the wife who had defiled the memory of her first and only true husband by remarrying, subconsciously associated with harlotry in the mind of the bereft son who hoped that both she and the debaucher Smith would burn? These are minor but provocative Freudian mysteries without definitive answers; nor could Newton have supplied them himself.

VI

By the mid-1670s Newton had resolved the centuries-old conflict between the followers of Athanasius and Arius to his personal satisfaction, allowing him to begin a new (but not entirely unrelated) chapter in his Biblical studies, the interpretation of prophecy. He wrestled with the subject until the day he died half a century later, continually revising the work eventually published in 1733, *Observations upon the Prophecies of Daniel, and the Apocalypse of St. John.* Like many of his contemporaries, Newton believed that prophecy concealed direct revelations of hidden truths that would reveal to men—very special men—the future course of history as set forth by the Creator from the beginning of time. He was especially drawn to Daniel of the Old Testament and John of the New because "the language of prophetic writings was symbolic and hieroglyphical and their comprehension required a radically different method of interpretation."[50] To be even more explicit, the Book of Daniel and the Revelation of Saint John the Divine were for Newton the keys to the long lost *prisca theologia.* There is no doubt that he already believed himself the possessor of a significant part of that wisdom before reaching his mid-thirties. "Having searched after knowledge in ye prophetique scriptures, I have thought my self bound to communicate it for the benefit of others, remembering ye judgment of him who hid his talent in a napkin."[51] Keeping this in mind, one can more readily understand his fanatical preoccupation with seclusion and freedom from controversy. Was anyone else in the world simultaneously reading and decoding God's sacred word and the no less holy works of nature? Newton would one day reveal all, but in his own good time and on his own terms. Meanwhile, he must be left in peace so that his spider-mind, endowed with singular skills of analysis and synthesis, might spin the tautest, most symmetrical of intellectual webs.

Though he was conversant with all of Scripture, Newton's protracted focus on only two of the Bible's many prophets, Daniel and St. John, is illustrative of the fundamental skepticism with which he approached all subjects, including holy writ. He was especially leery of mystical seers like the miracle-working, acid-tongued Elijah, who allegedly raised a widow's son from the dead, survived on food provided by ravens in the wilderness, and departed the earth in a chariot of fire enveloped in a whirlwind. For Newton the true prophet was a supremely rational man. "Nothing would have been more alien [to his] conception of the ancient prophet than the distraught mystic running naked through the streets of Jerusalem that Voltaire later conjured up."[52] The few men chosen by God to receive His special revelation were usually known for their mental faculties as teachers and as visionary intellectuals. Daniel lived at the court of Babylon, where he interpreted King Nebuchadnezzar's dreams, while John was thought to have been the disciple "whom Jesus loved" and to whose care he committed the Virgin Mary in his dying moments.[53]

While the mind of the prophet was perfectly lucid, the enigmatic language, full of symbolism and imagery, in which he expressed himself was not, and for good reason. Prophecy is concerned only with things to come. "I would not have any discouraged by the difficulty & ill success that men have hitherto met with these attempts" at interpretation, Newton wrote. "This is nothing but what ought to have been. For it was revealed to Daniel that the prophecies concerning ye last should be closed up & sealed until ye time of ye end: but then ye wise should understand and knowledge should be increased." Indeed, the longer prophecies concerning the apocalypse remain abscure, the nearer the time "in which they are to be made manifest. If they are never to be understood, to what end did God reveal them?" Obviously, "he did it for ye edification of ye church," though "not [for] all that call themselves Christians, but a remnant, a few scattered persons which God hath chosen."[54] Newton illustrated by example the inherent danger of neglecting prophecy. The means by which the Jews were to recognize their Messiah had been foretold in the Old Testament. But they had ignored Scripture and so brought down upon themselves the wrath of God. A chillingly clear parallel suggested itself. Unless those Christians, both Catholic and Protestant, who had been seduced by the false trinitarian doctrine quickly mended their idolatrous ways, the number of lost souls entering the gates of Hell must multiply at an exponential rate. Yet there is precious little evidence to indicate that this consummate Puritan thought the prospect avoidable, or for that matter even cared, at least on an emotional, evangelistic level. He looked down on lesser mortals from a vastly superior position; his own soul, after all, was safe. But if this were true, how were the ignorant and spiritually imperiled to know what to do when perhaps the only man in Christendom who did not see through a glass darkly refused to budge from his shuttered chambers?

No better proof exists of Newton's predominantly intellectual approach to theology than his carefully drafted "Rules for interpreting ye words & language in Scripture." As he would not confuse that part of his work in science which was certain with that part which was hypothetical, neither would he substitute private conjecture for the language of the prophets. One must "observe diligently the consent of Scriptures & analogy of the prophetique stile & . . . reject those interpretations where this is not duely observed." Thus it would be wrong to interpret the "Beast" spoken of in Revelation and other Books of the Bible as "some great voice," because according to the style and tenor of Scripture a Beast signifies a "body politique & some times a single person wch heads that body, . . . and there is no ground . . . for any other interpretation." One must also assign only a single meaning to each passage of Scripture. If two meanings—"literall and mystical"—seem equally plausible, "he is obliged to believe no more then in general that one of them is genuine untill he meet with some motive to prefer one side." Although it might not seem so to the uninitiated, the Bible was consistent from beginning to end. If a prophetic symbol applied in a given way in one part, one could apply it with assurance in the same way in any other part. Most critical of all, however, is the requirement of simplicity; and here Newton even more consciously merged his philosophies of science and religion.

> Truth is ever to be found in simplicity, & not in ye multiplicity & confusion of things. As ye world, wch to ye naked eye exhibits the greatest variety of objects, appears very simple in its internall constitution when surveyed by a philosophic understanding, & so much ye simpler by how much the better it is understood, so it is in these visions. It is ye perfection of God's works that they are all done wth ye greatest simplicity. He is ye God of order & not of confusion. And therefore as they that would understand ye frame of ye world must indeavour to reduce their knowledge to all possible simplicity, so it must be in seeking to understand these visions. And they that shall do otherwise do not onely make sure never to understand them, but derogate from ye perfection of ye prophecy.

Newton, it is clear, addressed the Bible with the same directness as he did nature. The heretical Arian had much in common with the heretical Socinian, who would tolerate no mysteries, especially those contrary to reason. He even issued the very warning to potential critics that he had given to those who protested the hypothesis of light: "Hence if any man shall contend that my Construction of ye Apocalyps is uncertain,... he is not to be regarded unlesse he shall show wherein what I have done may be mended."[55]

Needless to say, Newton not only lived up to but far exceeded the high standard he had established in his self-imposed rules for interpreting Scripture. Believing that there had been a prophetic language common to the ancient Near East, he drew upon writings beyond those of the Bible, including the work of Artemidorus, a Hellenistic prototype of Freud, known for his interpretation of dreams.[56] Nor, of course, were the prophetic figures slighted in any way. Newton compiled a catalog detailing, insofar as was possible, the lives of seventy biblical seers and followed it with an essay, called "The Proof," in which he presented extensive evidence of what their individual prophecies meant.[57] Like any reputable scholar, he showed a deep concern for the very sources over which he labored. The Scriptures contained the word of God, but that word had often been corrupted, intentionally or otherwise, by those who had their own religious axes to grind. John's admonition ran continuously through his mind: "If any man shall add unto these things, God shall add unto him the plagues that are written in this book. And if any man shall take away from the words of the book of this prophecy, God shall take away his part of the book of life." Taking no chances, Newton plodded through Revelation, verse by verse, collating some twenty different manuscript and published accounts of the work.[58] He also compiled a dictionary of historical, political, and religious events that conformed to the symbols and images evoked in prophetic literature. If these exercises seem familiar, there is good reason. The chemical dictionary and "Index Chemicus" (not to mention the philosophical notebook) share the same paternity and purpose. Once the exact meanings of alchemical and prophetic terminology were established, the respective pages of the Books of Nature and Scripture must fall into place. And whatever knowledge of an ever present, ever active God was revealed to man in one was in harmony with what was vouchsafed to him in the other. The Bible, no less than nature, was for Newton a "perpetuall circulatory worker," enshrining universal truth spoken at different times from the mouths of divinely inspired but very different men.

Every period of Christian history has given birth to new, often bizarre, interpretations of the Apocalypse. An engimatic work at best, rich in imagery and obscure symbolism, it consists mainly of visions that compass the defeat of evil and persecution and the supreme triumph of God and His martyrs. Whoever John may have been, he was possessed of a majestic style and deft command of Scripture, for he constantly alludes to Old Testament prophecy, especially that of Daniel, Isaiah, and Ezekiel. Although the language is esoteric, John's plan is carefully laid out and depends heavily on recurrent patterns of sevens: the letters to seven churches in Asia Minor; the opening, by a powerful angel, of the seven seals on a book in the hand of God, the first four of which loose Apocalypse's Four Horsemen on a sinful world; the blowing of seven trumpets by angels standing before the throne of God; the seven visions, including the battle between St. Michael and Satan; the seven-headed dragon, and the rising out of the sea of the seven-headed Beast with ten crowned horns; the seven plagues; and the appearance of the scarlet-clad whore of Babylon, riding a scarlet beast with seven heads. The one obvious goal of Revelation was to give heart to persecuted Christians, but beyond this no consensus has been reached. What, then, did Isaac Newton make of John's sublime visions?

Revelation's author foresaw the mass abandonment of the true apostolic faith for the worship of Satan's surrogate, a beast thought by Newton to represent the great apostasy, trinitarianism, and its bastard offspring, Roman Catholicism. As a natural philosopher he instinctively sought safety in numbers, not the human but the mathematical kind. By calculating the length of the Beast's reign one could accurately predict the time of Satan's defeat, the last judgment, and the founding of the New Jerusalem. John spoke of a woman "clothed with the sun" who had fled into the wilderness where God had prepared a place for her. Here she was to be sheltered and fed for "a thousand two hundred *and* three score days." Protestant theologians, including Henry More and Joseph Mede, believed that each of those days signified a year in human terms; thus the apostasy would run its course in 1,260 years. Since it began with the triumph of Roman Catholicism during the reign of Theodosius the Great, at the end of the fourth century, true Christianity must be restored no later than about 1700. William Whiston, for one, argued that the prophecies of John were on the verge of being made manifest: The "beast was near destruction, and his demise would commence with the sounding of the seventh trumpet."[59] Few Anglican theologians demurred. It was no longer the Jews who were God's chosen people, but the English.

Newton was prepared to accept the 1,260 years as the length of the Antichrist's reign, but he rejected both the generally accepted date of its institution and the identification of the Church of England with true Christianity. Anglicans were trinitarians, and trinitarianism was the greatest apostasy of all.[60] Those who embraced it also worshiped the Beast. Besides, what evidence was there that a higher moral order had established itself in England? If anything, morality had declined during the Restoration, and nowhere more precipitously than at his own university, where aspirants to the clergy led remarkably dissolute lives. Surely this reprobate lot was not to be entrusted with the mission of leading the masses into the anticipated millennium. His silver leonine mane must have shaken in dismay when he read the prophet's few unequivocal words: The "fearful, and unbelieving, and the abominable, and murders, and whoremongers, and sorcerers, and idolators, and all liars, shall have their part in the lake which burneth with fire and brimstone: which is the second death." Murderers excepted, those at Cambridge who fitted this description would have filled a good-sized lake of fire.

No, the new order was not yet at hand; and Newton could prove this by numerical calculation, but only after undertaking an exact correlation of prophecy and history. His rule thirteen for interpreting Scripture provides the key: "To interpret sacred Prophecies of ye most considerable things & actions of those times to wch they are applied. For it would be weakness in an Historian whilst he writes of obscurer actions to let slip the greater."[61] This much-employed technique is familiar to contemporary Biblical scholars as typology. It simply means Newton believed that St. John, like all legitimately inspired prophets, predicted and prefigured subsequent historical events. As can be gathered from his statement, Newton was largely uninterested in what might be termed the "small change" of history. Certainly he was anything but enthralled with the day-to-day happenings of his time, especially those of the Cambridge years, observing in their mundane passage no evidence of imminent prophetic fulfillment.

Like every man, Newton was his own historian, and he concluded that the Beast would fall not in 1700 or thereabouts, but around the year 1867. To make this determination, he again plunged into the murky waters of the early Middle Ages, again with the same relish displayed during his prior investigation of Arianism. His efforts again bore fruit. The beginning of the 1,260 years spoken of by John had not come in Theodosius's reign after all. Instead, Newton settled on the year 607, when worship of the Antichrist reached its peak.[62] Each of the idolatrous kingdoms, represented by the ten horns of the Beast, must pass away before the sealed saints who denied Satan could resume their rightful place at the head of the apostolic church, "saints not defiled," in John's words, "with women; for they are virgins." The ten kingdoms were identified in the following order: the Vandals and Alans of Africa and Spain, the Suevians of Spain, the Visigoths, the Alans of Gallia, the Franks, the Britons, the Huns, the Lombards, and Ravenna. The eleventh and last horn was the Church of Rome itself. Daniel, interpreting Nebuchadnezzar's dream of the monster, had foretold the same historical events in the Old Testament, thus establishing the foundations of world history. His prophecy of seventy weeks also proved to Newton that Christ's coming had been accurately anticipated long before the momentous event. In sum, the respective visions of John and Daniel were without contradiction and so demonstrated the consistency of the Testaments, both Old and New.

The prophet John had asked, "Who is worthy to open the book, and to loose the seals thereof?" Newton had answered without reservation, "I am." God had spoken to him with equal directness on a related matter: "To him that overcometh I will give . . . him a white stone, and in the stone a new name written, which no man knoweth saving he that receiveth *it*." Newton wrote such a name, of all places, at the bottom of his most secret alchemical manuscripts, "Jeova sanctus unus." It must have reminded him of another passage from Revelation, even more simple and profound, which most clearly mirrored his belief in the unity of matter and spirit: "I am the Alpha and the Omega, the beginning and the end, the first and the last."

Chapter Eleven

A Pitfall in Eden

He pleases all the world, but cannot please himself.
Nicholas Boileau-Despréaux

I

If Wickins's departure from Trinity created an emotional vacuum in Newton's life, there is little indication that he sought to fill it by seeking the companionship of others. "He always kept close to his studyes," wrote Humphrey, "very rarely went a visiting, & had as few Visitors, excepting 2 or 3 Persons, Mr Ellis of Keyes, Mr Lougham of Trinity, & Mr Vigani, a Chemist, in Whose Company he took much Delight and Pleasure."[1] John Ellis, a Fellow of Caius College, had established an excellent reputation as a tutor. Elected Master in 1703, he was knighted on the same day as Newton in 1705. The "Lougham" to whom Humphrey referred was in fact John Laughton, the college chaplain and librarian. Perhaps the best measure of the depth of Newton's friendship with the two scholars is the absence of any correspondence between them after he left the University. His relationship with Vigani is rather more intriguing, but again the information is scant and mostly secondhand.

A native of Verona, Italy, the the chemist John Francis Vigani (his Anglicized name) settled in Cambridge between 1680 and 1682. With no official standing in the University, Vigani was permitted to teach an informal course in his specialty. The Italian must have been surprised and not a little pleased to encounter someone of Newton's erudition in his field, while the Lucasian Professor, as indicated by his amanuensis, was no less delighted to make Vigani's acquaintance. Newton's only direct reference to Vigani is contained in the partial draft of a letter in which he referred to the expatriate as one "who had been here performing a course of Chymistry to several of o^r University much to their satisfaction."[2] This Vigani continued to do unofficially for twenty years, until Cambridge finally honored him with the title of Professor of Chemistry but no stipend. Why it should have taken so long is difficult to say, but the delay may have been related to the question of his trustworthiness as a coreligionist. In any case, a moment of indiscretion in the presence of his puritanical friend resulted in an instant and irreparable breach. Vigani told Newton "a loose story about a Nun," for which he was never to be forgiven.[3]

Nearly three years had passed since Hooke and Newton, like reformed alcoholics, mutually pledged to refrain from raising any issue that might "kindle coal" between them. They had kept their word by corresponding very little during the interim, exchanging only a few trivial and obsequiously polite notes that did nothing to advance the cause of science—or of mutual respect. In November 1679 Hooke, perhaps in response to pressure from his colleagues at the Royal Society, sought to draw Newton out. The Secretary alluded to Newton's productive correspondence with Hooke's predecessor and nemesis, Henry Oldenburg: "I hope therefore that you will please to continue your former favours to the Society by communicating what shall occur to you that is philosophicall." Hooke went on to assure Newton that any future correspondence "shall be noe otherwise farther imparted or disposed of then you yourself shall prescribe." He apologized once more for their past differences and reasserted his position that scientific matters "shoud not be the occasion of Enmity—tis not with me I am sure." To prove that he nurtured no lingering animus Hooke laid his professional reputation on the line. "I shall take it as a great favour if you shall please to communicate by Letter your objections against any hypothesis or opinion of mine." Hooke suggested a few specific topics on which Newton might care to comment. He especially wanted to know if Newton had given any thought to Hooke's own theory that the motions of planets are compounded of a tangential motion and an attractive motion toward the sun.[4]

The intriguing idea that the planets are continually attracted to the sun as they revolve around it was first broached by Hooke in a paper he read before the Royal Society in 1666.[5] He never ceased to meditate upon the concept, and it resurfaced in both his *Attempt to Prove the Motion of the Earth* (1674) and the *Lectiones Cutlerianae* (1678). The biographical portraitist John Aubrey, Hooke's intimate friend, unabashedly called it "the greatest Discovery in Nature that ever was since the World's Creation." But Aubrey's hyperbole lost a good deal of its luster when he added, "I wish he had writt plainer, and afforded a little more paper." Still, Aubrey, who wrote his brief account of Hooke's life in the early 1690s, was convinced that Newton had been guilty of plagiarism—that "the whole coelasticall theory, concerning which Mr. Newton made a demonstration," was first intimated to him "from Mr. Hooke."[6] (Even though they corresponded, Aubrey pointedly left Newton out of his widely read biographical sketches.) Since this charge has cropped up periodically in one form or another over the centuries, it behooves us to have a clear understanding of precisely what Hooke said on the subject:

> I shall explaine a systeme of the world, differing in many particulars from any yet known, answering in all things to the common rules of mechanicall motions. This depends upon 3 suppositions; first, that all coelastiall bodys whatsoever have an attractive or gravitating power towards their own centers, whereby they attract not only their own parts, and keep them from flying from them, as we may observe the Earth to doe, but that they doe also attract all the other coelestiall bodys that are within the sphere of their activity, and consequently that not only the Sun and the Moon have an influence upon the body and motion of the Earth, and the Earth upon them, but that Mercury also, Venus, Mars, Saturne, and Jupiter, by their attractive powers have a considerable influence upon its motion, as, in the same manner, the corresponding attractive power of the Earth hath a considerable influence upon every one of their motions also. The second supposition is this, that all bodys whatsoever, that are putt into direct and simple motion will soe continue to move forwards in a straight line, till they are by some other effectuall powers deflected and bent into a motion describing a circle, ellipsis, or some other uncompounded curve line. The third supposition is, that these attractive powers are soe much the more powerfull in operating, by how much nearer the body wrought upon is to their own centers.[7]

Hooke's subsequent argument, based on the above statement, that he had evolved the theory of universal gravitation well in advance of Newton seems dubious to the point of denial. Hooke had said nothing about *every* particle of matter in the solar system attracting every other particle; nor had he extended this unenunciated principle to the distant stars, which Newton was soon to do. But even supposing Hooke's words were interpreted in the most generous possible light, one insuperable obstacle remained. Hooke admitted to not having "yet experimentally verified" his hypothesis of planetary attraction. The brilliantly intuitive mechanical theory could be carried no further by its originator, who lacked the mathematical powers needed for tackling the problem. The fact is, Hooke was in no position to write "plainer" or to fill "a little more paper" on the subject as Aubrey had wished. He had already stated virtually everything he knew or, we should say, suspected.

Hooke's letter was waiting for Newton when he returned from Lincolnshire on November 27, 1679, following Hannah's death and the settlement of her estate; he took Hooke seriously enough to write a reply the following day. His first inclination was to maintain his self-imposed moratorium on philosophical correspondence. He pleaded ignorance, having had no time "to study or mind any thing else but Countrey affairs." He also wanted to avoid the kind of wrangling that had arisen from Oldenburg's insistence that he communicate his discoveries to the scholarly world. "I hope it will not be interpreted out of any unkindness to you or ye R. Society," he wrote somewhat disingenuously, "that I am backward in engaging my self in these matters." As for Hooke's theory of planetary motion, Newton had not heard of it "yt I remember."[8] Perhaps not, but he had not forgotten other details of the Cutlerian Lecture as evidenced by a reference to it later in the letter. Be that as it may, Newton, when exiled from Cambridge during the plague, had at least tentatively entertained thoughts along the lines of Hooke's, or so Pemberton, Whiston, and Newton himself later maintained. To raise that issue now, however, would be tantamount to reentering the cauldron of controversy from which he had so recently extricated himself.

Following this lengthy disclaimer Newton, as he sarcastically informed Edmond Halley six years later, decided to "complement" Hooke by "sweeten[ing] my answer."[9] He drew a diagram of what he characterized in his reply to Hooke as "a fansy of my own about discovering the earth's diurnal motion." Proponents of the Copernican theory that the earth rotates on its axis had repeatedly met with the timeworn but dogged objection that a falling body would be left behind as the earth spins beneath it. Such an object should land to the west of where it was released, if, indeed, the planet rotates. The fallacy of this argument, as Newton was aware, is that the tangential velocity of an object dropped from the top of a high tower is greater than at the tower's base; hence the object ought to land slightly to the east, if the friction of the air is discounted. Hoping to demonstrate this principle, he sketched the path of such a body during a theoretical fall.

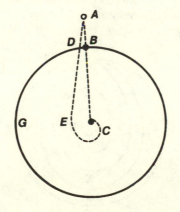

FIGURE 11-1.   Newton's diagram of a body falling toward a rotating earth (redrawn from *The Correspondence of Isaac Newton*, ed. by H. W. Turnbull, J. F. Scott, A. R. Hall, and Laura Tilling [Cambridge, England, 1959–77], vol. II, p. 301).

"The advance of ye body from ye perpendicular eastward," he observed, "will in a descent of but 20 or 30 yards be very small & yet I am apt to think it may be enough to determin the matter of fact." Newton's prescribed trajectory showed the body moving in a spiral ending at the earth's center. "I could with pleasure," he concluded, "heare & answer any Objections made against any notions of mine in a transient [oral] discourse for a divertisement." But Hooke, in language evocative of Newton's recently renewed contact with his agrarian roots, was cautioned to expect nothing more: "[M]y affection to Philosophy being worn out, so that I am almost as little concerned about it as one tradesman uses to be about another man's trade or a country man about learning, I must acknowledge my self avers from spending that time in writing about it wch I think I can spend otherwise to my own content."[10] To put the matter bluntly, Newton wanted to hear nothing further from Hooke, whom he trusted no more now than ever. The Lucasian Professor still nursed a fair-sized grudge against the Royal Society Secretary for his too hastily mounted attack on the theory of light. That grudge was about to wax fat at a single sitting, making the best a lasting enemy of the good.

Newton had made a serious mistake, which Hooke was quick to catch. Donning kid gloves and adding more than a little sweetening of his own, he replied to Newton on December 9: "I doe not despare of you at all for I find by your letter you doe sometimes for your divertisements spend an Hower or soe, in conversing . . . and I would never wish any thing more from a Person of your ability: I hate Drudges . . . at any thing [for] they produce nought but Molas or chymeras some what with out life or Sole." Hooke had read Newton's letter before the Royal Society (something he had pledged not to do without express permission) and was pleased to report that "most of those that were present," including Christopher Wren, John Hoskins,

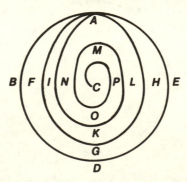

FIGURE 11–2.   Hooke's diagram of a body falling toward a rotating earth (redrawn from *The Correspondence of Isaac Newton*, ed. by H. W. Turnbull, J. F. Scott, A. R. Hall, and Laura Tilling [Cambridge, England, 1959–77], vol. II, p. 305).

and Thomas Henshaw, agreed with Newton's conclusion that a body let fall from a great height would land "Eastwards of the perpendicular and not to the Westward of it as most have hitherto Imagined." Then the rub: "[M]y theory of circular motion makes me suppose that it would be very differ[ent] and nothing at all akin to a spirall but rather a kind [of] Elleptueid."[11] Newton had not recognized that a body dropped under his stated conditions obeys the principles of orbital motion. Furthermore, the deviation from the vertical would be precisely to the east only if his experiment were performed at the equator; in the more northerly latitude of London it would be more to the south than the east. In his attempt to evade Hooke's question concerning planetary motion Newton had inadvertently raised the issue in a different and more subtle guise. Unlike Newton, Hooke quite properly treated the problem as one of orbital motion, referring to it as "my Theory of Circular motions compounded by a Direct motion and an attractive one to a Center."[12] He had correctly plotted an ellipse.

Hooke left Newton flat-footed with his mouth agape, but it would be only a matter of time before he regained his granite composure. For the moment Newton swallowed his bitter medicine, if only in part. "I agree wth you," he replied on December 13 (a Friday?), "yt ye body in our latitude will fall more to ye south then ye east if ye height it falls from be any thing great." He nevertheless refused to yield on the question of the ellipse, even as he abandoned the spiral trajectory contained in his previous letter: "[I]f its gravity be supposed Uniform it will not descend in a spiral to ye very center but circulate wth an alternate ascent & descent made by it's *vis centrifuga* & gravity alternately overballancing one another."[13] Hooke's insight again shines through. Whereas Newton now viewed circular motion as the product of opposing forces (gravity and centrifugal motion), which produce a state of equilibrium, Hooke thought of a central attraction continually diverting the planet from its natural rectilinear motion. Newton concluded his latest letter (also read before the Royal Society) with the obligatory mock apology for having bothered Hooke "wth this second scribble" and a hardly more sincere invitation to "correction" at Hooke's leave. Thus the way was grudgingly left open for the Secretary to inform Newton that he was wrong a second time, which Hooke promptly did, presumably with considerable pleasure.

Hooke's biographer has argued that her subject was not one to sustain a grudge. Perhaps this is so, but such a view calls for some rationalizing when we recall Hooke's lingering enmity toward Oldenburg and Huygens, to say nothing of his acute embarrassment at Newton's hands not once but twice before the Royal Society a few years earlier. Hooke now realized that he had achieved a rare psychological advantage over his inscrutable nemesis, who was clearly off balance, and his sense of exhilaration was heightened by the knowledge that Newton must soon own up to his error. So exultant was Hooke over Newton's inability to shake his planetary theory that on January

4, two days before responding to Newton's second letter, Hooke wrote of my "perfect Theory of Heavens" in his *Diary*. He was spreading the news by word of mouth as well, for on January 26 he noted a conversation with "Sir Chr. Wren about Planetary motion."[14] It is little wonder that Aubrey, bombarded by such inflated rhetoric, became hopelessly biased on Hooke's behalf. Yet Hooke's was anything but the campaign of intellectual annihilation for which Newton became famous. Even the deaths of his enemies— Hooke, Flamsteed, and Leibniz—did not quell the assault. So long as Newton lived they lived.

Hooke's third letter reached Cambridge during the second week of January 1680. This time he wasted little ink in getting to the point: "Your calculation of the Curve by a body attracted by an aequall power at all Distances from the center . . . is right," Hooke noted. "But my supposition is that the Attraction always is in a duplicate proportion to the Distance from the Center Reciprocall, and Consequently that the velocity will be in a subduplicate proportion to the Attraction and Consequently as Kepler Supposes Reciprocall to the Distance."[15] Not only would Hooke lay claim to the discovery of universal gravitation, he later cited this convoluted passage as proof that credit for setting forth the inverse square law defining planetary motion was due him as well. Hooke was again tantalizingly close, but his limited knowledge of mathematics once more barred the way. Suffice it to say that the force law here proposed, based on Galileo's formula for uniformly accelerated motion, conflicts with the alleged velocity relation erroneously derived from Kepler. Besides, Newton had discovered the inverse square relationship many years before by substituting the values from Kepler's third law into his own equation for centrifugal force. The incorrect value taken for a single degree of latitude had proved a major stumbling block, however, causing him to lose interest and turn to other matters.

To be twice corrected by Hooke (or by any individual for that matter) was more than Newton could bear; he enveloped himself in a hermetic cocoon of silence, refusing to write even a single letter to anyone for more than a year. Hooke, who was as nervous as Newton was vindictive, immediately sensed trouble and hastened to avert another rupture. "I can now assure you," he wrote on January 17, "that . . . the Experiment is very certain and that It will prove a Demonstration of the Diurnall motion of the earth as you have very happily intimated." The only remaining problem was to determine the properties of the curve that a body must follow when acted upon by an inverse square force. "I doubt not but that by your excellent method you will easily find out what that Curve must be, and its proprietys, and suggest a physical Reason of this proportion."[16]

One may argue, as Newton did in a fit of pique in 1686, that Hooke's was an easy guess, for Kepler had demonstrated that elliptical orbits fitted the observations of the planetary movements. But try as he might, Newton—whatever the nature of his private thoughts while watching the descent of the legendary apple—could not change the fact that it was Hooke who recognized that the shape of a planet's orbit is determined by an attractive force from the sun varying with the distance, that the fundamental problem of astronomical physics was the mathematical derivation of the orbit from the law of attraction. As the first to focus clearly upon this problem of orbital motion Hooke was Newton's mentor, and Newton admitted as much to Halley: Hooke's "correcting my Spiral occasioned my finding ye Theorem by wch I afterward examined ye Ellipes."[17] Two copies of that theorem survive, one in Newton's hand, the other in that of John Locke's amanuensis, Sylvanus Brownover. The original was composed before the *Principia*, most probably during the early months of 1680.[18] Employing his method of limits and infinitesimals, Newton was first able to demonstrate that the force of gravity varies inversely as the square of the distance at the apsides of an ellipse and then extend this relationship to every point on the curve. He had solved the mechanics of orbital motion for a planet circling the sun, an intellectual feat whose magnitude can hardly be overstressed. For the first time in history the cat's cradle of astronomical problems that had baffled natural philosophers since the days of Babylon were capable of theoretical solution. The controlled power of mathematics in the hands of genius promised nothing less than to remake the very universe itself. While Hooke had been the lightning bug, Newton was the lightning, a virtuoso of the *fait accompli*.

II

Robert Hooke got even less than his usual fitful amount of rest during the month of December 1680. But this particular round of sleeplessness bore no relation to the frequent bouts of insomnia associated with his legendary hypochondria or to his most recent skirmish with Newton. Hooke's *Diary* reveals the cause in a nutshell:

12th. Comet appeared with a very long blaze.
18th. The Comet at 7 appeared 90° and about 2 broad.
30th. Sat up all night about Comet.[19]

Astronomer Royal John Flamsteed was rather less matter-of-fact when he described the celestial visitor in a letter to James Crompton, Fellow of Jesus College, Cambridge, on December 15: "This night I have seen ye tail again wch is 50 degrees in length. I believe scarce a larger hath ever been seen 'tis above a Moon broad its motion decreases in Switftness, and I believe we shall see it longer than any hath been seen of late." Then Flamsteed, the scientist, caught himself: "When I have got some more observations of it I shall bee able to tell you how long it will last and where it will pass. At present I dare not pretend."[20] On December 24 Edmond Halley wrote to Hooke from Paris about the comet "after the most unpleasant journey that you can imagine, having been 40 hours between Dover and Calais with wind enough." Yet so brilliant were the celestial pyrotechnics that the dual obstacles of a pitching ship and racing clouds could not deny the eager young astronomer a view.[21] (Little did Halley suspect that a similar object— soon to guarantee him immortality—was fast closing on the sun to keep its only rendezvous in the normal life span of a man.)

Waste Book at the ready, Newton too waited and watched. He began to record information about the comet's tail on December 12, the very date of Hooke's first written observation. Newton eventually compiled a log containing detailed figures of twenty-one separate sightings between mid-December 1680 and February 5, 1681, including the precise times of his observations and the comet's ever shifting ascensions, declinations, latitudes, and longitudes.[22] Hampered by myopia, he employed a single concave lens to sharpen the comet's image, then graduated to a telescope fitted with a micrometer as the object moved deeper into the solar system, finally to disappear altogether early in March. (While Newton's observational techniques had grown more sophisticated over the years, his was a vigil reminiscent of the one that had caused him to become "much disoriented" during a winter of solitary comet-watching in 1664.) Still not content, he assigned himself the arduous task of collecting every recorded observation of the comet, whether from Venice, the East Indies, Paris, or colonial Maryland, where Arthur Storer, standing on the banks of the Patuxent River, saw it "Forme like a Sword streameing from ye Horizon."[23]

During the month of November 1680, a spectacular comet, a seeming twin of its December sister, had been observed moving toward the sun. Halley, who became obsessed with the phonomenon, had watched it from his residence in London, while an animated Flamsteed tracked its movements from the Royal Observatory in Greenwich. Newton also logged its passage with multiple entries in the *Waste Book*.[24] The appearance of two such similar objects within so brief a time span was most unusual, but it would have been a serious violation of conventional scientific wisdom to argue that the comet that had approached the sun from one direction was the same object seen leaving it from the other. No, they had to be different bodies moving in nearly parallel but clearly opposite directions. Johannes Kepler had contended that comets travel roughly in straight lines. The most elaborate adaptation of Kepler's rectilinear theory appeared in the treatise *Cometographia* (1668) by Johannes Hevelius, the eminent astronomer of Danzig. Newton had read the works of both men and, like almost everyone else, was erroneously convinced that the November comet, in a form of cosmic suicide, had either disintegrated on impact with the sun or, hidden from view by the sun's impenetrable glare, had headed outward into the abyss, never to be seen again. Not so, protested Flamsteed, the crusty Astronomer Royal. Comets are not, as characterized by ancient and medieval men, alien bodies that randomly violate the sanctity of the solar system, conforming to no principles save their own. Rather, they must follow the same rules as other objects in the universe. Comets reverse their direction in the vicinity of the sun, which is precisely what occurred between late November and early December. The comet of 1680 was just that, not two bodies but one.

An anxious Flamsteed hoped to have his theory vindicated by Newton, whose work with the reflecting telescope and experiments on light had made a profound impression on the Astronomer Royal. But this was not the only part of Newton's reputation to have preceded him. Apparently forewarned by a member of the cognoscenti about the Lucasian Professor's extreme touchiness, Flamsteed chose to broach his unorthodox theory via his Cambridge friend and correspondent, James Crompton. Newton accepted this channel of communication and never wrote directly to Flamsteed on the subject. Brilliant though his insight was, Flamsteed felt the need to develop a physics on which to ground his theory, and this led to its undoing. He argued that the comet's abrupt change of direction could be explained on the basis of a mysterious reversal of the sun's magnetic action. The comet was first drawn toward the sun like metal to a giant magnet. Then, after the comet entered the sun's vortex, this force somehow reversed itself, radically altering the flaming missile's trajectory and saving it from almost certain destruction. Flamsteed further imagined that, once the comet had been deflected, its magnetic pole would continue to point away from the sun, just as a hurtling bullet was thought (incorrectly) always to keep the same side forward. Nor had the comet circled the sun; it had turned back short of the sun.

Although not as far-fetched as Jonathan Swift's satirical proposals (made after a visit to the Royal Society) to propagate a breed of naked sheep and to extract sunbeams from cucumbers, Flamsteed's physics was badly muddled, costing him any chance of winning Newton's approval of his cometary theory, at least for the time being. And while Newton adopted a civil, almost cordial, tone, his manifold and incisive objections proved intellectually devastating. For example: "The instance of a bullet shot out of a cannon & keeping ye same side forward may be a tradition of ye Gunners, but I do not see how it can consist wth ye laws of motion, & therefore dare venture to say that upon a fair trial 'twill not succeed excepting sometimes by accident."[25] One historian has cited this argument as a benchmark in the social evolution of science. "Such an utterance marks the point at which the science of mechanics emancipates itself from the empiricism of craftsmen and enough confidence is felt in its laws to make predictions about the outcome of yet untried experiments."[26]

In a letter to Halley written during this same period, Flamsteed presented his idea of what a comet is made of—"it may have been some planet belonging formerly to another vortex now ruined: for Worlds may die as well as men: that its naturall motion being destroyed its body is broke & the humid parts swim over ye rest." Newton, who had received a similar description from Flamsteed, could at least agree on this point. But his analysis was far more cogent; indeed, by making proper allowances for the archaic style, it could well be inserted into the latest textbook on introductory astronomy: "ye tayle of ye Comet is a thin vapour, that it rises from ye atmosphere about ye head, that ye action of ye Sun's light conduces to raise it, that it shines not by its own light but only by reflexion of ye Sun's light."[27] More to the point, however, was Newton's refusal to accept Flamsteed's indentification of the two comets as one, even though he could have reconciled the hypothesis with a superior physics of his own. The recent exchange with Hooke had enabled him to define mathematically the orbital dynamics of a planet. But since Newton did not for the present apply the same physics to a comet, it seems certain that the idea of universal gravitation had progressed no farther than the intuitive stage, if that far. Where comets came from he knew not; nor did he yet consider the beautiful but bothersome trespassers amenable to the laws governing the solar system. Still, Flamsteed had struck a responsive chord. Newton continued collecting data on the subject, even turning oral historian by questioning a Cambridge undergraduate at some length about his recollections of the November comet's location in the firmament. Less than two years later, as if by cosmic ordination, another equally brilliant comet appeared in the early morning sky.

Hours before dawn in the early autumn of 1682, Edmond Halley left a warm bed and lissome bride to keep a rendezvous with a mistress of irresistible charms. Halley softly climbed the stairs leading to the private observatory in his house at Islington, then a village on London's northern outskirts. Armed with telescope and sextant, he intended to continue his "darling employment" of studying the moon as a means of determining longitude. What Halley saw caused him to forget the project temporarily. On the horizon, illuminated by the slowly rising sun, was the comet that would one day bear his name. Back in Cambridge Newton, for the third time in less than two years, had already opened the *Waste Book* to the section on comets, recording his first observations of the newly sighted object in August and September.[28] He found that its motion was retrograde and therefore contrary to that of the comet of 1680. This time there would be no second look, for the object had already circled the sun and was headed outward into space.

Never one to compile data as an end in itself, Newton had continued his silent pursuit of answers to the questions on comets raised by Flamsteed, repeating the cycle set into motion by Hooke's query on planetary orbits in 1679. Three manuscripts survive in which Newton tried to prove that the comet of 1680 had moved along a rectilinear path at a uniform velocity, the standard theory based on Kepler's analysis in *De Cometis* (1619).[29] Halley had attempted this same proof and, like Newton, was disturbed to find that the actual motion was neither truly straight nor constant in speed.[30] As always, Newton had sought to exhaust the most likely possibilities before turning to the more conjectural. Yet even before he carried out these computations, something about Flamsteed's single comet theory instinctively appealed to him. In a draft of a second letter to the Astronomer Royal via Crompton, Newton raised the possibility that the comet was "directed by the Sun's magnetism as well as attracted." Rather than turning in front of the sun because of a reversal of its magnetic attraction, as Flamsteed had proposed, the comet circled the sun, "the *vis centrifuga* . . . overpowering the attraction & forcing the Comet . . . to begin to recede from ye Sun." Newton even attempted a partial sketch of the orbit he envisioned.[31] But the letter he sent to Crompton was minus this passage, his first on cometary dynamics, and therefore seemed uncompromisingly critical of Flamsteed's theory.

At some point in 1681 or 1682 (one cannot be absolutely certain), Newton concluded that while Flamsteed's physics was hopeless, the astronomer definitely had the right idea after all. The unsatisfactory calculations of a rectilinear trajectory doubtless supplied some impetus; his breakthrough on the orbital dynamics of the planets supplied far more. Newton may have derived additional stimulus from a quite unexpected direction. Giovanni Domenico Cassini, the Italian astronomer chosen by Colbert to head the Royal Observatory in Paris, published his *Observations on the Comet* in the spring of 1681. A copy of the work had reached the Royal Society by May 19, and Hooke gave a full review of its contents at the meeting of June 8, "together with some animadversions on that astronomer's theories."[32] One of the most interesting of those theories held that not only were the comets of 1680–81 one and the same, but that this was the very object seen by Tycho Brahe in 1577; its revolution consisted of a giant circle whose dimensions and exact position Cassini had not yet calculated. Halley, who was with Cassini in Paris at the time, rejected the orbital hypothesis, but he nonetheless wrote to Hooke about it in May. Later, when Halley tackled the question of cometary motion in all its daunting intricacies, he remembered Cassini well. Perhaps Newton did too. He owned Cassini's work, though its exact date of purchase is unknown.[33] Word of the theory might have also come from one of the contacts supplying him with observational data, such as John Ellis or a Mr. Bainbridge, who were both mentioned by Newton as collaborators in observing the comet of 1680–81.[34]

However much Flamsteed and Cassini may have stimulated Newton's thoughts on comets, the most provocative writing on the subject was to be found in a small treatise simply titled *Cometa*. Its author was none other than the effervescent and ubiquitous Robert Hooke. While Newton made no mention of the work in his correspondence until much later, his *Waste*

Book notes stand as irrefutable proof that he subjected its contents to a rigorous examination about 1681.[35] Hooke had closely observed the comets of 1664 and 1665, and he also spent much time probing data collected by other astronomers. He had probably carried his ideas about as far as he ever would by 1666, but the extensive notes were readied for publication only after the brief appearance of another comet in 1677. Hooke never resolved to his own satisfaction the path a comet follows, but he nonetheless gave free rein to his powerful scientific imagination. "What kind of motion was it carried with?" he asked, "whether in a straight or bended line? and if bended, whether in a circular or other curve, an elliptical or other compounded line, whether the convex or concave side of that curve were turned towards the earth?" And "whether in any of these lines it moved equal or unequal spaces in equal times?" Hooke was particularly suspicious of the rectilinear theory as applied to the comet of 1664. Its movement, he thought, seemed to describe a circle, or perhaps an ellipse. The physical explanation might at first "seem pretty difficult," he admitted, "but this is no more than is usually supposed in all the planets." Is it not possible that "a kind of gravitation throughout the whole Vortice or *Coelum* of the Sun, by which the planets are attended," has a similar effect on comets?[36] Imagine the impact of such fertile conjecture on Newton, who had recently mastered the orbital mechanics of the solar system. Standing on far firmer ground than Hooke, he alone was in a position to bring comets into his new but expanding family of explicable mechanical phenomena.

Spurred on by the appearance of Halley's Comet, Newton put the curved orbit theory to the test. He knew there were four possible trajectories—a circle, an ellipse, a parabola, or a hyperbola—each of which can be obtained by slicing through a cone at different angles. Unlike the ellipse, the last two do not catch their own tail but begin and end at distances from the sun too vast to calculate. This would be the track of an object having but a single encounter with the solar system before rushing away forever. If, on the other hand, a comet should return, as Cassini predicted, its orbit must be elliptical, like those of the planets. Newton's calculations from this period have survived and show that he plotted both types of trajectories. He had further concluded that the sun and comets each show evidence of gravitation toward their centers, which decreases as the square of the distance. The sun's gravitation is by far the greater, and therefore the sharpest point of curvature in the orbit occurs when the comet passes nearest it; what astronomers refer to as the perihelion.[37] Here was substantial proof that orbital dynamics was not unique to the visible planets but a mathematical commandment of universal physical creation. Surely this revelation, coming as it did on the heels of his previous discovery, transported Newton to the purest state of intellectual ecstasy. Yet even in Eden there are pitfalls.

III

When one assesses the flow of Newton's thoughts between 1675 and 1682 a single word comes automatically to mind—convergence. From one direction came the new physics and the ability to solve the orbital mechanics of a planet or a comet; from a second came the alchemy and the promising hypothesis that all of nature, from the tiniest particle to the most grandiose star, is bound together by an elastic medium called ether; and from a third came the belief in an omnipotent, ever present Creator who has chosen a handful from each generation to share knowlege of which few humans even dream. But if, as Shakespeare once observed, the course of true love runs not smoothly, the metaphor applies equally well to the life of the mind. The biographer, as cartographer of inner space, must be sensitive to subtle shifts in those major intellectual currents that affect the contours of his subject's perception of the world. The first intimations of just such a shift occurred in Newton's thinking at some point toward the end of this period, when almost everything he wanted seemed within his grasp. True, he was destined to learn far more about the universe during the next few years, but that knowledge, however important it might be to him and to posterity, could never totally compensate for the dawning realization that the quest for a perfect transcendent unity must ultimately fail.

In the "Hypothesis of Light" Newton had postulated the existence of a universal ether as the agent by which the various forces acting on matter are transmitted. More subtle than air, which presses only on the outer surface of an object, ether penetrates to the interior and pushes against the walls of every hidden pore. Newton supported this view by recounting an experiment he had first performed in the mid-1660s while drafting the *Quaestiones*. He found that a pendulum's motion in a "glasse exhausted of air" decays at almost the same rate as it does outside the vessel, "no inconsiderable argument" for the presence of a resisting medium inside the sealed container.[38] The recent exchange with Hooke had opened the door to a radically different and most disturbing prospect, however. Planetary orbits apparently were not determined by tiny particles of matter in motion but by great forces reaching across vast stretches of theoretically empty space. If this was true, the same principle of action at a distance might also apply to the short-range attractions in evidence during the alchemical experiments. The earlier work with the pendulum obviously demanded closer scrutiny.

Newton fashioned a large, rounded bob from wood, which he suspended from a steel hook by a thread 11 feet long. He drew it aside 6 feet, marked the spot, and let it swing freely, not in a vacuum but in the open air. The places to which it returned on its next few oscillations were carefully noted. Newton then weighed the hollow bob, including the air inside, the thread that went around it, plus half the remaining thread. He next filled the sphere with various metals and calculated its weight as seventy-eight times greater than when empty. Care was also taken to adjust the string to compensate for the stretching caused by the increased load. "Then drawing aside the pendulum to the place first marked,' he wrote, "and letting it go, I reckoned about 77 oscillations before the box returned to the second mark, and as many afterwards before it came to the third mark, and as many after that before it came to the fourth mark."[39] According to the mechanical theory of ether, the bob was now heavier because the subtle matter penetrated its surface and pressed down upon the metal inside. This should have caused the ether to resist the motion of oscillation, bringing the pendulum to the prescribed marks in several fewer swings. It had not. The ether either was present in quantities too small to have any physical effect or, more plausibly, simply never existed. Newton misplaced the paper on which he recorded the data, but the lucidity with which he was able to recall, when composing the *Principia* a few years later, the demonstration that unsettled his *idée fixe* speaks for itself: It was for him another *experimentum crucis*. With the evaporation of this invisible mechanism he was forced to come to grips anew with the concepts of attraction and repulsion. From the early 1680s onward he began to think of forces by which bodies and individual particles act on each other at a distance without direct contact. Newton would define such a force on the cosmic plane, but in 1693 his sanity was jeopardized during a final desperate and unsuccessful attempt to extend it to the diminutive realm of chemical phenomena, the one great intellectual failure of his life—or so he told himself.

The only scientific correspondence initiated by Newton, that with Boyle on alchemy, continued into the 1680s, although few of the letters survive. Of those that do, the best-known is a lengthy missive drafted by Newton on February 28, 1679. "The truth is," he began on a familiar note of false modesty, "my notions about things of this kind are so indigested yt I am not well satisfied my self in them."[40] He then proceeded to expand in masterly detail on the theory of ether first announced in the "Hypothesis of Light," fairly conclusive proof that the pendulum experiment was still to be performed. With Newton, however, one can never be absolutely certain, for every card, especially an ace, was kept extremely close to the vest. The suppressed draft of his letter to Flamsteed suggesting that the sun's gravity might shape the orbits of comets is a perfect example. If he were wrong, his embarrassment would be private rather than public, preserving his expanding aura of infallibility. If right, which was usually the case, it gave him absolute control, the freedom to pick and choose the precise moment of disclosure, as would occur in 1684 when Halley made his famous visit to Cambridge.

Chapter Twelve

The Most Perfect Mechanic of All

Doing easily what others find difficult is talent; doing what is impossible for talent is genius.

Henri-Frédéric Amiel

I

On a bitter Wednesday in January 1684, Edmond Halley decided to brave the winter elements by going up to London from Islington. That afternoon, presumably over drinks before a roaring fire at their favorite inn, he entered into a lively conversation with Robert Hooke and Sir Christopher Wren. The subject, Halley recalled vividly in a letter to Newton dated June 29, 1686, was celestial motion. The astronomer was anxious to get his friends' reaction to his new theory that the inward force of attraction between the planets and the sun must decrease in inverse proportion to the square of the distance between them. Though Halley never said so, it seems likely that his pronouncement was met by concerted laughter, leaving him to wonder in his embarrassment just where he had blundered. His confusion would have been momentary, however, for "Mr Hook affirmed that upon that principle all the Laws of the celestiall motions were to be demonstrated." Wren made it unanimous, claiming that he, too, had reached the same conclusion independently, an assertion later supported by Newton.[1] The problem, of course, lay in finding the means of proving it. To stimulate the inquiry, "Sr Christopher," who candidly admitted to his own lack of success, "sd that he would give Mr Hooke or me 2 months time to bring him

281

a convincing demonstration thereof." The winner, should there be one, would receive two rewards—the intangible gift of honor and a present from Wren of a book worth forty shillings. Impetuous as always, Hooke laid preemptive claim to the modest spoils by declaring himself the victor, "but . . . he would conceale [the solution] for some time that others triing and failing, might know how to value it, when he should make it public." Halley remembered that Wren "was little satisfied that he could do it," and rightly so. Hooke's own diary reveals that he had conversed with Wren on the subject of orbital mechanics as early as January 1680, while he was engaged in correspondence with Newton on the same matter. If, as must have seemed likely to Wren, Hooke's mathematical thinking had failed to mature in the interim, this latest in an endless spate of claims could be chalked up to fustian, the seventeenth-century equivalent of braggadocio.

The month of March came and went, and with it the deadline set by Wren. Spring turned to summer, and still Hooke remained mute, leading Halley to observe later: "I do not yet find in that particular he has been as good as his word."[2] Finally, in August 1684, Halley, long since having lost patience with Hooke, cast an anxious eye in the direction of Cambridge and made a fateful decision: He would go there and beard the lion in his den.

In contrast to Newton, who had attended a backwater grammar school and made his way in the University by waiting tables and running errands, the young Halley enjoyed almost every advantage money could buy. Born in 1656 at the family country house in the village of Haggerston, Edmond was named after his father, a wealthy landowner, salter, and soapmaker in nearby London. According to Aubrey, the boy was taught writing and arithmetic at home by his father's apprentice. Anxious that his son should have the benefit of a formal education, the Senior Halley later enrolled him in St. Paul's School and afterward, at the age of seventeen, in Queen's College, Oxford. Like Tycho Brahe, Halley possessed an unusual knowledge of the heavens at an early age. "From my tenderst youth I gave myself over to the consideration of Astronomy," he was to write in the preface of his famous *Catalogue of the Southern Stars*. One Moxton, the globemaker at St. Paul's, told Aubrey that if a star were misplaced on the globe the youth would soon find it.[3] Halley left an equally favorable impression of himself at Oxford. The antiquary Anthony à Wood observed that "he not only excelled in every branch of classical learning, but was particularly taken notice of for the extraordinary advances he had made at the same time in mathematics."[4] Halley fulfilled his promise early. Leaving Oxford in 1676, he sailed to the lonely island of St. Helena in the South Atlantic, today remembered as the site of Napoleon Bonaparte's final exile and death nearly a century and a half later. Halley's two-year stay proved rather more productive than the little General's. He catalogued the stars of the Southern Hemisphere with an accuracy and completeness never before attempted and discovered a star cluster in Centarus for good measure.

Charles II applauded his work, and the Royal Society enthusiastically elected him Fellow in 1678. His earliest portrait, painted when he was in his late twenties, shows Halley as he first appeared to Newton: tall, dark-eyed, soft of face, a presence pleasing to almost everyone.

There is no record of any correspondence between Halley and Newton before 1686. However, Newton's *Waste Book* refers to a meeting with Halley (c. 1682) at which he pumped the young astronomer dry of every detail he could remember concerning the comet of 1680.[5] Perhaps a letter informing Newton of Halley's impending visit is lost. Yet their recollections of the meeting (which are admittedly less than perfect) suggest that Newton was caught off guard, an indication that Halley scouted his quarry well and rejected written notification in the hope that more candor would result from a face-to-face exchange. If so, Halley made no shrewder decision in his life, for Newton, the recipient of many nettling letters but few distinguished visitors, was both flattered and disarmed. Abraham de Moivre later retold the dramatic story as he said Newton had related it to him:

> In 1684 Dr Halley came to visit him at Cambridge, after they had been some time together, the D^r asked him what he thought the curve would be that would be described by the planets supposing the force of attraction towards the sun to be reciprocal to the square of their distance from it. Sr Isaac replied immediately that it would be an Ellipsis, the Doctor struck with joy & amazement asked him how he knew it, why saith he I have calculated it, whereupon Dr Halley asked him for his calculation without any farther delay, Sr Isaac looked among his papers but could not find it, but he promised him to renew it, & send it.[6]

However gratified Newton may have been by Halley's spontaneous display of emotion, something more profound moved him to take the ebullient astronomer, whom he barely knew, into his confidence. Halley had come to Cambridge with an awareness of the inverse square principle. He wanted more than the oral confirmation already supplied by Hooke and Wren. What Halley truly sought was a clear-cut mathematical demonstration of its dynamics, which to Isaac Newton made all the difference. Implicit in Halley's request was the recognition that in the new science the making of hypotheses, no matter how brilliant, must never be confused with scientific truth itself. The "calculated" result is everything, a reality that Hooke never fully fathomed. In addressing Hooke's claim that Newton had stolen from him, the eighteenth-century French scientist Alexis Claude Clairaut astutely observed, "what a distance there is between a truth that is glimpsed and a truth that is demonstrated."[7] To Clairaut's concise and penetrating insight Newton would have added not a single word.

Time and circumstance also conspired to Halley's considerable advantage. At twenty-eight, he was already a well-published natural philosopher of the first rank, but of a different generation from Newton and the gifted men against whom the Cambridge virtuoso entered the lists when his vulnerable center ceased to hold. Halley also possessed *savoir-faire*, an indispensable requirement for dealing productively with the great man. Despite numerous instances of provocation, his treatment of Newton was unwavering, ever polite and deferential from their first encounter to the last letter, a relationship spanning more than four decades. In later years Halley liked to think of himself as Ulysses, who had persuaded a temporarily unnerved but valiant Achilles to cross the Hellespont and meet his fate head on.[8] While Halley doubtless embellished the facts, he clearly approached the solitary genius during one of those rare moments when, sated from hunting and feeding alone on the endless plain of his private thought, Newton proved receptive to an outside stimulus.

Days passed, then weeks, and finally months, during which Halley surely experienced the same helpless frustration that was the lot of Oldenburg and Collins more than a decade before. From what little Halley could tell, Newton had forgotten his pledge. Perhaps, in moments of despair, Halley thought he was on a fool's errand. The temptation to write to Cambridge must have bordered on the unendurable, but the youthful astronomer showed a wisdom beyond his years by somehow staying his pen. He would one day learn the cause for the delay, which Newton related to de Moivre:

> Sr Isaac in order to make good his promise fell to work again but he could not come to that conclusion wch he thought he had before examined with care, however he attempted a new way which tho^u longer than the first, brought him again to his former conclusion, then he examined carefully what might be the reason why the calculation he had undertaken before did not prove right, & . . . he made both his calculations agree together.[9]

The delay, which totaled three months, resulted from more than Newton's prudent recalculations, which one of his mathematical gifts could have completed in substantially less time. He spent much of those three months working on a nine-page treatise whose origin may have predated Halley's visit. Newton titled the work *De motu corporum in gyrum* (*On the Motion of Revolving Bodies*). In November Edward Paget, a Fellow of Trinity whom Newton had recently nominated for the position of Mathematical Master of London's Christ's Hospital, delivered a copy of the paper to Halley. Suffice it to say that the astronomer was little prepared for this bolt from the blue, which not only contained a solution to the original problem but far more—the mathematical seeds of a general science of dynamics.

Halley was fully capable of appreciating the magnitude of what he held in his hands, but he had to be certain that his personal relationship with Newton rested on firm ground. The test would come when he requested permission to inform the Royal Society of *De Motu*. In this instance the idea of writing to Newton never crossed Halley's mind, as it undoubtedly had in advance of their previous meeting. He made directly for Cambridge early in December, his thoughts racing far ahead of the coach that carried him toward the moment of truth. From a rational standpoint there was good reason for optimism—why else would Newton have even bothered to send him the paper? But across this bright prospect drifted the baleful clouds of a bleaker reality formed from the accounts of those who, like Hooke, had spoken of Newton's flashing temper and Janus face. Halley did not dawdle in Cambridge after receiving an answer; he returned to London in time to attend the Royal Society meeting of December 10, which marked the beginning of Samuel Pepys's two-year tenure as President. The Fellows were confronted by a lengthy agenda, and by the time Halley, the last to speak, rose to take the floor, a number of those present had slipped into that comfortable torpor which invariably attends such occasions, thus forcing them to improvise when they attempted to recall the historic moment years later. "Mr. Halley," the minutes read, "gave an account that he had lately seen Mr. Newton at Cambridge, who had showed him a curious treatise, *De Motu*; which, upon Mr. Halley's desire, was, he said, promised to be sent to the Society to be entered upon their register." The Chair "desired" Halley "to put Mr. Newton in mind of his promise for the securing [of] his invention to himself . . . till he could be at leisure to publish it." [10] Paget, who was also present, was requested to assist Halley in this effort, and they accomplished their task early in 1685. [11] Newton had given his answer to Halley, and Halley had dutifully communicated it to the scholarly world—proof, after all, that the reclusive and intractable don wanted to taste immortality while still among the living.

II

To describe Isaac Newton as he reached his forty-second birthday on December 25, 1684 (O.S.), one must of necessity invoke the very scientific symbols that captured and held his imagination for nearly three-quarters of a century. His mind functioned like a powerful achromatic lens, collecting disparately tinctured rays of thought and bringing them to a single unblemished focus. Like some Promethean hero of old, Newton believed his dazzling trajectory from obscurity to greatness was prefigured in legend and folklore. His birth on Christmas Day convinced him that he was a prodigy of nature with a divinely ordained mission. (Halley, who was anything but pious, merely reconfirmed what Newton already believed in the ode he composed and prefixed to the *Principia*: "Nearer the gods no mortal may approach.") As is often the lot of the inexplicably gifted, Newton paid an awesome price for his riches. Just as he labored successfully to remove the scoria from his crucibles, so also did he work with equal diligence, if less favorable results, to purge the blackness from his soul. Newton's psychic defenses were frightfully thin; he nurtured grudges and hoarded imaginary slights like so much alchemist's gold, making it virtually impossible to establish lasting human relationships. Yet a flaw like unsociability often contributes more to great accomplishments than a virtue. That extraordinarily private but gravid mind was about to fulfill the high promise of a long and perfect gestation, but only after eighteen months of intellectual labor performed at the outermost limits of human capacity.

Newton's belief that God was on his side never led him into that fatal, and to a Puritan damning, illusion that God would do the work. He withdrew farther than usual from the world around him, as if in conscious anticipation of the great struggle ahead. Of the handful of letters he wrote in 1684 only two survive. Both were written in December and relate directly to the business at hand. Halley, who was burdened with family matters until August, may have hoped to avoid his first visit to Cambridge by approaching Newton during one of the latter's infrequent visits to London, but the opportunity never arose. Newton did not stir from the University for so much as a day between May 1683 and March 1685.[12] While he may have initially thought of De Motu as an end in itself, the creative flow loosed during its composition soon relegated the little treatise to a brilliant beginning. "Now that I am upon this subject," he wrote to Flamsteed in January 1685, "I would gladly know ye bottom of it before I publish my papers."[13] Measured though his words were, they indicate that the controller had already relinquished control. Newton had achieved a level of autonomic creative euphoria in which all things seemed possible to him, and to a remarkable extent they truly were.

Newton's brief but earnest exchange of letters with the Astronomer Royal, beginning late in December 1684 and ending exactly one month later, provides significant insight into the revolution as it was unfolding in his thoughts. Hungry for observational data that only Flamsteed could supply, Newton bypassed Crompton and wrote directly to the Royal Observatory, making excellent use of the recently instituted London Penny Post. To get on Flamsteed's good side, he had instructed Paget to deliver a copy of De Motu to Greenwich, but the transfer was unpredictably delayed until very near the end of this exchange by a combination of bad weather and Paget's contraction of ague.[14] Fortunately, it was the thought that counted with Flamsteed in this instance: "[Y]ou can not propose more readily then you shall be willingly answered," he wrote on December 27, and the Lucasian Professor found the Astronomer Royal to be as good as his word, a glaring contrast to Flamsteed's later contrariety, which Newton's complex and insistent demands did nothing to ameliorate.

Newton first wanted to know the exact position of the two fixed stars in the foot of the constellation Perseus. The comet of 1680, as viewed from earth, had passed between them, and Flamsteed, who immediately forwarded all the data at his disposal, accurately surmised that Newton was seriously rethinking the conventional comet theory. "If you will give me leave to guesse at your designe," he replied on January 5, 1685, "I believe you are endeavoring to define yt comet described in ye aether from your Theory of motion."[15] One week later Newton confirmed that "I do intend to determin ye lines described by ye Comets of 1664 & 1680 according to ye principles of motion observed by ye Planets, & should be glad of your help as to those places of ye latter."[16] Still, in light of their previous correspondence on the subject Flamsteed must have been somewhat taken aback to learn just how far Newton had advanced this "theory" when *De Motu* finally reached his hands later in the month.

The second and more revealing issue raised by Newton was one that far exceeded the confines of *De Motu*, indeed seemed so radical to Flamsteed that he expressed grave doubts reminiscent of Newton's reaction to his own cometary theory of 1681. Newton asked Flamsteed for information about the respective velocities of Saturn and Jupiter as they enter conjunction—that is, share the same celestial longitude. Saturn's orbit as defined by Kepler's tables, he believed, was too small. "This planet so oft as he is in conjunction with Jupiter ought (by reason of Jupiters action upon him) to run beyond his orbit about one or two of ye suns semidiameters or a little more & almost all the rest of his motion to run as much or more within it." Here was positive proof that the inverse square attraction between the sun and the planets and the sun and certain recently observed comets was yielding in Newton's mind to a far grander conception: the idea of a universal power of attraction between all celestial bodies, including the earth. "I can not conceave of any impression made by ye one planet . . . can disturbe ye motion of the other," Flamsteed wrote, captive as he was of the theory of ether; "it seems unlikely such small bodies as they are compared with ye Sun, the largest & most vigorous Magnet of our systeme, should have any influence upon each other at so great a distance." Newton, he cautioned, would do well to consider the fact that the largest magnets have no influence either on one another or on a needle at a distance of 100 yards. Yet Flamsteed was simultaneously able to confirm that the orbits of Saturn did indeed vary to a degree consistent with Newton's calculations. "Your information about ye error of Keplers tables . . . has eased me of several scruples," Newton replied. "I was apt to suspect there might be some cause or other unknown to me."[17] As far as he was concerned, Flamsteed's objections, which for the most part would have been his own no more than a few months earlier, were not worth the time and paper an answer required. If not a camel, then at least a man had passed through the eye of the Astronomer Royal's needle.

Either satisfied for the time being with the data placed at his disposal or perhaps distracted by an abrupt shift in the country's political winds, Newton addressed no further letters to Greenwich for several months. From mid-January to September 1685, when be broke his silence to request more information about the comet of 1680, only two people were favored by brief messages from his otherwise occupied pen. Dr. William Briggs, an oculist with whom Newton had exchanged letters in 1682, received the Lucasian Professor's favorable appraisal of his work, which was prefixed (with Newton's permission) to the Latin edition of Briggs's *Theory of Vision.* And on February 23 Newton wrote to Aston of Paget's hope, "when last with us," that a philosophical society be established at Cambridge. Both More and Montague favored the idea, but Newton reported that the business was "dasht" for want of persons willing to undertake experiments, himself included. "I should be very ready to concurre . . . so far as I can doe it without engaging the loss of my own time."[18] His current preoccupation speaks for itself, while the inability of others to carry on with little more than his moral support does nothing to enliven the wan portrait of Cambridge's malnourished scientific spirit. To complete the equally pallid record of all that is known of Newton's day-to-day activities during the first eight months of 1685, one has only to note two visits to Lincolnshire of about a fortnight each, the first in April, the second in June.[19]

Though normally little affected by external events, Newton, owing to unforeseen political developments, labored under greater tension than usual early in the year, as did all of England. It is even possible that he suffered a temporary loss of concentration, which would account in part for the lapse of his productive correspondence with Flamsteed. As Thomas Lord Bruce, attendant to King Charles II, later remembered, the sudden extinction on Sunday night, February 1, of a huge candle held by a page "where no wind was to be found" portended all that followed. The King, who spent the evening in the chambers of his mistress, later returned to his own apartments, where he passed an uncharacteristically restless, dream-filled night. Six days later, in his fifty-fifth year, Charles II died, his body shrunken and pale from the remorseless dosing, purging, and bleeding prescribed by the royal physicians (some fifty-eight drugs and twenty-four ounces of blood), his shaven head and feet blistered from the liberal application of red-hot irons.[20] Some doubtless looked upon this torture as a fit preparation for hell, and history has certainly done little to honor Halifax's epitaph: "Let his royal ashes then lie soft upon him, and cover him from harsh and unkind censures."

Newton did not fall in with Halifax's sentimentality, but if he were honest with himself, he had more reason than most to mourn the King's death. It was Charles who had lent his support to the founding of the Royal Society and who warmly praised the very device that won Newton election to it, the reflecting telescope. More important, Charles had waived the statutory requirement that Newton take holy orders to retain the Lucasian Professorship. True, Charles's Catholic sympathies could not have sat well with Newton, but the King, bowing to duty, had discreetly refused conversion until two days before the inevitable, rumors to the contrary notwithstanding. Now the crown rested on the head of Charles's overbearing brother, James II, a practicing Catholic. Knowing what we do of Newton's uncompromising hatred for the Church of Rome, it must have seemed to him that the beast of Revelation, having once been expelled by Trinity's royal founder, had again been loosed upon the land.

There was little to be done but wait and watch. The University, as dictated by tradition, met in full academic regalia at the public schools and from there wound its many-threaded way to the market place to proclaim the new King on February 9. Bells pealed all day long, and each college held a feast. Even the town's prisoners were treated to an extra ration of bread. Night was ushered in with bonfires and the quaffing of spirits, a further excuse (as if one were needed) to evade the demands of scholarly life. The University later published the obligatory verses of mourning and celebration, which few, save possibly James's anxious political advisers, bothered to read. On the surface little had changed, at least for the time being. In July Monmouth's pitiful attempt to wrest the crown from his uncle came to a swift and ignominious end, and the traitor paid with his head after being taken prisoner while cowering in a ditch. The Duke's picture, painted in honor of his election to the Chancellorship of Cambridge in 1674, was duly burned, and a grace was passed by the Senate requiring that the offender's name be stricken from all catalogs of University officers, an enforced rewriting of history as ancient as civilization itself.[21]

III

In 1675 Sir Christopher Wren laid the first foundation block of St. Paul's Cathedral; aging but still vital, he was present thirty-five years later when his son and namesake set the final stone of seventeenth-century England's architectural masterpiece. In keeping with the spirit of the age the simple yet eloquent memorial inscription on the sanctuary's wall is in Latin. In translation: "If you would seek his monument look around you." The words may be incomprehensible to most of the nonclassically educated visitors who file past them by the thousands every day, but Wren's genius becomes immediately apparent to all. St. Paul's massive Baroque dome, pierced at the crown to afford a view of the lantern above, first isolates the secular pilgrim from his fellows, then reduces him to an inconsequential entity in the midst of surpassing grandeur. Newton contemporaneously built his masterwork, the cathedral of modern science, out of materials gathered from a different quarry. Still, those who have read and understood the *Principia* have experienced a transcendence of the mind and spirit not so very different from that generated by entering a great and sacred edifice for the first time.

While Wren labored for all occupations and classes by designing scores of elegant yet utilitarian buildings in the wake of London's Great Fire,

Newton spoke only to an intellectual elite, whose numbers he sought to reduce to an absolute minimum by every means at his disposal. Shunning Galileo's innovative use of the vernacular, he chose to cast his revolutionary treatise in classical Latin. This decision, of itself, was hardly obscurantist; indeed, one can argue that it facilitated the dissemination of his work among the Continental virtuosi. At the same time, an English edition would have been welcomed by Newton's countrymen, especially after his fame was secure, making him a part of his nation's history as well as its science. Although he lived forty years more and published three editions of the *Principia*, Newton never undertook an English translation. Nor did he authorize anyone else to do so until the year before his death, if then.[22] (In contrast, the first edition of the *Opticks* was published in English in 1704, and a Latin edition, prepared at Newton's suggestion by Samuel Clarke, was issued just two years later).

Even more telling than the language used was his intent. The heart of the *Principia* is Book III, which bears the title "System of the World." In the introduction the author claims to have seriously considered framing the Book "in a popular method that it might be read by many." Still, Newton could not persuade himself to trust in human nature, particularly his own. Affecting a style of glacial remoteness, he made no concessions to the reader: "To prevent the disputes which might arise upon such accounts" by those who "could not easily discern the strength of the consequences, nor lay aside the prejudices to which they had been many years accustomed ... I chose to reduce the substances of this Book into the form of Propositions (in the mathematical way) which should be read by those only who had first made themselves masters of the principles established in the preceeding Books."[23] Many years later, freed of the constraints imposed by the printed word, Newton boasted to his friend the Reverend Dr. Derham that he had made the *Principia* "abstruse" intentionally "to avoid being bated by little smatterers in mathematics."[24] Such hypersensitivity is strongly reminiscent of Copernicus's blunt pronouncement to potential readers at the beginning of the long-delayed *De revolutionibus*: "for mathematicians only." Conditioned in part to this feeling by his earlier clashes with Hooke and the English Jesuits, Newton was reminded regularly and painfully of the futility of making himself understood by any but the mathematically and scientifically learned. Once a week he stood in the semidarkness of a deserted lecture hall literally talking to the walls, a living nightmare about which Martin Luther and Mark Twain only dreamed. He lectured from *De Motu* and the *Principia* between 1684 and 1687, taking his revenge by never bothering to repeat the material even when he had had no audience the previous week. Those few students who put in an occasional appearance were deeply confused and gave evidence of their alienation by inattention, as Whiston, a product of the system (if one can call it that) later testified. After the *Principia* was published, Newton passed such a perplexed undergraduate on the street. The student's comment serves as a fitting epitaph for a failed professorial career: "There goes the man that writt a book that neither he nor any body else understands."[25]

And what of the mathematics employed in the *Principia*? Newton, the inventor of the calculus, seemingly shunned modern analysis by clothing his masterwork in the garb of classical geometry. Before the relevant papers became readily available to scholars, a rather recent development, it was generally held that Newton had made extensive use of the calculus but had then recast the problems in the mathematical language of the ancients, composing, in effect, not one version of the *Principia* but two. The source of this misconception is Newton himself. During the calculus controversy, when he was attempting to establish his claim to priority, Newton wrote an anonymous, self-serving article for the *Philosophical Transactions*. "By the help of the new *Analysis* Mr. Newton found out most of the Propositions in his *Principia Philosophiae*: but because the Ancients for making things certain admitted nothing into Geometry before it was demonstrated synthetically, he demonstrated the Propositions synthetically, that the Systeme of the Heavens might be founded upon good Geometry."[26] If in fact Newton had drafted such an extensive set of papers with which to support his claim, he never produced them. Nor have they turned up in the manuscript materials, which only adds to the suspicion that they never existed. Had he been more forthright, he would have simply admitted to his preference for classical geometry on the grounds that it was more elegant than the analytic algorithms of the fluxional calculus, and to his belief that it had enabled the ancients to discover what he was only rediscovering some two millennia later.

Insofar as the *Principia* was concerned, Newton need not have raised the spurious question of prior demonstrations; instead, he should have encouraged mathematicians studying the work to seek out, in one scholar's words, "the thought patterns of the calculus behind the façade of classical geometry,"[27] which were readily apparent. By virtue of his singular analytical powers, Newton alone proved capable of devising a reliable method whereby the principles of ancient geometry could be extended beyond circles and straight lines to what he described as curved lines generated by continuous flow. His belief in the *prisca sapientia* never wavered, and by utilizing the mathematical tool available to Euclid, Pythagoras, Apollonius, and Archimedes he hoped to be included in their company. "He that works with less accuracy is an imperfect mechanic," Newton wrote in the *Principia's* revealing introductory paragraph, "and if any would work with perfect accuracy, he would be the most perfect mechanic of all."[28] As usually happened, Newton was destined to have his way, for this is precisely how posterity has chosen to remember him.

The *Principia* consists of three books, the first of which sets forth a general dynamics for bodies operating in the idealized conditions of no friction and no resistance. Book II, which appears not to have been a part of Newton's original conception, is concerned primarily with the motion of fluids and the effect of friction on the motions of solid bodies suspended in a fluid medium. It was here, by mathematical demonstration, that Newton settled once and for all the nagging question of the Cartesian vortex, proving to both himself and the scientific world that the tourbillion plays no part in the new physics. But it is Book III that most concerns us, for, as Newton wrote, "in the third I derive from the celestial phenomena the forces of gravity with which bodies tend to the sun and the several planets. Then from these forces, by other propositions which are also mathematical, I deduce the motions of the planets, the comets, the moon, and the sea."[29] Here the mark of genius left its imprint upon almost every page; here the grandeur of the *Principia* was made manifest.

The essential problem confronting Newton was to formulate a dynamics capable of unifying terrestrial and celestial motion into a single coherent system. Of the many pieces to the giant puzzle confronting him, the most crucial parts were Kepler's three laws of planetary motion and Galileo's laws of the motions of bodies on earth. Despite certain points of affinity, these essential components hardly seemed compatible. For example, the forces that drove the planets in their Keplerian orbits appeared not to apply on earth. And, vice versa, Galileo's terrestrial laws of motion and falling bodies apparently had no significant bearing on the movements of the planets, their satellites, or comets. According to Kepler, the planets circled the sun in ellipses, while Galileo, harking back to the ancients, said they moved in circles. Kepler believed that the planets were driven along by invisible "spokes" (the *anima motrix*) issuing from the sun, while Galileo, again under the influence of classical science, believed they were not driven at all, because circular motion was self-perpetuating. To connect (and correct) the work of his two illustrious predecessors Newton formulated a dynamics centered on his three laws of motion, which appear in Book I and are put to further use in Book III.

According to Newton's first law, "Every body continues in its state of rest, or of uniform motion in a right [straight] line, unless it is compelled to change that state by forces impressed upon it." Galileo had actually been the first to formulate this principle, which Newton, taking up where the Italian had left off, recast and integrated into his own system, never claiming that it was his alone. It is simply a statement that force is needed to change the motion of a body. Thus if no force acts on a body it will continue to move uniformly in both speed and direction. The old idea, supported by Kepler, among others, held that a force was needed to maintain motion when in fact just the contrary is true. As the first law asserts, a force is required to alter or destroy it. Leave a moving body, like a planet, alone, free from friction, and it will go on eternally. What is more, the force that produces a planet's curved path is not a propelling but a deflecting one, as the second law holds.

Newton's second law states, "The change of motion [of a body] is proportional to the motive force impressed; and is made in the direction of the right [straight] line in which that force is impressed."[30] In this instance Newton borrowed from Galileo the concept that the acceleration of a body's motion could be taken as a measure of the effect produced by one body on another, or the measure of forces acting upon such a body. But, as was characteristic of his genius, Newton carried this idea a giant step farther by introducing the concept of "centripetal" force. He had coined the term in *De Motu* and defined it as "seeking the center" to distinguish it from Christiaan Huygens's word "centrifugal," which means to flee the center. For purposes of illustration, one need only whirl a weight round and round on an elastic tether, as children are wont to do. Increase the speed and the elastic stretches more. In pulling outward the circling mass exerts centrifugal force; the hand at the center pulls also, thus generating sufficient centripetal force to keep the mock planet in orbit. It has been observed that no single concept better characterized the *Principia*, a book whose primary objective centers on the delineation of inward-pulling forces as they determine orbital motion.[31] It was at this juncture that Newton's indebtedness to Hooke became most apparent, for it was the latter's correction of his diagram of a falling body that put Newton on the trail of centripetal force. Of nearly equal importance was Newton's introduction of another term into the expanding lexicon of seventeenth-century physics—mass, the measure of a body's resistance to acceleration. Experimenting with pendulums, he was able to show that mass—the quantity of matter in a body irrespective of its volume or of any forces acting on it—is proportional to, but different from, its weight. Furthermore mass, which is dynamically fundamental and measurable, remains constant regardless of a body's position or motion with respect to other bodies, concepts virtually as alien to Kepler, Galileo, and Descartes as they had been to the ancients.

In contrast to the first two laws of motion, Newton's third law was completely his own and involved the concept of force, a principle little understood by his contemporaries. It states: "To every action there is always opposed an equal reaction: or, the mutual action of two bodies upon each other are always equal, and directed to contrary parts." Newton, as if attempting to explain this principle to a simple man of the soil, briefly evoked the imagery associated with his agrarian origins, providing the reader with a rare glimpse of the normally well-concealed yeoman's son: "If you press a stone with your finger, the finger is also pressed by the stone." But who, it must have occurred to him as he paused to consider the thought process of an unschooled rustic, would be so foolish as to press on a stone with his finger? He cast about for a more credible example and found one in his memories of a ponderous beast: "If a horse draws a stone tied to a rope, the horse (If I may say so) will be equally drawn back towards the stone; for the distended rope . . . will draw the horse as much towards the stone as it does towards the horse, and will obstruct the progress of one as much as it ad-

vances that of the other."[32] Newton quickly despaired of such prosaic illustrations, however, for on deeper reflection they doubtless seemed rather
trite to him. From this point onward he opted for more erudite demonstrations, leaving only the barest hint of how he might have addressed an audience of laymen. As in religion, true understanding was beyond the grasp
of the ordinary individual. The third law, like the others, was used to explain phenomena of cosmic importance, to demonstrate, for example, that
the moon pulls on the earth with the same force with which the earth pulls
on the moon. This is also true of the earth and an apple, except that in this
instance the forces exerted cause the apple to change its position visibly,
while the earth, because of its immensely greater mass, seems totally unaffected. Through a rigorous application of his laws of motion Newton
precisely defined for the first time the concepts of mass, inertia, and force,
and painstakingly documented their relationship to acceleration and velocity. The modern science of dynamics at last had its legitimate founder.

Newton, we know, formulated the inverse square law well before he
wrote the *Principia*, applying it with a considerable measure of success to
the attraction between the sun and the planets. According to his third law,
however, gravitation could no longer be thought of as an isolated
phenomenon peculiar to the central body of the solar system. It must apply
to every planet, moon, comet, and star in the universe, as profound a
thought as has ever crossed a sentient mind. Picard's accurate measure of
the earth was now available to him, so the obstacle that had apparently
blocked his progress twenty years earlier was removed. But before he could
resolve the question of universal gravitation to his scientific satisfaction, a
further obstacle remained, namely, whether one spherical body would act
gravitationally toward another body as if all its mass were concentrated at
its center. To illustrate, consider once again the example of the earth attracting the apple. Imagine for a moment that the entire earth is divided
into small particles (Newton's corpuscles, if you will) and that each interior
particle of the planet attracts the particles of the apple near its surface. The
problem is that these attractions are not all of the same intensity, because
some particles are at a much greater distance from the outside than the
others. Moreover, the directions of these attractions necessarily vary.
Generally speaking, however, those particles nearest the apple will pull immensely harder than those on the other side of the earth, but there will be
far fewer of them. The question is this: "Will the extra number of bits pulling at a distance just compensate for their greater distance, so that the pull
of all the bits add up to give the same results as if all the mass of the earth
were at one point at the earth's center?" The answer is yes, as Newton,
through his work with homogeneous spherical shells, triumphantly
demonstrated in Section XII of Book I. He first proved that a corpuscle
placed outside a sphere is "attracted toward the centre of the sphere with a
force inversely proportional to the square of its distance from the centre."

Then, in Proposition LXXV, he reached the general conclusion that any two such spheres will be attracted toward each other as if their masses were concentrated at their centers. Newton himself was awed by this accomplishment:

> I have now explained the two principal cases of attractions; to wit, when the centripetal forces decrease as the square of the ratio of the distances, or increase in a simple ratio of the distances, causing the bodies in both cases to revolve in conic sections, and composing spherical bodies whose centripetal forces observe the same law of increase or decrease in the recess from the centre as the forces of the particles themselves do; which is very remarkable.[33]

A series of elegant mathematical theorems follow, including one designed to determine the force with which a corpuscle placed inside a sphere may be "attracted toward any segment of that sphere whatsoever. " In Section XIII, with a brilliant application of mathematics not fully revealed to the reader, Newton extended the treatment of the attractions to nonspherical solids, fulfilling his promise "to comprehend and determine them all by one general method."

Finally, having established the principle that the force laws apply to all bodies, great and small, Newton concluded Book I with a section on the motion of minute particles. He recounted some of his earlier optical experiments and noted that light behaves in a manner quite similar to corpuscles when in the proximity of large bodies: "those rays which in their passage come nearest to the bodies are most inflected . . . and those which pass at greater distances are less inflected." Did this mean, then, that light was composed of corpuscles after all? Newton was more firmly convinced of this than ever, but he had been harshly criticized by Hooke for having said as much in 1672. The evidence, while promising, remained inconclusive. "Therefore because of the analogy there is between the propagation of the rays of light and the motion of bodies," he wrote ever so cautiously, "I thought it not amiss to add . . . Propositions for optical uses; not at all considering the nature of the rays of light, or inquiring whether they are bodies or not; but only determining the curves of bodies which are extremely like the curves of the rays."[34]

Up to the time of Descartes no natural philosopher had formulated a purely mechanical explanation of motion. The Aristotelian explanation, which attributed to bodies an inner striving toward their natural places, had persisted even through the Copernican Revolution. In contrast, Descartes's vortex theory made the planetary motions purely mechanical, separating mind and matter into two distinct orders of being. But while the vortex theory explained physical phenomena plausibly enough in broad outline, it proved of no practical use when it came to detailed calculations. It could not, for example, account for Kepler's planetary laws, although a number of astronomers still clung to the hope that a reconciliation of such glaring discrepancies might be effected one day. Newton was no longer among them, nor had he been for many years. He undeniably had been influenced by the Cartesian spirit, but Book II of the *Principia* constituted an unqualified assault on Descartes and the many advocates of the French philosopher's system.

Newton first challenged the belief that space is filled with great whirlpools of frictional fluid by proving that circular motion in a fluid medium would, under the action of gravity, decay into a spiral path and ultimately cause the planet to crash into the sun. He had discovered no evidence of instability in the planetary orbits; therefore, space must be empty and free of the friction central to the Cartesian hypothesis. In Proposition LII Newton offered conclusive proof that a vortex cannot account for Kepler's laws:

> If a solid sphere, in an uniform and infinite fluid, revolves about an axis given in position with an uniform motion, and the fluid be forced round by only this impulse of the sphere; and every part of the fluid continues uniformly in its motion: I say, that the periodic times of the parts of the fluid are as the squares of their distances from the centre of the sphere.

The mathematical demonstration of this proposition, along with the numerous corollaries Newton derived from it, constituted nothing less than the *coup de grâce* for Cartesianism as applied to celestial mechanics. "The hypothesis of vortices is ultimately irreconcilable with astronomical phenomena," he wrote with the confidence that only genius can inspire, "and rather serves to perplex than explain heavenly motions. How these motions are performed in free spaces without vortices, may be understood by the first Book; and I shall now more fully treat of it in the following Book."[35] Even had Book III of the *Principia* been aborted by its temperamental author, which for a time Halley saw as a distinct possibility, Newton had already demolished what remained of the beautifully self-contained, geocentric world of medieval Christian theology. Everything that had once appeared large and mighty shrank to the smallest significance in a universe where the earth no longer occupied the center of all existence. Concepts such as "eternity" and "infinity," which had once had only religious significance, now related directly to time and space.

IV

Books I and II of the *Principia* were progressing well by the spring of 1686, but as yet Newton had no publisher for his thriving brainchild. The scene now shifted to London, where Halley, Newton's chief exponent before the Royal Society, relinquished his Fellowship in January. It has long been argued that the recent death of Halley's prosperous father had resulted in a significant decline in the astronomer's income. The elder Halley had left no will. Protracted litigation over the estate ensued between the son and his stepmother, who married a third time. This reversal of fortune supposedly compelled Halley to seek the office of clerk to the Royal Society at the modest salary of £50 per annum. But a case can also be made that money was a secondary consideration and that Halley sought the position to keep in closer touch with scientific developments. Whatever his motives, he was elected by a majority of the thirty-eight Fellows present at the meeting of January 27, 1686, outballoting the physicians Denis Papin and Hans Sloane. The terms of the appointment not only required that Halley vacate his Fellowship but that he be single and childless, and "constantly lodge in the college, where the Society meeteth." The latter conditions were waived, however, probably in recognition of his superior intellectual gifts ("he shall be completely keen in mathematics and experimental philosophy").[36] The Fellows were not disappointed; satisfaction with Halley's performance was expressed on several occasions, especially in connection with his editorship of the *Philosophical Transactions*, the private enterprise initiated by the Society's first Secretary, Henry Oldenburg.

Whether in applying for the office of clerk Halley also sought to maneuver himself into a position of influence concerning the publication of Newton's manuscript is difficult to say. We do know that about this time Newton felt the urge, rare for him, to have his work evaluated by a person of advanced scientific understanding. Notes with suggested revisions, a number of which Newton undertook, survive among his papers. The handwriting, while not absolutely conclusive, strongly suggests that their author was Halley, as does additional evidence of a more circumstantial nature.[37] The astronomer had last seen Newton more than a year before in Cambridge. So far as is known, they had exchanged no letters during the interim. It would have had to be a stunning experience for Halley to read a manuscript that had evolved from a brilliant sketch of orbital mechanics (*De Motu*) into a revolutionary treatise whose central precept was a law of force uniting every particle of matter in the cosmos. Besides showing evidence of haste, the handwriting in which the notes are set down looks somewhat more uneven than Halley's usually formal script. Total composure, after all, could hardly be expected of one who has been asked to comment privately on a work of immortality in the making.

The likelihood that Halley served as the unnamed reviewer is strengthened by the fact that he read a paper of his own based on Newton's theory of gravitation at the Royal Society meeting of April 21. Halley applied the principle of attraction to the flight of projectiles, skillfully demonstrating how to set up and fire a mortar at a given target, no matter the terrain.[38] Ironically, he had chosen to make the first practical application of an elegant theoretical principle a potentially destructive one.

At the next meeting, a week later, the following note, almost certainly written by Halley, was entered in the Society minutes:

> Dr. Vincent [Senior Fellow of Clare Hall] presented to the Society a manuscript treatise intitled, *Philosophiae Naturalis principia mathematica*, and dedicated to the Society by Mr. Isaac Newton, wherein he gives a mathematical demonstration of the Copernican hypothesis as proposed by Kepler, and makes out all the phaenomena of the celestial motions by the only supposition of a gravitation towards the center of the sun decreasing as the squares of the distances therefrom reciprocally.
>
> It was ordered, that a letter of thanks be written to Mr. Newton; and that the printing of his book be referred to the consideration of the council; and that in the mean time the book be put into the hands of Mr. Halley, to make a report thereof to the council.[39]

Newton, who in the past had been waited upon so often by so many, now tasted of his own acrid medicine. Book I of his masterpiece was held in abeyance for lack of an officer to preside at the Council. Pepys, a surpassing diarist but a miserably lackluster President, was attending James II, while the Vice Presidents were in the country taking advantage of the fine spring weather. Three weeks passed during which Halley must have grown increasingly restive. Up to then his association with Newton had proved fruitful beyond imagination, and now was hardly the moment to tempt fate by provoking the savant's sulfurous temper. At the same time Halley realized that he could just as easily jeopardize the project by making indiscreet demands of his superiors. Well respected though he was, he harbored no illusions regarding his status. As the new clerk he had been instructed to sit "uncovered [wigless] at the lower end of the table" during Council meetings.[40]

When the Council failed to convene for its third regularly scheduled meeting in a row, Halley finally decided to take the initiative, or so it would appear. Whether or not he was aided by other Newton supporters is impossible to say, but the interest of Paget and Aston can hardly be ruled out. At the meeting of May 19, with Pepys still absent and Vice President Joseph Williamson in the Chair, it was ordered that "Mr. Newton's *Philosophiae naturalis principia mathematica* be printed forthwith in quarto in a fair letter; and that a letter be written to him to signify the Society's resolution, and to desire his opinion as to the print, volume, cuts, & c."[41] Halley could relax for the time being, but he was well aware that the Council might yet take away what—according to the Royal Society Charter— had not been legally within the Fellows' prerogative to bestow.

Three days later, on May 22, Halley wrote to Newton about the decision. He referred to the manuscript as "Your Incomparable treatise," apologized for the regrettable but unavoidable delay, and assured the author of the "Great Honor you do [the Royal Society] by your Dedication." He outlined the terms of publication, including the promise that all expenses would be underwritten by the Society. "I am intrusted to look after the printing, and will take care that it shall be performed as well as possible, . . . what you signifie as your desire shall be punctually observed."

Newton, as he had every reason to anticipate, received all that he had hoped for. But Halley, through no fault of his own, was forced to include a less than welcome bonus:

> There is one thing more that I ought to informe you of, viz, that Mr. Hook has some pretensions upon the invention of ye rule of the decrease of Gravity, being reciprocally as the squares of the distances from the center. He sais you had the notion from him, though he owns the Demonstration of Curves generated therby to be wholly your own; how much of this is so, you know best, as likewise what you have to do in this matter, only Mr. Hook seems to expect you should make some mention of him, in the preface, which, it is possible, you may see reason to praefix. I must beg your pardon that it is I, that sent you this account, but I thought it my duty to let you know it, that so you may act accordingly; being in myself fully satisfied, that nothing but the greatest Candour imaginable, is to be expected from a person, who of all men has the least need to borrow reputation.

Fearing that he had already said too much, Halley hurriedly brought the strained communication to a more positive conclusion: "When I shall have received your directions, the printing shall be pushed on with all expedition, which therefore I entreate you to send me, as soon as you may be."[42]

Newton's reply was not long in coming. It contained not one word of thanks nor any instructions concerning the printing. Instead, he sought to clarify his position vis-à-vis Hooke, "for I desire that a good understanding may be kept between us. In the papers in your hands there is noe one proposition to which he can pretend, & soe I had noe occasion of mentioning him there." Halley could rest assured, however, that Hooke, "and others" would receive the credit due them in another book, which was as yet unfinished. To prove that the inverse square law was his own, Newton encouraged Halley to query Sir Christopher Wren about "a visit Dr Done and I gave him at his Lodgings" where we "discoursed of this Problem of Determining the Hevenly motions upon philosophicall principles." Newton could not recall the exact date of the meeting but was certain that it had taken place a year or two before his exchange of letters with Hooke in 1679.[43]

Halley, from all indications, had expected the worst. Thus Newton's firm but tempered reply could not have pleased him more, and he moved to take immediate advantage of the situation. A week later, on June 7, he sent Newton a proof sheet of the *Principia*'s first printed page and the following comment: "[I]f you have any objection; it shall be altered: and if you approve it, wee will proceed." Halley also urged Newton to complete what remained of the "system of the world," for this "is what will render it acceptable to all Naturalists, as well as Mathematiciens; and much advance the sale of ye book."[44]

That Halley should have raised the issue of sales appeal is reflective of something more than an enlightened concern for the dissemination of original ideas. Although its President remained absent, the Council of the Royal Society had finally convened on June 2, its first meeting in nearly six weeks. The membership agreed with the Fellows' resolution that Newton's book would be printed and that Halley should see it through the press. But when it came to the question of shouldering the considerable expense the Council balked. Halley was asked to print it "at his own charge; which he engaged to do." The reason for this decision is hardly a mystery. In 1685 the Society had agreed to finance the publication of the late naturalist Francis Willoughby's *De Historia Piscium* (*The History of Fishes*). By April 21, 1686, the project's expenses had ballooned to £360, and the bills were still coming in. So straitened had the Society's always limited finances become that Halley received fifty copies of *De Historia Piscium* as an honorarium for measuring a degree of latitude, instead of the £50 promised him. In July 1687 the Council ordered that Halley should receive fifty additional copies of the same treatise in lieu of his salary, plus twenty more copies in settlement of wages in arrears from the previous year. Hooke, who was presented with a similar proposition, wisely asked for six months to consider the acceptance of such a payment, hoping, no doubt, that the organization's fortunes would improve in the meantime.[45] (He was eventually paid in cash.) In December Papin also received a small school of *Fishes* for the "good services" he had rendered. If Halley had indeed taken the post of clerk with the idea of shoring up his deteriorating finances, he would have done better by opting for debtors' prison, where shelter, however rude, was at least provided gratis. It is difficult, in light of these circumstances, not to sympathize with Halley's hope that Newton would seek the largest possible readership for his demanding treatise.

Having dodged one bullet after another, Halley could at last look forward to bringing the task at hand to a smooth and reasonably swift conclusion. But while he had been occupied with matters pertaining to publication, Newton smoldered in grandiose isolation back in Cambridge. Like the moon whose movements and powers he so ingeniously delineated, Newton had his permanently dark side. During the previous three weeks his pique, obeying some fierce inner compulsion, had gradually been transformed into an irrepressible rage. Hooke, his inferior, had committed another act of intellectual lese majesty, defiling Newton's supreme moment on Olympus and tarnishing the unblemished image of himself Newton sought to project. The inevitable conflagration erupted on June 20, and its withering fury, which surpassed that of any previous outburst, sent the unsuspecting Halley reeling. Dispensing with all pleasantries, Newton tore into his old nemesis, kindling coal by the carload:

> Mr Hooke could not from my Letters [1679] wch were about Projectiles & ye regions descending hence to ye center conclude me ignorant of ye Theory of ye Heavens. That what he told me of ye duplicate proportion [the inverse square law] was erroneous, namely that it reacht down from hence to ye center of ye

earth. That it is not candid to require me now to confess my self in print then ignorant of ye duplicate proportion in ye heavens for no other reason but because he had told it me in the case of projectiles & so upon mistaken grounds accused me of that ignorance. That in my answer to his first letter I refused his correspondence, told him I had laid Philosophy aside, sent him only ye experimt of Projectiles (rather shortly hinted then carefully described) in complement to sweeten my Answer, expected to heare no further from him, could scarce perswade my self to answer his second letter, did not answer his third, was upon other things, thought no further of philosophical matters then his letters put me upon it, & therefore may be allowed not to have had my thoughts of that kind about me so well at that time.

To undermine Hooke's contention further, Newton scoured his papers for evidence that the inverse square relation was his long before it was claimed by anyone else. He found it in a letter to Oldenburg "that's above fifteen years ago." As for the concept of gravity, Halley should refer to the "Hypothesis of Light" penned in 1672. "I hope I shall not be urged to declare in print that I understood not ye obvious mathematical conditions of my own Hypothesis."

The more Newton wrote the more damning his accusations—and the more ominous his threats. He had recently decided that the *Principia* should consist of three books rather than two, but "the third I now designe to suppress. Philosophy is such an impertinently litigious lady that a man had as good be engaged in Law suits [which Newton was] as have to do with her. I found it so formerly & now I no sooner come neare her again but she gives me warning." Standing alone, the first two books "will not so well beare ye title of *Philosophiae naturalis Principia Mathematica*." Better to call the work *De motu corporum libri duo.* "But upon second thoughts I retain ye former title. Twill help ye sale of ye book wch I ought not to diminish now tis yours."[46] One wonders how Halley managed to contain his gratitude.

The letter, signed and perhaps even sealed, was waiting for the London carrier when Newton received word, probably from Paget, that Hooke was making "a great stir pretending I had all from him & desiring they [the Royal Society Fellows] would see that he had justice done him." Angrier than ever, he penned a vehement postscript, longer than the letter itself. "Now is not this very fine?" he fumed. "Mathematicians that find out, settle & do all the business must content themselves with being nothing but calculators & drudges & another that does nothing but pretend & grasp at all things must carry away all the invention as well of those that were to follow him as of those that went before." Hooke "has erred in the invention he pretends to & his error is ye cause of all the stirr he makes." Newton would have given much to be able to take back the letter written in such haste after his mother's death. While he never expressed his complete feelings on the subject, he returned to it once more, an indication that he believed Hooke had chosen to query him at that most vulnerable moment in the perverse hope of provoking a blunder. Blunder he did, but through no fault of Hooke's:

> Should a man who thinks himself knowing, & loves to shew it in correcting & instructing others, come to you when you are busy, & notwithstanding your excuse, press discoveries upon you & through his own mistakes correct you & multiply discoveries & then make this use of it, to boast that he taught you all he spake & oblige you to acknowledge it & cry out injury & injustice if you do not, I believe you would think him a man of a strange unsociable temper. Mr Hooks letters in several respects abounded too much wth that humor wch Helvius & other complain of & therefore he may do well in time to consider whether after this new provocation I be much more bound (in doing him that justice he claims) to make an honourable mention of him in print, especially since this is ye third time that he has given me trouble in this kind.[47]

Newton made good his threat. He went over his entire manuscript and expunged reference after reference to the sickly London virtuoso, though Hooke's name did remain in a few places, most notably in the section on comets in Book III, where Newton was forced to retain certain of his observations. No matter what Newton did, however, the memory of having once been corrected by Hooke never ceased to smart. Nearly a quarter of a century later he settled upon a different if no less tortured rationalization, attempting to explain his error away by informing Abraham de Moivre that he had been the victim of a "negligent stroke" of his pen. With his capacity to rationalize, Newton took care of every objection by spinning ever new figments to resolve contradictions in the old. He lived partly in a dream world created by imagination and vanity, suspended between distant points of truth. Such fantasies no doubt flourish and expand in the vacuity of a hermit life. To him they were fully coherent and rational.

By now Halley must have been asking himself a question that had frequently occurred to Henry Oldenburg: "Why me?" Yet notwithstanding the difficulties he had already faced and overcome, he realized that this was no time for an indulgent wallow in self-pity; there was simply too much at stake. The beast in Newton must be soothed somehow. Halley sagaciously decided to employ a combination of reason and flattery. "I am heartily sorry," he replied on June 29, "that in this matter, wherein all mankind ought to acknowleg their obligations to you, you should meet with any thing that should give you disquiet, or that any disgust should make you think of desisting in your pretensions to a Lady, whose favours you have so much reason to boast of." But, Newton must remember, "Tis not shee but your Rivalls enviing your happiness that endeavor to disturb your quiet enjoyment, which when you consider, I hope you will see cause to alter your former Resolution of suppressing your third book." Halley had conferred with the Fellows on the matter, and they "are very much troubled at it, and that this unlucky business should have happened." Halley had also honored

Newton's request and discussed the subject with Wren. Sir Christopher assured him that what Newton wrote was true. But Halley went a step further. He had a frank exchange with Hooke: "I have plainly told him, that unless he produce a . . . demonstration, and let the world be a judge of it, neither I nor anyone can believe [his claim]."

Newton should also know that Hooke's claim to the inverse square law "has been represented in worse colours than it ought; for he neither made publick application to the Society for Justice, nor pretended you had all from him." Hooke had become upset when Newton's manuscript was presented by Dr. Vincent at the meeting of April 28. Sir John Hoskins, Hooke's "particular friend" was in the Chair, but in praising Newton's work Hoskins made no mention of what Hooke had supposedly confided to the acting President of his own discoveries, "upon which they two, who till then were the most inseparable cronies . . . are utterly fallen out." After the meeting, when the Fellows had gathered as usual at their favorite coffeehouse, Hooke had argued "that he gave you the first hint of this invention," but his protestations were to no avail. "I found that they were all of the opinion, that nothing thereof appearing in print, nor on the Books of the Society, you ought to be considered as the Inventer; and if in truth he knew it before you, he ought not to blame any but himself; for having taken no more care to secure a discovery, which he puts so much value on." Having set forth the facts in the most diplomatic manner possible, Halley implored the Moses of modern science not to withhold any of the tablets containing his laws: "Sr I must now again beg you not to let your resentments run so high, as to deprive us of your third book . . . which will undoubtedly render it [the *Principia*] acceptable to those that will call themselves philosophers without Mathematicks, which are by much the greater number." [48]

Halley had done his best. He now waited to learn the outcome of his efforts and the fate of the greatest scientific manuscript ever written. Two weeks later, on July 14, Newton drafted a reply. "I am very sensible of ye great kindness of ye Gentlemen of your Society to me, far beyond wt I could ever expect or desire & know how to distinguish between their favour & anothers humour." Halley's clarification of Hooke's position, which "was in some respects misrepresented to me," caused Newton to wish that "I had spared ye Postscript in my last." Neither an apology nor an expression of heartfelt contrition, it was merely Newton's way of retaining the support of the Fellows, who were already on his side, of "sweetening" his answer, so to speak. Like the phototropic moth to which the deadly flame is irresistible, Newton needlessly justified himself to Halley once more by trotting out the same evidence of priority contained in their previous correspondence. He also sent a revised scholium demonstrating that an inverse square force results when a planet's periods and times correspond to Kepler's third law, adding, parenthetically, "(as our countrymen Wren, Halley & Hooke have also severally concluded)." The placing of Hooke's name last was no accident, and Halley did his best to lessen the slight by reversing the order of Hooke's name and his own in the published version. In the end Newton had taken back with one hand what little he had so disingenuously given with the other: "And now having sincerely told you ye case between Mr Hooke & me I hope I shall be free for ye future from ye prejudice of his Letters."[49]

Halley surely wished for the same vis-à-vis Newton, but a fortnight later the astronomer was the recipient of a final blast of indignation from the direction of Cambridge. The previous day Newton had "unexpectedly [or so he said] struck upon a copy of ye Letter [July 21, 1673] I told you of to Huygenius." He quoted a lengthy passage "concerning the *vis centrifuga*," which proved "yt I . . . had an eye upon ye forces of ye Planets, knowing how to compare them by the proportions of their periodical revolutions & distances from ye center they move about." A victim of his own paranoia, he could also prove that "it is authentic," for "tis in ye hand of one Mr John Wickins who was then my chamber fellow & is now Parson of Stoke Edith neare Monmouth." Newton continued:

> In short as these things compared together shew that I was before Mr Hooke in what he pretends to have been my Master so I learnt nothing by his letters but this that bodies fall not only to ye east but also in our latitude to ye south. In ye rest his correcting & informing me was to be complain'd of. And thô his correcting my Spiral occasioned my finding ye Theorem by wch I afterward examined ye Ellipsis; yet am I not beholden to him for any light into yt business but only for ye diversion he gave me from my other studies to think on these things & for his dogmaticalnes in writing as if he had found ye motion in ye Ellipsis, wch inclined me to try it after I saw by what method it was done.[50]

Realizing that with Newton almost anything was possible, Halley chose not to raise the vexed issue of Book III. The threat to suppress it had not been repeated, but neither had it been withdrawn. At best it was a cautiously optimistic, at worst a confused and downcast Halley who proceeded with the editing and printing of the manuscript pages already in his possession.

For all his great capacity for anger and vituperation, there is little reason to believe that Newton would have "multilated" or "castrated" his progeny, as some scholars have suggested. And while he may have temporarily convinced himself that the *Principia*'s final book would not be set in type during his lifetime, his investment, both emotional and intellectual, was much too heavy to allow him to turn back, no matter how tempted be may have been. Completely in the grip of his inexorable creative powers, Newton had entered a period when his will, his likes, and his dislikes were of secondary consequence. He was behaving like one dominated by an irresistible force, and it would not release him until it had carried him on to the accomplishment of the supreme intellectual feat of his life. Humphrey Newton was intimidated by his employer and saw little but the outer man, but his verbal portrait, drawn from the most creative years, is too vivid to be discounted:

I never saw him take any Recreation or Pastime, either in Riding out to take ye Air, Walking, Bowling, or any other Exercise whatever, Thinking all Hours lost, y^t was not spent in his Studyes, to wch he kept so close, y^t he seldom left his Chamber. . . . So intent, so serious . . . y^t he eat very sparingly, nay, oftimes he has forget to eat at all, so y^t going into his Chamber, I have found his mess untouch'd, of wch when I have reminded him, would reply Have I; & then making to ye Table, would eat a bit or two standing, for I cannot say, I ever saw Him sit at Table by himself. . . . He very rarely went to Bed, till 2 or 3 of ye Clock, sometimes not till 5 or 6, lying about 4 or five hours.

The fastidiousness of a self-conscious young don had mellowed into the comfortable sloth of an eccentric pillar of the academic establishment:

He very rarely went to Dine in ye Hall unless upon some Publick Days, & then, if He has not been minded, would go very carelessly, wth shoes down at Heels, Stockins unty'd, Surplice on, & his Head scarcely combed. . . . I can't say I ever saw him wear a Night-Gown, but his wearing Cloathes, that he put off at Night, at Night do I say, yea rather towards ye Morning, he put on again at his Rising. . . . In a Morning he seem'd to be as much refresh'd with his few hours Sleep, as though he had taken a whole Night's Rest. He kept neither Dog nor Cat in his Chamber, wch made well for ye old Woman, his Bedmaker, . . . for in a morning she was sometimes found both Dinner & Supper scarcely tasted of, w^{ch} [she] has very pleasantly & mumpingly gone away with.

It seems rather foolish to attempt a precise definition of "normal behavior" or, more to the point, to debate whether there is such a thing as a "normal" human being. There are, however, certain broad standards of conduct into which most of our actions fit, and these are usually described as being normal or socially acceptable. As surely as the person born with six fingers or the calf with two heads, Isaac Newton was a mutant, seeming, as often as not, more a phenomenon than a man. It can be legitimately argued, given the weight of historical evidence, that he was hardly a conventional figure in any sense of the word. His intellect was too profound, his capacity for rage too great, his desire for seclusion from the outside world too obsessive, his passion for original thought and scholarship too exclusive. He was the incarnation of the abstracted thinking machine that some scientists predict humankind may become through the natural process of evolution and others, impatient in their desire to move among a race of intellectual giants, hope to create by the administration of powerful mind-altering drugs. Humphrey provided a rare glimpse of that machine in operation:

At some seldome Times when he design'd to dine in ye Hall [he] would turn left hand [instead of heading almost straight ahead across the Great Court as he should have], and go out into ye street, where making a Stop, when he found his mistake, would hastily turn back & then sometimes instead of going into ye Hall, would return to his Chamber again. . . . In his Garden . . . he would, at some seldome Times, take a short Walk or two, not enduring to see a Weed in it [he had a gardener]. . . . When he has some Times taken a turn or two has made a sudden stand, turn'd himself about, run up ye Stairs like another Archimides, with an eureka fall to write on his Desk standing, without giving himself the Leasure to draw a chair to sit down on.

And, as would be expected of a machine, Humphrey said he had heard Newton laugh but once: "Twas upon occasion of asking a friend to whom he had lent Euclid to read, what progress he had made in that author, and how he liked him?" The friend answered with a question of his own: "What use and benefit in life that study would be to him? upon which Sr Isaac was very merry."[51]

If Newton, the most perfect mechanic of all, found little cause for levity during this deeply taxing period, the same can be said of his woefully over-matched rival. Robert Hooke's diary lasped during the mid-1680s, but it is known that his intense dislike of Newton poisoned his dwindling reserve of good cheer for years to come. After seeing Newton at Halley's home on February 15, 1689, Hooke wrote that he "vainly pretended claim yet acknowledged my information. Interest has no conscience." A year later, while dining with friends, Hooke could scarcely contain his delight on hearing Newton referred to as a "knave."[52]

Whether Hooke's untenable cry of plagiary actually resulted in an interruption of Newton's writing is impossible to know. Judging from the satisfying regularity with which the manuscript pages were piling up, however, if an interruption did occur it must have been extremely brief. Book II, Newton later informed Halley, was completed in the autumn of 1686, but he did not forward it to London until the following March. He blamed the delay on "being told (thô not truly) that upon new differences in ye R. Society you had left your secretaries place."[53]

The difference to which Newton referred first surfaced at the Council meeting of November 29. "It was resolved, that there is a necessity of a new election of a clerk in place of Mr. Halley; and that it be put to a ballot, whether he be continued or not." On January 5, 1687, the Council appointed a committee of seven, of which Hooke was a member, to inspect the books to "see if Mr. Halley had performed his duty." One month later Sir John Hoskins read the committee's findings—"said books and papers [are] in a very good condition."[54] Whether or not someone was after Halley's scalp is uncertain, but it has been suggested that as Newton's agent he was the object of an aborted conspiracy hatched by a vengeful Hooke and his small circle of supporters. Whatever the truth of the matter, Halley hastened to assure Newton that all was well, although in rather noncommittal terms: "I am sorry the Societie should be represented to you so unsteady as to fall so frequently into varience, but there is no such thing; . . . I serve them to their satisfaction, though 6 of 38, last generall Election day, did their endeavour to have put me by."[55] One thing seems clear: If Hooke had made a bid to get at Newton through Halley, the latter was not about to dwell on it in his correspondence with Newton, for to do so would only risk stirring up the emotional fires Halley had tried so desperately to bank.

With Book II in his hands, Halley employed a second printer to set the manuscript while the first finished Book I. He informed Newton that with any luck the task would be complete in about seven weeks. The question of the final book now loomed larger than ever. Nothing had been said about Book III since Newton's threat to suppress it nine months earlier. The moment of truth at hand, Halley broached the matter gingerly:

> You mention in this second, your third Book *de Systemate mundi*, which from such firm principles, as in the preceding you have laid down, cannot chuse but give universall satisfaction; if this be likewise ready, and not too long to be got printed by the same time, and you think fit to send it; I will endeavour by a third hand to get it all done togather, being resolved to engage upon no other business till such time as all is done: desiring herby to clear my self from all imputations of negligence, in a business, wherein I am much rejoyced to be any wais concerned in handing to the world that that all future ages will admire."[56]

Unbeknownst to Halley, Book III was almost ready for his scrutiny. On April 5 he reported to Newton that the last part of "your divine Treatise" had reached him the previous day: The "world will pride it self to have a subject capable of penetrating so far into the abstrusest secrets of Nature, and exalting humane Reason to so sublime a pitch by this utmost effort of the mind."[57] As per the instructions accompanying the precious cargo, Halley sent his communication in care of "Mr Parish, Rector of Coulsterworth in Lincolnshire." Newton, who had written his 550-page treatise in the incredible space of eighteen months, had gone home for a richly deserved rest.

V

In Proposition VII of Book III Newton announced the most celebrated of his scientific discoveries, the law of universal gravitation: "That there is a power of gravity pertaining to all bodies, proportional to the several quantities of matter which they contain."[58] But it is the common textbook definition with which most of us are familiar: every particle of matter attracts every other particle with a force proportional to the product of the masses and inversely proportional to the square of the distances between them. With this single law of physics Isaac Newton "democratized" the universe, as it were, by laying permanently to rest the concept of a hierarchical dominance among the celestial bodies, an idea that had appealed as strongly to the mystical Kepler as to the practical-minded Aristotle. In the seemingly infinite universe envisioned by Newton, no one body is more important than any other. All matter—from the largest of stars to the tiniest of corpuscles—obeys the same invariant principle. This being established,

Newton was in a position to explain how his extended conception of gravitation tied together astronomical phenomena that for untold centuries had perplexed the finest minds in the history of scientific thought, "to admit," according to his first rule of reasoning in philosophy, "no more causes of natural things than such as are both true and sufficient to explain their appearances."[59]

Ironically, the very law that promised to expand and clarify man's understanding of the cosmos presented its discoverer with innumerable difficulties when he attempted to apply it to planetary, lunar, and cometary motions. To speak or to write in the abstract of the law of universal gravitation is one thing; to demonstrate it in concrete physical terms is quite another. For contradictory though it seems, the very appeal of Newton's law—its universal applicability—also constituted the greatest drawback to its clear-cut demonstration and acceptance by men of the late seventeenth century. Consider, for example, the problem Newton faced when he sought to calculate the orbit of Saturn. Had it been only a matter of solving the comparatively simple problem of two mutually attracting bodies, Newton could have rather easily calculated the planet's orbital motion on the basis of the mathematical formula contained in *De Motu*. But, as he revealed in his correspondence with Flamsteed, the problem is made far more complex because, as Saturn circles the sun, it is also influenced by the attraction of other bodies, most notably its neighboring planet Jupiter, the largest in our solar system. Although such secondary attractions are minor when compared with the pull of the sun, which contains many times more matter than all the planets and their satellites combined, they still create minor deviations or perturbations from the true orbit, which must be taken into account by the physicist. Even Newton, armed with the calculus and blessed with analytical skills par excellence, was unable to resolve this problem in more than a general fashion. Yet his revolutionary investigation of planetary perturbations pointed the way for later generations of astronomers: A more refined application of Newton's method led to the discovery of the planet Neptune in 1846.

Of more immediate concern to Newton were the irregularities in the moon's motion, certain elements of which had been known to close observers of the skies for millennia. While the moon is controlled by the pull of the earth, the enormous mass of the sun, even at its far greater distance, visibly disturbs the normal lunar orbit. Unlike the perturbations of the planets, those of the moon are both more numerous and more profound, making lunar theory one of the most complicated and difficult in astronomy. By employing a complex system of calculations Newton succeeded in accounting for the satellite's major perturbations; he also discovered other irregularities not previously observed, and indicated the existence of similar aberrations in the motions of the moons of Jupiter and Saturn. As a corollary to his calculations of the motions of the planets and their satellites, Newton also demonstrated how to compute the masses of the planets and the sun from the mass of the earth. His results, understandably, were not as refined as those yielded by later methods, but they proved remarkably accurate for his time. For example, Newton computed the earth's density to be between five and six times that of water. The accepted figure is almost exactly five and one-half.

One of the more interesting deductions offered by Newton in the *Principia* concerns his belief that the planets, including the earth, are oblate bodies: "That the axes of the planets are less than the diameters drawn perpendicular to the axes." In other words, the planets are flattened somewhat at their poles and bulge to some degree at their equators, much like a balloon when gently pressed between one's palms. He suspected this because a rapidly spinning body composed of yielding material tends to swell at its circumference and shrink along its axis. What is more, the seas are not concentrated at the equator, as they would be if the earth were a perfect sphere. Thus, a plastic spheroid the size of the earth, held together by the force of gravity and rotating once a day, can be shown to have an equatorial diameter a few miles greater than its polar diameter. To confirm his hypothesis he drew upon the distance measurements of Norwood and Picard and also conducted a series of experiments with pendulums and columns of water, the details of which need not be related here. He further calculated the oblateness of Jupiter, which, owing to its more rapid axial rotation, must be considerably greater than that of earth. In 1691 Cassini's telescopic observations of the giant planet provided the requisite visual confirmation of Newton's theory.[60]

Through his study of oblate bodies Newton made an additional and more significant discovery. A sphere, as he had demonstrated in Book I, attracts another body as if its mass were concentrated at its center, but the same is not quite true of a spheroid. This means that the intensity of the earth's gravitational field will not be exactly the same everywhere. As a spheroid it pulls and is in turn pulled by the moon and sun with a slightly uncentric attraction, meaning that the line of pull is strongest at the equatorial bulge, where its matter is most concentrated. What Newton was dealing with, in effect, was a giant top slightly overloaded on one side, so that gravity acts on it asymmetrically. This causes the planet's axis to change its angle of rotation very slowly, so as to trace out the shape of a cone in the heavens, a movement familiar to astronomers as "the precession of the equinoxes." The phenomenon was first observed by Hipparchus, a Hellenistic astronomer of the late second century B.C., but its significance had continued to elude the best of minds. Copernicus, no inconsiderable thinker himself, devoted a good deal of attention to it in *De revolutionibus*, but, like his bemused predecessors, he failed to shed much practical light on the phenomenon. This is not surprising, for astronomical observation had not been going on long enough to permit the measurement of more than a small fraction of the complete axial cycle. Newton undertook the calculation of this conical motion, which he correctly ascribed to the slightly asymmetrical attraction of the moon. He found that it takes some 26,000 years for the earth's axis to complete the cone. "A calculation of this kind," the British scholar E. N. da C. Andrade wrote, "is not the same kind of thing as suggesting that an apple falls because the earth pulls it!"[61] Nor was it the kind of answer Hooke could possibly have arrived at on his own, his rich intuitive powers notwithstanding. "Chance," Pasteur once remarked, "favors only the mind which is prepared."

Of all the great mysteries that confounded astronomers through the ages, none proved less tractable and more frustrating than the eternal fluctuation of the seas. Kepler came the closest to a correct explanation of tidal motion by attributing it to magnetic forces generated by the sun, moon, and earth, a concept he borrowed from the English natural philosopher William Gilbert. But Galileo had dismissed this hypothesis as simply one more example of Kepler's superstitious preoccupation with occult forces. He thought the tides resulted from the combined effects of the earth's rotation on its axis and its movement around the sun, thus linking the phenomenon to purely mechanical principles without having to appeal to any such force as the attraction of the moon on the waters. Descartes, it will be remembered, had his own views on the subject: The French philosopher believed that when the moon approaches earth its vortex presses down upon the planet's waters and hence produces a low tide beneath it. A high tide thus would result on the areas of the earth's surface relatively free of the moon's vortical pressure.

Newton, in one of the *Principia*'s most incisive yet simply stated propositions, wrote that "the flux and reflux of the sea arise from the actions of the sun and moon."[62] By applying the law of gravitation to the phenomenon, he discovered that the power of attraction is greater on the waters facing the attracting body than on the earth as a whole, and greater on the earth as a whole than on the waters on the opposite side. Since the moon, while in perigee, exerts the more powerful force, the main tides are caused by the earth's satellite. Its chief effect is to raise a pair of waves or oceanic humps, of tremendous area, causing them to travel round the earth once in a lunar day, or a little less than twenty-five hours. The sun causes a similar but less pronounced pair of waves to form and circle the planet once in a solar day of twenty-four hours. These two pairs of waves periodically overtake each other, producing the spring and neap tides. The spring tides, the more marked of the two, occur at the point of syzygy, that is, when the sun, moon, and earth are in a line, thus exerting a maximum gravitational pull on the earth's surface. The neap tides result when the sun and the moon, both attracting the waters of the seas, pull at right angles to each other. These alternating alignments occur every two weeks. While Newton's figures were not sufficiently refined to enable him to construct tables of the precise time or height of a tide at any given place, they did provide a masterly explanation of the main details of tidal action and of the general characteristics of the tides themselves. Yet another quantum advance in scientific understanding was his to claim.

With the possible exception of lunar theory, which caused his head to ache so badly that it often "kept him awake," nothing in the *Principia* gave Newton more trouble than his attempt to demonstrate the influence of gravity on comets. Though he had previously formulated a workable theory of cometary motion, he admitted to Flamsteed, in a letter dated September 19, 1685, that "I have not yet computed ye orbit of a [specific] comet but am now going about it: & taking that of 1680 into fresh consideration, it seems very probable that those of November & December were ye same comet." Flamsteed doubly underlined this sentence, which he obviously saw as a vindication of his own single-comet theory, and wrote in the margin: "he would not grant it before see his letter of 1681." As he had the previous December, Newton asked Flamsteed to send him additional data, something the Astronomer Royal was again pleased to do. Newton was still wrestling with the problem as late as June 1686, when he informed Halley that the "third [book] wants ye Theory of Comets."[63]

At some point during the next few months—precisely when is uncertain—the breakthrough came. "Comets," Newton wrote, "move in some of the conic sections, having their foci in the centre of the sun, and by radii drawn to the sun describe areas proportional to the times." Their orbits "will be so near to parabolas, that parabolas may be used for them without sensible error."[64] The proof of the theory, which he had embraced since 1684, came when he successfully plotted the curve of the great comet of 1680–81. At long last comets were no longer the vaporous exhalations from earth the ancients thought them to be, nor were they phenomena that inexplicably appeared and then just as mysteriously disappeared, never to be seen again. Every comet seen from earth must return at regular intervals unless its velocity is great enough to overcome the gravitational force exerted upon it by the sun.

In years to come Halley, employing Newton's theory, would plot the orbit of the brilliant comet of 1682. Then, through a painstaking search of past records worthy of his mentor, he determined that there had been reports of similar sightings in 1531 and 1607, or about once every seventy-five years. Could not these sightings, Halley conjectured, indicate successive appearances of the same object, rather than the regularly spaced arrival of three unrelated comets? He calculated the orbit on the assumption that if in fact this was one and the same comet, it would next appear in 1758, give or take a year. The streaming body, which now bears Halley's name, was first observed on Christmas Day, 1758 (the anniversary of Newton's birth), by an amateur astronomer named George Palitsch. Halley's confirmation of Newton's reduction to rule of the complex movements of comets, and their inclusion with the planets in the category of bodies orbiting the sun, may be justly regarded as one of the most profound discoveries first announced in the *Principia*.

One of the central objections raised by those who in the beginning strongly opposed Copernican theory was that it required a universe of such immense proportions as to challenge the credulity of rational men, not to mention the teachings of the Church. Though what Koyré called "the world-bubble" had swelled considerably during the century and one-half since the death of the reluctant revolutionary, Newton was the first natural philosopher to establish a true idea of the distances separating the celestial bodies, especially the stars. His calculations indicated that they must be hundreds of times more remote than Saturn, then the most distant planet known. Were it otherwise, they would either fall into the sun or swing into orbit around it. Moreover, at such great distances the stars would not be visible by reflected light as are the planets; they must be self-luminous bodies like the sun. And if the stars are suns like our own, they too must act as centers of other planetary systems. Newton expanded upon his view in a letter to his friend Dr. Richard Bentley written in December 1692:

> [I]t seems to me, that if the matter of our Sun & Planets & all ye matter in the Universe was eavenly scattered throughout all the heavens, & every particle had an innate gravity towards all the rest & the whole space throughout wch this matter was scattered was but finite: the matter on ye outside of this space would by its gravity tend towards all ye matter on the inside & by consequence fall down to ye middle of the whole space & there compose one spherical mass. But if the matter was eavenly diffused through an infinite space, it would never convene into one mass but some of it convene into one mass & some into another so as to make an infinite number of great masses scattered at great distances from one to another throughout all yt infinite space.[65]

Newton's universe, when stripped of metaphysical considerations, as stripped it would be, is an infinite void of which only an infinitesimal part is occupied by unattached material bodies moving freely through the boundless and bottomless abyss, a colossal machine made up of components whose only attributes are position, extension, and mass. Life and the sensate world have no effect upon it and are banished, à la Descartes, from its rigorously mechanical operations. Humankind's only contact with the objective world is limited to observing and interpreting its manifold phenomena. And yet, for all its lack of feeling, Newton's universe is a precise, harmonious, and rationally ordered whole. Mathematical law binds each particle of matter to every other particle, barring the gate to chaos and disunity. By flinging gravity across the infinite void, he was able to unite physics and astronomy in a single science of matter in motion, fulfilling the dream of Pythagoras, Kepler, and countless others in between. And even though Newton was unable to discover a demonstrable principle with which actually to explain the phenomenon of gravitation, the laws he formulated provided convincing proof that man inhabits a preeminently orderly world. We remember and honor him today not for providing us with ultimate answers to the most profound scientific questions but because, in apprehending the Pythagorean power by which number holds sway above the flux, Isaac Newton contributed more than any other individual of the modern age to the establishment and acceptance of a rational world view.

VI

On July 5, 1687, three months to the day after Halley wrote that Book III was safely in his hands, the Royal Society clerk drafted what must have been an even more satisfying communication to both the sender and the recipient:

> I have at length brought your Book to an end, and hope it will please you. the last errata came just in time to be inserted. I will present from you the books you desire to the R. Society, Mr Boyle, Mr. Pagit, Mr Flamsteed and if there be any elce in town that you design to gratifie that way; and I have sent you to bestow on your friends in the University 20 Copies, which I entreat you to accept. In the same parcell you will receive 40 more, wch, having no acquaintance in Cambridge, I must entreat you to put into the hands of one or more of your ablest Booksellers to dispose of them.

Halley set the price "of them bound in Calves leather and lettered" at nine shillings and those in "Quires" (or sheets) at six, but he was willing to reduce the price of the latter to five shillings if payment was made in cash or "at some short time." Indeed, "I am contented to lett them go halves with me, rather than have your excellent Work smothered by their combinations."[66] Halley's was a most generous concession, especially when one recalls that he was virtually drowning in copies of *De Historia Piscium,* the less than satisfactory substitute for his own modest clerk's salary. Though the Royal Society itself was not the publisher, Halley again set personal interest aside by nurturing the fiction that it had served as such. In July 1686 Samuel Pepys, who was then President, signed the imprimateur. Thus the *Principia* appeared under the Society's license, and Pepys, whose acquaintance with the new science was superficial at best, saw his name displayed just as prominently on the title page as that of the author.

It seems fair to characterize Newton's treatment of Halley throughout this long and often tortuous ordeal as polite but reserved, lacking in genuine warmth and cordiality. If Newton ever acknowledged in private his debt to the gifted virtuoso who had encouraged and upheld him at every turn, we do not know about it. He did, however, pay public tribute to his richly deserving publisher in the *Principia*'s Preface:

> In the publication of this work the most acute and universally learned Mr. *Edmund Halley* not only assisted me in correcting the errors of the press and

preparing the geometrical figures, but it was through his solicitations that it came to be published; for when he had obtained of me my demonstrations of the figure of the celestial orbits, he continually pressed me to communicate the same to the *Royal Society*, who, afterwards, by their kind encouragement and entreaties, engaged me to think of publishing them.

For his part Halley prefixed a number of verses to the *Principia*, which he titled "Ode to Newton." His opinion of the treatise and its author is best expressed in the following lines:

> *Then ye who now on heavenly nectar fare,*
> *Come celebrate with me in song the name*
> *Of Newton, to the Muses dear; for he*
> *Unlocked the hidden treasuries of Truth:*
> *So richly through his mind had Phoebus cast*
> *The radiance of his own divinity.*
> *Nearer the gods no mortal may approach.*

Newton was surely flattered but perhaps also a little embarrassed by Halley's enthusiastic praise. As if to compensate for his publisher's zeal, he wrote an excessively modest appraisal of his accomplishments some years later: "If I have done ye Publick any service . . . 'tis due to nothing but industry & a patient thought."[67]

Chapter Thirteen

"Go Your Way, and Sin No More"

My hair is grey, but not with years . . .

Byron

I

Newton was well aware of what he had wrought. Physics and astronomy were to be removed at last from the smothering ecclesiastical mantle that the exigencies of an earlier age had demanded. He knew also that, having placed natural philosophy on a revolutionary but firm foundation, his life would never be the same again. He had spent most of the past quarter-century at Cambridge in a reclusive state of meditative ecstasy, revealing his thoughts to others only on occasion. When he did, it was always with the deepest trepidation, for he felt compelled to ward off every criticism, no matter how petty. Now the world at large was to know the name of Isaac Newton, and its bearer was helpless to control what it might think of him.

Either too shy or too aloof to distribute personally the twenty presentation copies of the *Principia* provided him by Halley, Newton assigned Humphrey the task of making the rounds of his colleagues and masters of the colleges. Few, if any, of them were in a position to comment intelligently on the masterpiece; "some," Humphrey later recalled, "(particularly D^r. Babington of Trinity) said that they might study seven years, before they understood anything of it."[1] How Newton reacted when informed of this rather backhanded compliment Humphrey did not say; perhaps a smile flitted across his normally austere visage, as when an unidentified friend had

315

asked him what possible benefit in life the study of Euclid might confer. Humphrey was probably also charged with placing the remaining forty copies with local booksellers, and one suspects that he encountered some difficulties. The daunting contents would surely have precluded the cash discount arrangement preferred by Halley. A special appeal may have been made to Newton's stationer, who must have been delighted, if somewhat mystified, by the Lucasian Professor's frequent purchase of large quantities of imported writing paper. At last he found out what use Newton had made of it.[2]

Halley, who was extolling the *Principia*'s virtues in his scientific correspondence well before its publication, wasted no time in bringing the finished work to the attention of the influential. In July 1687, the month of publication, he sent a copy to James II. Fearing that the King would never be able to master the book's contents on his own, Halley drafted and enclosed a popular account, "being sencible," he wrote diplomatically, "of the little leisure wch. care of the Publick leaves to Princes." Also aware that James had once been Lord High Admiral of the British fleet, Halley concentrated on the theory of the tides in the hope of stimulating royal interest. He concluded with an offer to come before the King to answer questions about anything "I have omitted."[3] His Majesty was deeply preoccupied by less abstruse matters, however, and he found Halley's proposal of a private audience easy to resist. Had not James's more scientifically oriented predecessor died two years before, the outcome might have been different, for Charles II was familiar with the names of both Halley and Newton.

In the final analysis, it was not the opinion of kings or princes that truly mattered but the judgment of Newton's peers, the virtuosi, and more particularly what Darwin called the "great guns." Halley once more prepared the ground by composing a lengthy unsigned review for the *Philosophical Transactions*. Better equipped to present a synopsis of the *Principia*'s contents than anyone save the author himself, Halley left no doubt that Newton's book constituted a turning point, a hinge in the history of science. The diligent reader would learn of a new dynamics capable of explaining the phenomena of planetary motion, tidal action, and the movements of the comets; "it may be justly said," the review concluded, "that so many and so Valuable *Philosophical Truths*, as are herein discovered and put past Dispute, were never yet owing to the Capacity and Industry of any one Man."[4]

Halley, one might argue, could not afford to do anything but wax enthusiastic. He was, after all, Newton's publisher and, what is more, a publisher encumbered by mounting debt. As the book's fortunes went, so went his. Independent confirmation of Halley's assessment was not long in arriving, however. From Scotland David Gregory, Professor of Mathematics at Edinburgh, wrote Newton on September 2: "Having seen and read your book I think my self oblidged to give you my most hearty thanks for having been at pains to teach the world that which I never expected any man should have knowne." Such were Newton's contributions in physics and mathematics, "that you justlie deserve the admiration of the best Geometers and Naturalists, in this and all succeeding ages."[5] Newton, it seems, chose not to reply, but neither was the young mathematician forgotten. Gregory became Savilian Professor of Astronomy at Oxford in 1691 on the dual recommendation of Newton and Flamsteed.

So scarce yet in demand was the *Principia* north of the border that John Craig, Gregory's former pupil, loaned a copy in his possession to the mathematician Colin Campbell, "tho I have not yet perused on[e] quarter of it, and have an unsatiable desire to know the wholl." Campbell was forewarned by Gregory that the going would not be easy: "I beleeve Newton will take you up the first month you have him." The month turned into a year, and Craig was finally compelled to recall the volume.[6] The Scots were not the only mathematicians experiencing difficulties with the text. Gilbert Clerke, the author of several minor mathematical and theological studies, confessed to Newton in September 1687 that "I doe not as yet well understand so much as your first three sections, for wch you do not require yt a man should be a *mathematicé doctus*."[7] Meanwhile, a young teacher of mathematics named Abraham de Moivre, who like Gregory was destined to become a disciple of Newton, sought to master the master by stuffing his pockets with pages torn from the *Principia*, which he read while traveling from one pupil's home to another.[8]

Though it was not published in full until 1690, John Locke's finest work, *An Essay Concerning Human Understanding*, was all but completed in 1687, the year of the *Principia*'s publication. Its author, suspected of radicalism by the government of Charles II, hand long since fled to Holland, where he now set himself the task of mastering Newton's treatise. Locke had studied medicine at Christ Church, Oxford, from whose records his name was stricken by royal mandate, and his high regard for experimental philosophy exerted a strong influence upon his philosophical thought and method. Locke's unsigned review of the *Principia*, which appeared in the March 1688 issue of the *Bibliothèque Universelle*, was the first to be published on the Continent. No mathematician himself, he asked Christiaan Huygens for an opinion as to the veracity of the quantitative demonstrations. "[B]eing told he might depend upon their Certainty; he took them for granted, and carefully examined the Reasonings and *Corollaries* drawn from them, became Master of all the Physics, and was fully convinced of the great Discoveries contained in the Book."[9] Judging by the contents of his review, however, Locke's mastery of the work was less than complete. Any reader of Locke's account would have found it extremely difficult to grasp Newton's central proposition that the motion of all matter is determined by the force of gravity.[10] The philosopher's tenuous hold on the treatise's contents is further underscored by the survival (in copied form) of a lengthy but somewhat simplified version of the proof that elliptical orbits require an inverse square force, the original of which Newton drafted for Locke shortly after the latter's return to England aboard the same ship carrying Mary Stuart, wife of the triumphant Prince William of Orange.[11]

By far the longest and most laudatory of the continental reviews appeared during the summer of 1688 in the *Acta Eruditorum*, published in Leipzig. Its author remains a mystery, but his grasp of the *Principia*'s physics and mathematics was complete, thus extending knowledge of Newton's work to intellectual circles beyond those reached by the *Philosophical Transactions*. While there was much in the review with which he did not agree, Leibniz, who was having his own special problems with Newton, could not suppress his admiration for his rival. "This account," he wrote to Otto Mencke, the editor of the journal, "I have read eagerly and with much enjoyment, although it is far removed from my present lines of thought. That remarkable man is one of the few who have advanced the frontiers of the sciences." Leibniz enclosed three short papers, all published in the *Acta* in 1689, on the subjects of refraction, motion through a resisting medium, and orbital dynamics, claiming priority in these areas. "Newton's work stimulated me to allow these notes, for what they are worth, to appear, so that sparks of truth should be struck out by the clash and sifting of arguments."[12] Sparks would indeed fly, but truth, unfortunately, was not one of their key elements.

In the meantime, Hooke unsuccessfully continued his dogged pursuit of the justice he believed had been denied him.[13] Despite his contentiousness and overinflated sense of his own accomplishments, one cannot but sympathize with this tortured human soul. The tale of Hooke's failure—like that of Newton's triumph—is the stuff of high drama, something the Greeks, with their unwavering belief in fate, would have relished to the full. If nothing else, the claims of Leibniz and Hooke again illustrate that there is always a desire, after a grand intellectual feat, to merge one's own account with that of the victor, forgetting past hesitations and inconsistencies, as though the path of discovery contained no gaping pitfalls or great stumbling blocks. When his two antagonists died, Newton felt he had no reparations to make.

The last of the reviews appeared in the August 1688 issue of the *Journal des Sçavans*, the official publication of the *Académie Royal des Sciences*. It too was printed anonymously, and the author's identity has never been ascertained. A strange mixture of harsh criticism and lavish praise, the review, obviously written by a disciple of Descartes, unjustifiably attacked Newton's "Physique" on the grounds that it was purely hypothetical, when in truth he presented positive proof that the motions of the celestial bodies are as he described them. There is no evidence to indicate that Newton responded directly to this charge, but it was guaranteed to enrage him, much as had the attacks on his theory of light. At the same time, while rejecting the concept of attraction at a distance, the reviewer called his mechanics "the most perfect . . . one can imagine," a seeming echo of the thoughts expressed by Newton in the *Principia*'s introduction.

Still, perfection meant one thing to the anonymous reviewer of the book and quite another to its author. So overpowering were Newton's arguments that the caveat he inserted in the preface was little noticed by his awed readership:

> I wish we could derive the rest of the phenomena of Nature by the same kind of reasoning from mechanical principles, for I am induced by many reasons to suspect that they may all depend upon certain forces by which the particles of bodies, by some causes hitherto unknown, are either mutually impelled towards one another, and cohere in regular figures, or are repelled and recede from one another. These forces being unknown, philosophers have hitherto attempted the search of Nature in vain; but I hope the principles here laid down will afford some light either to this or some truer methods of philosophy.[14]

Despite having reduced the theory of universal gravitation to mathematical rule, Newton could never rest content so long as he lacked the knowledge of what gravity *is*, and the ability to reconcile it with the diverse natural phenomena that had commanded his attention since 1663.

Among the papers that never reached Halley are two versions of the suppressed *Conclusio*, originally intended for publication in the first edition of the *Principia*. "Hitherto I have explained the System of this visible world, as far as concerns the greater motions which can easily be detected," Newton began. "Whatever reasoning holds for greater motions, should hold for lesser ones as well. The former depend upon the greater attractive forces of larger bodies, and I suspect that the latter depend upon the lesser forces, as yet unobserved of insensible particles."[15] The *Conclusio* was abandoned just short of completion, and for obvious reasons. The revision of natural philosophy contained in the *Principia*, radical though it was, rested on a sound mathematical foundation; the theory of matter did not. To present it at that time, even in hypothetical form, would open wide the door to attack and possibly jeopardize the acceptance of the dynamics, concerning which Newton harbored not the slightest doubt. The ultimate synthesis of cosmic and microscopic phenomena would have to await further alchemical experimentation, which Newton returned to with a vengeance once the *Principia* was in press. Indeed, it was almost as if his masterpiece, once completed, seemed to him little more than a bothersome detour along the path of a far profounder quest, just as Einstein would one day view relativity as a bend on the road to a unified field theory. Humphrey Newton, who never encountered anyone remotely similar to the man whose name he shared, thought he was witnessing the supernatural in action, and perhaps in a sense he was. In England his employer had undergone an apotheosis of the kind usually reserved for martyrs or the participants in miracles. This transformation was delayed somewhat on the Continent, but by the mid-1690s many of the more astute European virtuosi had joined the ranks of the true believers. The Marquis de l'Hôpital, a noted French mathematician, inquired of Dr. John Arbuthnot, a prominent member of the Royal Society: "does [Newton] eat & drink & sleep. is he like other men?" The French savant was much surprised when Arbuthnot replied that Newton "conversed chearfully with his friends, assumed nothing & put himself upon a level with all mankind."[16]

II

When James II was young he had outshone his older brother Charles; he was thought handsome, and he possessed all of the social graces, including the ability to speak French fluently. Brought up in exile in Catholic courts by a Roman Catholic mother who was ever obedient to the Pope, James despised his brother's tendency to compromise on religious matters and reacted by openly espousing the creed of Rome. His Protestant grandfather on Henrietta Maria's side, Henry of Navarre, later King Henry IV of France, had considered Paris well worth the price of a Mass. A century later James was to be congratulated by the Bishop of Rheims for renouncing his three kingdoms (England, Scotland, and Ireland) for a Mass.[17] Of course, it had not happened quite that way. Fearful of suffering the fate of Charles I, his executed father, James fled London in 1688, abandoning his crown forever.

Fear also breeds ruthlessness, and James was certain in his own mind that his father's death at Cromwell's hands could have been prevented had stronger resolve and less toleration been exercised. Having crushed two rebellions and suspended a recalcitrant Parliament, the King was now determined to forge ahead with his plan to grant Roman Catholics freedom of worship as well as the right to enter into the service of the state and the universities. Such a policy would have made most Englishmen uneasy under any circumstances, but when played out against a backdrop of vicious religious persecution on the Continent it was certain to provide a hostile reaction.

In October 1685 Louis XIV revoked the Edict of Nantes, which had extended legal protection to French Protestants, or Huguenots, in 1598. Huguenot churches were razed, their schools closed, and dragoons were quartered in the most influential Protestant households to guarantee conformity. Word swept across England of official murders, the torture of women as well as men, and the forcible conversion of newborn babies. Although James was not a party to these deeds—nor did he condone them—he was tarred with the same brush, in the public view, as the Sun King. English Catholics themselves were sorely troubled lest they be caught up in a backlash against their coreligionist's excesses abroad.

A letter addressed to Dr. John Peachell, Master of Magdalene College and Vice Chancellor of the University, reached Cambridge on Ash Wednesday, February 9, 1687. It began on a disarming note—"Trusty and well-beloved we greet you"—but the rest of the letter could scarcely have been more ominous. What Peachell, whose most distinguishing feature was a bulbous red nose that betrayed his tippling, received was a mandate that one Father Alban Francis, a Benedictine monk, be admitted to the degree of Master of Arts "without obliging him to performe the Exercises requisite thereunto . . . and without administering unto him any Oath or Oaths whatsoever."[18] When he heard the news, Newton was in the final throes of giving life to his brainchild. He nevertheless found time to pronounce judgment on the King's order "Those yt Councell'd his Maj. to disoblige ye Univ. cannot be his true friends," he wrote to an unidentified correspondent on February 19. "The Vice-chancellor cannot by Law admit one to yt degree, unless he take ye Oaths of Supre[macy] & allegiance wch are cojoyn'd by 3 or 4 Statutes." A similar attempt to have an aging Catholic nonjuror named Popham appointed a pensioner of the Charterhouse, Newton noted, had been resisted by its Master, Thomas Burnet, "& ther's no more said of it."[19] Could men of conscience at the universities do less, if England was to be saved from the wiles of the rampant Whore of Babylon?

But the Charterhouse was neither Oxford nor Cambridge, and the King meant to have his way with these influential institutions, whatever the cost. At Oxford Obadiah Walker, a Roman Catholic, had already been appointed Master of University College and told that he could turn the chamber of his choice into an oratory, or private sanctuary of prayer. Within the year James would order the Fellows at Oxford to accept as their President one Samuel Parker, the crypto-Catholic Bishop of the city. Their refusal to comply so enraged the King that he "treated them with foul language" during a visit to the University. When they still insisted on upholding their oath, James, speechless with anger, saw to it that twenty-five of their number were dismissed. Even this drastic action did not quell the resistance. When Parker, who was elected by a single vote, sought to take up residence in the President's quarters, the doors had to be broken down in order to obtain the keys, presumably dispensing with the need for them in the process. Closer to home, Cambridge's Sidney Sussex, a quintessential Puritan foundation dating from the early seventeenth century, was forced to accept as its Master the Catholic Joshua Basset, a Bachelor of Divinity from Caius. Derided by one of his successors as "a mongrell Papist," Basset was exempted by mandate from taking the master's oath to "detest and abhor Popery."[20] The ground must have fairly trembled when Basset first celebrated Mass in his Lodge. Still, the despised Master never quite dared to make good his threat to have the "Popish Service" performed in the Chapel.

Meanwhile Peachell, an innocent victim of circumstances, was sick with worry. He hurriedly drafted a letter to the Duke of Albemarle, Chancellor of Cambridge and a member of court, begging him to intercede with the King. Albemarle complied with Peachell's request, but James remained adamant. The Chancellor suggested that a petition drafted and signed by the University as a body might prove more effective. Albemarle's strategy was rejected, however, on the grounds that it "might look tulmultuary." The more acceptable course would have entailed the preparation of a grace or petition to be considered by the University Senate's two houses, but that avenue was blocked by Basset's recent appointment. The new Catholic Master of Sidney Sussex was also a member of the six-man caput whose unanimous agreement was required before a petition could be considered by the Senate at large. As has happened more than once, the violation of law could not be effectively resisted because of the existence of an inhibiting statute.

The authorities finally decided to dispatch a messenger from each house to London in the hope that this more "respectful way" would make the desired impression on the King. James knew what to expect well before the petitions reached his court. Word from Cambridge had already arrived via Alban Francis himself, who, immediately upon learning of the Senate's decision, "took horse for London, to represent at Whitehall what had been done." Both the King and the Secretary of State, Lord Sunderland, refused to accept the petitions in person. Once they were delivered, the King dispatched another mandate exactly like the first, but for the addition of this chilling clause: "to do it [disobey] at our own peril." The second mandate was read in the Senate on Thursday, March 11, by which time Newton, who temporarily overcame his hermit instincts, had become publicly and, it would appear, passionately involved.

Why Newton, a heretic who rejected the doctrine of the Trinity, the fundamental tenet of orthodox Christianity, chose to speak out on this particular occasion seems clear. Religion, even more than the science that made him famous, was his *raison de être*, Roman Catholicism his *bête noire*. True, the Church of England also embraced the trinitarian credo, but Newton's investigation of church history had persuaded him that the papacy had invented and perpetuated it through the willful perversion of the Scriptures. Thus to his way of thinking heresy, however deplorable, was not necessarily of one stripe. James's recently launched attempt to pack the University with his coreligionists was of the lowest order of moral leprosy; it had to be excised at once lest the malignancy spread. Newton had turned to the Book of Proverbs for solace in the past, and there is no reason to suppose that he did not seek consolation in the words ascribed to Solomon during this time of trial: "The wicked flee when no man pursueth: but the righteous are bold as a lion."[21]

On a less exalted level, Newton's decision to oppose the King was fraught with fewer risks than might appear at first glance. Whatever the outcome, he would not have to submit to a conscience-violating oath, something he had previously avoided by securing a royal dispensation exempting him from taking holy orders. Hence his opposition to granting Francis a degree would simply appear that of an insubordinate but orthodox Anglican professor, a consideration of which he was well aware. Moreover, by shunning ordination he had long since closed the door on further preferment, so that he risked less than those who viewed the crown as a future source of patronage.

The anxiety generated by the prospect of blighted careers proved motive enough for the Senate to thrust Newton into the breach. He, along with the Public Orator John Billers, a Fellow of St. John's, was chosen by the nonregent house to meet with Peachell for the purpose of informing the Vice Chancellor that the King's second mandate could not, in good conscience, be obeyed. Representatives from the University were again dispatched to London. This time, accompanied by Albemarle, they were admitted into the King's presence, but hardly in the manner they would have liked. Their brief audience, during which James spoke not a word to the academics, took place in the passage leading from the royal bedchamber. Albemarle told the King of their mission, and he responded by accepting their letter before hastily departing for his regular evening visit with the Queen Dowager. The representatives were then taken by the Duke to Sunderland's bedside, where the Secretary of State accepted a second letter and entertained their oral arguments. A month later, on April 9, this preliminary jousting was brought to an abrupt halt. The offended King, who was not known for his patience, commanded Vice Chancellor Peachell and a delegation from the University to appear before the newly reconstituted Commission for Ecclesiastical Causes for having refused to comply with the letters mandate in behalf of Alban Francis.

The Senate convened on April 11 to select its representatives. By this time Newton had applied the finishing touches to the *Principia* and had left for his home in Lincolnshire. Out of sight, he was anything but out of mind, for, along with Humphrey Babington and six others, he was chosen to serve. But when he returned about a week later he found himself in the minority, if Conduitt's account of the events can be believed. The pressures to temporize being considerable, the delegation had virtually agreed on a plan drawn up by the Chancellor of nearby Ely: The University would yield to the King on the matter of Father Francis provided that his case not be used as a legal precedent in the future. Newton, "disliking it arose from the table & took 2 or 3 turns & said to the Beadle ... this is giving up the question." Urged to elaborate on his thoughts, "he returned to the table & told them his mind."[22] The sight of a normally taciturn Newton fervently arguing a cause not only was riveting but had an impact sufficient to turn the tide. Those assembled had seriously misjudged their colleague with the melancholy countenance, and they resolved to stand their ground with him. Such was their respect for Newton that if they were not already doing so, the Fellows would soon be going out of their way to avoid disturbing the "odd" diagrams he was disposed to sketch in the gravel walks with a stick.

So unruly was the crowd that gathered in the chambers of the Ecclesiastical Commission on April 21 that the summons which brought the Cambridge delegates to London had to be read a second time. The University had its share of supporters among the throng, but a majority of those present had come for one reason only—to enjoy the performance of George, Lord Jeffreys, head of the seven Commissioners and a fearsome throwback to the type that had once presided over the inquisitorial Star Chamber. While Jeffreys's legal learning was limited, owing largely to a proclivity for the tavern rather than the Temple during his student days, his talent in cross-examinations was almost legendary. Ruthless, coarse, and frequently not averse to intimidating those he interrogated by resorting to gallows humor, one peer characterized his conduct as more that of a "jack-pudding" than a judge. Charles II once observed that Jeffreys "had no learning, no sense, no manners, and more impudence than ten carted street walkers," but this did not prevent the King from promoting him to the office of Lord Chief Justice. James soon found in him a servant after his own heart and raised him to the peerage at the unusually early age of thirty-eight. Jeffreys did not disappoint. He presided over the "bloody assizes" in the north following Monmouth's doomed rebellion, passing the death sentence on scores and selling many more into slavery. The good judge made himself rich in the process by exchanging promises of leniency for the money and property of certain rebels and their friends. Besides knowing Jeffreys by reputation, Newton may have had occasion to wait upon him in the most literal sense many years before. Jeffreys had been admitted a pensioner of Trinity College in March 1662, while Isaac was a sizar, but the future jurist left Cambridge for London and the Inner Temple a year later without taking a degree.

Jeffreys addressed all of his questions to the alcoholic Peachell, whose nerve had failed long before he set foot in the Council Chambers. When asked why he had not obeyed the King's command, Peachell replied, "My lords, I am a plain man, not used to appear before such an honourable assembly, and if I should answer hastily, it may be I might speak something indecent or unsafe, which I should be afterwards sorry for." He begged a recess to prepare an answer "as may be both for our safety, and your lordships honour." A smiling and contemptuous Jeffreys immediately seized the advantage and resorted to his favorite tactic of humiliation by mordant wit: "Why, Mr. Vice-Chancellor, as for your own safety, my lords are willing you should take all the care you can; but for what concerns our honour, do . . . not trouble yourself; we are able to consult that, without any interposition of your's."[23]

A delay was nevertheless granted. On April 27 the delegates again assembled in the Council Chamber, where Peachell read the lengthy document containing the statutory objections to James's mandate. The Commissioners, well aware that they had no substantive legal case against the University, were content to let Jeffreys continue his bullying ways. When William Cook, Doctor of Civil Law, came to the quailing Peachell's defense, the Lord Chancellor issued a stinging rebuff: "Nay, good Doctor, you never were Vice-Chancellor yet; when you are, we may consider you." George Stanhope, a Fellow King's College and future Dean of Canterbury, also tried to intercede and met with similar disdain: "Nay, look you now, that young gentleman expects to be Vice-Chancellor too; when you are, Sir, you may speak, but till then it will become you to forbear."[24] Having been instrumental in preparing the University's written defense and rallying support for it, Newton uttered not a single word. However, he gave vent to his barely controlled anger at the bottom of one of several drafts composed in answer to the Commissioners' mandamus: "A mixture of Papist & Protestants in ye same University can neither subsist happily nor long together. And if ye fountains once be dryed up ye streams hitherto diffused thence throughout ye Nation must soon fall of. Tis not their preferments... but their religion & Church wch men of conscience are concerned for."[25] The paragraph containing these blunt sentiments never found its way into the official reply, a further indication that he thought long and hard about just how far he might safely go.

Mercifully, the judicial axe finally fell on Peachell during the session of May 7. He was removed from the office of Vice Chancellor and suspended from the Mastership of Magdalene. Jeffreys also announced that the Commissioners would meet with the rest of the delegates for the fourth and final time the following Thursday, when "we shall take them into consideration."

No one knows how Newton and his colleagues passed the five trying days while waiting to learn if their academic careers were at an end. It was doubtless a period of considerable soul-searching, and their long robes surely disguised some wobbly knees as they confronted the redoubtable Jeffreys on May 12. "Gentlemen," the Lord Chancellor began, "You cannot but be sensible, and so must all the world, how pernicious and obstinate the university has shown themselves in refusing the king's command... that ought to be obeyed." Though their guilt was not of the same degree as Peachell's, "my lords understand very well the sly insinuations in your paper." In the future the best course will be a "ready obedience." "Therefore I shall say to you what scripture says, and rather because most of you are divines; Go your way, and sin no more, lest a worse thing come unto you."[26] Considering what had befallen Peachell, theirs was a light, if humiliating, public spanking. Of course, there was not the slightest doubt in Newton's mind as to who the real sinner was. Neither in Newton's triumph nor in his tribulation had that person—James II, King of England—come to know him by name.

Having emerged unscathed from the taut encounter with the Ecclesiastical Commission, Newton returned to the University and the low profile that was the hallmark of his academic career. His usually fitful correspondence dwindled to less than a trickle during the next eighteen months; in all matters but one, it sheds no light on his private concerns. The running skirmish with his perenially uncooperative tenants, the Storers, had erupted into a battle during Newton's most recent visit to Woolsthorpe on Lady Day. He had walked the deteriorating property with his childhood friend and, while admittedly handicapped by nearsightedness, discovered much that offended his Calvinist sense of order and duty, not to mention his Puritan purse. Then, when he had attempted to collect the rent, the crafty Storer demanded that 30s. be deducted for boards his son had obtained to shore up gutters that were rotting from neglect. "I put it to him whether he could honestly affirm yt ye boards were worth so much. He answered he could not, but he hoped I would not stand with him for a small matter. To wch I presently answered yt I would not stand with him and so remitted 30s." Angry with himself over his temporary failure of resolve, Newton returned to the leasehold three or four times before departing Woolsthorpe, making careful mental notes of the decrepit state of the buildings and fences. Looking for the slightest provocation to file suit, he found it when Storer's outspoken son Oliver "disparaged the living ... in my hearing, I being of opinion that he did it as well behind my back as before my face, to hinder me of tenants who might put me upon calling them to account for repairs." [27]

These charges, along with the damages Newton sought, were contained in a letter he wrote to an unidentified agent into whose hands he entrusted the matter. Nevertheless, Eduard Storer's charmed life as a tenant, while threatened, did not yet come to its well-deserved end. Newton grudgingly drew upon his nearly exhausted reservoir of goodwill one last time. Perhaps thinking of Eduard's kind mother and stepfather, not to mention the sprightly sister he still visited when in Grantham, Newton promised, "If Mr. Storer will send me a satisfactory answer to my last [letter], I'll endeavour to make a final end in my next." [28] The surviving record shows no further mention of the subject, hence we are permitted to surmise that Newton was finally satisfied. As it turned out, however, he never quite managed to bring his tenants into line. When he died, one tenant owed him £60 for three years' rent while another was in arrears by two and a half years for a smaller unspecified sum. [29]

Another member of his dwindling circle of more than passing acquaintances died in September 1687. "I am going on a long journey," Dr. Henry More told his nurse at the end, "where I shall change these for better possessions." The saintly theologian of Christ's was buried, as he had wished, by torchlight on a Sunday. We may assume that his fellow Granthamite was among the mourners and was touched to receive one of the funeral rings bequeathed by his philosophical mentor. It was a good time for one so aged and sensitive as More to go. Toward the end he had lived in perpetual fear that James II's unbending policy regarding Catholicism might lead to his suspension, as it had to that of his good friend Dr. John Sharp, the once powerful Bishop of London.

Later that same month Newton took only his third pupil since his appointment as Lucasian Professor seventeen years before. He was also to be Newton's last. Robert Sacheverell is distinguishable from his shadowy predecessors—St. Leger Scroope and George Markham—only because of his talented and influential father, William Sacheverell, who has been described as "the brain" in the House of Commons under Charles II. A Whig like Newton, the elder Sacheverell alone dared challenge openly James's right of succession. His political career had since gone into predictable eclipse, but the time was fast approaching when he and his son's new tutor would become allies in the Convention Parliament. As far as Robert's instruction is concerned, Newton seems to have approached it with the same lack of interest he had shown toward private tuition in the past. An even more important connection was the one that Newton had already forged with Charles Montague. This swiftly rising star had recently decided to leave the University for the world of political and social affairs. Montague, who was among those who supported the letter of invitation to the Prince of Orange, would shortly join Newton and Sacheverell as members of the majority party at Westminster.

Memories of the Civil War were still vivid; so as long as James's opponents could expect him to be succeeded by one of his two Protestant daughters, they were inclined to put up with his arbitrary rule. But on Trinity Sunday, June 10, his second wife was delivered of a son, an irony that must have given Newton pause. Christened James Francis Edward (Edward after the Black Prince, who was also born on Trinity Sunday), the infant would become known to history as "the Old Pretender." News of the birth reached Cambridge the following day: The bells of Great St. Mary's were rung, and there was a bonfire on Market Hill, where soldiers stationed in the city formed ranks and fired the obligatory salvos. Here, as elsewhere throughout the land, the joy was largely ersatz, manufactured by order of the proud father. Barring an act of God, a Catholic succession to the throne now seemed certain, which in the words of one contemporary observer "gave the greatest agonies imaginable to the generality of the kingdom." [30]

Once exiled, James had all the time in the world to contemplate the errors of a failed political strategy. Had Machiavelli been able to analyze the events of 1688 from the grave, he would have singled out James's misplaced trust in William of Orange as the deposed King's Achilles' heel. James could not but have agreed. No one was more surprised than he to learn, after it was too late, that Prince William had been considering an invasion of England ever since April, when he told two influential English visitors in The Hague that he would "come and rescue" the nation, if invited. The birth of the Prince of Wales virtually guaranteed such an invitation would be forthcoming. It was dated June 30 and signed by six peers and the Bishop of London.

When James learned late in September that an amphibious expedition was being prepared to cross the North Sea from Holland, the King literally panicked, proving himself even less steadfast than the father and brother whose lack of resolution he so often decried. In a desparate effort to rally support, he did an about-face, promising to do nothing "inimical" to the Church of England in the future. At Cambridge a shamefaced Peachell, who had seen his name affixed to the gate of Magdalene for disobedience, was restored to the mastership of his college in October, while at Sidney Sussex the uncompromising Fellows were given permission to rid themselves of the Popish Basset and replace him with a master of their own choosing. Alban Francis, the proximate cause of so much sectarian strife, never did receive his mandated degree, and the matter was dropped without further discussion.

These and a hundred similar attempts to mend fences came too late. On November third and fourth, an expectant crowd, braving the chill of late autumn, gathered on Dover's white cliffs to watch William's six hundred vessels pass through the Straits. Those with the sharpest eyes may have been able to make out the Prince's banner fluttering on the masthead of his flagship: *Religione et Libertate—Je Maintiendrai*. A day later the future constitutional monarch landed unmolested in Torbay. It is said that James heard the grim news while sitting for a portrait by Godfrey Kneller, commissioned by Samuel Pepys. Nosebleeds, treated by more bleeding, and insomnia, treated by opiates, preceded the King's ignominious flight from London in December. He was carried by yacht to a point on the French coast near Calais, much to the relief of his anxious Queen, Mary of Modena. James stepped ashore on what was Christmas Day in England, the forty-sixth birthday of Isaac Newton.

Newton's defiance of recognized authority was not soon forgotten, and the reputation he gained during his struggle with the Ecclesiastical Commission took on added luster once the *Principia* was in print. Thus when the new government issued writs for the election of a Convention Parliament to establish a legal foundation for the revolution, Cambridge's Lucasian Professor—along with fellow Whig Sir Robert Sawyer—was chosen to serve by the University Senate on January 15, 1689. The balloting was very close: Sawyer, a noted lawyer and old friend of Pepys, received 125 votes; Newton, 122; and Edward Finch, an M.A. of Christ's College, 117.[31] A week later, at daybreak, Newton took his seat in the House of Commons, where he served until its dissolution the following February.

The distinguished natural philosopher turned backbencher could hardly have been more pleased with himself. Possessed of a keen sense of history, he sought to capture this rare moment in time by commissioning a portrait by Kneller, the German expatriate and principal painter of the day. Kneller's is the first and most appealing of the many likenesses we have of Newton, executed when the subject was at the height of his powers and poised on the threshold of international fame. The silver hair is thick and flowing, the myopic eyes somewhat protuberant but piercing, the angular chin deeply cleft, the mouth sensuous and delicately formed. The long thin fingers of the right hand, which extend from beneath an academic gown, are suggestive of a virtuoso of quite another kind, a performer of music rather than a revolutionary choreographer of matter's deterministic dance. In sum, Kneller's midlife portrait is a foreshadowing of that famous Jovian look borne by the aged Newton, who shared with Einstein the aspect of one present at the Creation.

While the painting betrays no hint of Newton's fearsome vindicative side, at that time it was by no means at rest. His revenge was exacted through the publication of a pamphlet in 1689 titled: *The Cambridge Case . . . for refusing to admit Alban Francis, a Benedictine Monk, to the degree of Master of Arts, without taking Oaths*. With the assistance of his amanuensis, he had compiled a record of the hearings and documents pertaining to the recent dispute. Three manuscript versions survive; all vary to some degree from the published account, and one of these is in Humphrey's hand.[32] Indeed, most of our information surrounding the events comes from this pamphlet, and there is little question that Newton was the primary motivating force behind its composition and printing.

The new member of Parliament could hardly have shed tears over the recent fate of George Jeffreys. When James fled London, the Lord Chancellor had disguised himself as a common sailor and hid on a ship anchored off Wapping. But the tedious wait for a favorable change of wind proved too much for the inveterate tippler, who foolishly went ashore and was recognized in the Red Cow tavern by a man he had once treated ill in court. Jeffreys was arrested and removed, at his own request, to the Tower, where he died of natural causes exacerbated by alcoholism in April 1689, at the age of forty-one. He was buried in the Tower chapel next to Monmouth, whose followers he had persecuted and sent to the gallows by the score.

III

Newton first set eyes on William Henry, Prince of Orange, only two days after his election to Parliament, when he and some fellow Whigs dined with the future King on January 17. With his elongated face, hawkish nose, and sharply curved brows, William was far from attractive. He had just turned thirty-eight the day before he landed in England, but his health was so poor that it gave him the appearance of an aging man. As a professional warrior, and a courageous one at that, he projected a natural authority, a trait strengthened by his general contempt for the leading English politicians, whom he referred to in private as "weak" of mind, "blockheads," or simply "mad." But William was also much more intelligent than his predecessor and a shrewd judge of character as well. When he was advised to consult Newton on a matter of politics later in the year, he is said to have declined on the grounds that Newton "is merely a great philosopher."[33]

Newton's first impression of William is not known, but he voted with the majority who declared the English throne vacant and marched, in mid-February, in the brilliant procession through the capital after William and his shy wife, Mary, the Protestant daughter of James II, were proclaimed monarchs by Parliament. The pageantry of the coronation followed two months later, in April, and Newton was there. He feasted with the "Parliament men" in the Exchequer Chamber and received a medal of gold, which Evelyn valued at "five & fourty shill[ings]."[34] One side bore the likenesses of the King and Queen; on the other side Jupiter was frozen in time as he cast a thunderbolt at Phaëthon, the exiled James's mythical surrogate, who nearly destroyed the universe by fire.

In truth, the enthronement marked the beginning rather than the end of the revolution. The Whig-dominated Parliament now asserted itself by passing a host of laws designed to safeguard the rights of Englishmen and to protect the legislative authority from executive usurpation. First came an act requiring that monies be appropriated on an annual basis, thus sharply curtailing the crown's freedom of action. This measure was followed by the Toleration Act, which extended religious freedom to all Christians except Catholics and Arians (which would include the antitrinitarian Newton). Most important, Parliament enacted the famous Bill of Rights on December 16, 1688. A model for the first ten amendments to the United States Constitution and the French Declaration of the Rights of Man and Citizen, it affirmed the right of Englishmen to petition their government for a redress of grievances and forbade excessive bail, exorbitant fines, and cruel punishments. Moreover, it prohibited the king from suspending laws or levying taxes without the express consent of Parliament. The theory of divine right monarchy had been dealt a mortal blow. Never again would any crowned head in Britain be able to defy the legislative branch as the Stuarts had done, and almost none, save the inept George III, even dared try. Isaac Newton had participated in what is surely the most crucial session of Parliament in English history. Having reduced the motion of matter to a single universal principle, he witnessed the no less revolutionary beginnings of constitutional government, illustrating once again that the lines of history often converge in the lives of special individuals. Yet the surviving records of the deliberations contain not a single reference to his name. According to an oft-repeated anecdote, the silent spectator at these historic debates spoke but once, when he asked an usher to close a window because of a chilling draft.

As with his refusal to address Jeffreys and the Ecclesiastical Commission, Newton's decision to remain silent on the floor of Parliament should not be mistaken for indifference or inactivity. Montague, who was moving ever closer to William III, kept in close touch on the important political issues, while Newton's relations with Sacheverell, one of Montague's co-leaders in the Commons, could not have been better.[35] However, as was true of many in this age of incipient political parties, Newton was more concerned with parochial than with national affairs. He perceived his primary role as one of effecting a smooth transition of the University from the Stuart to the Orange regime.

On February 12, 1689, Newton wrote a letter to John Covel, the new Vice Chancellor of Cambridge, announcing the proclamation of William and Mary by Parliament. He enclosed the form to be used by the University in its proclamation of the constitutional monarchs and expressed the hope that Cambridge "would so compose themselves as to perform ye solemnity wth a seasonable decorum because I take it to be in their interest to set ye best face upon things they can, after ye example of ye London Divines."[36] Covel agreed. One week later delegates from the University assembled to hear the public reading of their pledge of "all faith and true Allegiance" to the new order. Afterward they dined with officials of the city and supplied the King's soldiers with liberal quantities of drink.[37]

But there was a problem, and it lay in that always sensitive realm of oath-taking. Despite the fact that the academic community had been outraged over its harsh treatment at the hands of James II, a number of high church academics steadfastly refused to violate their consciences by renouncing the oath of allegiance they had sworn to the last of the Stuarts. In his next letter to Covel Newton referred to them as "persons of less understanding" and proceeded to set forth arguments in favor of pledging fidelity to the King: "[B]y Law we are Free men notwithstanding those oaths." His colleagues had to realize that obedience was passing from crown to Parliament, which must be trusted to protect its citizens from future roguery by the executive. He also reminded the Vice Chancellor that a book of verses celebrating the settlement was in order. "If you do it at all, ye sooner ye better." John Montague, Master of Trinity College, and his counterpart at Peterhouse, Joseph Beaumont, were among those chosen to crank out the fawning offerings.[38]

No one felt more keenly than Newton the pangs of conscience that arise from the taking of a morally repugnant pledge. Nor could he truly blame those who were suspicious of the new government, for they were not a party to its operations as he was. Hence when Covel wrote that the nonjurors refused to budge, Newton adopted a different strategy. "I perused it [the proposed oath] a week before it was brought into ye House . . . & found nothing in it for imposing [it] on all preferments." Only those seeking their first preferment or candidates for a higher position would have to comply. Conscience and the law were not at loggerheads, after all. But this illusion was quickly shattered when the House of Lords sent down a far sterner and less equivocal bill. Newton and his allies in the House of Commons were defeated by about fifty votes, and in May he did his duty by drafting a joint letter with Sir Robert Sawyer, informing Covel that the new oath should be administered to all and telling him how to go about it.[39] At St. John's, the center of resistance, some twenty Fellows refused to forswear their fealty to the royal exile, including John Billers, who had stood shoulder to shoulder with Newton before Jeffreys. Even so, few sanctions were brought to bear. Most of the holdouts retained their preferments until they died in the next century.

The thirteen letters Newton sent to Covel between February 12 and May 15, 1689, contain almost everything that we know of his official activities during this politically dramatic period, which, in point of fact, is very little. Brief, legalistic, and downright dull, they reveal even less about his personal life. Newton did mention that he had taken rooms in the house of one Mr. More, conveniently located at the west end of Westminster Abbey in the section now known as Broad Sanctuary. Early in May he contracted "a cold and bastard Pleurisy" (perhaps from not asking soon enough that the window near his seat in Parliament be closed) and was confined to his quarters for an unspecified period.[40] He seems to have recovered sufficiently to participate in his third installation of a new chancellor of the University, however. On May 30 Charles Seymour, Duke of Somerset, succeeded Albemarle, who had recently died in the far-off colony of Jamaica, over which he had been appointed governor by James. Like many others at Cambridge, Newton had favored Dr. William Sancroft, Archbishop of Canterbury, for the chancellorship.[41] But Sancroft removed himself from consideration, pleading advanced age and infirmity. It proved to be a fortunate decision so far as Newton was concerned. In 1691 Sancroft refused to take the oaths to the King, and, with six of his bishops and about four hundred of the parochial clergy, the prelate was reluctantly deprived of his office and living by William III.

The following month Newton had the opportunity to make amends for his previous ill treatment (via correspondence) of a most distinguished colleague. Christiaan Huygens arrived in London in June to visit his brother Constantyn, a royal favorite, who had accompanied the newly crowned King on the invasion from Holland. The two giants met for the first time at the Royal Society on June 12, and indications are that Huygens bore no grudge against Newton for the caustic denunciation he had suffered at the latter's hands in 1673. Huygens had read the *Principia* immediately upon publication, and his assurance to Locke that its mathematics was unimpeachable is proof of his profound regard for Newton's mind. The Dutch savant was far less enthusiastic over the dynamics, particularly the principle of attraction at a distance, but we can assume that he kept such major criticisms to a minimum, since he later parted company with Newton on amicable terms. Huygens presented a summary of his forthcoming *Traité de la Lumière* or *Treatise of Light* and *Discourse on the Cause of Gravity*. Newton then made some observations on the double refraction of Iceland spar. Each committed some significant errors of which the other was aware, the result of selecting topics in which his fellow scientist was better versed.[42]

On July 9 Newton visited Huygens at Hampton Court, where the natural philosopher was staying with his brother. His was something more than a mere courtesy call, however. Early the next morning the two set out for London. They were accompanied by John Hampden, one of the leading Whigs in Parliament, and Nicolas Fatio de Duillier, a young and gifted Swiss virtuoso about whom more will be heard shortly. Newton's friends had persuaded him that he should seek the office of Provost of King's College, whose current occupant was at death's door. Cambridge's new Chancellor, the Duke of Somerset, was also consulted, and we may be quite certain that Montague, too, did his utmost to incline the King in Newton's direction. William finally consented and issued a mandamus announcing the prospective appointment. Unfortunately, the King had not received the best advice on the matter, for the college statutes required that its Provost be both in holy orders and a current or former Fellow of the institution. Because Newton met neither qualification, the decision aroused considerable opposition at the August 24 meeting of the King's Council. An embarrassed William was too well schooled by the example of his imperious predecessor to repeat James's fatal mistake. He quickly withdrew his approval.[43] Temporarily stymied but undaunted, the friends of Isaac Newton did not allow the matter of his patronage to lapse for long. More important, Newton himself took an active role in this continuing quest for preferment, a telling indication that he—a man of influential political connections and high intellectual standing—was no longer willing to suffer the recurring insult of speaking into the void of a dim lecture hall, forsaken by uncaring students. If the year in London had not turned him into a social lion, it had at least wrought a subtle transformation of character. He left Cambridge a solitary man and returned a private one. Assuming a familiar role would prove much more difficult than the professor had anticipated; by now he was subconsciously defying the gravity of a deeply cloistered past.

Newton was in the capital during all of 1689, except for a few weeks in September and October, when Parliament was adjourned. William, who continued the royal tradition of attending the races at Newmarket, paid a visit to nearby Cambridge on October 7. Alderman Samuel Newton, a long-time observer of such events, thought the occasion less ostentatious than the visits of Charles and James. For one thing, the King was preceded by no mace-bearer; for another, his progress through the narrow, muddy streets was not announced by kettledrums.[44] He took the time to view the Wren Library, the still unfinished but splendid inanimate legacy of Isaac Barrow's tenure at Trinity College. Perhaps he also acknowledged Barrow's even more illustrious living legacy, a crimson-clad professor of mathematics and natural philosophy. In any case, this subdued pageant was to be Newton's last as a political personage until 1701. William prorogued the historic Convention Parliament on January 27, 1690, and dissolved it the following month. Sensing a Tory sweep, Newton shrewdly decided not to stand for reelection. He signed the college redit book on February 4 and returned to his silent chambers without causing a ripple.

Chapter Fourteen

Cul-de-Sac

We have sought for firm ground and found none. The deeper we penetrate, the more restless becomes the universe; all is rushing about and vibrating in a wild dance.

Max Born

I

Although the greater part of London was ravaged by fire in 1666, John Evelyn, in his diary entry of October 4, 1683, un–self-consciously described his newly reconstructed home as "the most august Cittie in the world."[1] Edward Ward, a popular writer of the late seventeenth century, observed of Cambridge, in contrast, that its abominably dirty streets could not have been better suited for a scavenger. The town was full of meager countenances that had studied themselves into a "Hypocondriack Melancholy." The rakishly thoughtless were also in evidence, young men designed by nature to grind mustard or pick mushrooms for some nobleman's kitchen, but whose parents, in defiance of destiny, were resolved to make scholars of them. The sight that rankled Ward most, however, was one of arrogant neophytes in their late teens "Struggling along...in new Gown and Cassock, as if they had receiv'd Orders about two hours before, and were the next Morning to have Institution and Induction, to become the hopeful Guide of a whole Parish."[2] Repeated encounters with individuals such as these had also taken their toll on Newton, making it easier for him to rationalize a significant change of course. He had not set foot in a lecture hall for well over a year, and there is no indication that he returned to the podium after serving in Parliament. Still, he continued to hold the Lucasian

Professorship for eleven more years, the last five of which were spent in London.

While the return to Cambridge was not easy, the onetime sizar could take heart in the knowledge that his counsel and company were valued by the local establishment, enabling him to settle accounts with the past, if only in part. Each year from 1689 to 1701, when he resigned both his fellowship and chair, Newton was among the commissioners appointed to oversee the collection in Cambridge of aids or taxes voted to the crown and Parliament. His fellow commissioners included, among others, the Mayor and Alderman of Cambridge, the Vice Chancellor of the University, and assorted Doctors of Civil Law and Divinity.[3] Another measure of his prominence is to be found in David Loggan's magnificent *Cantabrigia Illustrata,* published in 1690. Several of the plates were dedicated by the author–engraver to men of substance and influence: Thomas Tenison, the future Archbishop of Canterbury; the Duke of Lauderdale; Baron Guilford; Humphrey Babington; the Earl of Westmoreland; the Bishops of Ely and Lincoln; and Isaac Newton. Loggan succinctly described Newton's major achievements, but probably only after consulting him in advance: "Lucasian Professor of Mathematics, Fellow of Trinity, Fellow of the Royal Society, Very Accomplished Mathematician, Philosopher, Chemist." (As the author of the *Principia Mathematica,* Newton would naturally have wanted the only modifier to precede the source of his widening fame.) Ironically, he also appeared in the volume as the patron of the print of that Anglican bastion Great St. Mary's Church, adding yet another subtle layer to the false cocoon of orthodoxy that masked an abiding heresy.

The satisfaction he derived from this type of public recognition was occasionally tempered by a distasteful reminder of things past. During a visit to Cambridge in 1690, Archbishop Sancroft found it necessary to rebuke Peachell, the restored Master of Magdalene College, for "drunkenness and other loose conduct." Peachell promised to repent and began his penance in earnest by fasting for four days. When he then attempted to take food his system, ravaged by alcohol, failed to respond, and the helpless reprobate eventually starved to death.[4] Only Newton's hatred of Roman Catholicism had permitted him to stand in public beside one notorious debauchee who was being censored by another. If he attended the funeral, which seems unlikely, it was solely for the sake of appearance.

An important part of the compensation Newton derived from the year spent in London was his ongoing correspondence with two distinguished intellectuals, the political philosopher John Locke and Nicolas Fatio de Duillier, a gifted but arrogant Swiss mathematician whose central role in thwarting a Bourbon plot to kidnap the Prince of Orange in 1686 elevated him to a position of glittering celebrity at the new court. As has been noted, Locke became familiar with the *Principia* well before he was introduced to its author early in 1689, shortly after he returned from exile in Holland.

Echoing the enthusiasm of the review he had written for the *Bibliothèque Universelle,* Locke paid further homage to "the incomparable Mr. Newton" in the opening pages of his own masterpiece, *An Essay Concerning Human Understanding,* placing him (as diplomatically as possible) above the other "Master-Builders" of the age—Boyle, Sydenham, and Huygens.[5] The architect of the mechanical world view gave Locke every hope that he might discover the relationship between universal physical laws and the workings of government and society, thus bringing the world of men in line with the universe of machine. For surely the Creator applied to human beings and social institutions inviolable principles similar to those governing the planets and the stars. Religion, however important it might be to the individual, could no longer serve as the basis of public activity. Reason born of natural law was the way of the future, the way out of chaos and sectarian strife. In Locke Newton also found one of the very few who merited his implicit trust, a man who had chosen exile over moral and political compromise, a man who shared a substantial part of his larger vision of things—past, present, and future. Together they would literally monopolize the attention of those educated in the West for the next century and more, shaping the thought of educators, philosophers, scientists, and political leaders alike.

The occasion of their first meeting is unknown, but it probably took place at Lord Pembroke's weekly salon in London, where the repatriated Locke became the center of polite society. As would have been natural, Locke seems to have taken the initiative by questioning Newton about certain of the mathematical demonstrations in the *Principia.* Newton was flattered by Locke's desire to know more; he reciprocated by drafting a somewhat simplified demonstration that the planets move in ellipses. He also took the unusual and, for him, remarkably generous step of preparing an annotated copy of the *Principia* as a gift for his newfound friend, making certain that the corrections he planned for a future edition were included. Since both men were Puritans at heart, believing that science and religion were inseparable, it seemed only a matter of time before theology would become the dominant theme of their widening discourse.

Newton's return to Cambridge marked the beginning of a correspondence with Locke that continued until the political philosopher's death in 1704. The earliest extant letter, dated October 28, 1690, is in Newton's hand, but it is clear from the contents that it was preceded by others. He apologized for having delayed answering a letter from Locke, but "I staid [waited] to revise & send you ye papers wch you desire." Those revisions were not yet complete, and, like Henry Oldenburg before him, Locke would simply have to exercise patience while Newton resolved certain lingering doubts concerning his subject. More personal concerns were also vying for Newton's attention, though he cleverly concealed them behind a veil of nonchalance. The Earl of Monmouth, a strong supporter of the new regime, had been enlisted by Locke to help find a public post for Newton commensurate with his growing fame. By the autumn of 1690 Monmouth's efforts had been to no avail. "I am extremely much obliged to my Ld & Lady Monmouth for their kind remembrance of me & whether their design succeed or not must ever think my self obliged to be their humble servant." Newton concluded on an equally revealing note: "I suppose Mr Fatio is in Holland for I have heard nothing from him ye half year."[6]

Writing his cousin, Peter King, the year before his death, Locke observed, "Mr. Newton is really a very valuable man, not only for his wonderful skill in mathematics, but in divinity too, and his great knowledge in the Scriptures, wherein I know few his equals."[7] A substantial part of this complimentary assessment was no doubt based on the papers Newton had promised to send Locke earlier but which did not arrive until November 1690. At some unknown point in their conversations Locke had either asked, or Newton had volunteered, his thoughts on the doctrine of the Holy Trinity. Yielding once more to Locke's winsome manner, he had gone over his theological manuscripts, undertaken a series of revisions, and presented the result in the form of two letters bearing the unequivocal title *An Historical account of two notable corruptions of Scripture in a Letter to a Friend*. Not even John Wickins, his chamberfellow for two decades, had been privy to Newton's antitrinitarianism. Thus, the willingness with which he laid bare one of the deepest secrets of a deeply troubled conscience speaks eloquently of the rare trust Locke inspired in him. Long a champion of religious freedom himself, Locke had published several anonymous versions of his *Epistola de Tolerantia (Letter on Toleration)* on the Continent. Temporarily emboldened by this example and by Locke's considerable powers of persuasion, Newton amazingly consented to have the first of his heretical papers "done into French, [while] the other may stay till we see what successe the first will have." He added, even more incredibly, that "I may perhaps after it has gone abroad long enough in French put it forth in English."[8] Locke's *Epistola* had been translated into English in 1689. If Locke would serve as his liaison with a trustworthy publisher, Newton would cover the expenses. His identity, of course, must be kept secret, for exposure as an Arian would almost certainly result in his professional ruin and perpetual banishment from the circle of his peers.

Locke knew just the person to contact in Europe. He made a copy of the manuscript, which he then sent—without identifying its author—to his good friend the liberal thinker Jean Le Clerc, founder of the *Bibliothèque Universelle* and Professor of Moral Philosophy in the Remonstrants' Seminary at Amsterdam. Strongly opposed to rigid Calvinist doctrine, the Remonstrants hewed to a creed of universal grace by seeking harmony between the laws of God and human liberty. They were particularly averse to imposing their beliefs on others, a position with which Locke openly sympathized. Le Clerc agreed that the manuscript should be printed and offered to translate it into either Latin or French. He suggested, however, that Locke refer its author to Richard Simon's newly published *Historie Critique du Texte du Nouveau Testament.* Locke complied, and Newton, who did not yet know that his papers were in Holland, made several additions based on his reading of Simon. These he sent to Locke, who in turn forwarded them to Le Clerc.

It was certain that Newton would eventually come to his senses and realize that he had let himself be seduced by the heady wine of celebrity, not that the author of the *Principia* wasn't entitled to a certain measure of self-indulgence. And if his waxing ego enhanced his stature in the eyes of the gifted and even more famous Locke, what was the harm? But when Newton learned that his friend had truly taken him at his word, he panicked and hastened to contain the damage: "I was of [the] opinion my papers had lain still & am sorry to heare there is news about them. . . . Let me entreat you to stop their translation & impression so soon as you can for I designe to suppress them." He would naturally pay for any out-of-pocket expenses incurred by the publisher.[9] Le Clerc, who by now had puzzled out Newton's identity, was informed of the author's wishes and, though disappointed, complied without protest, thereby living up to his lofty philosophical principles. He deposited the papers in the Remonstrants' Library in Amsterdam, where they lay for the next half-century until published posthumously under Newton's name. Contrary to Newton's worst fears, Locke's opinion of him changed only to the extent that the philosopher vowed "to press him to nothing but what he is forward in himself to do" in the future.[10] Locke himself had recently experienced the discomfiture of having his authorship of the controversial *Epistola* become an open secret once it had been translated into English. He reacted by angrily scolding his Dutch confident, Philip van Limborch, for revealing to others "the secret you could not keep yourself,"[11] rather telling proof that Newton's refusal to publish had been wise after all.

The design of Lord and Lady Monmouth to have Newton made Comptroller of the Mint had come to nothing, though not, as he was about to convince himself, for their lack of trying. It was simply that his prospective patrons were Whigs, and for the time being the Tories held sway, an inconvenient political fact that Newton chose to ignore. He decided to try his luck in another pond by visiting Locke at Oates, the Essex estate of the latter's intimate friends, Sir Francis and Lady Masham. His stay of about a week took place in January 1691. Newton's letter thanking his hosts for their hospitality was most cordial: "You have obliged me by mentioning me to my friends at London, & I must thank both you & Lady Massam for your civilities at Oats & for not thinking yt I made a long stay there." In June Newton wrote more pointedly to Locke on the subject: "If ye scheme you have laid of managing the controuler's place at ye M.[Mint] will not give you the trouble of too large a letter, you will oblige me by it. I thank you heartily for your being so mindfull of me & ready to assist me with your interest."[12] By December, two years after he first sang "ye tune of King's College," Newton finally had a concrete prospect, but it was not the post he coveted at the Mint. An attempt was being made to obtain the Mastership of the Charterhouse for him. He was not impressed: "I thank you for putting me to mind," he wrote to Locke, "But I see nothing in it worth making a bustle for. Besides a Coach wch I consider not, its but £200 per an with a confinement to ye London air & to such a way of living as I am not in love with." He would only enter into serious competition "for a better place." Six weeks later Newton was crying conspiracy. Charles Montague, his most influential contact at court, "Upon an old grudge wch I thought had been worn out, is false to me, I have done with him & intend to sit still unless Ld Monmouth be still me friend."[13] He added that he had no prospect of seeing Locke again unless the philosopher was willing to visit him at Cambridge. Unfortunately, Locke's replies have not survived. Still, he evidently reassured Newton, as he would in the fast-approaching Black Year of 1693, that all was well. For when Newton next wrote to Locke at Oates, on February 16, 1692, he expressed his relief on hearing that Monmouth was still his friend. He added, however, that it was his intent not to give Locke or his Lordship "any further trouble" about a preferment. He also apologized for a related and personally embarrassing incident, "beg[ging] his Lordship's pardon for pressing into his company at the last time I saw him. I had not done it but that Mr Paulin prest me into the room."[14] Whether, as seems unlikely, he had truly committed a *faux pas* is less important than his fear of having done so, one more indication that he remained forever uncomfortable when in the company of his social superiors. Locke sought to reassure his neurotic friend further, first by visiting him at Cambridge in May 1692 and then by submitting a chapter from his *Third Letter on Toleration* for Newton's perusal. How difficult it was to reverse even a small part of the flow that had alienated the strange professor more and more from communion with other men.

In addition to religion and the question of Newton's preferment, natural philosophy played its part in the Newton–Locke correspondence. For example, Locke wanted to know if Newton had given any thought to Boyle's description of a man who, after staring at length into the sun, was able to recapture its brilliant image simply by thinking about it. Newton drafted a graphic reply, detailing the agony he had experienced as a young man when he, too, participated in this most dangerous of experimental games, having taken his cue from Boyle. However, he could furnish Locke with no easy answer to the question of causation, "wch I must confess is too hard a knot for me to untye."[15]

Robert Boyle was much on the minds of both men at this time. The great virtuoso's delicate health finally gave way in 1691, and he died on December 30, seven days after his beloved sister, Lady Ranelagh. Newton left for London the next day to attend the funeral.[16] He returned to Cambridge on January 21, 1692, and five days later added the following postscript to the letter accusing Montague of duplicity: "I understand Mr Boyle communicated his process about ye red earth & [mercury] to you as well as to me & before his death procured some of yt earth for his friends."[17] This single sentence launched a vigorous written exchange with Locke on the subject of alchemy, the greater part of which has long since disappeared.

Locke was somewhat taken aback by Newton's knowledge that Boyle had also entrusted him with his great secret. Where had Newton obtained his information? Locke wanted to know. Newton replied, "Mr Paulin told me you had writ for some of Mr Boyles red earth & by that I knew you had ye receipt."[18] What Newton did not tell Locke is that he had long been dubious of Boyle's alleged discovery of a special mercury capable of multiplying gold, especially if, as seemed likely, it was the same stuff of which Boyle had written in the *Philosophical Transactions* so many years before. Still, it was possible that Boyle's complete papers, which were now in Locke's hands, contained crucial details that he did not possess. Newton recalled that Boyle had always claimed he hadn't revealed everything to anyone. Newton soon asked for a sample of the red earth, and Locke obliged by sending a much larger quantity than he had expected. "For I desired only a specimen, having no inclination to prosecute ye process. For in good earnest I have no opinion of it." Yet he immediately contradicted himself by offering Locke his unqualified assistance, though "I feare I have lost ye first & third part [of Boyle's receipt] out of my pocket." Whether this was true or not, Locke responded just as Newton hoped he would by forwarding transcripts of two Boyle manuscripts. "If you desire the other[s] . . . I will send you them too," Locke promised.[19]

It was as Newton had originally thought: The process revealed to Locke was identical to the one he had obtained from Boyle. Newton replied on August 2, returning Locke's manuscript on toleration, while at the same time cautioning his friend against wasting his time on Boyle's unproven process. He concluded with a promise to make available to Locke an argument against multiplication "wch I could never find an answer to . . . & wch . . . if you will let me have your opinion about it, I will send you in my next."[20] Sadly, this intriguing letter, written at an important juncture in Newton's alchemical thought, is lost, as are all the others he exchanged with Locke on the subject.

II

Nicolas Fatio de Duillier remains a fascinating enigma, as he was to Isaac Newton, his most intimate acquaintance of the early 1690s. Fatio burst upon London society with the aura, almost, of a Magus, yet he eventually

wound up in the pillory at Charing Cross, an unmourned victim of his indiscreet espousal of heterodox theology.

Fatio, Newton's junior by twenty-two years, was born in Basel on
February 16, 1664, the seventh of twelve siblings, all of whom survived infancy and childhood. "My father," he wrote in an autobiographical letter
dating from the 1730s, "designed that I should study divinity; and accordingly having been instructed, both at home and at Geneva, in ye Latin and
Greek tongues, I spent two or three years in ye study of philosophy,
mathematicks, and astronomy; and began to learn ye Hebrew tongue and
go to the lessons of ye Divinity Professors."[21] But it was the will of
Catherine Barbaud, Fatio's doting mother, that eventually prevailed. In the
spring of 1682 Catherine sent her second and favorite son off to Paris, hoping that he would acquire the necessary polish for eventual employment at
some Protestant court in Germany. Fatio stayed in France for more than a
year, living well of the letter of unlimited credit provided him by his prosperous father as he mingled with the social and intellectual elite.

Though he was but eighteen, Fatio had already corresponded with
Domenico Cassini, head of the Royal Observatory in Paris. The noted
astronomer was much impressed by the youth's method of calculating the
earth's distance from the sun and his explanation of the form of Saturn's
rings. Before leaving the French capital, Fatio also told Cassini of his promising theory of zodiacal light. In the course of his later travels on the Continent he established strong ties with the most distinguished of natural
philosophers: Huygens in Amsterdam, Leibniz in Hanover, de l'Hôpital in
France, and the Bernoullis in his native Switzerland. It seemed to each of
these luminaries that here was a young man destined to make his mark,
both wide and deep. So it would soon seem to Newton as well.

The way to England proved strangely circuitous for Fatio. It began in, of
all places, the quiet garden of his father's estate of Duillier in the Swiss
province of Vaud in 1685. Fatio and a Piedmontese count named Fenil were
walking together one morning when, as Fatio later recalled, "he acquainted
me that he had written to Monsr de Louvois, and proposed to him to seize
the Prince of Orange, and deliver him into their [French] hands." What is
more, Fenil, a former officer in the army of Louis XIV, had received "a
most encouraging answer...written wth Monsr de Louvois own hand;
whose name being subscribed, I presently knew it to be written like y^t wch I
had seen at Paris." Fatio subsequently divulged the plot to a friend, the exiled English Bishop Gilbert Burnet, whom he accompanied to Holland for
the purpose of informing the unsuspecting William. In a generous display
of gratitude the Prince, whose advisers had certified Fatio's political and intellectual credentials with Huygens, resolved to create for him a private
mathematical professorship, complete with a house at The Hague and a
yearly stipend of twelve hundred florins. Tempting though this offer was,
the young virtuoso feared reprisals from both the French and the quick-

tempered Fenil, whose chance for instant fame and wealth he had dashed. Fatio asked for and received permission to go to London in the spring of 1687, resolving to stay "in England till the Prince of Orange was in full power of these Kingdoms."[22]

He arrived in England carrying a letter of introduction to the Royal Society from one Henri Justel, a then-noted Parisian scholar and natural philosopher. The document was read at the meeting of June 8, 1687, and Fatio was voted membership two weeks later, although he did not attend his first meeting until May 2, 1688, when he was formally admitted by signing the register. When Christiaan Huygens paid his celebrated visit to England the following year, Fatio took pleasure in escorting his Dutch friend about the capital. He was present at the Royal Society meeting of June 12, 1689, during which Newton and Huygens met for the first time. This may have served as the occasion for Fatio's introduction to Newton as well. When Huygens and Hampden accompanied Newton to London a month later in support of his petition to William for the provostship of King's College, Fatio was also on hand.

If one is to judge by the youthful portrait that hangs in Geneva's University Library, the Fatio Newton came to know was far from a handsome man. The countenance is dominated by a great Roman nose made even more prominent by a high, sweeping forehead, diminutive mouth, and rather sharply pointed chin. Only the eyes are compelling: Keen and intelligent, accentuated by steeply arched brows and high cheeckbones, they seem capable of transfixing anyone who might venture too near.

The powerful attraction experienced between the middle-aged bachelor and his younger colleague is perhaps best described as infatuation. The earliest surviving letter, dated October 1689, is in Newton's hand. He was about to return to London after the month-long Parlimentary recess. "I . . . should be very glad to be in ye same lodging wth you. I will bring my books & your letters with me." So close were they already that Newton felt free to criticize the great Robert Boyle, one of his few true models, for "conversing [on alchemy] wth all sorts of people & being in my opinion too open & too desirous of fame."[23] Such private candor, so rare for Newton, is paralleled only in his theological and alchemical correspondence with Locke, and even then he said nothing negative about Boyle until after his death. Fatio's reply is lost, but his feelings toward Newton were expressed in no uncertain terms in a November letter to Jean-Robert Chouet, his friend and onetime professor of philosophy at Geneva. How wonderful it would be, he rhapsodized, to become a permanent citizen of England and to inhabit the same dwelling as Newton, who was both the greatest mathematician ever to have lived and the most worthy gentlemen he had ever met. The Cartesian system was finished, no matter what others might think. "Its all over for vortices . . . which were only an empty imagination." What a pity that his old teacher was not sufficiently trained in mathematics so that he could understand and acknowledge this "pure truth" by himself. If only the money (one hundred thousand *écus*) were available, he, Nicolas Fatio de Duillier, would erect great and lasting monuments to his godlike friend. Posterity would then know that Newton had at least one contemporary who recognized his true worth.[24] Whether or not he was conscious of it, Fatio saw in Newton a mirror that perfectly reflected his own virtues and aspirations. Newton saw a similar reflection when he first gazed upon Fatio, but its vivid hues were destined to fade with time until only the barest of outlines remained, a hazy image as seen through a glass darkly.

The *Principia* was monument enough for Newton, who now nourished the more practical dream of landing a lucrative civil post. Aware of this design, Fatio thought of a way to be of service. "I did see Mr Lock above a week ago," he wrote on February 24, 1690, "and I desired that he should speak earnestly of you to Mylord Monmouth." He gently cautioned Newton not to get his hopes up, however, for "it is plain the King doth wholly give himself up to ye high Tory party." (It was sound advice, and Newton would have spared himself considerable mental anguish had he only taken it.) "Mr. Hampden and I had a design to go see you at Cambridge; but you send me word you will come hither, which I am very glad of." Fatio was expecting a copy of Huygens's *Traité de la Lumière* almost any day. A gifted linquist, he proposed to hold the copy until Newton's arrival: "It beeing writ in French you may perhaps choose rather to read it here with me."[25] Had this reminder of his difficulty with foreign languages come from any source other than Fatio, Newton would have thought it condescending. That he took no offense is a further measure of their closeness. It may be nothing more than coincidence but Fatio, who has been characterized as the "ape of Newton," employed many of the idyosyncratic writing techniques of his new friend and master. This was especially true during his later life, when he fully realized just how much he had lost.[26]

Newton set out for London on March 10 and passed the following month in the company of his young admirer. They had much to discuss in addition to Huygens's *Traité*. "My theory of Gravity," Fatio had written in his latest letter, "is now I think clear of objections, and I can doubt little but that it is the true one. You may better judge when you see it." He had expressed similar thoughts in a no less exuberant letter written to Huygens on the same day.[27] Taking his cue from the *Principia*, Fatio sought to elucidate the true physical cause of gravity, whereas Newton had confined himself to its mathematical demonstration. The theory, such as it was, held that ether, or a subtle matter like it, passes in great circular waves about the earth, driving objects by the regular impulse of its motion towards the planet's center. Fatio presented his hypothesis at the Royal Society meeting of February 26 and took care to obtain Newton's signature at the bottom of the paper when he arrived in London two weeks later. Halley also signed it, as did Huygens when Fatio visited Europe in 1690, unabashedly trumpeting its merits. Newton, he proudly proclaimed, "did not scruple to say *That there is but one possible Mechanical cause of Gravity, to wit that which I had found out.*" But honesty compelled him to add the following caveat: "Thô he would often seem to incline to think that Gravity had its Foundation only in the arbitrary will of God." Some time later, after passions had cooled, David Gregory learned that Fatio had indeed confused Newton's signature with his imprimatur: "Mr Newton and Mr Hally laugh at Mr Fatios manner of explaining gravity."[28] Later in life a broken and disgraced Fatio clung even more tenaciously to the illusion that he would one day share a place in history beside his idol. In 1735, nearly half a century after its composition, an embarrassed George Cheyne was reluctantly coaxed into adding his signature to this pitiful manuscript, which time and the scientific world had long since forgotten.[29]

Never at a loss for ideas, Fatio entertained more immediate plans for basking in Newton's luster. In a backhanded swipe at Halley he informed Huygens that he "knew of no one who so well and thoroughly understands a good part of [the *Principia*] as I do." (Huygens wrote sarcastically in the margin opposite this claim: "Happy Newton.") It was only natural that the great man would want him to oversee a second edition, for Newton was eager to correct his treatise on matters Fatio had told him about. Fatio also wanted to add a large amount of new material, a rather daunting prospect. Yet such an effort would be most worthwhile in the end, because his edition would be easier to read and comprehend than the original.[30] Little came of this intent, although Fatio was a gifted mathematician and worked closely with Newton in resolving several errors and ambiguities contained in the first edition. One senses that such conceited remarks were partly Fatio's way of salvaging a measure of his own identity while at the same time attempting to prove that he was truly worthy of the august company he kept. It may also explain why he curried Hooke's favor at the considerable risk of alienating his newly found alter ego. It did him little good. "Facio read his own hyp[othesis] of Gravity," Hooke noted in his *Diary* on February 26, 1690: "not sufficient." One week later he derisively labeled Fatio the "Perpet[ual] Motion man."[31]

Fatio sailed for the Netherlands in the spring, where he spent much of the next fifteen months in the company of Huygens. After hearing nothing from the young Swiss for "ye half year," Newton could suppress his anxiety no longer and inquired after his friend in an October 1690 letter to Locke. Coming from anyone else, this would seem a matter of little import, but Isaac Newton almost never asked about an acquaintance—no matter how learned or influential. Whether Locke took this for the muffled *cri de coeur* that it was is impossible to say because his reply, like so many others, is lost. Fatio returned to London early in September 1691. He wrote to Huygens on the eighth and mentioned that Newton would be arriving in the capital in a very few days.[32] It was just as Fatio said it would be. Newton left the

University on September 12 and returned exactly one week later.[33] So private was this reunion that their London friends were unaware that it had even taken place. The mathematician David Gregory, with whom Newton had recently visited in the city, wrote to inform him that "Mr Fatio hath been in the country some days."[34] Gregory composed his letter on October 10, three weeks after Newton's return from seeing Fatio.

The renewal of their friendship led to a revitalized correspondence, although none of the letters exchanged between the time of Fatio's return and November 1692 have survived. We do know that Fatio soon visited Cambridge and was shown some of Newton's most advanced mathematical papers. The young savant, who never seemed at a loss for words, informed Huygens that he was "frozen stiff" on seeing what Newton had accomplished."[35] Fatio paid at least one other visit to Newton's chambers, this in the late autumn of 1692. He drafted a chilling follow-up letter on November 17 that shook Newton to his very roots:

> I have Sir allmost no hopes of seeing you again. With coming from Cambridge I got such a grievous cold, which has fallen upon my lungs. Yesterday I had such a sudden sense as might probably have been caused upon my midriff/diaphragm by a breaking of an ulcer, or vomica, in the undermost part of the left lobe of my lungs. . . . I thank God my soul is extreamly quiet, in which you have had the chief hand. My head is something out of order, and I suspect will grow worse and worse. The Imperial powders, of which I have taken today four of the weakest sort, and one of the best sort, have proved quite unsignificant. . . . Were I in a lesser feaver I should tell You sir many things. If I am to depart this life I could wish my eldest brother, a man of extraordinary integrity, could succeed me in your friendship. As yet I have no Doctor.[36]

Fatio's offer of his brother's friendship to replace his own has been compared to the last desperate act of a dying woman: In choosing a successor for her husband she proves her love and eliminates a series of potential rivals.[37] Then again, it may simply have been a melodramatic attempt to assure himself that the genius of his age cared for him above all others, for why, if Fatio feared that he was dying, had he not called in a doctor? If indeed this was his hope, it was quickly fulfilled. Newton's hastily drafted reply summoned up a pathos unlike anything he had ever written:

> I . . . last night received your letter with wch how much I was affected I cannot express. Pray procure ye advice & Assistance of Physitians before it bee too late & if you want any money I will supply you. I rely upon ye character you give of your elder brother & if I find yt my acquaintance may be to his advantage I intend he shall have it, & I hope yt you may still live to bring it about, but for fear of ye worst let me know how I may send a letter &, if need be, a parcel to him. . . .
>
> Your most affectionate
> and faithfull friend to serve you
> Is Newton[38]

It had taken three days for Fatio's letter to reach Cambridge. Newton's emotional reply arrived in London the day after it was written, which suggests that he contracted the services of a special carrier in the hope of outracing death. He need not have bothered. "I hope the worst of my disease is over," Fatio wrote on the twenty-second. "My lungs are much better, tho' I have still a cold upon them. I think I had no Vomica, whatever that accident might be, which I have described to You."[39] He went on to trace the clinical aspects of his illness in tedious detail and followed this discussion with a lengthy (and superfluous) character sketch of his elder brother, proof in itself that the crisis, whether real or imagined, had quickly passed. The truth is, this temperamental, melancholy, hypochondriacal soul was blessed with a physical constitution the equal of his prodigious idol's, which he proved by living sixty-one years more.

Jean Alphonse Turretin, a Swiss theologian, visited Cambridge in January 1693, bringing Newton word that Fatio's cold still lingered. More upsetting was the news that Fatio might soon return to his native Switzerland. Newton again expressed his abiding concern in a letter of January 24: "I feare ye London air conduces to your indisposition & therefore wish you would remove hither so soon as ye weather will give you leave to take a journey. For I believe this air will agree with you better." Fatio replied that he must return home, for his mother had recently died, leaving him an inheritance. If the money proved sufficient he would seriously consider living permanently in England and "chiefly at Cambridge, and if You wish I should go there and have for that some other reasons than what barely relateth to my health and to the saving of charges [living expenses] I am ready to do so; But I could wish in that case You would be plain in your next letter."[40]

Their shared interest in Biblical prophecy surfaced in the same letter. "I am persuaded and as much as satisfyed," Fatio wrote, "that the book of Job, allmost all the Psalms and the book of proverbs and the history of the Creation was as many prophecys, relating most of them to our times and to times lately past or come." He believed that the serpent of Eden represented the Roman Empire, Adam the clergy, and "his wife the Church or people submitted to the clergy." Newton applauded Fatio's newly awakened interest in exegesis but cautioned him to guard against the wiles of irrational enthusiasm: "I am glad you have taken ye prophesies into consideration & believe there is much in what you say about them, but I fear you indulge too much in fansy in some things."[41] Had Fatio but heeded this sound warning he could have avoided the public censure that was to be his share.

The exchange of letters continued for the next few months. Newton had somehow convinced himself that Fatio was nearly destitute and sent him £14 in March, ostensibly to cover the cost of some relatively inconsequential purchases made during their collaboration on scientific matters. He also responded to Fatio's query concerning possible living arrangements: "The chamber next me is disposed of; but that which I was contriving was, that since your want of health would not give you leave to undertake your design for a subsistence at London, to make you such an allowance as might make your subsistence here easy to you."[42] If Newton had anything else in mind he could not bring himself to put it on paper. Fatio again flexed his emotional muscle by offering him hope: "I could wish Sir," he wrote on May 10, "to live all my life, or the greatest part of it, with you, if it was possible, and shall always be glad to any such methods to bring that to pass as shall not be chargeable to You and a burthen to Your estate or family." He added that Locke hoped they might both join him at Oates, where he was residing with Lady Masham. Fatio saw himself as the bait in this projected colloquium à trois. "Yet I think he means well & would have me to go there only that You may be the sooner inclinable to come."[43]

The last two letters of their extant correspondence from this period were written by Fatio, and both describe alchemical experiments about which there had been previous discussion. Fatio had been working with one "f. of L." (quite possibly a friend of Locke) on a process to purify mercury:

> When his [mercury] is ready he taketh some filings of pure [gold] and maketh a softish amalgam of them. These matters being put in a sealed egg in a sand heat do presently swell, and puff up, and grow black and in a matter of seven days go through the colours of the Philosophers. After which time there grows a heap of trees out of the matter, which trees change by degrees their colours. They begin now to be near their bottom of a copper colour or violet; the branches are gold and silver."[44]

Newton had described what was virtually the same process in the *Clavis* some two decades earlier. His living mercury also made gold swell, putrefy, and spring forth into sprouts and branches like those observed by Fatio. And, like Fatio, he had "recokon[ed] this a great secret in Alchemy." The youthful image of himself lingered but momentarily, however. There had been no real breakthrough after all, something Fatio was soon to learn from the master. Newton therefore had no good reason "to burn this letter after You have done with it," as Fatio had requested. Besides, to destroy anything wrought by his intimate friend would be to destroy a part of himself.

The last letter, dated May 18, 1693, bears all the marks of a manic-depressive personality. Fatio's inheritance had turned out to be most disappointing, but his health was much improved, thanks to a generous acquaintance whose secret remedy he had taken with "a wonderfull success." He was now considering becoming a physician, "which I think would not require above a year or two of my time." He would then go into partnership with the creator of the miraculous new elixir. "I could cure for nothing thousands of people and so make it known in a little while. After which it would be easie to raise a fortune by it." The medicine's only drawback was its emetic quality, but that, he assured Newton, could be overcome without difficulty. Of course, he must expend between £100 and £150 a year "for four years or more before this thing is established. And I dare not propose to any body nor to yourself," he continued disingenuously, "to be Partners in such a design because of ye many accidents, which could spoil it."[45] Newton must advise him on the proper course to take, preferably at a private meeting in London. Still under Fatio's seductive but apparently waning spell, Newton met with him in the capital two weeks later, and he seems to have made a second trip to London late in June.[46] What passed between them on these occasions we do not know, except that Fatio abandoned his grandiose dream of becoming a wealthy doctor and reluctantly admitted to Huygens, in September 1694, that he had not heard from Newton for more than seven months.[47]

Ever since his nineteenth year Newton had been confined to the company of males. So far as is known, the only woman he saw regularly was his aged housekeeper, who merrily (if that be the translation of "mumpingly") made off with his uneaten meals. Because he was barred by statute from taking a wife, his natural sex drive had to be either repressed or given release outside the conjugal estate. It seems most unlikely, given his unbending scruples, that he would have taken up with women of the streets, even though Cambridge and London afforded ample opportunity to do so with impunity. Voltaire, who loved to dwell on the salacious, was told by Dr. Richard Mead, Newton's physican and confidant in old age, that his patient had never violated chastity.[48] Thus, except for John Wickins and a broad streak of narcissism, Newton had been objectless in his adult affections until he met Fatio de Duillier. It is impossible to be certain how those affections were expressed during their private meetings, but Newton's moral code would have raised a strong barrier against an overt sexual liaison, if, indeed, it ever came to that. On the other hand, their correspondence—with its lavish praise, requited loneliness at separation, and melancholy swings of mood—bears haunting overtones of an ill-fated romance. The final break itself appears to have been prefigured in their agonizing desire to share the same chambers, a desire quite possibly overridden by the fear of what might happen if they were to attempt it.

Whatever Newton's actual sexual proclivities, there is nothing to indicate that they exerted a telling influence, either positive or negative, on his chosen lifework. Had he been a poet, playwright, artist, or novelist, we could search his creations for vital clues to such deep-seated feelings. But science, unlike literature and the arts, is neuter; emotion and passion become its mortal enemies when they are allowed to enter the arena of daily labor. And the fruit of that labor, exhilarating and beautiful though it can be, is expressed in the most dispassionate symbols known to literate society. Of this much alone we can be certain: In reaching out to an equally distressed spirit, Newton for once proved himself human in the most human of ways.

III

Between the time of Fatio's illness in September 1692 and the agonizing rent in their intensely emotional relationship the following summer, Newton willingly engaged in a brief but significant correspondence with another young disciple named Richard Bentley. Born in Yorkshire in January 1662, Bentley entered St. John's College, Cambridge, in the humble capacity of subsizar in 1676. Little is known of his university career, except that he showed early a strong taste and aptitude for the cultivation of ancient learning. Bentley received a bachelor's degree in 1680, and on leaving Cambridge he was appointed master of the grammar school of Spalding in Lincolnshire. That he had attended any of Newton's lectures, as conjectured by his biographer James Henry Monk, is highly doubtful, for there is no evidence of any interest in natural philosophy on Bentley's part before 1691. By then he had been made chaplain to Edward Stillingfleet, Bishop of Worcester, whose son he had tutored during the 1680s. In 1700 Bentley was appointed Master of Trinity College. He had the first observatory in Cambridge constructed on the roof of the Great Gate (it was later removed for aesthetic reasons) and established a chemical laboratory overlooking the bowling green. A consummate academician, heady administrator, and caustic wit, Bentley was ranked by G. M. Trevelyan as Trinity's greatest scholar next to Newton. But the history of his tenure is also the narrative of an endless series of quarrels and litigations, sparked by arrogance and rapacity, for which, it must be noted, he was fully as well known to his contemporaries as for his erudition.[49] Newton, it would seem, attracted young men not only of extraordinary talent but of temper to match.

Bentley, who was conversant with the writings of Locke, warmly embraced the concept that the existence of God could no longer be argued from the principle of innate ideas. He became the first influential divine to seize upon Newtonian philosophy as the natural ally of the rationalist theology championed by less doctrinaire Restoration churchmen. Eschewing the popular idea of an interventionist Deity who sometimes acts on nature from without, causing matter to behave in an "unnatural" manner, Bentley preferred to contemplate the vastness and harmony of the cosmos as the primary measure of God's design and presence. With this in mind he enlisted Newton's direction in gaining a better understanding of the *Principia* in the summer of 1691. Newton drew up a list of mathematicians for Bentley to consult and encouraged the theologian to concentrate on Book III of his treatise. Bentley elicited and received a similar but even more daunting set of instructions from the mathematician John Craig,[50] and that is where matters stood when Boyle died some six months later.

Fully cognizant that the end was fast approaching, Boyle, a man of deep and active piety, had taken steps to assure that his religious views would not perish along with the flesh. His last will and testament, which was drafted in the summer of 1691, contained a codicil bequeathing the sum of £50 a year "for proving the Christian Religion against notorious Infidels, *viz.* Atheists, Theists, Pagans, Jews, and Mahometans, not descending lower to any Controversies, that are among Christians themselves."[51] What most haunted the dying virtuoso was the specter of Thomas Hobbes's atheism and its growing appeal among the higher classes of English society. Boyle hoped to lay this most pernicious of spirits to rest by underwriting a series of eight sermons to be delivered annually in one of London's most prestigious churches. The choice of theologians, also to be made on an annual basis, was left in the hands of four trustees, the most prominent of whom were John Evelyn and Thomas Tenison. The four gathered on February 13, 1692, and, according to Evelyn's diary, "we made choice of one Mr. Bently, a Chaplain to the Bishop of Worchester."[52]

Bentley had been much in the public eye owing to the recent publication of his *Epistle to Dr. Mills*, an essay of enviable erudition and critical insight. This, added to the young theologian's influential connections with the Low or moderate faction of the Church of England, may have tipped the scales in his favor. But Bentley's inquiries about the best method of approaching the *Principia* from a layman's standpoint were made at the very time Boyle was writing his will, causing some scholars to wonder if Newton might have played an important role in Bentley's selection. The best evidence comes from an invitation sent by Pepys to Evelyn on Saturday morning, January 9, two days after Boyle's funeral at St. Martin's-in-the-Fields. "Pray lett Dr. Gale, Mr. Newton, and my self [Royal Society Fellows all] have the favour of your company to day, forasmuch as (Mr. Boyle being gone) we shall want your helpe in thinking of a man in England fitt to bee sett up after him for our Piereskius [a French savant]."[53] The trustees took formal action the following month, but the real selection of Bentley may have taken place already by way of a gentlemen's understanding reached over dinner and brandy at Pepys's.[54]

Whatever the actual circumstances surrounding his appointment, Bentley hewed closely to Boyle's wishes and collectively titled his sermons *A Confutation of Atheism*, the first of which was delivered at St. Martin's in March 1692. The young man excited deep public interest, drawing crowds that far surpassed the expectations of his patrons. Dealing first with the role of religion in society, he skillfully led up to the conclusion, delivered in the culminating lectures of October, November, and December, that a divine Providence exists as proved by the evidence of design implicit in the universal laws revealed by Newton. Before he committed his sermons to publication, however, Bentley wrote the *Principia*'s author for the purpose of clarifying a number of points. Newton responded by drafting four letters to Bentley between December 10, 1692, and February 25, 1693. He began with these now famous words: "When I wrote my treatise about our Systeme I had an eye upon such Principles as might work wth considering men for the beleife of a Deity & nothing can rejoyce me more than to find it usefull for that purpose." [55] It has been observed that this assertion should not be taken literally, because the *Principia*, at least in the first edition, offers scarcely any evidence of such an underlying presupposition. [56] Not only would Newton have scoffed at this suggestion, he would have thought it one of the greatest wrongs posterity could commit against him. The *Principia*, perhaps more than the rest of his works put together, stood in his eyes for the doctrine of God's existence and immutability—the absolute confutation of atheism by natural law that Bentley claimed it to be.

Newton launched his argument for design by setting forth his personal view of creation. In the beginning God endowed every particle "with an innate gravity towards all the rest" and scattered them evenly throughout the void. If space were finite, this mutual attraction would cause all matter to move toward the center, where it would form a single spherical mass. But if the universe were infinite, as he believed it to be, matter would be prevented from coalescing into one sphere—"some of it [would] convert into one mass & some into another so as to make an infinite number of great masses scattered at great distances from one to another throughout all yt infinite space." [57] Like laymen of generations to come, Bentley was awed by the magitude of Newton's conception and revised his manuscript to provide a more detailed account of it. So, too, was the mathematician David Gregory, who, in a partial attempt to digest the idea, wrote this incredulous note to himself in June 1693: "If light travels from the Sun to the Earth in an interval of 10 minutes as Newton and Huygens wish to think, it will need many days, months indeed for light from a fixed star to reach the Earth." [58] Newton saw further evidence of God's presence and planning in the orderly arrangement and regular movements of the planets. No natural cause alone was responsible for propelling these bodies in the same direction along the same plane. "And to compare & adjust all these things together in so great a variety of bodies argues that cause to be not blind & fortuitious, but very well skilled in Mechanicks & Geometry." Nor was it possible that cold bodies like the planets and flaming stars like our sun could have come into existence by the same mechanism. The fact that both types were to be found in nature, and that such an arrangement was essential for life, pointed to an omniscient Creator who had prevented matter from following a single purposeless pattern. He concluded his first and most revealing letter with an intriguing remark: "There is yet another argument for a Deity wch I take to be a very strong one, but till ye principles on wch tis grounded be better received I think it more advisable to let it sleep." [59] If, as seems doubtful, Newton ever committed to paper precisely what he had in mind, all trace of it has vanished. We can only assume that it was somehow related to his heterodox theology.

In subsequent letters Newton returned again and again to the thesis that an intelligent agent was responsible for framing the universe and continually directing the movements of its innumerable components through natural law. Nothing, including universal gravitation itself, must be permitted to obscure this reality. "You sometimes speak of gravity as essential & inherent to matter," he wrote somewhat admonishingly in the second letter, "pray do not ascribe that notion to me, for ye cause of gravity is what I do not pretend to know, & therefore would take more time to consider it." Several weeks later, in the fourth and final missive, Bentley received the fruit of that consideration: "Tis unconceivable that inanimate brute matter should (without ye mediation of something else wch is not material) operate upon & effect other matter without mutual contact. . . . Gravity must be caused by an agent acting constantly according to certain laws."[60] He could just as well have added that this agent, operating as it does at both the macrocosmic and microcosmic levels, would never come within the purview of the scientific method, no matter how rigorous and skilled its practitioners. For to assert the contrary would be to embrace the chimera that we can attain an absolute knowledge of phenomena, to believe that we can become God Himself—the first and, according to Scripture, the greatest blasphemy ever committed by humankind.

Despite his private and rather petty criticisms of Boyle toward the end, Newton knew well and greatly admired his melancholy peer. Together they shared the belief that the scriptural and the scientific are best understood when considered as part of unified view of reality. The book of nature was for them a kind of Third Testament, which merely reaffirmed what God had revealed to man in the Old and the New. Critical though Newton was of lesser minds for whom he deliberately made the *Principia* abstruse, he nonetheless chose to play an important role in the first public exposition of his philosophy. And one suspects that few things pleased him more than to attend the Boyle Lectures after he removed to London, where he sat in rapt attention while other young Newtonian disciples—John Harris, Samuel Clarke, William Derham, and William Whiston—extolled the spiritual and intellectual virtues of his handiwork.

While Newton was hardly besieged by acolytes who came to genuflect at the altar of his greatness, neither were Fatio and Bentley the only young men to be admitted into the ranks of his disciples in the early 1690s. Having twice failed to establish a correspondence with Newton during the previous decade, David Gregory, Professor of Mathematics at Edinburgh, finally caught up with his idol while visiting London in the summer of 1691. Newton was sufficiently impressed by the Scot to send him to Greenwich bearing a warm letter of introduction to John Flamsteed. "You will find him a very ingenious person & good Mathematician worth your acquaintance." He urged the Astronomer Royal to publish his projected catalog of the fixed stars as soon as possible, adding, "If you & I live not long enough [to complete our work], Mr Gregory and Mr Hally are young men."[61]

Newton had more on his mind than mere professional courtesy, however. Edward Bernard, Savilian Professor of Astronomy at Oxford, had recently resigned the post he had held since the early retirement of Christopher Wren in 1673. Anxious to have the chair filled by a scholar of his own choosing, Newton liked what he saw in Gregory and hoped that the influential Flamsteed would come to share that opinion. Indeed, Newton had already written a letter of recommendation to Oxford in support of Gregory's candidacy, calling his new acquaintance a great artist in mathematics and an ornament to his country.[62] In quest of a preferment himself, Newton had long since learned that nothing would be gained by the use of temperate language in a society that lived by the effusive and the rhetorical.

Two other prominent thinkers were also vying for the position: John Caswell, a mathematician-astronomer who held the post of Vice Principal to Hart Hall, Oxford, and Edmond Halley, to whom Newton owed a debt beyond repayment. Fearing his chances were rapidly dissolving, Gregory wrote to Newton on August 27, 1691, that "without more noise Mr Casswell will gett in ther." By November 7 his little remaining hope had turned to quiet desperation: "ther is nothing done about the Savilian professor yet. pray Sr, if ye did leave your opinion in that affair with any body here [London], tell me what it is that I may know how to behave in it if it come to be talkt of befor I goe down to Edenburgh." When Newton failed to take up the matter in his reply, Gregory resigned himself to the outcome: "Mr Caswell will defeat Mr Hally and me in that affair." But Caswell subsequently withdrew his name from consideration, supposedly, according to Flamsteed, at the behest of Halley, who had once studied under the departing Bernard.[63] Halley had also gained the enthusiastic support of the Royal Society and now seemed certain to win the appointment, barring a single impediment. The successful candidate had to be approved by the Church. The astronomer was already suspected in some quarters of heresy, a charge vigorously supported by Flamsteed, his erstwhile friend, who wrote Newton in February 1692 that if Halley and his "Infidel companions" were present at the Second Coming "they should not scape free ye calumnies of their venomous tongues." As a good Christian himself, Flamsteed hastened to add, "I hate his ill manners not the man."[64] (In another draft of this same letter, which was never sent, he had urged Newton to do what he could to block Halley's election on the grounds that he would "corrupt ye youth of ye University with his lewd discourse.")[65] Bishop Edward Stillingfleet, a man of piety and rigid orthodoxy, was one of the referees for the appointment. He sent his trusted chaplain, Richard Bentley, to interview Halley, and the meeting did not go well. Although Halley was more of a deist than an infidel, the Church could hardly have been expected to distinguish between the two, and his nomination was summarily rejected. Gregory, Newton's first choice owing to the religious controversy surrounding Halley, thus obtained the chair, although Fatio boasted that he could have gotten the post if he had desired it.[66] Some time later one of Gregory's countrymen is said to have made regular albeit rather mysterious visits to Mus's, the London coffeehouse frequented by Halley and his circle. When the man was finally asked the reason for his persistence, he replied, "I would fain see a man that has less religion than Dr. Gregory."[67]

There can be no doubt that Newton nurtured a degree of affection for talented young men who considered him their intellectual hero. Much of the correspondence that burgeoned, relatively speaking, after the *Principia*'s publication was directed to this small but dedicated elite. It began with Halley, who was replaced by Fatio, who, in turn, yielded to Gregory. In 1694 William Whiston, another young hopeful, made himself known to the master by submitting the manuscript of his *New Theory of the Earth* to Newton's critical eye. It seems to have won approval, for Whiston was secretly initiated into the fraternity of Arians and awarded the Lucasian chair following Newton's resignation in 1701. Three years later Newton made it possible for Halley to join Gregory at Oxford as Savilian Professor of Geometry, despite the fact that, as Catherine Conduitt told her husband, Newton occasionally became angry at Halley for making light of Christianity.[68] For his part, Richard Bentley never reconciled himself to Halley's seeming lack of religious principle. In 1734 the Master of Trinity College produced a widely circulated tract titled *The Analyst or a Discourse Addressed to an Infidel Mathematician*. Virtually everyone who read it took it for what it was, a pointed attack on Halley, who by then had taught mathematics at Oxford for thirty years.

IV

Nothing is known of the actions and events that directly preceded Newton's dark passage in September 1693, save this. A letter arrived late in August from his stepsister, Hannah Barton, wife of the Reverend Robert Barton of Brigstock in Northamptonshire. Its message was anything but welcome: "Dear Brother [,] My Dear Husband . . . has been very ill. . . . I find noe hopes of Cure but that hee lossis his flesh and strength very fast."[69] Confronted with impending widowhood, Hannah, who seems to have inherited her mother's obsession with financial security, sought her brother's advice on matters pertaining to the estate. There is no indication that Newton replied; indeed, he may have been psychologically incapable of doing so, at least for the present. For on September 13 he wrote the following letter to Samuel Pepys, breaking an ominous silence that had extended over the past several months.

Sir,
    Some time after Mr. Millington [a Fellow of Magdalene College, Cambridge] had delivered your message, he pressed me to see you the next time I went to

London. I was averse; but upon his pressing consented, before I considered what I did, for I am extremely troubled at the embroilment I am in, and have neither ate nor slept well this twelve month, nor have my former consistency of mind. I never designed to get anything by your interest, nor by King James's favour, but am now sensible that I must withdraw from your acquaintance, and see neither you nor the rest of my friends any more, if I may but leave them quietly. I beg your pardon for saying I would see you again, and rest your most humble and most obedient servant,

Is. Newton[70]

Pepys was not only deeply shaken by Newton's allegations, he was innocent on all counts. The former Secretary to the Admiralty had been forced to retire during the Glorious Revolution and was lately living in virtual seclusion while attempting to sort through his voluminous papers. If he possessed any influence at court it was of a negative and therefore useless kind, although Newton, in alluding to James II, may have imagined that time and history had turned back upon themselves, as when Alice passed through the looking glass. Nor had Pepys contacted Newton through John Millington. He had to be certain, nevertheless, that he was the innocent victim of some horrible misunderstanding. In an attempt to learn more he asked his nephew, John Jackson, to interview Millington at Cambridge. Not satisfied with Jackson's handling of the matter, Pepys wrote directly to Millington himself, expressing his fear that Newton was suffering from "a discomposure in head, or mind, or both." He begged Millington to learn the truth, "For I own too great an esteem for Mr Newton . . . to be able to let any doubt in me of this kind concerning him lie a moment uncleared." Millington assured Pepys that he had delivered no message to Newton of any kind. He took the added precaution of visiting Newton on the twenty-eighth. His host must have anticipated the reason for the call because Millington had said nothing before Newton confessed to having written Pepys "a very odd letter." He blamed it on "a distemper that much seized his head, and that kept him awake for about five nights together." He begged Pepys to forgive him, "He being very much ashamed." There was no reason, so far as Millington could tell, to suspect that Newton had taken permanent leave of his senses, though he observed that the natural philosopher remained under some small degree of melancholy. A deeply relieved Pepys thanked Millington for delivering him from the fear that he had contributed somehow to Newton's breakdown. He remained understandably chary of writing to the savant, however, and sagely left it up to Millington to express his regards and undiminished esteem.[71]

The unsettling letter Newton sent to Pepys was followed three days later, on September 16, by an even more disturbing and vengeful one to Locke. It was dispatched from the Bull Inn in Shoreditch, London. What Newton was doing there he did not say.

Sr

Being of opinion that you endeavoured to embroil me wth woemen & by other means I was so much affected with it as that when one told me you were sickly & would not live I answered twere better if you were dead. I desire you to forgive me this uncharitableness. For I am now satisfied that what you have done is just & beg your pardon for having hard thoughts of you for it & for representing that you struck at ye root of mortality in a principle you laid down in your book of Ideas & designed to pursue in another book & that I took you for a Hobbist. I beg your pardon also for saying or thinking that there was a designe to sell me an office, or to embroile me. I am

your most humble & most
unfortunate Servant
Is. Newton [72]

A most unfortunate servant indeed. A deluded Newton, now in his fifty-first year, accused his loyal admirers of conspiring against him, referred to conversations that never were, and sought to sever his hard-won and lately flourishing contacts with the outside world. Sick of mind, he was no less sick of heart.

If anything, Locke's handling of the volatile situation was even more delicate and compassionate than Pepys's, but then Locke enjoyed the advantage of having dealt in this emotionally inflated currency before. As we know, he was the recipient of the letter in which Newton charged Montague with being false to him and Monmouth of no longer being his friend. The political philosopher had hastened to smooth the troubled waters in February 1692, and he responded with equal understanding on this occasion: "[G]ive me leave to assure you that I am more ready to forgive you than you can be to desire it," he wrote in all sincerity on October 5, "and I do it soe freely and fully yt I wish for noe thing more than the opportunities to convince you yt I truly love & esteem you & yt I have still the same goodwill for you as if noe thing of this had happened." Locke had gone to Cambridge to recement their relationship following the previous outburst, and he was more than willing to meet with Newton at a place of his choosing to work out their imagined differences now. But by the time his letter reached Cambridge the most acute phase of the paranoia had passed, much to the relief of all concerned. The only explanation Newton could offer Locke for his bizarre conduct was a more detailed account of what Millington had learned a fortnight earlier:

The last winter by sleeping often by my fire I got an ill habit of sleeping & a distemper wch this summer has been epidemical put me further out of order, so that when I wrote to you I had not slept an hour a night for a fortnight together & for 5 nights together not a wink. I remember I wrote to you but what I said of your book I remember not. If you please to send me a transcript of that passage I will give you an account of it if I can.

There was nothing insincere about Newton's plea of forgetfulness. He had misplaced letters sent him by Leibniz and Otto Mencke several months earlier. When he finally felt up to drafting replies they contained uncharacteristically profuse apologies for the delay but not a word of the illness that had laid him low.[73]

The factors that impinged upon the mental equilibrium of Isaac Newton are many and virtually impossible to disentangle. No one can be certain—nearly three centuries after the event—as to which of them plunged him into the depths of psychic chaos. The following analysis, like others put forward by a wide range of scholars, is fashioned from deductions and arguments that at best attain to nothing better than probability.

One theory that became popular among Newton's contemporaries is that certain of his manuscripts were destroyed by fire, an experience so wrenching that he was unable to cope. Christiaan Huygens noted in his journal on May 29, 1694, that a Scot named Colm informed him that Newton had become deranged eighteen months previously, from either overwork or excessive grief at having his papers accidentally burned. His friends supposedly took him in hand, kept him secluded—"whether he liked it or not"—and nursed him back to the point where he could understand the *Principia* once more.[74] Abraham de la Pryme's diary account of a fire, which the undergraduate recorded in February 1693, seemed for a long time to lend credence to the story imparted to Huygens. However, we noted above that Newton suffered a considerable loss of papers by burning during the late 1670s, a tragedy attested to in his correspondence with Anthony Lucas, the English Jesuit. Moreover, Newton was engaged in a regular exchange of letters with Fatio and Bentley when de la Pryme made his notes. There is not the slightest hint of a catastrophe in their correspondence; in fact, Newton was pleased over Fatio's improving health and further buoyed by the possibility of their living together. And if, as Edleston has suggested, de la Pryme actually made this diary entry a year earlier, in February 1692, one searches the correspondence with Locke in vain for any mention of a conflagration. The eighteenth-century poet Thomas Maude added to the legend when he told of Newton's being called out of his study and foolishly leaving his dog alone with a burning candle. He supposedly returned to find that his pet had overturned the flame, which consumed irreplaceable parts of his lifework. The latest edition of *Bartlett's Familiar Quotations* still contains the heart-rending but spurious cry of discovery: "O Diamond! Diamond! thou little knowest the mischief done!"[75] Diamond has since been aptly described as a mythical beast, the creation, perhaps, of a bored but mischievously clever undergraduate and a far too credulous auditor.[76]

A quite recent and more enterprising theory holds that Newton slowly but surely poisoned himself by his constant alchemical experiments with heavy metals, most particularly mercury and lead. He had carried out hundreds of experiments over the past quarter-century, inhaling toxic vapors, tasting every type of concoction imaginable, and handling dangerous compounds of arsenic, lead, mercury, and antimony with his bare hands. What is more, the general symptoms of his illness, as he described them to Pepys, Millington, and Locke, appear consistent in the main with those associated with mercury and lead poisoning: insomnia, heightened irritability, loss of appetite, paranoia, and amnesia. Newton may have further exposed himself to toxic compounds in the form of homemade medicines, which he used with some regularity. Unable to test this theory by exhuming Newton's remains from his tomb in Westminister Abbey, the researchers P. E. Spargo and C. A. Pounds obtained samples of Newton's hair from locks in the possession of the Portsmouth family and Trinity College Library. Spectrographic analysis revealed the presence of unusually high concentrations of lead and mercury, leading them to conclude that Newton's "so-called 'nervous breakdown' of 1692/93 was due principally to poisoning by the metals which he used so frequently and with such cavalier disregard for his own safety in his chemical experiments."[77]

One need hardly observe that any attempt to reach a definitive medical diagnosis across a gulf of centuries is fraught with as many risks as psychoanalysis by historical fiat. To begin with, Newton's description of his malady is lacking in detail, nor did he himself see a cause-and-effect relationship between his illness and alchemical experimentation. On the contrary, he thought himself the victim of a "distemper" of "epidemical" proportions. While it is clear that the term meant many things to people of the seventeenth century, distemper was hardly associated with mercury or lead poisoning. This would have been especially true in Newton's case, for, as the most accomplished alchemist of his day, he knew full well the considerable risks associated with the tasting of metallic compounds and breathing their fumes in an unventilated room. His jest to Wickens that his hair had turned silver prematurely from working with mercury tells us that much. Knowing the hazards, against which he admittedly took few if any precautions, he would also have known the symptoms, many of which were carefully documented by Paracelsus in the sixteenth century. Among the most prominent is a tremor affecting the hands, a condition once endemic among felt hatters who used the metal in their trade, and immortalized in literature by the shy, stuttering Oxford mathematician Charles Lutwidge Dodgson, whose pen name was Lewis Carroll. One might reasonably expect Newton's handwriting to reflect such a condition at the time of his breakdown, but it does not. Ulceration of the gums and a loosening of the teeth are also symptoms common to mercury poisoning, yet Newton died, at the age of eighty-four, with all of his secondary teeth but one, rather remarkable for a man of his day. Numerous other pronounced symptoms received no mention, whether by Newton himself or by those who shared his company: jaundice, emaciation, flaking skin, darkened nails, premature aging, weight loss, and lethargy, let alone visual disturbances, convulsions, coma, and paralysis. Uremia attendant on kidney failure is one of the common consequences of mercury poisoning, but the record is no less mute on symptoms of this condition. Most important of all, perhaps, is Newton's exceedingly rapid and seemingly complete recovery at a time when alchemy continued to occupy him more than any other study. Should there not have been additional breakdowns, given his prolonged addiction to the fire and the crucible? The only other mental disturbance to be documented with certainty occurred in 1664, well before alchemy had entered his life, when he suffered from exhaustion and disorientation brought on by observing comets the night long. Thus, tempting though the poisoning thesis is, the medical and historical evidence is far from conclusive. (We do not even know when the hair containing the metal traces was cut from Newton's head or the conditions under which it has been preserved.) The most one can say after weighing the evidence is that metal poisoning may have played a part in bringing on an illness whose absolute cause will always remain open to question.

To have postulated a scientifically based theory in order to explain Isaac Newton's breakdown during a single year of his life is to ignore the many interlocking components and distinctive traits that make up the whole person. For example, despite much evidence to the contrary, two proponents of the poisoning theory have written that the breakdown of 1693 occurred during a period devoid "of any extraordinary pressures."[78] Even Spargo and Pounds, whose arguments for poisoning are more thoroughly researched and therefore more persuasive, overlook the fact that Newton was still laboring at a pace similar to that which accompanied the *Principia*'s composition; nor do they have anything to say concerning Newton's emotionally charged break with Fatio de Duillier or his continuing disappointment and frustration at not being honored with an influential civil post. These and certain related factors are deserving of closer scrutiny.

The Newton we have come to know was famous for the clarity and persuasiveness of his scientific reasoning, the profundity of the questions he addressed to nature. No conundrum seemed too knotty for him to untangle, no phenomenon too obscure to ponder. The cold blood of stolid yeomen, lawyers, and Puritan clergymen coursed through his veins; his logic was unsullied by human emotion. And then came Fatio, to whom he was drawn as to a sympathetic powder. Never in his adult life had he experienced such a broad and frightening range of emotions, rejoicing in anticipation of their every meeting, despairing when Fatio left for Europe and then, after his return, hesitated to join him at Cambridge permanently. When the young man became ill Newton also started to ail. He informed Pepys that his insomnia and poor appetite had lasted a year. Significantly, word of Fatio's sickness arrived in a letter dated November 17, 1692, while Newton's letter to Pepys announcing his own infirmity is dated September

13, 1693. And though it is only speculation, Newton's mysterious visit to London from where he wrote an equally strange letter to Locke could well have been the occasion of a final shattering rendezvous with Fatio. It has been suggested that the guilt Newton experienced over this ill-fated attachment was simply too great to be contained *ad infinitum*. Accusing Locke of attempting to embroil him in sexual improprieties was perhaps Newton's way of disguising a deeply troubled conscience, hiding the bitter truth from himself. Coveting Fatio, whom Locke allegedly used as bait in an unsuccessful attempt to lure Newton to Oates, caused him to sin in thought, which to an inveterate Puritan was tantamount to deed.[79] Newton eventually righted himself, but Fatio, who was neither as intellectually gifted nor as emotionally resilient as his idol, never fully recovered his equilibrium.[80]

Also of consequence was Newton's difficulty in making his existence known to the nonscientific world. He had succeeded in convincing himself that the position he sought at the Mint was an entitlement of which he was being wrongfully deprived, an extension of his conspiracy theory of history. To make matters worse, the wounded lion had groveled like a whimpering sizar before lesser mortals, like Monmouth, in order to obtain what was his by right of accomplishment. Others seemed in control of his destiny, an unacceptable violation of his obsession with absolute freedom of action. It grated even more when he thought of such members of his pride as Paget, Bentley, and Gregory, who had profited so handsomely by virtue of his favor. He may have also harbored a secret nostalgia for the surprisingly pleasant social life and recognition that was his while he served in the Convention Parliament. This major source of disappointment and irritation, which had surfaced repeatedly in the correspondence from 1690 onward, is most conspicuous in the anguished letters to Locke and Pepys. Equally revealing is the fact that Millington, who met with a recomposed Newton shortly thereafter, wrote Pepys that *"it was a sign of how much it [learning] is looked after, when such a person as Mr. Newton lies so neglected by those in power."*[81] Millington's decision to underscore this particular passage strongly suggests that Newton, who had forgotten much else, remained deeply aggrieved over what seemed to him a never ending wrong—which apparently had precious little to do with chronic exposure to toxic compounds and their vapors.

All of these factors were played out against a background of rentless intellectual activity. When no appointment was forthcoming, Newton sat down at his desk one last time in the hope of restoring his connection with some divine source, a proven cure for rejection and disappointment in the past. The last of his prodigious reserves were thrown into a final assault in an effort to achieve an even more comprehensive understanding of the natural order. His attitude toward unfinished business was revealed in a May 1694 letter to Nathaniel Hawes, Treasurer of Christ's Hospital in London: "A Vulgar Mechanick can practice what he has been taught or seen done, but if he is in error, he knows not how to find it out and correct it; . . . Whereas he that is able to reason nimbly and judiciously about figure, force, and motion, is never at rest till he gets over every rub."[82]

As always, there was an element in Newton's thought that overstepped the boundries of logic and rational analysis. He had ever looked on ancient alchemy and mathematics as a sign of the divine or quasi-divine powers of the human mind. God thundered and spoke in corpuscles, conic sections, and differentials; and Moses, Pythagoras, and Plato were privy somehow to that divine language. That such thoughts obsessed him at the time of his breakdown is beyond question. Fatio was amazed early in 1692, when Newton told him that the ancients knew the law of gravitation announced in the *Principia,* and the young Swiss hastened to inform Huygens of this supposedly privileged communication.[83] David Gregory, who spent several days with Newton in May 1694, recorded the substance of their conversations in the form of memoranda. Newton had recently completed his brilliant mathematical treatise *De quadratura,* which introduced the now familiar dot notations for fluxions, and he expressed the belief that its contents were known to the Greeks, who had destroyed all evidence of algebraic analysis in favor of more elegant geometrical proofs. Indeed, much of his time in the early 1690s was spent reinterpreting the rich legacy of classical geometry, most particularly the works of Apollonius and Euclid.[84] Gregory also noted that Newton believed his natural philosophy was most consistent with the teaching of Thales, while the Egyptian Thoth "was a believer in the Copernican system," a key element of the *prisca sapientia.*[85]

Many of the highly creative look with greater fondness on what they pursue than on what they possess, on what they intend to do than on what they have done. Newton had spent the past three decades of his life tracing the mazelike ridges in the fingerprint of God, only to discover a *cul-de-sac* at their center that barred any prospect of effecting the grand synthesis of his dreams. He had, in Stukeley's words, "unfolded the economy of the macrocosm," but the secret of the microcosm and the forces concealed by it still eluded him. Newton's work in alchemy, which suffered immeasurably from a dearth of lawgiving predecessors like Kepler and Galileo, reached its febrile climax, if not quite its end, at the very moment of his breakdown, when he metaphorically immolated himself on Vulcan's altar. Working at the ragged edge in the spring and summer 1693, he completed five chapters of a manuscript titled *Praxis,* his projected magnum opus on the multiplication of metals. For a brief while the secret of the ages seemed within his grasp, as it had some twenty years earlier when gold mixed with a purified mercury sprouted, swelled, and branched on being exposed to fire. But as before, disillusionment quickly set in, at first manifesting itself in canceled passages and finally in the complete abandonment of the work.[86] For the first time in his life Newton had to face up to something he had always known but never wanted fully to admit: that there exists in the world nothing physical to account for the world, nothing to account for the drive of elemental particles to become flesh and bone, nothing to account for the primordial balance between undulating sea and steadfast shore, hurtling planet and anchored star. Having forever banished certain of nature's mysteries, he had also exposed the imperfections as well as the limitations of mechanical philosophy, a science that had failed him for the first time, as men had so often done before. Sadly, the wonder of his achievement was permanently blemished. The lengthening shadows temporarily enveloped him, obliterating the line between dream and reality. In a crushing fall of arrogance, Isaac Newton lost the secret confidence of God, shattering his self-myth and stemming the wellsprings of his creative imagination. Gone forever was the visionary's near madness that kept him scribbling out equations and watching over crucibles with a religious determination so carefully hidden beneath a placid exterior. The mind was still intact, but the spirit was no longer willing. He who had attempted to rise above human art and industry longed to be shed of Cambridge lest someone chance to discover his terrible secret.

V

Having satisfied himself that Newton was in full possession of his faculties once again, Samuel Pepys decided to seal their reconciliation by requesting a favor of his friend in November. Thomas Neale, M.P., who was both Master of the Mint and Groom-Porter to the Royal Household, had consulted Pepys about the odds on winning a newly instituted lottery of Neale's design. The question was this: If a man is given a chance to roll dice, is he more likely to roll one six with six dice, two sixes with twelve, three sixes with eighteen, or will the odds in each instance be the same? The calculations in probability theory were not particularly difficult, and Newton assured Pepys that the odds clearly favored the player shooting for a single six. His solution seemed paradoxical to Pepys, however, and Newton patiently drafted another letter spelling it out in greater detail.[87] In all, he wrote Pepys three letters on the subject, where in more normal circumstances one would have sufficed. Newton was obviously anxious to dispel any lingering doubts Pepys may have had concerning the state of his mind. Perhaps he was also hopeful of making a favorable impression on Neale, whose operation at the Mint he had longed to join ever since 1691.

Unfortunately, word of Newton's rapid recovery spread less quickly than did certain wild rumors born of his temporary distraction. As already noted, Huygens heard that Newton's madness lasted eighteen months and that he had to be protected by his friends from doing injury to himself. "There is a man," he wrote to his brother Constantyn on May 24, 1694, "lost and so to speak dead for research, so I believe, which is deplorable."

Huygens supplied Leibniz with the same information and must have been surprised and considerably relieved to learn from the reply that his intelligence had been inaccurate. "I am very happy that I received information of the cure of Mr. Newton," Leibniz replied, "at the same time that I first heard of his illness, which, without doubt, must have been most alarming. It is to men like Newton and yourself, sir, that I desire a long life and good health, in preference to others, whose loss would not be great, speaking comparatively."[88] Still the gossip persisted, both at home and abroad, growing ever more grotesque in the retelling. In the spring of 1695 an incensed John Wallis received from Johann Sturm, a German virtuoso, an account of how Newton was "so disturbed in mind . . . as to be reduced to very ill circumstances."[89] Wallis wasted no time in drafting a forceful rebuttal, for Newton had suddenly become a national treasure whose reputation must be protected at all costs. Far worse, John Flamsteed, in one of his more classic examples of poor taste, wrote Newton in February 7, 1695: "The day after I received your Last Mr Hanway brought me News from London yt you were dead but I shewed him your letter which proved the contrary." What makes this story even more puzzling is the fact that John Hanway had recently graduated B.A. from Newton's own Trinity College. Hanway also told Flamsteed that the normally circumspect Christopher Wren was the source of his false information. When the Astronomer Royal visited London some two months later, he again heard the rumor of Newton's death and reported the news with the same indelicacy as before.[90] Perhaps such stories were the twisted outgrowth of a letter written to the Royal Society by Huygens in October 1694, urging that Newton be persuaded to publish a second edition of the *Principia* along with his other discoveries, "lest by his death they should happen to be lost."[91] In neither instance did Newton acknowledge Flamsteed's gaucherie, but one can safely assume that he did not receive reports of his own death with the same aplomb as Mark Twain.

The mask of optimism donned by Newton for the benefit of Gregory and his other disciples belied the painful legacy of the breakdown of 1693. He now realized that in establishing the laws of the universe he had reached the pinnacle of his genius, that the flood of revolutionary ideas had been permanently checked. One is reminded of Bertrand Russell's autobiographical account of his own struggle, between 1902 and 1910, to complete the *Principia Mathematica,* which he coauthored with Whitehead. "I persisted, and in the end the work was finished," Russell wrote, "but my intellect never quite recovered from the strain. I have ever been since definitely less capable of dealing with difficult abstractions than I was before. This is part, though by no means the whole of the reason for the change in the nature of my work."[92] Newton, too, felt the instinctive urge to undertake other labors in the face of a slow but accelerating decline in his ability to evolve new techniques of solution and to absorb fresh findings.[93] How prescient he had been as a young man when he observed: "The boyling blood of youth puts ye spirits upon too much motion or else causet too many spirits. But cold age makes ye brain either two dry to move roundly through or else is defective of Spirits."[94] Until called upon for other duty by his King, he could do little more than amend, reform, and restate principles that had been set forth in the past.

It was for such a purpose that he, along with Gregory, visited Flamsteed at the Royal Observatory on September 1, 1694. Newton was anxious to obtain accurate observations of the moon from the feisty Astronomer Royal in order to improve upon the dynamically based lunar theory sketched out in the *Principia*. According to Gregory's memorandum of the visit, Newton was in high spirits: "He believes that the theory of the Moon is within his grasp." To find its exact position he will need but "five or six equations."[95] Flamsteed provided his illustrious guest with some fifty lunar positions reduced to a synopsis and promised him one hundred more observations in the near future, the distillate of twenty years' hard labor. There were conditions, however, as Flamsteed later recalled in a memorandum of his own: "that what ever emandations of ye Theory he derived from them should be imparted to me before any other."[96] Neither was Newton to reveal the observations themselves to anyone, for Flamsteed planned to include them in his long-projected masterpiece, a catalog of the fixed stars.

Flamsteed, England's first Astronomer Royal, had been appointed to the position by Charles II in March 1675 at the age of twenty-eight. His father, Stephan, was a fairly prosperous maltster from Derby; his mother, Mary, the daughter of an ironmonger, died when John was but three years old. The youth attended the Derby free school until his fourteenth year, when a summer cold developed into a severe rheumatic affliction which eventually forced his withdrawal. He was unable to walk for several months, and his health remained ever delicate thereafter, a condition frequently attested to in his later correspondence. Left largely to his own devices, the adolescent began to instruct himself in the details of astronomical science. A turning point came late in 1669, when the young astronomer submitted a small work on lunar observations to the Royal Society. He soon became engaged in an extensive correspondence with Henry Oldenburg and John Collins, who in 1670 presented Flamsteed to Sir Jonas Moore, Master of the Royal Ordnance. Flamsteed emerged from the meeting with a micrometer and Moore's promise of some good telescope lenses, equipment which enabled him to initiate his serious observational work. Moore later persuaded Charles II to appoint Flamsteed his "astronomical observer" at a salary of £100 per annum. In June 1675 the King further decreed that a royal observatory be constructed according to the design of Sir Christopher Wren. The great architect chose a site atop a slope in Greenwich Park, and the building was hastily constructed at a cost of £520, a sum realized by the sale of old and decayed gunpowder. (Some 690 barrels of the stuff were sold to one Polycarpus Wharton for 40s. a barrel. Wharton, an enterprising soul, allegedly made it serviceable again and sold it back to the Government for £4 a barrel.) Flamsteed, reluctantly bowing to custom, cast the new observatory's horoscope after the laying of the foundation stone at 3:14 P.M. on August 10, 1675. Hardly known for his sense of humor, he nevertheless added a witty inscription to the document: *Risum teneatis amici* (May this keep you laughing, my friends.)[97] The Astronomer Royal occupied the new edifice the following July. He was accompanied by two assistants—Thomas Smith, an educated man whose wages Flamsteed paid out of his own meager salary, and Cuthbert Denton, a flunky described by his superior as a "silly, surly, labourer" from the Tower.

Flamsteed's financial problems were severely compounded by the fact that his new observatory was virtually devoid of scientific instruments, which Charles, for reasons unknown, failed to supply. Desperate for more income, the dour and sickly virtuoso, who had taken an M.A. at Cambridge by letters-patent, sought and obtained a church living at Burstow, an inconvenient 20 miles from the Royal Observatory. He was further obliged to accept private pupils, some 140 of whom trekked to Greenwich over the years. (In contrast Newton, who was rather handsomely rewarded to do nothing but teach, had only three such pupils, not one of whom graduated from Cambridge.) In spite of these multiple disadvantages Flamsteed achieved remarkable results. He gradually built up an excellent collection of instruments and had executed well over twenty thousand observations by the time Newton paid his call.

That Flamsteed stood in awe of his august visitor is beyond doubt. The Astronomer Royal drafted a second memorandum in February 1695. "I study not for present applause," he wrote. "Mr Ns approbation is more to me then the cry of all the Ignorant in ye world." But neither had Flamsteed's hatred of Halley, which he first expressed to Newton in 1692, abated to any degree. Indeed, this same memorandum indicated that it was growing and had begun to color, however subconsciously, Flamsteed's perception of the great man: "about E.H. That he [Newton] is very much mistaken in him. that I never found any thing so considerable in him [Halley] as his craft & forehead. his art of filching from other people & makeing their workes his owne as I could give instances. but that I am resolved to have nothing to doe with him for peace sake."[98]

The irony is that Halley and Flamsteed had labored together over many of the same scientific questions during the 1670s. Although Hooke noted that the two had "fallen out" as early as January 1676, the main trouble between them seems to have started over some tables of tides published by Flamsteed in 1683. Halley found his friend to be in error on several counts, and Flamsteed took offense when Halley later published his contradictory data in the *Philosophical Transactions*. Flamsteed's professional jealously was doubtless exacerbated by a clash of personalities. Halley's gregarious nature, cavalier attitude toward orthodox Christianity, and "hit-and-run" approach to astronomy could not but offend the puritanically single-minded Flamsteed, who did all he could to frustrate Halley's hope of becoming Savilian Professor. Flamsteed rarely praised any third party and passed up few opportunities to criticize. In January 1693 Hooke wrote in his *Diary* that Flamsteed "rails at Hally." Hooke, of course, disliked Halley as well, but his opinion of Flamsteed, whose difficult childhood and painful climb to a position of prominence had been quite similar to his own and Newton's, was none too high either. He alternately described Flamsteed as "a conceited cocks comb" and "Ignorant impudent Asse."[99] For his part, Flamsteed wrote that he was "much troubled" by Hooke. "I know not how to deal with him."[100] He would soon be thinking the same—and worse—of Isaac Newton.

Thus began an exchange of some forty-two letters between September 1694 and January 1696. Flamsteed, the first to write, reiterated his promise to provide Newton with any observations he required. "When Mr H. shews himself as Candid as other men I shall be as free to him as I was the first seven years of our Acquaintance. when I refused him nothing yt he desired." Until that day arrived, however, Newton must keep his promise never to impart either the observations or any deductions he might derive from them, except, of course to Flamsteed himself. Newton replied on October seventh. He had already compared several of Flamsteed's observations with his own preliminary calculations, "& now I have satisfied my self that by both together the Moons theory may be reduced to a good degree of exactness perhaps to ye exactness of two or three minutes."[101] Flamsteed, who wanted nothing so much as to be accepted by Newton as a collaborator and peer, was intoxicated by what he read. He underlined this passage and one other that "begged" fifty further observations. As regards Halley, Newton wrote not a word.

Flamsteed's reply of October 11 contained news of certain unexpected developments. Having recently been denied access to Flamsteed's observations by Newton, Halley, who also decided to try his luck with lunar theory, appealed directly to Flamsteed, only to be sharply rebuffed: "this I perceived was your aequation & told him so. he was silent." But Halley had persisted. Soon thereafter he arrived in Greenwich with an unidentified friend: "I was surprised at it & took the occasion of mindeing him of his disingenious behavior in several particulars which he bore because he could not excuse it." After venting his spleen, Flamsteed showed Halley his synopses and "suffered him to take a very few notes." Newton hastened to assure his suspicious friend that Halley was not behaving in a surreptitious manner. He had kept Newton fully appraised of his intentions, which were anything but dishonorable. Indeed, Newton was pleased to learn of the new dialogue, however strained, and expressed his hope that it would result in a renewal of an old friendship.[102] It was now Flamsteed who declined to comment on the subject.

Meanwhile, Newton's demands for additional data grew ever more insistent, while the Astronomer Royal, who did everything within reason to comply, was stretched to the limit by his many professional responsibilities. In November Newton complained that a number of "faulty observations" were hindering his progress. Flamsteed, who had been incapacitated for several days by a severe cold, patiently rechecked his calculations and admitted to some errors. But he also informed Newton that most of the data in question were quite accurate, the first indication that something was seriously amiss.[103] As previously noted, no phenomenon dealt with in the *Principia* was more intractable than lunar theory. While Newton had succeeded in demonstrating in general terms some of the perturbations introduced by the presence of a third attracting body, the moon's precise trajectory continued to baffle him. The actual path of the satellite, which is pulled simultaneously by the earth and the sun, has been likened to the course of a drunkard, who in attempting to follow a prescribed path staggers from side to side at a varying rate of speed. What he did not know is that a full analytic solution of the three-body problem is impossible. Never, in all the many years that he labored, had Newton once complained of being overwhelmed by the task at hand. Only later did he confide to John Machin that "his head never ached but with his studies on the moon."[104]

Far harder on himself than on anyone else, Newton kept up the pressure. He attributed more faulty observations to Flamsteed, even as he exhorted the astronomer to supply him with scores of additional sightings. He also made the mistake of promising to pay Flamsteed for his labors. It proved a devastating blow to the ego of a proud, self-made man, who wanted only to be treated as an equal. Flamsteed bristled, but instead of attacking Newton he again made Halley the convenient object of his aggression. "I never took any thing of any for communicateing of my skill or paines, except of those who forced themselves upon me to devour my time ... pray therefore lay by any prejudicall thoughts of me, which may have crept into you by malitious suggestions" of a "malitious false freind." Contrary to Newton's hopes, the rift between the two astronomers had not been healed. From that point forward Flamsteed never passed up an opportunity to express his contempt for Halley, who displayed his superior breeding and common sense by refusing to enter the fray. Flamsteed was still smarting when he next wrote to Newton on December 10, and his self-righteous remarks struck much closer to home: "But I am displeased wth you not a little for offering to gratifie me for my paines either you know me not so well as I hoped you did or you have suffered your self to be possest with that Character which ye malice & envy of a person from whom I have deserved much better things has endeavord to fix on me." Flamsteed wanted but one thing from Newton: "All the return I can allow or ever expected from such persons wth whom I corresponded is only to have ye result of their Studies imparted as freely as I afford them the effect of mine or my paines."[105] The younger Newton would have drafted an outraged reply and summarily broken off the tortured correspondence. As it was, he needed Flamsteed's help more than he had anyone's in the past. Newton chose conciliation instead, a rare posture for him. "What you say about my having a mean opinion of you is a great mistake," he replied on December 20. "I have defended you when there has been occasion but never gave way to any insinuations against you." He had offered Flamsteed money only to avoid the suspicion of being ungrateful: "if you please to let all this pass and concur with me in promoting Astronomy I'le concur with you."[106] And so the vexed exchange continued, lurching forward on a zigzag course like nothing so much as the perturbing moon.

Newton soon began to press his earlier charge that Flamsteed's dilatory ways were threatening the entire enterprise. Assuming a magisterial tone, he warned the Astronomer Royal that only if his observations were published in conjunction with an accomplished lunar theory would Flamsteed be assured of a place in history:

> But if you publish them wthout such a Theory to recommend them, they will only be thrown into ye heap of ye Observations of former Astronomers till somebody shall arise that by perfecting ye Theory of ye Moon shall discover your Observations to be exacter than the rest. But when that shall be God knows: I fear not in your life time if I should dye before tis done. For I find this Theory so very intricate & the Theory of Gravity so necessary to it, that I am satisfied it will never be perfected but by somebody who understands ye Theory of gravity as well or better than I do.

Two months later, on April 23, Newton again chided Flamsteed for keeping him from achieving his objective. He had become even more irritable, evidence that the difficulty of the work was exacting its toll:

> When I set my self wholy to calculations . . . I can endure them & go through them well enough. But when I am about other things (as at present) I can neither fix to them with patience nor do them wthout errors. Which makes me let the Moons Theory alone at present with a designe to set to it again & go through it at once when I have your materials. I reccon it will prove a work of about three or four months & when I have done it once I would have done with it forever.[107]

The magnificent powers of concentration were clearly on the wane; without them his keen mind was like a knife without a whetstone. It penetrated a little less deeply with each successive stroke.

The tension was heightened by Flamsteed's ill-advised reporting of rumors that Newton had died. Added to this was his incessant sniping at Newton over the recent appointment of a new Mathematical Master at Christ's Hospital in London. The position had been held for the past several years by Newton's disciple Edward Paget, whose unslakable taste for wine had finally eclipsed his passion for the instruction of adolescent boys, forcing his resignation. The leading candidates for the post were John

Caswell, Flamsteed's choice for Savilian Professor at Oxford in 1692, and Samuel Newton, who ran a mathematical school at Wapping. While Samuel was not related to Isaac Newton, the latter wrote a recommendation on behalf of the candidate, which tipped the balance in Samuel's favor. Flamsteed, the main suporter of Caswell, was more than mildly upset by this turn of events, particularly because Gregory, Newton's choice for the Savilian Chair, had prevailed over Caswell in 1692. Paget's dissipation poured salt on this old wound, for the puritanical Flamsteed had backed his appointment in 1682 on Newton's recommendation. Flamsteed made no attempt to disguise his bitterness. He took the opportunity to examine the new Master and found him wanting in Latin and, what is worse, basic mathematical skills, the fundamental tools of his trade. "The school will decay under his as much I fear as it did formerly under Dr. Woods Managemt." Newton retaliated: "As for ye late election," he wrote on April 25, "it belongs not to me to enquire what made ye governours so much against Mr Caswell; but now Mr Newton is in, ye best way is to make the best of it." [108] Flamsteed's grievances, however indelicately stated, were not without foundation, and of this Newton was soon made aware by a third party. The Governors of Christ's Hospital had second thoughts about their new Master's fitness, and within six months of the appointment their representative, Nathaniel Hawes, wrote Newton of their concerns. Newton admitted that "I was almost a stranger to him when I recommended him," but he preferred to think that the real problem lay with certain of Caswell's jealous friends, namely Flamsteed. [109] The incompetent Samuel Newton kept his post until 1709, a measure of the influence of his renowned if distant patron.

Exhausted of patience with Flamsteed, who, in attempting to play the mathematician, had taken to sending him erroneous calculations of the sun's and moon's positions rather than the raw observations he demanded, Newton fairly exploded late in June:

> I want not your calculations but your Observations only. . . . If you like this proposal, then pray send me first your Observations for the years 1692 & I will get them calculated & send you a copy of the calculated places. But if you like it not, then I desire you would propose some other practicable method of supplying me wth Observations, or else let me know plainly that I must be content to lose all the time & pains I have hitherto taken about the Moons Theory & about the Table of refractions. [110]

This proved the unkindest cut of all. Where Flamsteed sought in Newton a colleague, Newton merely wanted the services of a dependable and submissive observer, a pure technician, and a bloodless one at that. In self-justification the astronomer drafted a memorandum on the back of Newton's letter, including a precise listing of the many observations he had already supplied. It concluded with a simple declaration: "I contend it." [111]

The worst was yet to come. As he had done in the past, Flamsteed reminded his obstinate friend that reliable lunar observations would have to await the completion of his star catalog, for the moon's position must be measured against those of the fixed stars. Moreover, it was rumored about London that Richard Bentley, "who I know not," had been complaining that the second edition of the *Principia* would lack a lunar theory "because I do'nt impart my observations to you." Newton must set the cleric straight, "that I *shall furnish you to your satisfaction in yt particular.*" [112] Flamsteed's prickly nature aside, nothing but the looming prospect of Newton's own failure could justify his abusive rejoinder of July 9:

> After I had helped you where you stuck in ye three great works, that of the Theory of Jupiters Satellites, that of your Catalogue of the fixt stars & that of calculating the Moons places from Observations, & in all these things freely communicated to you what was perfect in it's kinds (so far as I could make it) & of more value then many Observations & what (in one of them) cost me above two months hard labour wch I should never have undertaken but upon your account, & wch I told you I undertook that I might have something to return you for the Observations you then gave me hopes of, & yet when I had done saw no prospect of obteining them or of getting your Synopses rectified, I despaired of compassing ye Moons Theory, & had thoughts of giving it over as a thing impracticable & occasionally told a friend so who then made me a visit. But now you offer me those Observations wch you made before ye year 1690 I thankfully accept of your offer & will get as many of them computed as are sufficient for my purpose. [113]

Poor Flamsteed could only do what he had done before. "I was ill all this Summer," he wrote plaintively on the back of the letter, "& could not furnish him as I had done formerly he mistook my illness for design & wrote this hasty artificiall unkind arrogant letter." [114] How right Flamsteed was—and how forgotten.

Faced with the unenviable choice between personal pride and professional ostracism, the Astronomer Royal, who was still suffering from savage headaches, hastened to forward more raw data on July 13. He wrote again five days later, and even Newton must have experienced some pangs of conscience over Flamsteed's total capitulation:

> I have just cause to complaine of the stile & expressions of your last letter, they are not freindly but that you may know me not to be of yt quarrelsome humor I am represented by ye Clerk of ye Society [Halley]. I shall wave all save this expression *that what you communicated to me was of more value yn many observations.* I grant it. as Wier is of more worth then ye gould from which twas drawne: I gathered the gould melted refined & presented it to you sometimes unasked. I hope you valew not my paines ye less because they became yours so easily. I allow you to valew your own as high as you please & require no other reward for what assistance I sometimes afford you, but that I may now & then see some of your workmanship. & if that be not ready when I desire it or if you think it not fit to favor me wth it I can easily be contented. [115]

While he was in no mood to apologize, Newton was temporarily mollified, and he drafted a conciliatory reply. He assured Flamsteed that he was acting to squelch the baseless rumor that the responsibility for any delay in perfecting the lunar theory lay at Flamsteed's door. The latter's health was suddenly of concern to him as well. Perhaps Flamsteed would welcome a cure for headaches advocated by Dr. John Batteley, chaplain to Archbishop Sancroft. It required that the head be bound tightly with a garter until numbed: "For thereby his head was cooled by retarding the circulation of the blood."[116] Newton did not mention what Machin later learned—that he, too, was experiencing headaches from his work on lunar theory—much less whether he had tested Batteley's novel but plausible remedy on himself. Nor did he give Flamsteed any credit when he informed Machin of their quarrel. "Machin told me," Conduitt wrote, "that Flamsted said 'Sr I. worked with the oar he had dug—to wch Sr I. said if Flamsteed dug the oar he had made the gold ring.'"[117]

As is indicated by the dwindling correspondence, Newton suddenly began to lose interest in the moon late in the summer of 1695. He wrote no letters to Flamsteed in August and only one in September, his last on the subject. He had just returned from Lincolnshire and was about to leave on another journey whose destination he chose not to specify. "I have not yet got any time to thinke of ye theory of the Moon nor shall have leisure for it this month or above: wch I thought fitt to give you notice of that you may not wonder at my silence." But wonder Flamsteed naturally did. He waited four months and then wrote Newton on January 11, 1696: "I have been told ... that you have finished ye Theory of ye Moon *on uncontestable principles*. . . . Pray let me know how far you have proceeded."[118] Newton did not even have the courtesy to answer, perhaps in part because he had not succeeded in constructing a personnally satisfying theory of the moon's path, although he eventually refined its two major perturbations and discovered several lesser ones. When these findings were formally incorporated into the *Principia*'s second edition of 1713 Flamsteed received not a word of credit, and, like Hooke, he suffered the additional indignity of having certain other references to his name expunged. There appears to have been no further correspondence between the two men for three more years. When they next faced off over the question of Flamsteed's observations, all pretense at civility was abandoned. This time neither would ask for nor give any quarter.

VI

Where Newton spent the last two weeks of September 1695 is a matter of conjecture, but all signs point to London and a secret meeting with the man he had denounced to Locke for being false to him. The Whigs had regained control of Parliament, and Charles Montague, now one of their leading lights, was at last in a position to act on behalf of his touchy friend. Appointed Chancellor of the Exchequer in April 1694, Montague was sworn as a member of the Privy Council the following month. Determined to remedy the alarming depreciation of the currency, which had been reduced to a fraction of its face value by the nefarious practices of coining and clipping, he led the successful fight in Parliament for recoinage and the preservation of the existing monetary standard. Newton had no sooner returned to Cambridge than rumors began to circulate that he had already been appointed to a post. On November 26, 1695, Wallis wrote Halley from Oxford: "We are told here . . . he is made Master of the Mint: which, if so, I do congratulate to him." That same month Narcissus Luttrell, a chronicler of political developments in London, noted: "The King hath been pleased to conferr the place of Master of the mint, vacant by the decease of the late master, upon Dr Newton, Mathematical Professor at Cambridge."[119] Luttrell's intelligence was only partly correct, or else he subsequently confused his dates, for Thomas Neale, the Master, still had nearly four years to live.

Approached in strictest confidence by either Montague or one of his representatives, Newton went about his business as if an impending change of status was the farthest thing from his mind. Sara, the wife of his half-brother Benjamin Smith, became seriously ill in October. The family, as it had in the past, turned to him for advice. He suggested a remedy, and by mid-November Benjamin reported that all was well. "The effects have pved very successfull; for the swelling is verry much abated, and the blackness quite gone."[120] Part of Newton's cure for the unspecified malady included a diet of "Sowes," (*Sowens*), which is made by steeping oat husks in water until sour, straining and boiling the mixture until it attains the consistency of a light pudding, and then serving it with milk. Though relieved, he was painfully reminded of his widowed stepsister Hannah, whose dying husband he had been in no condition to help because of his own infirmity in 1693. He chose this moment to send her £100 and purchased an annuity for a like amount to be divided among her three children upon his own death.[121]

In the meantime the languishing correspondence with Flamsteed was replaced by a revitalized exchange with Halley, their first since the *Principia* was being readied for publication. The two met early in August, and although we have no information as to what they discussed, their letters dealt almost exclusively with the delineation of elliptical orbits for certain comets. Halley was onto something important, but he required Newton's help: "I must entreat you to procure for me of Mr Flamsteed what he has observed of the Comett of 1682, . . . for I am more and more confirmed that we have seen that Comett now three times, since ye Yeare 1531, he will not deny it you, though I know he will me."[122] Newton complied, and Flamsteed's data permitted Halley to carry out the orbital calculations of the comet destined to etch his name in the memory of posterity, which calls it to consciousness once every seventy-five years. On a more secretive note, Newton was simultaneously conducting what turned out to be the last of his alchemical experiments. On Tuesday, March 3, 1696, an unnamed adept from London paid him a visit during which they exchanged views on the current state of the art.[123]

By then Newton himself had gotten wind of the gossip that he was about to depart Cambridge for the Mint. He appealed to Halley on March 14 to help him refute such loose talk: "And if the rumour of preferment for me in the Mint should hereafter . . . be revived, I pray you would endeavour to obviate it by acquainting your friends that I neither put in for any place in the mint or would meddle with Mr Hoar's [the Comptroller's] place were it offered me."[124] In the strictest sense, what Newton wrote was true. Montague had doubtless approached him, rather than the other way round, and the position in question was that of Warden, as opposed to Master or Comptroller. In any case, Halley was spared the inconvenience of taking up the gauntlet on Newton's behalf. Just five days later, on March 19, Montague wrote the letter Newton had been waiting to receive for upwards of five years. "I am very glad that at last I can give you a good proof of my friendship, and the esteem the King has of your merits," he began on a note mindful of their past differences. "[T]he King has promised Me to make Mr Newton Warden of the Mint, the Office is the most proper for you 'tis the Chief Officer in the Mint, 'tis worth five or six hundred pounds per an, and has not much bus'nesse to require more attendance then you may spare." Montague promised to smooth the way by handling the paperwork if Newton would only visit London at his earliest convenience, "that I may carry you to kiss the Kings hand."[125] Newton did not even pause to draft a formal letter of acceptance; March 23 found him on the road to London and his long-awaited audience with William III.

Chapter Fifteen

A Morality Play

Here beginneth a treatise how the high Father of heaven sendeth Death to summon every creature to come and give account of their lives in this world, and is in manner of a moral play.

Prologue to *Everyman*

I

John Dryden, whom the historian Lord Thomas Macaulay labeled "an illustrious renegade" for having converted to the Roman Catholic faith, had fallen on hard times. With the accession of William III the poet paid dearly for his apostasy by forfeiting his laureateship and his annual pension of £200. Now in his declining years, Dryden sought to relieve his straitened circumstances by translating the works of Juvenal and Virgil. The task proved more tedious than he had imagined, and the financial rewards illusive. Jacob Tonson, his friend and publisher, was slow in paying. When he did pay it was in gold, a volatile currency which Dryden refused to accept. "I expect fifty pounds, in good silver," he wrote testily on October 29, 1695. "I am not obligd to take gold, neither will I." Tonson instructed his agent to pay Dryden in silver coin, but the poet expressed his intense displeasure once again: "none of the money [all shillings and sixpences] will go; for which reason I have sent it all back." Through the rampant practices of counterfeiting and clipping, false and much-depreciated coin had been substituted for the "good silver" Dryden demanded. No banker or tradesman would accept Tonson's money without discounting it by upwards of half its face value. It was with great reluctance that Dryden ultimately agreed to accept the gold guineas first tendered by his publisher,

375

but not without leveling a final blast of indignation: "Upon triall I find all of your trade are sharpers & you not more than others: therefore I have not wholly left you."[1]

The financial and monetary tribulations of John Dryden mirrored those of the nation at large, a fact he grudgingly acknowledged by not dismissing Tonson altogether. Biting references to the state of the currency also appeared in the correspondence and works of his fellow writers. The dramatist and future poet laureate Colley Cibber, who, like Newton, received his grammar school education at Grantham, supplied one of his characters, a witty young gentleman, with this line: "Virtue is as much debased as our money; and, faith, *Dei Gratia* is as hard to be found in a girl of sixteen as round the brim of an old shilling." Sir William Blackmore, a lackluster poet and court physician to William III, offered an even more sweeping indictment: Comparing libertine Restoration literature and drama to the country's debased coin, he argued that both should be thrown into the melting pot and restamped.[2]

The stereotype of the watery-eyed crone who tests every shilling she touches by running a gnarled finger round its edge, then clamps down on it with the blackened stumps that were once her teeth, has a sound basis in historical fact. Almost nothing could be purchased in late-seventeenth-century England without a squabble, or worse. "On a fair day or a market day the clamours, the reproaches, the taunts, the curses, were incessant: and it was well if no booth were overturned and no head broken."[3] Every buyer insisted on valuing his coins by denomination, every seller by weight. No entrepreneur would sign a contract for the delivery of his wares without first spelling out the quality as well as the quantity of the money he was to receive. As the size and consequent value of silver coins shrank, the price of virtually every commodity rose accordingly. By 1695 a shilling would go only as far as a sixpence had a decade before; little more than a loaf or two of rye bread could be purchased with the same currency that had once secured food enough for a family. Frustrations born of daily wrangling often culminated in weekly riots: On Saturday evenings employers normally paid for six days of hard labor in adulterated coin. Clipped shillings were bad enough, but counterfeit replicas of brass and other metals were also beginning to flood the market place. Tonson's attempt to discharge his debt by paying Dryden in clipped money had been preceded by the equally unsuccessful offer of forty brass shillings, and this between long-time friends. Such treatment of England's greatest living writer gives a hint of how desperate conditions must have been for the powerless masses.

The clipper's trade, which demanded more perseverance and daring than native skill, was an ancient if less than noble calling. It first became a serious threat to the monetary system of England during Plantaganet rule. In 1290 Edward I, dubbed "the English Justinian" by nineteenth-century historians, leveled the harshest of penalties, hanging, against those convicted of the crime. For good measure he expelled the Jews, who in the public mind were guilty to a man of currency manipulation and debasement, with the warning "never to return," which they did not until the mid-1700s. But, as soon became apparent, no social class or religion held a monopoly on clipping. Any individual so inclined could, with a pair of shears or a file, trim off or shave away small quantitites of silver from the edge of a coin. He than camouflaged his handiwork by rubbing the newly exposed metal with dirt or some other dulling agent before returning the debased money to circulation, where it was inevitably subjected to further mutilation by others.

This specialized form of criminal activity could thrive only so long as the old methods of coining by hand and hammer continued, which was the case in England until the end of 1662. A specific number of blanks, supposedly as uniform in size and weight as possible, were cut from a pound of bullion. These were flattened and worked into an approximately circular shape on an anvil. Each rounded blank was sandwiched between a die and a puncheon, which was struck with several hammer blows so that the double impression would become distinct. Excess metal was filed from the edges, and the freshly minted coin was ready for circulation. Unfortunately, the workmanship was often slipshod, as would be expected of such a repetitious exercise. It frequently happened that the stamp was laid on off-center, so that the impression ran off one side of the coin and left the opposite side blank. Until the reign of James I no fixed standard weight applied for each coin. What mattered was that the requisite number of coins be cut from a pound of metal, regardless of variations in the weight of each piece. A pamphlet published during the Commonwealth accused the moneyers of violating the new standard by producing shillings ranging from half to one and half times the proper weight.[4] Moreover, the moneyers were permitted a margin of error inappropriately called "the remedy," which was set at two pennyweights per pound of bullion, an allowance generous enough to discourage all but the crudest attempts at quality control.[5] Since every coin was irregular in shape to some extent and without a milled edge, the clipper and counterfeiter flourished. The greater part of the old hand-hewn coins, which comprised ninety-five percent or more of all money in circulation on the eve of Newton's departure for the Mint, were either lightweights or fakes forged from lead, copper, tin, and brass.

The French had long since devised a proven technology with which to combat these twin evils, but the English were slow in giving their approval to the French innovation. Pierre Blondeau, a gifted experimenter whose patron was none other than Cardinal Richelieu, the Machiavellian Chief Minister of Louis XIII, had perfected a relatively cheap and utilitarian method of producing coins by machine. The French Mint first employed the new technique on Christmas Eve, 1639, and in 1645 its proven superiority resulted in the abolition of the old hand method. In addition to its greater speed and efficiency, the primary advantage of machine production was the appearance on every coin of a firm, ridged edge. This technique, called milling, stopped the clipper in his tracks, for he could only deface an edged coin by paring away its rim, thus rendering it useless in the market place. As one observer wryly noted, milling "saved the life of many a criminal."[6]

In 1649 Blondeau crossed the Channel and laid his process before a committee of Parliament, which recommended its adoption by the Council of State. But the hackles of the moneyers, a powerful guild whose livelihood appeared threatened, were raised against the Frenchman, and Blondeau, after exchanging virulent pamphlets with his detractors, returned to Paris in disgust. Twelve years of embittered wrangling and further currency debasement followed before Charles II, on May 17, 1661, ordered that clipping and counterfeiting be stopped by the mechanical production of coins with grained and lettered edges. Blondeau was sent for in November, as were the Roettiers, a family of engravers with whom the King had become acquainted during his long exile. Tower Mint, an institution founded in the Middle Ages, was expanded, and the full-scale production of milled money began in December 1662.

Mechanization, as applied to any age, the late seventeenth-century included, is a relative term. Thus the recoinage in which Newton was about to become intimately involved still demanded a prodigious outlay of labor, both human and animal. The process began not at the Tower but at the Treasury, where gold, the most precious but least plentiful component of England's bimetallic system, was melted in small pots of earthenware over a charcoal fire, and silver in giant crucibles of iron which took a charge of a thousand pounds. After a few hours, the liquid bullion was ladled off into sand molds, where it congealed into small bars or "journeys," scarcely thicker than the coin to be cut from them. The moneyers, who were stationed at the Tower, then took possession of the bullion, which they passed three times between a pair of horizontal iron cylinders. These rollers were worked by gears attached to a capstan in the cellar below. Four horses provided the power required to squeeze down the gold and silver bars to the thickness of a coin, after which human muscle took over. Round blanks of fairly uniform size were next cut from the rolled ingots with a punch operated by two laborers. Since the punch was sharply angled to increase the force of each blow, the warped blanks had to be placed in the column of a drop press for straightening. The leftover scissel, almost equal in weight to the blanks themselves, was collected, weighed, and returned to the Treasury for remelting. Each blank was also weighed, and those containing too much silver or gold were filed across the face; lightweights were returned to the melters with the scissel.

The prepared disks then went to Blondeau, who, for a fee of £100 per annum and a handsome royalty, personally supervised the edging of every coin until his death. Though his machine had been described in print and was commonly shown to mint visitors in both Paris and London, the inventor insisted that every person who had regular access to it take an oath never to reveal the details of its operation, Newton being no exception. Worked by means of a handcrank, the ingenious devise forced the blank into a narrow space whose engraved metal edges impressed their design on its rim, lines of graining for small coins, an inscription for larger ones: *Deus et Tutamen* (Ornamentation and Safeguard). After milling, the blanks were reweighed and, if necessary, either filed once more or sent back to the cauldron. Those judged good enough were passed to the coining room for stamping. Though much larger, the coining presses resembled those used to cut the blanks from rolled bars. Each press had two or sometimes four horizontal arms weighted at the tips by up to three hundred pounds of lead. Ropes were attached to the ends of the arms and hauled on with as much force as possible by sweating laborers. This violent motion sent the die plunging against the blank inserted by a moneyer seated in a pit at the base of the press. He flicked the newly minted coin away with his middle finger and set a fresh blank with his index finger and thumb. Part of the price of abolishing the clipper was the clipped flesh of men, for it was a rare moneyer indeed who had not sacrificed one or more of his finger joints to the cause. (How many other workmen were accidentally brained by wandering absent-mindedly into one of the whirling, head-high leads is anyone's guess.) Newton, who delighted in reducing every possible operation to mathematical rule, calculated that new coins could be stamped at the rate of thirty per minute, or one every two seconds. The effort was exhausting, however, for the machines, which operated on a recoil principle, could not be powered by circling horses as the rolling presses were. A crew of seven worked each four-man press, and though each man was relieved about every twenty minutes, they remained at the arduous task only five hours in a normal working day of ten.[7]

When Dryden demanded good silver from Tonson, he was naturally referring to that coined by the new method, and therein lay the rub. The Mint turned out the finest milled pieces, of the best design available, only to see them disappear from circulation virtually on issue. Scholars generally agree that Sir Thomas Gresham did not author the economic law that bears his name, but the monetary policy of Restoration England stands as a perfect illustration of the principle that "bad money drives out good." The government had blundered in not calling in the old debased coins after 1662, so the new money speedily disappeared into the melting pots of the goldsmiths. They shipped their bullion to Holland and the East, where it commanded a higher price than that set by the Treasury. So widespread was this practice that the Privy Council ordered masters of packet boats to search their passengers' luggage for precious metals.[8] Hopton Haynes, a close associate of Newton at the Mint, estimated that between 1672 and

1696 the staggering sum of two million pounds had been lost in the illicit bullion trade, while William Lowndes, Secretary to the Treasury, claimed that only one circulating coin in two hundred was a product of machine coinage.[9] Conditions being what they were, why would anybody pay for goods with a new coin, containing a superior amount of silver, when an old lightweight would suffice? Except for the half million or so hammered coins produced during the first three years of Charles II's reign, all more or less clipped, most of the others—all badly mutilated—had been struck under James I, Charles I, and Cromwell. A fair number of ragged shillings and sixpences issued by Elizabeth also remained in circulation, as did a scattering of various denominations from the reign of Edward VI. By 1695 it was apparent that the old coins were not only growing scarcer in number but were eroding at an exponential rate in the garrets and basements of the Soho clippers and in the festering, slum-ridden warrens beyond. Meanwhile, unscrupulous moneyers continued copying the clipped shillings and issuing them, at a handsome profit, to poor people desperate for tender, legal or not. Dudley North, brother of John North, Trinity's addled Master from 1677 to 1680, called the situation ludicrous, "a perpetual Motion found out, whereby to Melt and Coyn without ceasing, and so to feed Goldsmiths and Coyners at the Publick Charge."[10]

The inevitable reckoning might have been put off a few more years by muddling through had the economy been spared the additional drain and runaway inflation brought on by a full-blown Continental war. In an effort to thwart the predatory foreign policy of Louis XIV, King William had taken an army to the Spanish Netherlands in 1691 and was constantly involved in campaigning until the conclusion of an uneasy peace by the Treaty of Ryswick in 1697. The national debt skyrocketed; a wave of fear and speculation similar, though less severe in degree, to the one that preceded the debacle of the South Sea Bubble in 1720 saw the guinea soar from 22s. in November 1694 to a peak of 30s. in June 1695. A troy ounce of silver bullion that yielded 62s. in coin rose to 75s. in the market during this same period.[11] The measures taken by the government to transfuse the hemorrhaging economy presented an unusual mixture of the conservative and the novel: Taxation was increased to the point of provoking civil unrest; loans were secured from anyone who would lend; and a tontine and a lottery of a million pounds each were authorized by Parliament at Montague's insistence. Finally, with Montague again in the lead, the government founded the Bank of England, which institutionalized the national debt. Still, bankruptcy loomed. It was widely held that an economic collapse would almost certainly result in another Stuart restoration, bringing with it a betrayal of the hard-won principles of constitutional monarchy. Further reform was deemed imperative, and in 1695 the government turned to certain of its leading citizens for counsel.

Discussion centered on the reforms proposed by William Lowndes in a

Report containing an Essay for the Amendment of the Silver Coins (1695).
Lowndes had entered the Treasury in 1675 at the age of twenty-seven and
was promoted to Secretary in April 1695. A respected and dedicated public
servant, he also represented the borough of Seaford in Parliament. The
new Secretary favored a recall of the debased silver coins with redemption
at face value. The cost, estimated at a million and a half pounds, would
probably have to be shouldered by the Treasury, although he hoped that it
could be divided somehow between the Treasury and the last holder of
each coin. New coins struck from the reminted silver would weigh the same
as the old, but the value of each denomination would be increased by 25
percent, thus bringing the price of English silver into line with that in the
European nations, particularly France and Spain. Such a reform would
have the added salutary effect of ridding the country once and for all of clippers and melters of coin. Indeed, Lowndes did not believe that recoinage
among these lines would harm anyone, at least in comparison to the
damage already done.

The chief critic of Lowndes's strategy was John Locke, who immediately published a rebuttal in a pamphlet titled *Further Considerations concerning Raising the Value of Money*. Locke, more astute as a political
theorist than as a monetary analyst, opposed recoinage on the grounds that
it was both unnecessary and much too costly for a country in the midst of a
great war. He recommended instead that the worn and clipped coins be
kept in circulation, but that on a given day they be reduced in value to a
level commensurate with their silver content, the loss—roughly 50
percent—to be absorbed by the holder. New silver coins should continue at
the weight, fineness, and value of the old standard. Newton, who was also
asked to comment, came down on the side of Lowndes. Like the Secretary
of the Treasury, he favored bringing the face value of silver coin into line
with the market price of bullion by increasing the value of milled currency
already in circulation. But he also opted for a 20 percent reduction in the
weight of silver coins issued in the future. He realized that prices would
tend to rise in the wake of this devaluation, but "Parliament may perhaps
think itself concerned in Honour to take some care" to protect the holders
of government annuities, of which he just happened to be one. Inflation
must be assaulted by the establishment of a Price Control Board for the
duration of King William's War, the name given by historians to the grinding European conflict. Lastly, he supported the gradual withdrawal of
clipped coins at the expense of the Treasury, beginning with those of the
oldest reigns and lowest denominations.[12] In the meantime Parliament,
which seemed marvelously immune to these recondite proposals, continued its pursuit of more facile solutions. It enacted a law granting £40 to
every person who informed on a clipper, while every clipper who informed
on two of his fellow criminals was entitled to a pardon, and every person
found in possession of silver filing or parings was to be branded on the

cheek with a red-hot iron. As Newton was about to discover in his new capacity as Warden of the Mint, many a silver-tongued rogue would have testified against the King himself for a guinea, let alone the irresistible sum of £40, while the guilty went free in droves for the paltry moral fee of bearing false witness. As for branding on the cheek, what did it matter to one who was subject to swinging from the gibbet for the conduct it was supposed to deter.

No professional economist himself, Montague finally persuaded a sluggish and divided Parliament to enact a series of measures based on the ideas of both Lowndes and Locke. It accepted Lowndes's view that the old coins should be immediately recalled and melted down into ingots for reminting. Locke, on the other hand, won his point that the new silver coins should be of the same weight and fineness as those struck by machine since 1662. It was also decided that the Treasury would bear the cost. Holders of hand-struck silver coins were given a grace period during which old money would be exchanged for new. Conterfeit coins made from silver filings in combination with lead or some other nonprecious metal were also exchangeable, but those of copper and brass were not. To help defray the cost, a new tax was enacted. The old hearth money tax, which required the assessor to enter almost every dwelling in the land, often at the risk of bodily harm, gave way to the less objectional window tax. Unlike hearths, windows could easily be counted from the outside, and the wealthy were certain to pay more than in the past. The only effective means of evasion involved the blocking-up of openings, which is precisely what occurred at Woolsthorpe. Whether Newton himself ordered this done to avoid paying part of the tax that helped support him at the Mint is not known. But whoever did so inadvertently preserved the etchings left by a gifted child on a stone windowsill in the very room where he was born, and which resemble nothing so much as the crude sundials he later carved on the outside walls of his mother's country home.

II

The legislation had already been passed and the Mint had just shifted into high gear when Newton was called to London by Montague in March 1696. Many years later the former Whig leader observed that Newton proved indispensable in carrying out the recoinage,[13] but at the time of his friend's appointment Montague clearly harbored no such thoughts. The office of Warden, like so many of those in the government, was a sinecure, pure and simple. With few exceptions, Newton's predecessors had left the day-to-day operations at the Mint to their salaried assistants while they pursued other interests, chiefly of an economic nature. Some did not even live in the city, setting foot in the Tower only a few times a month, and almost never in the summer, which God made for relaxation in the country. Montague expected little if any more from Newton. While he had exaggerated the income—"'tis worth five or six hundred pounds per An" (in truth it paid only four hundred)—he was sincere when he wrote that it "has not too much bus'ness to require more attendance than you may spare." In other words, Newton remained free to continue the full-time pursuit of his intellectual interests. Indeed no one, least of Montague, who was then serving as President of the Royal Society, would have expected him to quit the problems that were commanding the attention of the scientific world. Whether his benefactor even thought Newton should give up his residence in Cambridge seems a matter of conjecture.

G. K. Chesterton might well have had Newton in mind when he wrote: "The man who makes a vow makes an appointment with himself at some distant time or place." Many of those who do so live to curse their failure, yet Newton had kept his private pledge to attain the recognition usually denied a lowborn yeoman's son. While no man can vouch for another man's dreams, the journey to London and a royal audience was almost surely tinged with a certain bittersweet nostalgia as the accumulated images of the past thirty-five years drifted through his mind. He had gone up to Cambridge as a servant to those of rank and privilege and was at last departing its bucolic precincts as the most revered and learned natural philosopher in the world, a man now destined to serve his King, country, and God in a very different capacity. But Isaac Newton was also in flight for having placed a greater burden on science than it could reasonably bear, to say nothing of himself. He had failed in his attempt to prove the grand synthesis of the age: that a subtle and elastic ether or spirit is the causative agent for phenomena ranging from gravity, magnetism, and heat to electricity, optics, and the movement of muscles via the nerves of the body. With so much at stake Newton never looked upon alchemy, physics, and mathematics as discrete subjects to be pursued within finite limits of time; rather, they consituted the core of his existence—and his religion. A post in the bureaucracy might seem a woefully poor substitute for a thinker of his stature, but it was the best that was likely to come his way and at the same time fulfill his deep psychological needs. If not a challenge as described by Montague, Newton would make it a challenge all the same. The Mint would be no sinecure. If for some unforeseen reason things did not work out as he hoped, the Lucasian Professorship remained his to command.

Newton left for London on March 23, and the warrant for his appointment was drafted two days later.[14] He soon returned to Cambridge, where he set about packing trunks with his belongings and papers. Such was his haste that he removed only the most portable of his possessions, for, as previously noted, the chambers were shown to visitors long after his death with the boast that every relic was preserved to the minutest particular. April 20 saw Newton sign Trinity's Exit and Redit Book for the last time.

He climbed aboard the London coach and watched as the spires of the small fenlands town merged gently with the horizon. He would rarely return and only when it suited his purposes. Judging from the dearth of correspondence, the few friends he had made there were soon forgotten. Isaac Newton had at last passed through what he had so aptly called "the prime of my age for invention."

The London to which Newton came shared with Paris the distinction of being one of Europe's two most populous urban centers. The city and its immediate environs was home to some 750,000 souls, or about one-tenth of the English populace. Like Paris, London was not only the capital of a nation and seat of government, but a center of art and culture as well. And as a hub of commerce London, which in 1696 lay mainly on the north bank of the Thames, trailed only Amsterdam, which it was about to surpass. Stretching from Tower Hill to the Houses of Parliament in Westminster, the city grew up along a great boulevard connected by London Bridge, the only span across the river until the completion of another in 1750. The Thames, some 750 feet across and forever crowded with hundreds of ships and even greater numbers of smaller craft, provided a quicker and safer means of transportation than did the city streets, as the writings of Pepys, Evelyn, and Boswell so graphically attest. But in the late autumn and winter piercing mists and impenetrable fogs rolled off the polluted waters, mixed with the hovering smoke of countless sooty chimneys, and engulfed the streets in a brown toxic cloud, keeping all but the foolhardy and hard-pressed indoors.

More an emporium than a center of production, London saw its work done mainly by a rough-and-tumble population of cockneys—watermen, porters, dockmen, and day laborers—not to mention a more than ample representation of beggars, prostitutes, and professional criminals. This latter element lived largely unpoliced in the Liberties beyond the old city walls, where the powers of the Lord Mayor did not extend. They inhabited warrens of unsanitary hovels facing dark and dangerous passages choked with garbage and human excrement, pitched with impunity from the glass-less windows above. Here violence walked hand in hand with the crudest of pleasures. The legal age of consent for a female was twelve and remained so until late in the nineteenth century. Public floggings did little to discourage criminal conduct, and scarcely more did the frequent executions that drew vast crowds, proof that severity of punishment is no deterrent by itself. Most favored of all was the spectacle of hanging. On "Hanging Day" the condemned were borne the 2 miles from Newgate Prison to Tyburn, the place of execution on the northeast corner of Hyde Park. Laborers and shopkeepers temporarily abandoned their livelihoods to gawk and jeer at the passing victims, while the privileged paid handsomely to secure an unobstructed vantage point. The gallows appears to have been a permanent structure resting on three posts, hence the name "Tyburn's triple tree." Wooden galleries were erected nearby for the comfort of the eager spectators. John (Jack) Ketch, the infamous executioner of the day, also beheaded those convicted of state crimes at the Tower. The most ghastly form of death was exacted for treason. The condemned was hanged until nearly dead from strangulation, then taken down, disemboweled while still breathing, beheaded, and chopped into quarters for good measure. Death on the gallows was the penalty for coiners, and Warden of the Mint Newton provided the evidence required to accomplish this end on more than one occasion. Forgery was considered treasonable under the law, but those convicted of it were spared the agony of being drawn and quartered.

London also had a large middle class of respectable shopkeepers and artisans, the latter mostly engaged in the finishing trades. At the apex of this cosmopolitan society stood the wealthy merchants, bankers, and political officials who inhabited the city proper. While no one was wholly immune to the ubiquitous congestion and squalor, the middle and upper classes also knew a city where beauty, grace, and culture thrived. Christopher Wren designed the handsome buildings that occupied the sites left charred by the Great Fire a generation before. His nearly three score parish churches, adorned by elegant spires, gave London a unique and breathtaking skyline, especially when viewed from the Thames. Most dominant of all was St. Paul's Cathedral. On December 2, 1697, an official day of thanksgiving for the peace (temporary, as it turned out) secured by the Treaty of Ryswick, the choir was opened for services. Wren's son laid the highest and final stone on top of the lantern surmounting the great dome in 1710. Though the architect lived to see his masterpiece completed, at seventy-eight he was in no condition to make the long, exhausting climb to the heights above his beloved London. In addition to the churches, thousands of new homes had been erected, along with forty-four company halls and numerous public buildings such as the new Custom House. And when, at the end of the day, a gentleman sought good food, drink, and conversation, he could retire to any one of dozens of coffeehouses. Many became known for the specific interests they nurtured: literature, natural philosophy, politics, religion, agriculture, or shipping. In Lloyd's, for example, the concept of marine insurance was born; Hooke discussed astronomy with Flamsteed at Mus's, while at Will's, an establishment where writers sharpened their wits at each other's expense, Addison had his chair next to the fire in winter and near the sunniest window in summer.

London possessed no more forbidding structure, nor any more steeped in historical symbolism, than the massive limestone fortress located atop rising ground on the Thames's north bank. Viewed from without, the moat, still flooded in Newton's day, is bordered within by a soaring crenelated wall, broken by huge flanking towers located at intervals along its rim. Once inside this outer line of defense, the visitor immediately encounters a second tower-encrusted wall of similar construction, but of even greater height. Inside this are the several barracks, armories, and other buildings dating from the various reigns of English history. At the center of the sprawling 13-acre complex stands the lofty keep or donjon known as the White Tower, whose construction was commissioned by Gundulf, Bishop of Rochester, in the time of William the Conqueror. Its solid masonry walls, said to have been restored by Wren in the decade before his death, are 16 feet thick at the base. White Tower, the most famous component of the Tower of London, served as the court of Plantagenet kings. Flamsteed, when first appointed Astronomer Royal, made his observations from its ramparts until Greenwich Observatory was ready for use. English monarchs of post-Plantagenet times made other parts of the great fortress their seasonal home, a custom finally abandoned by James I, founder of the Stuart dynasty.

As a serious student of the past Newton, had the choice been his to make, could not have selected a more historic site at which to perform his daily labors. On May 2, less than a fortnight after arriving in London, he came to the Tower to take the pledge of secrecy required of every Mint official: "You shall swear that you will not reveal or discover to any person or persons whatsoever the new Invention of Rounding the money & making the edges of them with letters or graining or either of them directly or indirectly. So help you God."[15] Perhaps for the first time he also walked the entire length of the congested and foul quadrilateral street running between the Tower's inner and outer walls, for it was on this avenue that the Mint was located. As the Tower's precincts became increasingly familiar to him, their richly storied past must have come to mind with an immediacy unavailable from the printed page. Kings had been conceived and nurtured within these mighty walls. Elizabeth, while still a princess, had entered via Traitor's Gate and wept against their stones; her father's luckless wives and loyal servants had suffered far worse. Ann Boleyn, Catherine Howard, Thomas Cromwell, and Sir Thomas More all breathed their last on Tower Hill or Tower Green after suffering miserable confinements, the last, a future saint, admonishing the headsman: "Pluck up thy spirits, man, and be not afraid to do thy office. I am sorry my neck is short, therefore strike not awry." Though more highly principled than her lusting father, Elizabeth did not shrink from her duty as she saw it. The Queen's beloved Essex met his God on Tower Green, while Raleigh endured the first of three long imprisonments in the Tower at her hands.

If one simply wished to contemplate the distant sights, the view afforded from atop the Tower walls was guaranteed to transfix any but the most insensitive observer. Below lay the Pool of London, the capital's major port area before the construction of docks farther downstream in the nineteenth century. Depending upon the weather and time of year, as many as two thousand masted ships, the pride of a great merchant fleet, rocked gently at their moorings.

The hermit of Cambridge was in need of a new hermitage, and it is possible that the Tower provided him with one, at least temporarily. As an officer of the Mint, Newton was entitled to the perquisite of a house worth £40 a year, in addition to free stationery and coal. The dwelling assigned him stood with its back against the northeast corner of the Tower's outer wall; its front door opened onto the street housing the Mint. Across this narrow space loomed the inner wall, Brick Tower on the right and Jewel Tower on the left. The garden that was also his lay across the street next to the inner wall, at the base of Jewel Tower. Only a master gardener could have made it bloom, however, for the giant barriers that ran along its front and back prevented the sun's light from reaching the ground for much of the day. To make matters worse, work at the Mint was conducted in two shifts, the first beginning at four in the morning, the second ending at midnight. Even when the din of official activity ceased, drunken guards and employees, who often came to blows, could be counted upon to shatter the peace of what little remained of the night, for they too were housed in the Tower at the taxpayers' expense. Smoke from the huge, seldom banked forges hung in the dead air, which was kept from circulating by the Tower walls, while rivulets formed by condensation on the ramparts trickled down onto the ramshackle wooden buildings below, giving an even keener edge to the perpetual chill within. Pollution of another kind also pervaded the dank atmosphere. The waste generated by the Mint's many horses and three hundred workers not only constituted a health hazard, it raised a constant and barely tolerable stench. According to the Mint papers in the Public Records Office, the cost of hauling manure away during the recoinage was almost £700.[16] These conditions would hardly have appealed to one who in 1691 had refused Locke's offer of the Mastership of the Charterhouse partly because it entailed "a confinement to ye London air." Following a popular precedent, Newton soon obtained a tenant for the Warden's house or perhaps let it to his deputy free of charge, and made other living arrangements for himself. He is said to have resided briefly in the nearby Minorities at Haydon Square. By August he had purchased a home of his own in Jermyn Street in fashionable Westminster, where he passed the next decade. The residence, which no longer stands, was a good half-hour from the Mint, but it was close to Parliament, a strong indication of Newton's resolve to gain reelection as a representative of the University. He also spent much of his time at the nearby Treasury in Whitehall, where the old coins were received and melted down into bullion. Perhaps it was during this period that he acquired the famous sedan chair in which he rode through the city streets, an arm protruding from either side, after he had grown thick at the middle from a lifetime of sedentary ways. Whatever Newton's means of conveyance, his kaleidoscopic world had undergone a sudden shift, drastically reordering the human pageant on which he so somberly gazed.

III

Considering the magnitude of the change wrought in his life, Newton adapted to his new responsibilities, such as they were, with surprising ease and self-assurance. Prior to the Mint's reorganization in the 1660s the Warden had been its highest officer, the manager of its finances, the judge in every conflict touching the Mint, and the supervisor of the Master and all lesser employees. These powers were now in the domain of the Mint's two other senior officers, the Master—or Master Worker as he was sometimes called—and the Comptroller. As chief executive officer, the Master had ultimate control over the coinage, which he conducted under a contract with the crown. In addition to a yearly salary of £500 he received a royalty on every ounce of gold and silver coin that passed through the rumbling presses. Since he was also responsible for all Mint expenditures, it was the Master who subcontracted the actual work of the institution. The casting of raw materials into bullion was entrusted to anyone he chose at the lowest cost he could negotiate, but the conversion of the ingots into coin still went, by tradition, to the Company of Moneyers, a holdover from medieval times, when the guild system prevailed. The Comptroller served as a kind of "second check" on the Master. He kept a separate bullion account and was also in charge of procuring all Mint supplies and seeing to the repairs of its ramshackle quarters. By the time of Newton's arrival the Comptroller had assumed responsibility for the day-to-day business affairs of the Mint.[17]

The Warden's powers, once virtually absolute, were now vitiated and ill-defined. The little authority left to him after the Master stepped into his shoes had severely eroded during a generation of neglect by Newton's mostly lackadaisical predecessors. As the Warden gradually receded from the picture, his clerk assumed the little remaining responsibility—a clerk no longer sworn to obey the Warden as before, but sworn to obey the Lord High Treasurer at Whitehall. In such circumstances Newton would have to work long and hard just to obtain meaningful work. He was fully prepared to do exactly that. Many years later, after he had queried Newton concerning this aspect of his life, Stukeley wrote that when his friend came to London he made the choice to leave off "intense study in general."[18]

Newton's plan to revitalize the Wardenship was bolstered by certain internal conditions that worked in his favor. His arrival coincided with a quantum jump in the workload resulting from the recoinage. Hence, even had his fellow officers resisted the newcomer's attempts to strengthen his position at their expense, and for the most part they did not, at least in the beginning, they were left with little choice but to share their escalating burden with him. Equally important was the attitude of the Mint Master, Thomas Neale. Aptly characterized as "a promoter of speculative enterprises," Neale obtained his office in February 1685. He had previously been appointed Groom-Porter of the court for life by Charles II, which, in addition to the responsibility of overseeing the King's furnishings, required him to provide the devices necessary for gambling, namely cards and dice, wherever the King might be. And since Sir Edmond Hoyle's reference book of rules for card games and other indoor entertainments did not go to press until 1742, it fell to Neale to rule on all disputes that cropped up during play. Neale was not merely an arbiter of gambling disputes but a player for high stakes himself, and not always "according to Hoyle." He founded the Penny Post in the American colonies and also created various lotteries designed both to enrich himself and to swell the flagging Treasury. It was in connection with the mathematical odds on one such venture that Pepys wrote to Newton in 1693 in a genteel effort to reestablish communications after Newton's breakdown. Having made and lost fortunes more than once, Neale couldn't be bothered with the dulling routine associated with the Mastership. He was quite content to leave matters in the hands of Thomas Hall, his new assistant, whom a committee set up to inquire into charges of misconduct at the Mint later described as "a very careful and diligent Officer," deserving of a raise in pay.[19] Newton concurred in that assessment; the two men worked hand in glove for the duration of the recoinage, and when Hall died in 1718 Newton, along with Hopton Haynes, was appointed an executor of his estate.[20] Another of Neale's deputies, John Francis Fauquier, a naturalized Huguenot and a director of the Bank of England, also impressed the new Warden. When Newton assumed the Mastership in 1699 he retained Fauquier until the faithful deputy's death in 1726, the year before Newton's own. Desperate for the enlightened leadership Newton was anxious to provide, these and other conscientious Mint officials, including Haynes and Fowle, soon began to view him as their best hope of weathering the mounting crisis in their midst. For surely the man who had brought order to the cosmos could help set to rights the tangled affairs of mere mortals.

On May 6, only four days after taking the oath of secrecy, Newton, with Neale and Hall, cosigned his first official communication to the Treasury. The salaries of several officers and clerks were in arrears by £800, hardly a comforting thought for the new Warden who was waiting for his own first payday. But this was not the half of it. When added to the unpaid wages of the workmen and the cost of new machinery, including eight rolling mills and eleven presses, the Mint's current obligations stood at £8,148 and were rising by hundreds more every day.[21]

By then the Mint was fully in the throes of the great recoinage. The Treasury ordered double shifts instituted, extending the working hours to twenty a day, six days a week. Still, the river of old coin poured in at a rate that swamped the institution's overburdened facilities. One reason was the initial deadlines set by the government for the redemption of unmilled currency. After January 1, 1696, no clipped crowns or half-crowns were to pass except those in payment of taxes and loans to the King. Shillings might pass until February 13 and sixpences until March 2. By April 2 no clipped money of any denomination whatsoever could be accepted for the payment of a debt, whether public or private. Within days after this proclamation was handed down the Treasury had taken in coins valued at one-third of a million pounds, with more in transit or in the hands of collectors. The impact on the citizenry was immediate and decidedly negative. "The Parliament['s] wondrous Intent on ways to Reforme the Coin," Evelyn wrote scoffingly in his diary, "made much confusion among the people." [22] Public misunderstanding was accompanied by widespread inconvenience and deprivation. The new money was slow to enter circulation; when it did reach private hands it came only in a pitiful trickle. Evelyn observed that the scarcity of silver coins made it virtually impossible to carry on the smallest of transactions in the market place. "All was on trust," a commodity as scarce among the poor and dispossessed as the new coin itself. For the rest of the summer England subsisted largely on the barter economy abandoned in the twilight of the Middle Ages.

Yielding to the threat of civil unrest, Parliament repeatedly extended the period during which clipped money would be received by the Treasury or its agents. Merchants nevertheless feared that they would be left holding large quantities of unredeemable coins. They soon began accepting clipped money only at a discount, if at all, which led to daily arguments in every shop and stall. In May the Bank of England was forced to stop payments on its notes for lack of specie. Only the considerable oratorical skills of the Lord Mayor, who also happened to be the Bank's governor, persuaded a mob to disperse after it had already stoned the building. The Treasury finally closed its doors to the old money at its nominal value on the second Saturday in May. During the previous week its harried clerks worked without rest to satisfy the continuous stream of people seeking to exchange the last of their mutilated coins. On the final day an impatient throng surrounded the Treasury from dawn to midnight, prompting officials to bring in a contingent of armed guards. Heads were bloodied and men were trampled in the frenzy, which lasted into the early morning hours. If Newton happened to be inside he probably lost a night's sleep waiting for the angry crowd to weary and go home. In any event, he must have been duly impressed by these grave developments.

According to the contemporary observer Narcissus Luttrell, a total of £4,706,003 in adulterated coins had been taken into the Treasury between January 17 and June 24, 1696.[23] One-sixth of the money was estimated to be counterfeit. The remaining coins had been reduced by industrious clippers, who like the Mint employees were working round the clock at the end, to little more than half their proper weight.[24] Before it was over the entire cost of the recoinage may have been as high as £2,700,000, nearly double the figure estimated by Lowndes.[25] Yet even this staggering sum would have risen considerably if every citizen had been fully aware of what was happening. As it was, the upper classes, who paid property taxes and subscribed to government loans and annuities, mostly succeeded in unloading their bad money on the Treasury, while merchants and bankers profited handsomely by purchasing clipped money at a discount from the less fortunate, then redeeming it at face value. Those who had only a few shillings to their name were often ill informed or too uncaring to make their way to the Treasury or one of its agents before the final exchange deadline passed them by. No one knows the true extent of individual losses, which have been estimated at between £1,000,000 and £2,500,000.

Part of the plan to ease the currency shortage and thereby bolster public confidence in the government involved the establishment of five branch mints at scattered locations throughout the country: Exeter, Bristol, Norwich, Chester, and York. Newton's superior Neale had shown no more interest in carrying out this project than any other at the Mint, thus enabling the new Warden to claim much of this territory for himself. On June 8, little more than a month after the assumption of his new duties, Newton informed the Treasury of the progress already made: "All the Iron Work & Instruments which cannot be made in the Country are bespoke & in a good forwardness here, & the men that are sent down are to get the rest of the things done there." Certain of the Deputy Wardens had also been appointed, but other candidates were dragging their feet "till your Lordships are pleased to Order what their Allowance shall be."[26] His meaning was unmistakable: His superiors at the Treasury could count on him to do his duty and more, but he must have their backing at all times.

Newton felt sufficiently confident to reiterate that message in a different context a week later. The Warden's salary, he complained, was only £400 per annum with a house of £40 and a few minor perquisites. The Master Worker, on the other hand, received £500 with a house and "very great perquisites," a reference, no doubt, to the profits garnered by Neale from every troy pound of newly minted gold and silver coin. In sum, the Warden's compensation "is so small in respect ... of the other Officers of the Mint, as suffices not to support the authority of his Office." Moreover, he was in need of a deputy.[27] On one level this may be taken as a gibe at Montague for promising Newton more in the way of remuneration than the Treasury had been able to deliver. More important, however, it was part of Newton's grand strategy, which was never wholly realized, to restore his position to what Montague in his letter offering the post had claimed it to be—"Chief Officer in the Mint." Zealous for the sanctity of precedent, Newton soon began to ply the Treasury with lengthy, meticulously researched documents on the origins and previous extent of the Warden's once almost limitless authority. With the burden of history as his ally, Newton quickly became satisfied in his own mind that the incompetent

Neale and his predecessors had usurped the powers that were rightfully his, much as the true church had been sapped of its vitality by the officious trinitarians. The Mint had already become his secular religion.

While the Lords Commissioners of the Treasury saw fit to honor Newton's request for a deputy, they temporized on the matter of a raise in wages. It was finally decided to "give him a Consideration extraordinarily suitable to his troubles," but only at the end of the recoinage and in an amount "proporconable to the Increase allowd to the other officers."[28] The Commissioners obviously feared that giving in to Newton would raise a hue and cry among his colleagues, further unbalancing a budget already scandalously in the red. They were also hamstrung by a law passed in 1666 that limited total expenditures on Mint wages and the upkeep of buildings to £3,000 a year. Corners were already being cut to a maximum, as evidenced by the proliferation of iron dogs used to shore up sagging structures that otherwise seemed to continue standing only out of habit.

If Newton's enthusiasm was dampened by this decision he hid his disappointment well by immersing himself in the task of opening the temporary branch mints. York and Exeter began coining in mid-August; Bristol and Norwich came on line a month later. Coining at Chester, the most troublesome of the country mints, was delayed until October because of a last-minute decision to move there from Hereford. Newton, Neale, and Comptroller James Hoare each had the right to appoint deputies of their choice, and Newton now found himself in a position to settle at least part of a long-standing debt. In an undated letter that may have been written in or near August 1696, Edmond Halley informed him that, "I will waite on you at your lodgings to-morrow morning to discourse the other matter of serving you as your Deputy."[29] Working with Hoare, who actually made the appointment, Newton secured the Deputy Comptrollership in the Chester Mint for Halley at the annual wage of £90. Though technically under Hoare's authority, Halley left no doubt about the depth of his loyalty to Newton. When simmering internal strife at Chester erupted into full-blown bureaucratic warfare in 1697, Halley let him know that he had a staunch ally in the field: "I would urge you to nothing but what your great prudence shall think proper, since it is to your particular favour I owe this post which is my chiefest ambition to maintain worthily."[30]

The seeds of the running feud at Chester were apparently rooted in the conduct, or perhaps one should say the misconduct, of its Deputy Master, one Thomas Clarke. Like Neale back in London, Clarke was rarely on the premises, which meant that Halley and the devoted Deputy Warden, Thomas Weddell, were left to do most of the work, but without the necessary authority or expertise. Halley complained of this in a letter to Newton drafted on February 13, 1697: "Mr Clarke our Master Worker has been at London this two months, and left all his business (which has for this month kept us all fully employed) to our care, though we know not why we should charge our selves therwith, he not desiring it."[31]

Clarke's nonfeasance might have been tolerated had not Halley and Weddell also uncovered evidence of a conspiracy to defraud the government. The details remain somewhat sketchy, but it seems that Edward Lewis, Halley's assistant, and another clerk named Bowles were secretly inflating the value of old currency being turned in for recoining and then skimming off part of the difference for themselves. When Halley and Weddell confronted the absentee Clarke with their suspicions the Deputy Master surprised them by defending the dishonest clerks. Even worse, when Halley then dismissed Lewis, Clarke immediately rehired him as his subdeputy, doubtless because Clarke himself was a party to the criminal operation. Fearing that they might become implicated as well, Weddell and Halley sought to stymie Clarke by locking the coin in a safe. The feud quickly took on all the coloring of a comic opera. Clarke pretended to take offense at some remark Weddell made during an argument with Bowles and Lewis. He borrowed Bowles's sword and vowed to waylay Weddell on his way home from work, but nothing came of this threat. In another exchange Weddell spat in Clarke's face and was immediately challenged to a duel. Clarke arrived on the field of honor intentionally early and left before Weddell could put in an appearance. Not to be outdone, Lewis, who seems to have been the real troublemaker, threw an inkstand at Weddell and talked Clarke into bringing false charges against both Weddell and Halley. Writing to Newton in December 1697, Halley charged "that proud insolent fellow" Lewis with being "the principall Author of all the disturbances we have had at our Mint." If Newton could see his way clear to remove him, "all will be easie."[32] But Halley was already seeking a graceful way out of his predicament. He had written Hans Sloane at the Royal Society two months earlier: "I long to be delivered from the uneasiness I suffer here by ill company in my business, which at best is drudgery, but as we are in perpetuall feuds is intollerable."[33]

No stranger to perpetual feuding himself, Newton was now cast in the novel role of peacemaker. Recriminating letters poured into the Mint from Chester and were interspersed with visits by one or another of the aggrieved parties. An anxious Halley wrote on August 2, 1697, "Our Adversary Clark went on Saturday afternoon for London with a resolution to loade the Warden [Waddell] and my self, with all the Calumnys he can: You see what weapons they fight with, and stick at nothing to compass their ends." Halley begged Newton's protection and the right to be heard should legal proceedings result. He added, somewhat desperately it would seem, "I hope your potent friend Mr Montague will not forgett me if there should be occasion."[34] The Mint Board, which comprised the three chief officers, grew increasingly impatient with the incessant bickering and responded with biting communications of its own. "Wee believe both sides much in the wrong & resolve to come & hear it Our Selves, . . . for the Mint will not allow of the drawing of Swords, & assaulting any, nor ought such Language, Wee hear has been, be used any more amongst You."[35] The Board did not visit Chester as threatened, however, and the feud continued until the branch was closed in 1698. By then respect for authority had so eroded that the clerks at Chester had to be warned against affronting their superiors by wearing hats and swords, a right reserved for senior officials alone.[36] Clarke himself provided a fitting end to this picaresque story when he was foiled in an attempt to abscond with the payroll for the final quarter.

Not to be outdone, certain criminal-minded officials at the other branch mints did their level best to undermine the great recoinage. The Deputy Master of Norwich, a Dickensian character named Anthony Redhead, was convicted of theft, was sent to prison, and had all his property confiscated for good measure. Most interesting, perhaps, is the case of Isaac Hayes, Deputy Comptroller of Exeter. When the accounts of all the branch mints were finally audited in London, it was discovered that Hayes had embezzled £1,100. The law could not touch him, however, for the old thief was living well in Madagascar, where he had been fittingly dispatched by the government as a commissioner for negotiating with its pirates. Hayes died in 1701, and his uncollectable debt was subsequently written off by the Treasury. Theft on such a grand scale was complemented by rampant petty larceny, the cumulative effects of which could be just as damaging as those resulting from greater crimes, or even worse. An official in the Company of Moneyers complained that the use of five hundred workers in six widely scattered locations had resulted in theft of such magnitude that the guild was "hard put to it to make good their wastes and losses."[37] Nor was Chester the only provincial mint haunted by the specter of violence. A committee of Parliament learned that Robert Williamson, who was appointed to inspect the mint at York, "hath several times [been] abused" by Barton, the Deputy Comptroller. It was further learned that "there lieth dead very great Sums of hammered Money uncoined, by the Negligence of the Officers" at Norwich and York.[38]

Incompetence, dishonesty, infighting born of jealousy, and a general lack of discipline and expertise all combined to keep output at most of the branch mints to a minimum during 1697, the peak year of production. Only Bristol, which turned out £77,000 of silver coin in the month of June, held its own. The production of each of the other four ranged between £15,000 and £25,000 a month, and even this modest average declined rapidly toward the end of the year. In contrast the London Mint, which, as we are about to see, was also plagued by a series of internal crises, turned out £330,000 in June and £360,000 in July.[39] Such figures would never have been achieved had not Newton gradually succeeded in making himself Master in virtually everything but name.

IV

In the early 1660s, when Edward Slingsby occupied the Master's house at the Mint, his guests, who included Pepys and Evelyn, were entertained by the most accomplished musicians from Italy, Germany, and France. One frequent visitor likened this little enclave of culture to "a nest of singing birds," a favorite metaphor of Restoration writers. Sadly, the soft, harmonious echoes of voice and viol had long since faded into forgotten summer nights, to be replaced by the reverberating dissonance of huge presses striking on different beats, like maddening grandfather and cuckoo clocks in a Mephistophelean dream. Where there was sound there was also fury, for frayed tempers exploded time and again as the recoinage was pushed forward without regard for its mental and physical toll on individual men.

Thomas Neale had little stomach for dealing with conflicts arising from encroachment on Mint territory or resulting in breaches of the peace. He was content, therefore, when Newton asserted his claim to what had once been the Warden's clear-cut prerogative: representative and champion of the Mint as well as its chief disciplinarian. Troubles began when the Mint expanded into quarters previously occupied by the Tower garrison, whose members found themselves reduced to sleeping three in a bed, if sleep they could with the din of production continuing twenty hours a day. Tensions steadily mounted, and fisticuffs erupted between soldiers and civilian employees on the slightest of pretexts. Newton moved to protect "his men" by drafting a document exempting them from retaliation by one of the more hated elements of official chicanery, the press gang. It was only a matter of time before he made his first important enemy in Lord Lucas, Lieutentant of the Tower, who came to despise the stocky, silver-haired interloper with a square jaw and imperious eyes. They pawed and postured at each other for nearly a year before finally locking horns over a number of seriocomic incidents that took place during the spring and early summer of 1697.

One of the Tower's more permanent fixtures was a feisty but wizened and perfectly harmless curmudgeon appropriately dubbed "Old Rotter" by those who knew him best. John Roettiers had come to England from France with his father and two brothers in 1661 at the behest of Charles II, who made them chief engravers at the Mint. Of the four, old John was the only one left, and according to a Treasury Report of July 2, 1689, he had long since lost the use of his right hand "by the shrinking of the tendents," which supposedly rendered him unfit for further service.[40] Nevertheless, he continued to draw the lifetime salary of £450 a year promised him in 1669, which greatly nettled his unscrupulous and incompetent successor, Henry Harris, who was paid only £325 annually. What is more, Roettiers and his sons, as described by a committee of the House of Commons charged with investigating abuses at the Mint in 1697, "are not only violent Papists, and refuse to take the Oaths . . . yet they still continue in the House belonging to his Majesty's Chief-Graver." Finally, it was widely believed that Old Rotter's namesake, his son John, was a conspirator in a plot to assassinate William III and return James II to the throne of England. A warrant charging the young man with high treason was duly issued by none other than "the Honourable Lord Lucas," but the suspect fled from Justice.[41]

Lucas, concerned about Mint security, and Harris, who was simply avaricious, joined forces in March 1697 with the object of ridding the premises of the recalcitrant old papist. They were assisted by others, including Marks, the Tower barber, who made a preposterous claim to the Lords Justices: King James had landed secretly in England and was being sequestered in Roettiers's house in "a very rich bed." The Justices were rightly skeptical, but these were perilous times, what with the King at war in Europe and the economy on the verge of collapse. They considered the "eaziness of seizing the Tower, and of what consequence it would be that there was a free conflux of people thither, under pretence of seeing the arms or the Mint."[42] Preferring to err on the side of caution, they ordered Lucas to make a proper search of the engraver's house and Harris to gather such additional evidence as he might find. The Treasury was also directed to search past records for a precedent that would support Roettiers's dismissal and removal from the Mint. Lucas and his soldiers turned up no bed worthy of a royal personage, not even a Roman Catholic one, nor was Harris willing to expose himself to the considerable risks of manufacturing additional evidence against his perceived rival. They doubtless hoped the Treasury would finish their work for them, but it did not. Roettiers's dies and puncheons were seized, and that was all.

Meanwhile Lucas was called on the carpet by an outraged Newton. A champion of neither Roettiers nor his civil rights, the Warden harangued the Lord Lieutenant for failing to obtain official permission (meaning his) before conducting the search, which had taken place on territory off limits to members of the garrison. The Mint's security had been seriously compromised, Newton fumed, and legal precedent wantonly violated: "if the military searches without us should be now allowed . . . we cannot undertake any longer the charge of the Dyes & Puncheons & Marking Engines & other coyning Tools & of the Gold & Silver wch lyes scattered about in all the rooms." Public confidence, especially that of the merchant class, would be undermined. As for Harris, Newton had correctly sized him up within a matter of weeks: "we have hitherto endeavoured to engage him (contrary to his mind) to copy after Mr Roettier's Puncheons that the money may all be alike," he wrote to the Treasury on July 6. "And we most humbly submit . . . whether he shall be any more employed as Engraver . . . or allowed a maintenance here till . . . he shall shew such speciments of his Art . . . or return with his family to Brussels his native country."[43] Harris was not dismissed, as Newton and Thomas Molyneaux, the new Comptroller, urged. However, he was forced to employ a skilled engraver named John Croker and to pay him £175 a year out of his own pocket, a punishing figure that constituted more than half of Harris's income, proof that justice occasionally prevailed at the higher levels of an entrenched bureaucracy.

Now that the battle with Lucas was joined, the gentlemanly sparring gloves were replaced by the street fighter's brass knuckles. The pretext for a second and fiercer confrontation was a bizarre incident on a June night, when a drunken officer of the guards attempted to break into one of the Mint residences. He shattered a window and ran his sword through the coattails of a servant who came to investigate the commotion. A sentinel refused to intercede on behalf of the fleeing victim, and when a third party finally roused a corporal and some musketeers they listened passively while the perpetrator sought to persuade the witness to lie about what he saw. Newton charged Lucas with refusing to investigate the matter thoroughly: "instead of being told the truth [he] is only informed by the soldiers that our men get drunk & affront Centinals, & he has upon such informations ordered the Centinals to fire at us." He branded this a "bloddy discipline" which so terrified the workers that some had left the Mint "to ye neglect of ye Kings service & insecurity of the Treasure." Lucas, in good conscience, should restore the old practice of having soldiers patrol in groups strong enough to overpower rather than kill. While it was true that many were addicted to the "crime" of drink, "it was never yet made death."[44] Playing the role of protector to the drunk was hardly to Newton's liking, but in the rough-and-tumble environment of the Tower even he had to lower his moral sights a notch.

At the height of the recoinage Lucas blockaded the Mint gates against the entry of all food and drink, raising the prospect of a strike by the angered workers. The Lord Lieutenant claimed that the action was taken to punish Mint employees who were lining their pockets by supplying his soldiers with alcohol. A work stoppage was averted at the last minute, but the output of new coin temporarily declined by half. All this and much more was set forth in official documents and personal testimony addressed to the parliamentary committee charged with investigating the Mint. Revealingly, its members addressed Newton as "Doctor," thus conferring upon him their own peculiar form of honorary degree. No matter where he went or what he did Isaac Newton was now regarded as a living legend. Anyone who chose to question his integrity or judgment faced the daunting prospect of having to engage both the man and the myth.

Lucas, no shrinking violet himself, also had his story to tell, which he related in equally vivid detail before the same forum. It began with Philip Atherton, a laborer, who had beaten up a head constable in an attempt to free his imprisoned wife and then fled to the protective confines of the Mint, where Newton, as Warden, claimed jurisdiction over almost all civil and criminal matters.[45] Lucas had aided the authorities when they came to arrest Atherton, who was removed to New Prison over Newton's vehement and, according to Lucas, legally decrepit protestation. That same day a Tower guard had shown some visitors round the Mint and was stopped by the gate porter, who allegedly demanded a tip. The porter was insulted when offered copper coins rather than silver and lost even this modest of-fering when the guard slipped his party out by an unattended exit. The guard then returned to settle rather different accounts with the porter, whose son, Newton claimed, was seized by the throat. Lucas countered that his soldier was menaced by the porter's entire family, who attacked and beat him to the ground. At least both men could agree that a riot, or the next thing to it, ensued, causing "such a tumultuous concourse of people as rendered ye money unsafe wch was then coming down the street of the Mint in Trays."[46] But Lucas's most damning accusation was leveled at a Mint worker who became violent from too much alcohol. He charged down the narrow street at a gallop with two horses, drove a sentinel into a post, "and call'd him many foul names, without any manner of provocation." When the sentinel attempted to recover his ground, another Mint em-ployee named Thomas Fowle exhorted his co-worker "to dragg the Cen-tinell by the Eares from his post, & shoot him through the head."[47] For-tunately the "Lieutenant's lady," who observed the assault from an upper window, called the guards, or blood might have flowed. While the commit-tee that scrutinized their testimony sympathized more with Newton than with Lucas, the Lord Lieutenant emerged from the proceedings relatively unscathed. Tensions remained high for another year until the recoinage wound down, after which the two officers and their men were presented with fewer opportunities to cross each other.

Thomas Neale fared considerably less well in his dealings with Newton than did the combative Lucas. Early in 1697 the preoccupied Master began to have disquieting second thoughts about the ambitions nurtured by this once seemingly harmless recipient of Montague's political benefaction. A member of Parliament and of the committee appointed to investigate the Mint, Neale read with increasing alarm the documents that flowed from the Warden's pen. It was one thing for this junior officer to take on certain of the Master's responsibilities during his frequent absences but quite another for Newton to initiate multiple claims that they belonged to his of-fice by right, not to mention his success in cultivating the personal allegiance of Mint employees from the lowest level to the highest. The two came into direct conflict over the ongoing struggle at the Chester Mint. Halley, as was noted, pointed an accusing finger at his former clerk Edward Lewis and urged Newton to remove him as Clarke's assistant if he could. Clarke, in turn, looked to Neale for protection and received it in the form of a promise not to dismiss Lewis. Unable to settle their differences amicably, Newton and Neale, who proved a skillful infighter when the chips were down, put their respective cases before the Lords Commissioners of the

Treasury. After much deliberation they decided in favor of Newton and sent Lewis packing, in essence confirming that the rumored palace coup was indeed a *fait accompli*.[48] Newton, whose contempt for the profligate and debt-ridden Neale had become that of a Puritan censor for a drone, later described the Master as a gentleman "who was in debt & of a prodigal temper & by irregular practices insinuated himselfe into ye Office."[49] Whether or not Neale took solace from having been trampled by a unicorn rather than a plow horse is impossible to say. In one sense Newton almost always won the battles he waged. In another he almost always lost, for his victories—more than a few of which were Pyrrhic—fanned his self-righteousness, spurring him headlong into new conflicts. The whole cycle called up echoes of the lonely, hurt, suspicious boy, left at home to his own devices, trusting few, keeping potential friends at a distance. Yet through it all, from the sublime formulation of elegant mathematical laws to the mundane timing of striking coin presses, Newton never once lost sight of what he cherished above every other thing—a supreme sense of order and rationality. It was this drive for order, this belief that every problem and institution could be turned into a rational system at his command, that so deeply colored his thoughts and deeds.

V

Like the tributaries of a great river, Newton's energies flowed simultaneously from many directions. Hopton Haynes gave him much of the credit for raising the Mint's output to £100,000 a week in the summer of 1696. The key to Newton's success lay in his absolute mastery of the entire production process, which he accomplished within weeks of his arrival in London. Among his early Mint papers is a document drafted in 1697 bearing the simple title, "Observations concerning the Mint." Each major step in the coining process is methodically described and meticulously analyzed under its appropriate heading: "Of the Assays," "Of the Melting," and "Of the making the Moneys." He calculated the speed of melting and the exact weights and kinds of materials needed to make the new ingots. The milling and stamping machines received no less careful scrutiny, as did the skills of the men required to work them. "Two Mills with 4 Millers, 12 horses two Horskeepers, 3 Cutters, 2 Flatters, 8 sizers one Nealer, thre Blanchers, two Markers, two Presses with fourteen labourers to pull at them can coyn after the rate of a thousand weight or 3000 *lib* [pounds] of money per diem."[50] Newton's long experience as an alchemist served him well. Whereas others would have been deeply satisfied with the sheer quantity of coin pouring from the great presses, he demanded—and got—the highest possible standard of quality, a reform which outlived him by centuries. Unwilling to rely solely on his formidable powers of observation, the Warden, able with his

hands as well as his head, practiced the assaying of gold and silver on old
coin until he mastered the ancient art. (An assayer's furnace and a touchstone said to have been his are still kept in the Royal Mint.) Some of the less
conscientious workers doubtless resented Newton for his "meddling,"
which required that they work all the harder, but the best among them
could take renewed pride in labors raised to unexampled heights by their
strange, seemingly all-knowing superior.

As when he had approached unfamiliar territory in the past, Newton
began purchasing books on economics, trade, and currency. Favoring quality over large numbers, he acquired works by some of the foremost thinkers
of the day: Francis Brewster, Josiah Child, John Locke, William Lowndes,
Pierre Daniel Huet, Jean Boizard, and others. He especially valued a collection of 180 official documents issued by the French between 1689 and
1701. Bound together in vellum under the title *Cour des Monnoyes*, many of
the pages bear unmistakable signs of dog-earing, that familiar Newton
trademark.[51] While familiarizing himself with the writings of others, he
composed the equivalent of several volumes of his own. If anything, the
compulsion to fill reams of blank paper had only grown more obsessive with
the passing years, and Newton now commanded a stable of amanuenses
ready to do his bidding. Five large volumes of Mint papers survive, and
Conduitt noted that his wife's uncle burned entire boxes of others (and who
knows what else) not long before he died.[52] Multiple copies of everything
from preliminary memoranda to official letters and reports abound. In
several cases an early and unrelated paper has been used for the drafting of
theological arguments or mathematical calculations, proof that Newton
always kept most of what he wrote very close at hand. The written word, no
matter if redundant or banal, to him represented a kind of truth that
thought or speech alone could never match.

If laggards winced when the Warden made his unscheduled rounds,
neither did he endear himself to certain government contractors whose
patriotism was eclipsed by greed during this time of national upheaval.
After conducting his study of the costs involved in the melting and refining
process, Newton found the bid of nearly $12\frac{1}{2}d.$ per troy pound submitted by
the financiers Peter Floyer and Charles Shales completely out of line. In
February 1697 he informed the Treasury that the two could accomplish the
task for $7\frac{1}{2}d.$ and still turn a fair profit—if fair they wished to be. Two
months earlier a master melter named Jonathan Ambrose had attempted to
float a similar scheme, but a watchful Newton had exposed him before the
Lords Commissioners with equal zeal.[53]

By the summer of 1697 Newton had come into direct conflict with virtually everyone who threatened what he believed was his rightful domain as
Warden: the Master, the Lord Lieutenant of the Tower, the Comptroller,
the Corporation of Moneyers, and plotting financiers. Almost from the

beginning he lobbied long and hard for a revision of the Mint charter, as is illustrated by his report to the Treasury on "The State of the Mint" in 1696:

> And thus the Wardens Authority which was designed to keep the three sorts of Ministers in their Duty to ye King & his people, being baffled & rejected & thereby the Government of ye Mint being in a manner dissolved those Ministers act as they please for turning the Mint to their several advantages. Nor do I see any remedy more proper & more easy then by restoring the ancient constitution.[54]

This and successive petitions were met by a pregnant official silence, for the Treasury tacitly approved of Newton's gradual consolidation and exercise of administrative power. Nor can one discount the possibility that Montague, who was First Lord, actively encouraged Newton from behind the scenes. By the time the recoinage had run its course in the summer of 1698, Newton held all but Neale's official title. Then, within little more than a year, the Master of the Mint was dead. With Neale's death in December 1699 his office passed to his impatient rival. These events coincided with the new Master's fifty-seventh birthday: Newton's star had danced once more.

VI

Before Newton could savor the pleasures of total victory, however, he faced a final challenge from a colorful rogue whose rare but squandered gifts nearly proved a match for his own. Unlike a number of lesser criminals, William Chaloner has been denied his rightful place in the *Dictionary of National Biography*, for this more mature and cunning form of the fictional Artful Dodger long succeeded in making fools of the highest government officials, while simultaneously masterminding a notorious ring of clippers and counterfeiters. In addition to this archvillian, Newton met, interrogated, and gave evidence against enough rascals and social misfits to fill a hundred works by Dickens. And like so many of the novelist's memorable characters, the real-life "bad guys" often proved more interesting than the good. The great mind that had so recently plotted the orbits of comets across the firmament now set about mastering the argot employed by the children of Cain, whose shadowy movements Newton attempted to trace through the dark warrens of the vast London underworld. His kaleidoscope suddenly shifted again. The Puritan who had once banished Vigani from his inner circle for telling a joke about a nun was now prepared to sacrifice his innocence in the service of the crown.[55]

Unlike many of his administrative powers, which he cleverly pilfered from lackadaisical colleagues, the Warden came by his charge to pursue and

prosecute coiners and clippers honestly enough. Indeed, apart from certain
ceremonial duties the office of Warden was largely judicial in nature. Once
strictly the responsibility of sheriffs, the apprehension and conviction of
false moneyers—a capital crime—had been drawn into the Mint shortly
after the Restoration. Newton's predecessors had delegated this task, like
almost all others, to their respective clerks, who were hampered by a lack of
funds and a proper supporting staff. Except for the clerk's salary, which was
paid by the Mint, all other expenses were to be covered by the proceeds
derived from the offenders' confiscated properties, which in most cases
amounted to relatively little.

Newton, who succeeded in expanding the authority of his office on virtually every front, initially sought to escape from this, his one area of clear-
cut responsibility. While he coveted recognition, he had always evaded the
public eye. Now he feared that his hard-won reputation was being sacrificed to the notoriety arising from his prosecution of criminals, many of
whom became objects of widespread public sympathy. As previously noted,
a new statutory reward of £40 for each conviction went to the informant
who testified successfully against the accused. Newton complained to the
Treasury in the summer of 1696 that this inducement "has now made
Courts of Justice & Juries so averse from believing witnesses & Sheriffs so
inclinable to impannel bad Juries that my Agents & Witnesses are discouraged & tired out by want of success & by the reproach of prosecuting and
swearing for money. And this vilifying of my Agents and Witnesses," he
continued, "is a reflexion upon me which has gravelled me & must in time
impair & perhaps wear out & ruin my credit." From the other side, "I am
exposed to the calumnies of as many Coyners & Newgate Sollicitors as I examine or admit to talk with me if they can but find friends to believe & encourage them in their false reports & oaths & combinations against me."
The work was not only threatening but downright "dangerous" and lacking
in "reward or encouragement." Realizing that the flow of such events was
beyond the control of any single individual, he concluded with the request
that this responsibility be turned over to the King's Attorney and Solicitor
General, who "are best able to go through it."[56]

After due consideration Newton's objections were overridden by the
Treasury, which was being pressured by the Lords Justices, who served as
the board of regents during William's absence abroad, to remove the
backlog of pending criminal cases and to investigate serious charges of corruption in the Mint itself. The only concession offered Newton was the appointment of an additional clerk, Christopher Ellis, who joined the
Warden's staff in September.[57] Sensing that further protest would fall on
deaf ears, Newton accepted the outcome and advaced with all the energy at
his disposal, holding nothing in reserve. To finance his operations he revived old Mint claims to confiscated goods held by the sheriffs of London
and won his case before a none-too-pleased Treasury, which was forced to

reimburse the sheriffs from other funds. These sources of revenue dried up quickly, however, forcing Newton to petition his niggardly superiors every time he incurred some new debt. He received only between £200 and £300 per year, in addition to the revenues derived from confiscated property, and this was spent on everything from lodging and transportation to disguises for his undercover agents—"paid [£5] to Humphrey Hall to buy him a suit of clothes to qualify him for conversing with a gang of Coiners of note."[58] Nor was the Warden averse to supporting the cause with funds of his own when the potential rewards seemed worth the cost: He spent several pounds on being made a justice of the peace in all of the seven home counties in an effort to strengthen his hand against the criminal element.[59]

The recoinage soon stopped the clippers in their tracks, but counterfeiters were still plying their old craft, though with greater difficulty than before. It was on this pernicious element that Newton focused most of his attention. His central part in this unfamiliar calling is richly documented at the Public Records Office in the surviving Depositions Book covering the years 1697–1704. His method was as straightforward as it was effective. A would-be informant or accused felon was brought before him at the Mint, in chains if necessary. The subject was cross-examined before witnesses, after which Newton either wrote or dictated to an amanuensis a summary of the proceedings. The deponent's statement was then read back to him, after which he was allowed to amend it as he saw fit before signing or, if illiterate as most were, making his mark. Newton's frequent use of the first person pronoun suggests that he took an active part in the long interrogations, as does his signature attesting to his presence at some two hundred sessions. This same volume contains copies of the many letters he received from distraught prisoners begging intervention or, when the death sentence had been pronounced, mercy. As one would expect, this Warden was little inclined to spare the rod (or, more precisely, the gibbet), unless the petitioner could provide information certain to secure additional convictions. The concept of rehabilitation was as foreign to Isaac Newton's thinking as to seventeenth-century thought in general: Criminals, like dogs, he once observed, always return to their vomit.[60]

Newton had hardly exaggerated when he informed the Treasury that his pursuit of counterfeiters involved an element of personal danger. Samuel Bond, a chirurgeon who treated men confined to Newgate Prison, swore on September 16, 1698, that one Francis Ball complained bitterly of the Warden for his severity against coiners: "Damne my blood I had been out before now but for him, and Whitfield who was also there in person made answer y[t] the Warden ... was a Rogue and if ever King James came againe he would shoot him [Newton] and the s[d] Ball made answer God dam my blood so will I." Whitfield was angered because Newton "had troubled his wife for a little bitt of clipping" found in her coat, and because the Chief Justices refused to grant him bail on a charge of gilding Spanish coins "y[t] was not the business of the Warden of the Mint but of the Spanish Embassad."[61] The previous year Newton had received a less explicit threat from a government employee named Gibbons, who permitted a confederate to counterfeit stamped paper in his home. "Notice was taken of his threatening the Warden of the Mint, that he would be revenged from the taking of examinations about him."[62] If Newton was intimidated by these and similar pronouncements, the record does not reveal it. Not only was he present at the Mint for the purpose of conducting interrogations at least 127 days between June 1698 and December 1699, but the Warden also applied for and received £120 as reimbursement for "coach-hire & at Taverns & Prisons & other places of all wch it is not possible for me to make account on oath."[63] The world into which Newton entered in his pursuit of counterfeiters was that immortalized by John Gay in *The Beggar's Opera*. His many contacts, any one of whom might have done him in as he moved about the city, included murderers, thieves, prostitutes, beggars, and perjurers—the socially dispossessed and the mentally deranged. "He learned of bribes to cut off evidence, of secret hideaways in the country, of clandestine meetings in garrets and taverns. He became conversant with men and women of many aliases."[64] Yet even London's sprawling precincts could not contain him. As Newgate began to fill and the list of convictions mounted, many coiners left the capital in flight from their seemingly possessed pursuer, who threatened not only their livelihood but their very lives. While some slipped through the contracting net, others, like Oedipus, ran to their fate rather than away from it. On June 9, 1698, Newton visited the Treasury to bring it up to date on the battle against counterfeiting in Yorkshire. He had previously paid Charles Maris £44 for services rendered in Worcestershire and Shropshire, while Bodenham Rewse (later a turnkey at Newgate) received £34 for prosecutions in five western counties. In March 1697 Joseph Williams and Caleb Clark received £40 for arrests and convictions at Worcester and Exeter.[65] Newton and his broad network of agents and informers seemed to be everywhere and nowhere at the same time.

He went after the powerful and the powerless with equal fervor, drawing no distinction between classes or between major and petty crimes. Among his most important victories was the arrest and conviction of Captain William Wintour, a leading citizen of Gloucestershire, who first provided financial backing for clippers and then recouped his investment many times over by receiving their wares, which he conveniently smuggled abroad. At the opposite end of the scale were the victims of lives spent beyond the pale of polite society. The most pitiful were widowed or abandoned women, for whom crime afforded a rare source of independent income in the seventeenth century. An informant named Julian Tuffin accused one Ann Pillsbury of passing a bad sixpence. Ann and her young daughter were arrested on Newton's orders and subjected to body searches at the Mint. The "informant found wrapt in a paper in the said Girle 5d worth of farthings & 4 six pennies two of which were counterfeit ones." In a similar instance a suspect named Elizabeth Pilkington was told by a woman informant at the Mint that she "must raise her bottom." This search yielded a napkin containing false coins.[66]

Hopton Haynes admitted that Newton's agents were "very scandalously mercenary," which was tantamount to an admission by Newton himself, for he was clearly behind the drafting of Haynes's mostly laudatory but essentially accurate *Brief Memoires*.[67] These men often intercepted the confiscated property of those they had accused and let others off in lieu of bribes, all without the consent of their superior, who condemned such practices but usually could do little about them.[68] Some, like Samuel Wilson, whom Newton eventually caught, were also professional blackmailers. Having sold some stolen shilling dies for £5, Wilson then obtained a warrant for the buyer's arrest, which he used as leverage to extort funds from his victim for eighteen months.[69] While Newton himself stayed within the law, the moral level of those with whom he dealt on a daily basis soon convinced him that its provisions had to be bent—if not broken—in the service of a higher morality. Depending on the circumstances, he would prosecute or relent, approve bail or have a man clapped in irons, threaten reprisals or promise leniency in exchange for cooperation in netting larger fish. The King pardoned the convicted counterfeiter Edward Jones on Newton's recommendation, but the Warden turned his back on the pleas of the Cambridge coiners Thomas Stadler and his wife, both of whom gave false evidence against others at a trial in Bury.[70] As betrayal led to betrayal, he had only his instincts to guide him. For the most part they seem to have served him well. According to surviving records, which are by no means complete owing to selective burning, he prosecuted twenty-eight coiners successfully. He paid for the discovery and prosecution of dozens more by others. In all, it has been estimated that Newton played a part in the pursuit of well over a hundred counterfeiters before gradually washing his hands of the dirty business after he became Master.[71]

Among the more curious and colorful publications of the late seventeenth century is a twelve-page pamphlet titled *Guzman Redivivus. A Short View of the Life of Will. Chaloner, The Notorious Coyner, who was Executed at Tyburn on Wednesday the 22nd of March, 1698/9.* Published anonymously in 1699, the work's title derives from Guzman de Alfarache, the central character of a popular work called *The Rogue*. Besides the arch-villain and one or two of his accomplices, the only other person mentioned by name in the pamphlet is "that Worthy Gentleman *Isaac Newton* Esq; Warden of his Majesties Mint." In 1697 Chaloner accused Newton of "Abuses and Cheats" before a committee of Parliament, "who upon a full hearing of the matter, dismissed the said Gentleman with the Honor due to his Merit, and *Chaloner* with the Character he deserv'd."[72] Like Doyle's

Sherlock Holmes, Newton had long been in pursuit of his Moriarty, and in the process collected every available scrap of information concerning this mastermind of the London underworld. Though one cannot prove it, the inclusion of hitherto obscure details, the accompanying air of tendentious morality, and the flattering reference to his name all suggest that a triumphant Isaac Newton was behind the composition and publication of *Guzman Redivivus*. If so, it was not the first time he had resorted to such tactics, nor would it be the last. Alban Francis had fallen prey to Newton's retributory phamphleteering in 1689, and Leibniz would suffer a similar onslaught in 1713 with the publication of *Commercium epistolicum*.

It was about the time of Newton's election to Parliament in 1689 that Chaloner also arrived in London, where he set off on the path destined to cross that of the future Warden of the Mint. A natural-born swindler, he first made his way in the hostile city by fashioning worthless tin watches and hawking them in the streets for whatever credulous buyers would pay. He soon gained an accomplice, and together they "set up for Piss-Pot Prophets, or Quack-Doctors." It was not long before Chaloner worked his wiles on his partner, who agreed to act as his servant while he impersonated a learned doctor, "pretending to tell silly Wenches what Husbands they should have, discovering Stol'n Goods, &c." The turning point in his career came when, after being implicated in a robbery and fleeing his lodgings, he bribed a japanner to teach him the finisher's art. With this knowledge he undertook the gilding of coins, which in turn brought him to the deeper study of metals and the eventual counterfeiting of guineas and French pistoles. Within two years' time William Chaloner was a rich and happy man, save for one thing; "he wanted nothing but a *Phillis*, (for a Coyner, you must know, is as rarely to be found without a Harlot, as a Sea Captain's Wife without a Gallant;) and as the Devil would have it, a *Phillis* he found." He abandoned his good wife and several children for a mistress, who reputedly asked and received her mother's blessing when she first went to bed with her spark.[73]

Chaloner's ring of accomplices swelled as he turned out coins by the thousands, but in 1692 one Blackford, an important cog in this criminal machine, was arrested and condemned to death. In a futile effort to save himself Blackford indentified Chaloner as his superior, forcing the latter to close up shop in London until the gallows finished off his erstwhile friend. Not satisfied with the prospect of continuing his operations in the relative safety of the country, Chaloner "began to peep abroad again." Ingenious and possessed of panache, he concocted a plan designed to rehabilitate his reputation and curry favor among prominent officials of the new constitutional monarchy. He prevailed upon some printers to copy certain declarations of the exiled James II. These were not for public distribution, he assured them, for that would be high treason, but for some discreet gentlemen in the country, who were willing to reward them well. Ordered by Chaloner to bring the materials to the Blue-Pots in the Haymarket, where he had invited them to supper, the printers were greeted instead by King William's musketeers. While the dupes cooled their heels in Newgate, Chaloner collected a staggering reward of £1,000 and was later heard to brag of how he had "funned" the King. Shortly thereafter yet another chapter was written in his double design of serving and cheating the nation. After counterfeiting several hundred-pound notes on the Bank of England, he turned the fake money in, along with a number of his expendable underlings, and collected what must have seemed a rather disappointing reward of £200.[74] Meanwhile Chaloner had boldly returned to his old trade with the boast that he had the wit and enough money to escape the law, for none but poor fellows and fools were hanged.

If the acquisition of money and material goods alone had been Chaloner's ultimate ambition, he might have rested content with his comfortable home in Kensington, a fine dinner service of plate, and the wardrobe of a gentleman, all of which were noted with meticulous care in Newton's records. But wealth and the false respectability it purchased seemed less important to this high-stakes gambler than the playing of the game. Chaloner loved the grand design—the plotting, the trickery, the cheating and fraud he called funning, the manipulation of pawns, but most of all the intellectual sleight-of-hand that pitted his wits against the combined talents and resources of officialdom. When confronted with evidence linking him to major crimes, he consistently wriggled free by resorting to the big lie, a strategem he jokingly referred to as "bubbling." By 1697, when he first crossed swords with the Warden of the Mint, Chaloner was nothing short of effervescent.

In February Newton received a terse communiqué ordering him to appear before the newly constituted Parliamentary Committee on the Miscarriages of the Mint. Its members were especially concerned about allegations raised by Chaloner. He charged that certain Mint officials were turning out false coin at their place of employment, even as they acted in concert with professional criminals. Furthermore, the Mint staff was composed of too many self-centered specialists, a condition that discouraged communication and led to gross inefficiency. The institution was in need of an officer distinguished by all-round knowledge of coining. Lastly, Chaloner requested that the Committee let him demonstrate his revolutionary coining techniques, which, if adopted, were guaranteed to put counterfeiters out of business once and for all. Chaloner claimed to be writing a book on the subject, which would soon be available to the Committee should it care to consider his views in greater depth. His primary objective, as he later bragged to a confederate named Thomas Holloway, was to gain an appointment in the establishment from which he could supply coining tools for use outside the Tower. It was a virtuoso performance, bolstered by the prior composition of three learned pamphlets addressed to

Parliament as part of his elaborate scheme to build a façade of respectability.

The Committee, which included such highly regarded monetary experts as Montague, Lowndes, and Neale, ordered Newton to provide Chaloner with the tools he required for a demonstration of his improved method of coining.[75] Newton refused, arguing that it would violate his oath of secrecy, because one of the recommended changes involved the use of Blondeau's edging machine. (The Warden clearly thought Chaloner's motives suspect, especially because the high-flying rogue had approached Newton earlier about making Holloway his special assistant for the pursuit of coiners.) Instead, he minted experimental coins to Chaloner's specifications and presented them to the Committee for its inspection. He also provided documentation to show what he (and Chaloner) had known all along: The proposed changes would be much too costly for adoption on a wide scale. The Committee, much to Newton's chagrin, did not see things his way: "That undeniable Demonstrations have been given ... by Mr. *Wm. Chaloner*, That there is a better, securer, and more effectual, Way ... to prevent either Casting or Counterfeiting of the milled Money."[76] Insult was added to injury when the Committee further recommended that legislation be drafted barring Mint officials from holding office for life. With this, the lines of battle were drawn for a classic confrontation between the Devil's disciple and the steadfast servant of the Lord.

Chaloner, who received no monetary reward from the Commons Committee, regrouped his forces and inaugurated "a new and dangerous way of coining" at Egham, some distance from London. Newton's diligent pursuit of coiners enabled him to declare that many had fled beyond the reach of his agents. They were spurred on, no doubt, by the passage of a new law proposed by the Warden, making the mere possession of coining tools a capital offense. It was not long, however, before Chaloner hatched yet another scheme by means of which he aimed to fun the government out of £1,000. Assisted by one Aubrey Price, he concocted a preposterous story of a Jacobite plot to seize Dover Castle in the summer of 1697. The two declared themselves willing to risk life and limb in the service of the crown by posing as secret agents for the purpose of compiling a list of suspected traitors. But Chaloner had made the ultimately fatal mistake of going to the same well once too often, and when the skeptical Lords Justices heard his charges in June they ordered Newton to supply them with more information about the background of this "tongue-padding" gentleman. Still smarting from the wounds he had sustained during their recent set-to, Newton enthusiastically concentrated most of his energies on building a legal case against his enemy. He was not yet ready to move when he received orders to arrest Chaloner and Price in September. Each informed on the other, but the most they could be convicted of was a misdemeanor. Chaloner must have taunted the man he later referred to as "that old Dogg the

Warden," for Newton vowed to hang his prisoner "for discovering to Parliament the mismanagement of the Mint."[77] Meanwhile Newton appeared before the Lords Justices on five separate occasions to discuss Chaloner's case. The highest legal authority in the land agreed that he was much too dangerous to set free on bail and cynically ordered Newton to proceed against him, even if it meant bending the law to secure a conviction.

Although the government was now on his side, Newton's hope of laying Chaloner low before his time was dealt a serious blow in the spring of 1698. Thomas Holloway, until now a confederate of Chaloner, had turned informer, supplying Newton with the evidence needed to condemn England's most accomplished coiner. But Chaloner somehow got to Holloway, who, for a bribe of £20 (which was never paid), fled to Scotland. On hearing the good news, the joyful prisoner "said a fart for ye world," after which he was set free for lack of evidence against him. It was at this point that Newton began drafting the first of four scathing accounts of Chaloner's wickedness, each many thousands of words in length.

Cool and cunning as a cat, Chaloner embarked on the last of his criminal lives, the engraving of a copper plate with which he printed forged malt lottery tickets, a new form of paper currency. The authorities soon learned of the operation, but no trace could be found of Chaloner or his copper plate. A reward of £50 was finally advertised in October 1698, and the master forger was betrayed and arrested in a matter of days. Early November found him back in the familiar confines of Newgate. This particular crime had nothing to do with the Mint, but Newton showed an unusually avid interest in it nonetheless. He collected and transcribed all the evidence about it and had copies made of Chaloner's long letters of protest to government officials. The investigation gradually broadened to include Chaloner's coinage crimes. Part of Newton's strategy entailed what is now commonly referred to as entrapment. He employed three men, whom he had sent to Newgate for crimes, to gain the suspect's confidence and pass on his criminal admissions. One of them was the notorious John Whitfield, who had earlier cursed the Warden and threatened his life. The broken counterfeiter had changed his tune when he wrote to Newton on January 25, 1699:

> May I humbly begg the favour of that you will be pleased to remember your promise which was to my wife. Every day I have thought it a month hoping to have the honour to have seen you either in passing or in repassing to the Tower. It lys in your power to admitt me to bayle So good S^r be pleased to let me know whether I shall send to you or you will be pleased to call here.

Two weeks later Whitfield begged Newton to come to Newgate's Dogg Tavern for the purpose of furthering their conspiracy—"we are now at a stand by shamming lunaticks and the more for want of the honour of your presence for one $\frac{1}{2}$ hour."[78] Newton's net was beginning to close. The hated name of Chaloner now flowed daily from his pen.

Of the many turncoats who had lately joined the Warden's ranks, none proved more damaging to his foe than Thomas Holloway, the suborned witness who was taken into custody after returning from Scotland. Ill-used by his one-time leader, Holloway, with the encouragement of his embittered wife, told all. Word of this omnious turn of events reached Chaloner in his cell, rattling him for the first time. As before, he attempted to undercut his accusers, but on this occasion he wrote directly to his tireless pursuer:

> I have been close Prisoner 11 weeks and no friend sufferd to come near me but my little child I am not guilty of any Crime, and why am I so strictly confined I do not know I doubt Sr You are greatly displeased with me abo[u]t the late business in Parliamt but if you know the truth you would not be angry with me for it was brought in by some persons agt my desire.
>
> Sr, I presume you are satisfied what ill men Peers and your Holloway are who wrongfully brought me into a great deal of trouble to excuse their Villanys Wherefor I begg you will not continue your displeasure agt me for I have suffered very much so I wholy throw my self upon your great Goodness.[79]

Ever more desperate missives followed. In one he claimed that his skills were not sufficiently well developed to counterfeit successfully, and this from a man who so admired his own handiwork that he cringed at the thought of striking coins from his freshly carved dies. In another he protested that "he never could work at the Goldsmith's Trade," though he was known to have fashioned false guineas by the thousands. The man truly responsible for the engraving was none other than Patrick Coffee, who, like Holloway, had turned against him.[80] When these ploys failed, Chaloner feigned insanity, perhaps part of the shamming lunacy alluded to by Whitfield: "That he was sometimes Delirious; he was continually raving that the Devil was come for him, and such frightful Whimseys."[81]

The ending, which had been so slow in coming, lasted but three weeks. Chaloner went on trial for high treason on March 2, 1699. Newton's several witnesses, all of whom were felons themselves, were allowed to recall the events of the past six years with virtually no regard for specifics as to time and place. The jury found the accused guilty the next day, and he was sentenced to be drawn on a sledge from Newgate to Tyburn and hanged. The King himself heard Chaloner's petition for a pardon on March 17, but he had no real chance for a reprieve, let alone legal absolution for his countless sins. He wrote one of his last letters to his great adversary, for whom he had developed a grudging respect. "I am going to be murthered allthough perhaps you may think not but tis true I shall be murdered the worst of all murders that is in the face of Justice unless I am rescued by your mercifull hands."[82] Newton, who could have spared him, if only for the moment, drafted no reply. His avenues of appeal exhausted, Chaloner, in a final salute to the victor, had his copper plate delivered to the Warden the day before he died. During his passage from Newgate to the gallows in Hyde Park on March 22, Chaloner is said to have lost his nerve and cried out to the assembled throng that he was murdered by perjury and injustice, while simultaneously cursing the name of Holloway, who kept his own rendezvous with the hangman a short time later. Though lessons abound in this true-life morality play, it seems only fitting that its anonymous narrator be allow the final words, which are likely as much Isaac Newton's as his own: "Thus liv'd and thus dy'd a Man, who had he squar'd his Talent by the Rules of Justice and integrity, might have been useful to the commonwealth: But as he follow'd only the Dictates of Vice, was as a rotten Member cut off."[83]

Chapter Sixteen

"Your Very Loving Unkle"

Her pure, and eloquent blood
Spoke in her cheeks, and so distinctly wrought,
That one might almost say, her body thought.

John Donne

I

At Cambridge Newton had been content with the slow progress of the days and seasons, the sense of isolation whose continuity was rarely interrupted by dramatic events. Whether he liked it or not, London and the Mint had transformed him into something of a man of action, a direct participant in the flow of those significant occurrences called history. In June 1697 a young Russian left home for Western Europe after having a seal engraved for himself with the inscription: "I am a pupil and need to be taught." Though he stood 6 feet, 7 inches, a height that in the seventeenth century would have made his skeleton the prize of every scientific society in existence, he intended to travel incognito, the better to protect himself from the hated formality and ceremony that inevitably attached to one of his position. After a leisurely stay of five months in Holland, where he met with a warring William III at Utrecht, he sailed for England to study English shipbuilding techniques and other forms of technology that might benefit his scientifically backward homeland. Arrangements were made to house the distinguished visitor and his party at Sayes Court, the splendid estate of John Evelyn in Deptford. From there the young man would have ready access to the shipyards, where arrangements had been discreetly made for his employment as a laborer. He was also taken to the Tower of London to

412

observe the recoinage, which was in full spate. It was there that Newton met him, this most progressive yet most authoritarian of all Europe's rulers. The Warden was informed by his kinsman Sir John Newton on February 5, 1698, that the strange visitor had specifically requested a meeting with the author of the *Principia Mathematica*: "The Czar intends to be here tomorrow before 12 and I thought my self obliged to signify to You he likewise expects to see You here. I have taken all possible care to have things in readiness & have not time to add more."[1] Exactly what passed between Peter the Great, Tsar of all the Russias, and Isaac Newton, author of the majestic law of universal gravitation, we do not know. Montague, who was then Chancellor of the Exchequer, was also present, but he probably deferred to the more knowledgeable object of his patronage when it came to explaining the fine points of minting new coin. We do know that Peter was sufficiently impressed to return. Two years later he initiated reforms in the Russian coinage based on those shown him at the Tower.

The Tsar also visited Greenwich Observatory, where he discussed mathematics with the duly impressed Astronomer Royal, and later showed up at England's main cannon factory, the Woolrich Arsenal. Peter discovered a compatible spirit in Romney, Master of the Ordnance, with whom he shared a delight in artillery and fireworks. Indeed, it would seem that about the only person who did not take pleasure in Peter's sojourn was Evelyn, for whom the high honor of having his estate occupied by foreign royalty became a lasting source of private anguish. The diarist had spent forty-five years planning and laying out the magnificent grounds at Sayes Court, which was vacated and redecorated for Peter and his companions. The first clue that something was seriously amiss came in the form of a letter from one of Evelyn's servants: "There is a house full of people, and right nasty."[2] Not until the Tsar and his party had left did the full extent of their wildly destructive behavior become evident. The carpets were so befouled with ink and grease that the floors beneath them had to be replaced. Windows were smashed in nearly every room, causing the occupants to reduce some fifty elegant chairs to kindling in an effort to stave off the chill winter winds. Featherbeds and their coverings were shredded "as if by wild animals," and twenty pictures and paintings were punctured repeatedly during frequent sessions of indoor target practice. The wanton damage inflicted on the lawn and gardens was no less severe and proved considerably more difficult to mend. Almost nothing, including a beautiful holly hedge, 400 feet long, 9 feet high, and 5 feet thick, had been spared. (Neighbors observed that the Russians played a game with three wheelbarrows, which were unknown back home; they placed a man in each one, the revoltingly dirty Peter included, and raced headlong into the hedges.) Evelyn finally called out Wren, the Royal Surveyor, and London, the Royal Gardener, who jointly recommended a settlement by the crown of £350, which it eventually paid.[3]

The day of Peter's arrival in England, as noted in Evelyn's diary, coincided with the unleashing of a far greater destructive force than a scurvy pack of Russian revelers, however crude: "White-hall burnt to the ground, nothing but the walls & ruins left."[4] The conflagration began about four in the afternoon of January 4 and raged out of control until the following morning. Its origins are still debated, but it was blamed at the time on a laundress who was said to have left a fire unattended while drying clothes in an upper room. Living nearby, Newton must have watched the shooting flames and the panicked citizenry with a sense of foreboding, for he doubtless remembered the Great Fire, which destroyed the city when he was a young man and whose grotesque effects he had observed during his first visit to London. Owing to the fact that much of his time was spent at the Treasury located on its sprawling grounds, it is possible, though by no means certain, that he had been inside the doomed palace's very walls when the alarm was first sounded.

Before leaving England, Peter sat for the obligatory portrait by Kneller, who, as one of a long line of court painters, performed the functions since taken over by official government photographers. The Tsar's contemporaries thought it a remarkable likeness, and the original still hangs in the King's Gallery of Kensington Palace, to which the court removed after the devastating fire. Shortly thereafter, in 1702, Newton sat for his second portrait by Kneller. In contrast to the earlier rendering, the flowing silver hair has yielded to the dark curled wig of a man of substance and position. The austere angular visage has been softened by the acquisition of more flesh, including a second chin. The eyes are just as protuberant and imperious as they were in 1689, however, and the long nose just as prominent. A loose covering robe has replaced the academic gown, which Newton rarely if ever donned again. In later portraits he is occasionally arrayed in fine velvets and rich brocades, the metamorphosis from professor to bureaucrat and capitalist complete.

If Conduitt's reminiscences are accurate, Peter the Great was not the only foreign ruler to take a direct interest in Newton. In the spring of 1698 Jacques Cassini, son of astronomer Giovanni Domenico Cassini, spent considerable time in London. Conduitt later reported that Cassini offered Newton a large pension on behalf of King Louis XIV.[5] Newton respectfully declined, though on what grounds Conduitt did not say—nor did he have to. The French Academy of Science was then being reorganized, and the offer almost certainly had to do with an appointment to full membership, which, on the average, was worth some 1,500 livres a year, a substantial amount but not enough to tempt the uncompromisingly anti-Catholic Newton. Thus snubbed, Louis chose not to include Newton's name on his list of three nominees to the strictly honorary position of foreign associate the following year. The Academy itself proved more forebearing than the King. It made Newton a foreign associate in February 1699, along with Roemer, Hartsoeker, and the brothers Bernoulli.

From Johann Bernoulli, mathematics professor at Basel, Newton had received an intriguing communication on January 29, 1697. The letter contained two challenge problems, one of which had been published the previous June in the *Acta eruditorum*. No one had forwarded a satisfactory solution within the prescribed time limit of six months, although Leibniz had solved the first problem and requested that the deadline be extended to offer all contestants an equal chance. Bernoulli consented and added the second problem before writing the *Philosophical Transactions* and *Journal des sçavans*, in addition to Wallis, Newton, and certain mathematicians on the Continent. "Let him who can, grasp the reward we have prepared, . . . a reward which is not a sum of gold or silver, for that is a reward by which only mean and mercenary minds are hired, minds from which we expect nothing either meritorious or fruitful for knowledge." Rather, "we offer a prize suited to a man of free birth, a prize composed of the honour, the praise and the applause with which we mean . . . to celebrate the perspicacity of our great Apollo-like seer."[6] Bernoulli's motives were not quite so exalted as his rhetoric. Neither, for that matter, were those of Leibniz. Both were eager to prove to their own satisfaction that a certain highly regarded English mathematician was deficient when it came to a command of the new differential calculus. Newton's silence during the previous six months was an encouraging sign, but they needed to be certain before laying claim to the very laurels they were publicly, but disingenuously, offering to others.

Known to mathematicians as the brachistochrone, the first of Bernoulli's twin challenges required the successful contestant to determine the curve in which a heavy body falling under the force of its own weight will descend most rapidly from a given point to another given point. The second problem required him to find a curve, having this same property, such that the sum of any two segments of a straight line drawn to intercept it, and raised to any power, will remain constant. While he was loath to risk involvement in such matters, Newton must have felt that his reputation as England's preeminent mathematician was on the line. This, in fact, was the real challenge, and he responded without hesitation.

According to his niece Catherine Barton, whose testimony was incorporated into her future husband's collection of anecdotes, Newton opened Bernoulli's letter late in the afternoon: "Sr I. N. was in the midst of the hurry of the great recoinage & did not come home till four from the Tower very much tired, but did not sleep till he had solved it wch was by 4 in the morning."[7] Though he might well have rested up before drafting a formal reply, he immediately set down the answers to both problems in a long letter to Montague, President of the Royal Society. It is dated January 30, one day after Newton received Bernoulli's challenge.[8] The solutions were published anonymously in the *Philosophical Transactions* for February 1697 and also read before the Royal Society. At Paris, Varignon, who received the problem of the brachistochrone in May, admitted that he was baffled by it, nor did he know of any mathematician capable of its solution.

His countryman l'Hôpital praised Bernoulli for his ingenuity in devising the "curious and pretty" problem but pleaded ignorance, "for physics embarrasses me." Six months later the Marquis forwarded a fake solution, and then only after conferring with the mathematician Joseph Sauveur. Wallis wrote nothing at all, while David Gregory, who spent two frustrating months conducting fruitless calculations, finally called on Newton for assistance, received it, and was forced to query him again.[9] In point of fact the answers to Bernoulli's problems had been well within Newton's grasp since 1684, if not earlier. The fact that he required all of twelve hours to solve them has been attributed to his total break with things mathematical for at least a year.[10] Rusty or not, Newton had accomplished in one night's labor what few other mathematicians of his generation were able to achieve over a lifetime. Bernoulli, despite his ulterior motives, could not hide his admiration. He had not the slightest doubt as to the identity of the man whose elegant solutions reached him via the *Philosophical Transactions*. As he wrote in a letter to Basnage de Beauval, the precision and quality bore the distinctive mark of Isaac Newton: "*ex ungue Leonem*"—"from the claws of the Lion."[11]

Instead of buoying Newton's spirits, the episode left a permanently bitter taste in his mouth. He dreamed of one day confounding the foreign mathematicians as they had attempted to confound him. "Newton had meant in his turn to propound to Bernoulli and Leibniz a problem about the path of a projectile when the resistance varies as the square of the speed," Gregory wrote in February 1698, "a problem which Leibniz had solved in a way not entirely satisfactory, in the *Leipzig Acts.*"[12] In the meantime a chastened and embarrassed Leibniz had written to the Royal Society disclaiming any responsibility for the challenge problem.[13] Ultimately it was Flamsteed who paid the price for Newton's lingering pique. An innocent reference to his lunar theory by the Astronomer Royal was answered with a broadside in January 1699: "I do not love to be printed upon every occasion much less to be dunned & teezed by foreigners about Mathematical things or to be thought by our own people to be *trifling* away my time about them when I should be about ye Kings business."[14] How much better it might have been—if only for Flamsteed—had Newton made good his threat to best Bernoulli and Leibniz at their own game.

II

His protestations notwithstanding, the King's business, though pressing indeed, was not the only business that occupied Newton's time during the recoinage. Early in July 1697 he conducted an examination of five youths enrolled in the Mathematical School at Christ's Hospital. If the boys were shaken at the prospect of being questioned by the world's most accomplished natural philosopher, they gave no evidence of it, for Newton pronounced them qualified to go to sea as apprentices. (Could it be that in such relatively inconsequential matters he was an easy mark?) He was present at the institution twice more during the year when educational issues were discussed.[15]

His rising value as a patron was also exploited, but not always successfully. David Gregory, who had been appointed Savillian Professor of Astronomy largely through the instrumentality of Newton, was hoping to obtain a position in London consistent with his professorship. He discussed the matter with his friend during a visit to the capital in 1697 and broached it again in his letter of December 23: "No doubt it will be thought necessary to instruct that young Prince in Mathematicks, and to take your advice in the choice of the tutor therein."[16] The prince to whom Gregory referred was William, Duke of Gloucester, the eight-year-old son of Anne, the future queen. Whether, as Gregory surmised, Newton's thoughts on the prince's education were solicited by his mother or her advisers is not known, but Gregory did not secure the coveted post. It was just as well from Newton's perspective, for Flamsteed also nurtured hopes of being appointed William's tutor. When the Astronomer Royal got wind of Gregory's ambitions he wrote Newton an indignant letter reminiscent of those bemoaning other lost opportunities: "I am told Dr Gregory is scheming to be Tutor of Mathematics to the Duke of Gloscester. *Which place I was told some Moneths agone . . . was designed for me.*" Incredibly, Flamsteed then went on as if the appointment actually meant very little to him: "I doe not thinke the Duke yet fit for a Mathematical tutor or that he will be this four or five yeares I hate flattery & shall not goe to Court on this account till I am sent for." Yet he closed with another swipe, also underscored, at Gregory's *"treacherous behavior."*[17] Flamsteed waited in vain for his call to court. The frail prince died at the end of July 1700, shortly after reaching his eleventh birthday.

At the same time Halley, who was mired in the bureaucratic feuding at Chester, hoped Newton could extricate him from the thankless post originally obtained through the Warden's influence. It proved a rather embarrassing situation for Newton, who sought to aid his erstwhile publisher by recommending him for two teaching vacancies with the King's Engineers. Halley, perhaps out of loyalty to Newton but more likely for financial considerations, remained at the Chester Mint until its closing a year later. Newton's role of patron manifested itself in a somewhat different manner in 1699. Along with Montague, Sloane, Aston, and some other members of the Royal Society, he underwrote the publication of 120 copies of Edward Lhuyd's *Lithophylacii britannici ichonographia.*[18]

Despite its now easy accessibility, the Royal Society did not play a prominent role in Newton's life during his years as Warden, a measure of his commitment to the recoinage and the suppression of crime. The Society

honored him by election to the Council in 1697, but the Journal Book indicates that he attended none of the Council meetings, nor did he appear at
any of the regular Wednesday meetings of the Society as a whole. It was the
established custom of the Warden, Master, and Comptroller of the Mint to
meet every Wednesday and Saturday during part of the year, although their
deputies were permitted to attend in their place. After being chosen President of the Royal Society Newton eventually changed its weekly meetings
to Thursday to avoid this personally inconvenient clash of dates. Not until
April 19, 1699, after the feverish pace at the Mint had slackened somewhat,
did he finally put in an appearance to report on a mathematical work by the
Spaniard Hugo de Omerique. He returned in August with a sextant, "being
the old instrument mended of some faults." Hooke characteristically laid
claim to the refinements at the next meeting on October 25. Newton, who
had been elected to the Council a second time, chose not to return that
year, perhaps in part because he was determined to avoid conflict with his
old nemesis.[19] A temperamental loner, he joined none of London's
prestigious and often risqué social clubs, any number of which would have
welcomed the renowned savant with open arms. Samuel Johnson's incisive
description of Sir John Hawkins fits Newton equally well: hard-working,
grave, and straitlaced, he was "a most *unclubbale* man."[20] Neither, it seems,
could he be drawn out for regular discourse on a higher plane. Richard
Bentley's attempt to form a private circle of London's intellectual elite late
in the 1690s came to naught. Its projected membership, in addition to the
rising young clergyman, was to include Wren, Evelyn, Locke, and Newton.
Never once, so far as is known, did they gather in Bentley's quarters at St.
James's Palace, where he was serving as Keeper of the King's Library.[21]

Newton's prolonged distraction from matters scientific did not apply to
the nagging question of the moon's irregularities. Flamsteed met with him
in August 1697 and followed up the London visit with a cordial letter of "no
great need" in which he expressed a willingness to correct earlier lunar
observations in hopes of better serving his friend, a significant gesture of
goodwill considering his harsh treatment at Newton's hands two years
before. Having received no reply, the Astronomer Royal wrote an equally
warm letter on December 10, declaring himself "fully satisfied yt all ye
Celestiall bodyes are moved by the laws of Gravity."[22] Flamsteed did not
yet realize it, but Newton was beyond placating. David Gregory, in another
of his memoranda, dated July 1698, wrote: "On account of Flamsteed's
irascibility the theory of the Moon will not be brought to a conclusion, nor
will there be any mention of Flamsteed, nevertheless [Newton] will complete to within four minutes what he would have completed to two, had
Flamsteed supplied his observations."[23] The problem from Newton's perspective, as before, was one not only of numbers but of quality. Much of the
data forwarded by Flamsteed proved useless, the result of erroneous
calculations undertaken by Flamsteed's assistant, James Hodgson. When he

learned what had happened, Flamsteed labored hard to make amends, but he, in turn, committed errors of his own, which vexed Newton the more. To make matters worse, Flamsteed, ever the credulous victim of rumors, feared that his tight-lipped acquaintance was keeping critical discoveries from him in violation of their long-standing pledge of mutual candor. Writing to the mathematician John Colson in October 1698, Flamsteed inquired about a report obtained from a servant that Newton told "you he . . . had perfected the Theory of ye Moon *from Mr Halleys Observations* and imparted it to him wth Leave to publish it." Flamsteed was still fretting over the matter two months later when he wrote Newton: "In your letter you say these corrections will Answer all my observations within 10 minutes. Mr Halley boasts that those you have given him will represent them within 2 or 3 Nearer."[24]

The ever predictable boiling point was reached soon thereafter. When Newton visited Greenwich on December 4, 1698, to collect some corrected lunar observations, Flamsteed wrote that "he is reserved to me contrary to his promise, I lie under no obligation to be open to him."[25] While put out with Flamsteed for supplying him with inaccurate data, Newton had a more immediate cause for anger, or so he had convinced himself. John Wallis was preparing the third and final volume of his mathematical works for the press, and he had requested from Flamsteed his observations on the phenomenon called stellar parallax, which Wallis was anxious to print. The Astronomer Royal was hesitant to publish anything at that time, but, he later informed Newton, "I wrote my letter to Dr. Wallis . . . to silence some busy people yt are allwayes askeing, *why I did not print?"*[26] Without dreaming that he would offend anyone, least of all Newton, Flamsteed included a paragraph relating to his past activities regarding lunar theory:

> I had become closely associated with Mr Newton, at that time the learned Professor of Mathematics at the University of Cambridge, to whom I had given 150 places of the Moon, deduced from my observations, her places as computed from my tables, and I had promised him similar ones for the future as I obtained them, together with the elements of my calculation in due order, for the improvement of the Horroccian theory of the Moon, in which matter I hope he will have the success comparable with his expectations.[27]

David Gregory, who had no love for Flamsteed, read the astronomer's paper while it was en route to Wallis and immediately informed Newton of the reference to his name. (Flamsteed later charged that Gregory, his rival for the post of tutor to the Duke of Gloucester, did so out of naked self-interest, and he may well have been right.) When Flamsteed returned home from a visit to London the night of December 28, a letter from Wallis was waiting. The mathematician had just received a communication "from one in London, which desires me not to print any paragraph of your letter which speaks of your giving to Mr Newton Observations of the Moon."

Wallis refused to name his correspondent, stating only that "He is a friend to both of you."[28] Flamsteed did not have to be informed, as he later was by Newton, that Gregory was the "friend" to whom Wallis alluded, for he angrily wrote in the margin that Gregory was no friend of his. After toying with the idea of calling on Newton in London, Flamsteed wrote to him instead. Newton replied in less than a week:

> Upon hearing occasionally that you had sent a letter to Dr Wallis about ye fixt starrs to be printed & that you had mentioned me therein with respect to ye Theory of ye Moon I was concerned to be publickly brought upon ye stage about what perhaps will never be fitted for ye publick & thereby the world *put into an expectation of what perhaps they are never like to have.* I do not love to be printed upon every occasion much less to be dunned & teezed by forreigners about Mathematical things or to be thought by our own people to be *trifling* away my time about them when I should be about ye Kings business. And therefore I desired Dr Gregory to write to Dr Wallis against printing that clause wch related to that Theory & mentioned me about it. You may let the world know if you please how well you are stored wth observations of all sorts & what calculations you have made towards rectifying the Theories of ye heavenly motions: But there may *be cases* wherein your friends should not be published without their leave. And therefore I hope you will so order the matter that I may not on this occasion be brought upon the stage.[29]

However irritable in temper Flamsteed may have been, he had acted with dignity and moderation throughout this bleak episode. Maintaining his composure, he replied to Newton's devastating assault on January 10, 1699. "I did not think I cou'd have disobliged you, by telling the world yt the Kings Observatory had furnished you with 150 places of the Moon." After all, Newton himself had spoken of the matter to others on several occasions. Nevertheless, he would instruct Wallis to delete the offending paragraph if it meant the preservation of "a reale friendship" with an ingenious man.[30] Flamsteed kept his promise, but Newton refused to be moved. Four years passed before he addressed another letter to the wounded denizen of Greenwich Observatory.

III

In December 1677 twenty-five-year-old Hannah Smith became the second wife of Robert Barton, a widower and clergyman from the Northamptonshire village of Bridgstock. Her half-brother, Isaac Newton, Lucasian Professor of Mathematics at Cambridge, cosigned the marriage articles, as did other members of the family, including their mother, Hannah Smith.[31] Little is known of the bridegroom, except that his forebears had possessed estates in Northamptonshire for generations and were related to a number of the county's most respected families. The Bartons' fortunes had gradually declined, however, making Hannah's inheritance of £500 from her late father, the Reverend Barnabas Smith, a welcome addition to the meager assets of a struggling country vicar. (When her mother died suddenly in some kind of epidemic a year and a half later, Hannah inherited £300 more.)[32] Like her mother, she bore her cleric husband three children, two girls and a boy: Hannah, who died in childhood; Robert, a reprobate army officer who was killed during Hill's ill-starred expedition against Quebec; and Catherine, a talented and vivacious charmer who lived in the London home of her stepuncle for several years before her marriage to John Conduitt in August 1717.

The exact time of Catherine's arrival at Newton's Jermyn Street house has been much debated but never established with any certainty. She turned seventeen in 1696, the year Newton acquired the property, and he may have sent for her to preside over the domestic arrangements upon moving in. Catherine's account of her uncle's receipt and resolution of Bernoulli's challenge problems in January 1697 would tend to support this view. On the other hand, the details may well have come from Newton himself, only to be passed on by his admiring relation many years later. Be that as it may, there is no disputing the fact that she had taken up residence with her uncle at some point before the summer of 1700. The only surviving letter from Newton to Catherine is dated August 5 of that year, when she had gone into the country to recuperate from the smallpox. Newton's affection for his niece, then twenty, is palpable, his tone solicitous and familiar. He is indebted to the Gyres for their ministrations to Catherine during her convalescence and proposes a simple and extremely mild palliative to relieve her lingering fever. He is also anxious to know if her delicate face has been permanently marked:

> I had your two letters & am glad ye air agrees wth you & th[ough the] fever is loath to leave you yet I hope it abates, & yt ye [re]mains of ye small pox are dropping off apace. Sr Joseph [Tily] is leaving Mr Tolls house & its probable I may succeed him [. I] intend to send you some wine by the next Carrier wch [I] beg the favour of Mr Gyre & his Lady to accept of. My Lady Norris thinks you forget your promis of writing to her, & wants [a] letter from you. Pray let me know by your next how your f[ace is] and if your fevour be going. Perhaps warm milk from ye Cow may [help] to abate it.[33]

Newton signed himself "Your very loving Unkle." He would be a grateful one as well, if we may judge from later descriptions of Catherine's beauty. Not a single scar defaced that most pleasing countenance.

Catherine Barton appears to have been the one relative of Newton who was gifted beyond the ordinary. And with the exception of his mother, who died in the year of Catherine's birth, she was the only one to influence his life to any degree. It has often been written that Newton personally supervised her education, but there is no documentation to support this dubious claim. Perhaps he did direct her reading for a time, while seeing to it that she was properly schooled in the etiquette required of a prominent citizen's niece and keeper of his household ("My Lady Norris thinks you forgot your promis of writing to her"). But since she could not have arrived in London before her seventeenth birthday at the earliest, Catherine would already have received the formal instruction considered proper for a young woman of her day. While she possessed a mind sufficiently keen to attract and hold the attention of talented men such as Jonathan Swift, she, like the rest of her sex, was most often judged by the narrow and inflexible standards of a male-dominated society. Catherine was not found wanting. Those who were lavish in their praise of her, and they were many, tended to focus on Miss Barton's grace, her wit, but most especially her extraordinary physical beauty.

Her uncle's friend and patron Charles Montague, who was created Baron Halifax in 1700, became her fervent admirer after she had taken up residence in Jermyn Street, though exactly when and under what circumstances are not known. Halifax was a leading member of the Kit-Kat Club, a fashionable Whig Association named after one Christopher (Kit) Cate in whose home it regularly assembled. It was the custom of members to inscribe verses to prominent beauties on toasting glasses with a diamond before drinking to their charms. When it was Halifax's turn in 1703, he is said to have composed the following lines:

> Beauty and Wit strove each in vain
> To vanquish Bacchus and his Train;
> But Barton with successful Charms
> From both their Quivers drew her Arms;
> The roving God his Sway resigns,
> And awfully submits his Vines.

Not content with a single verse of little merit, Halifax allegedly wrote two others, neither of which is any more distinguished than the first, perhaps because he honored half a dozen women in the same fashion that year, thus spreading his modest literary talents too thin. The following rhyme, also attributed to his pen, is of interest mainly because it invokes imagery of the Mint, possibly a self-conscious bow in the direction of the censorious, silver-haired figure in the background:

> Stampt with her reigning Charms, this Standard Glass
> Shall current through the Realms of Bacchus pass;
> Full fraught with beauty shall new Flames impart,
> And mint her shining Image on the Heart.[34]

It was writing of this quality that prompted Samuel Johnson to remark: "It would now be esteemed no honour...to be told, that, in strains either familiar or solemn, he sings like Montague."[35]

Swift, too, was much taken with the young woman when he met her in 1710, though for rather more platonic reasons. An excellent conversationalist and enthusiastic gossip who loved a risqué tale as much as the future Dean of St. Patrick's, Catherine became a cherished friend of whose company he could never get enough. Her name appears frequently in *Journal to Stella,* the title of the collected letters Swift addressed to Ester Johnson (Stella) and Rebecca Dingley. "I was this morning to see Mrs. Barton,"* he wrote in April 1711, "I love her better than any body here, and see her seldomer. Why really now, so it often happens in the world."[36] Rémond de Monmort, a French bureaucrat, fell head over heels for Catherine during a brief visit to London a few years later, notwithstanding the fact that there was a Madame de Monmort back home. "Ever since I beheld her," the smitten Monmort confided by letter to Brook Taylor of the Royal Society, "I have adored her not only for her great beauty but for her lively and refined wit. I believe there is no danger in betraying me to her. If I had the good fortune to be near her, I would henceforth become as awkward as I was the first time we met."[37] More than a century and a half later, Newton's biographer Sir David Brewster, who possessed a now lost photograph of Catherine's vanished miniature, wrote to a friend of her stunning, almost mesmerizing beauty.[38]

Aside from the admiring verses carved on the toasting glasses of the Kit-Kat Club, we hear nothing of Catherine between the summer of 1700 and April 12, 1706, when Halifax added a codicil to the will he had made two days before. It specified that all his jewels and £3,000 go to Miss Barton "as a small Token of the great Love and Affection I have long had for her." Six months later Halifax added another bequest by purchasing for her, in Newton's name, a lifetime annuity of £200 a year.[39] Voltaire, who admired Newton and first drew popular attention to the theory of gravitation on the Continent, got a whiff of the scandal resulting from this more than generous act when he visited England at the time of the natural philosopher's death in 1727. It would have been asking too much of the iconoclastic savant to overlook such a juicy morsel, which, of course, he did not: "I thought in my youth that Newton made his fortune by his merit. I supposed that the Court and the city of London named him Master of the Mint by acclamation. No such thing. Isaac Newton had a very charming niece, Madame Conduitt, who made a conquest of the minister Halifax. Fluxions and gravitation would have been of no use without a pretty niece."[40] There was more. On February 1, 1713, two years before he died, Halifax replaced the 1706 codicil with a second, whose terms were nothing short of munificent. It began with a bequest of £100 to Isaac Newton "as a Mark of the great Honour and Esteem I have for so Great a Man." The bequest to Catherine had risen to £5,000 and a life interest in the rangership and lodge

* Mrs. was the common form of address for a single woman at this time.

of Bushy Park, just to the north of Hampton Court. To make certain that she would be able to maintain the house and sprawling grounds, Halifax also granted her a life interest in Apscourt Manor, his country estate in Surrey, together with all the rents and profits accruing from the lucrative property. "These gifts and Legacies, I leave to her as a Token of the sincere Love, Affection, and Esteem I have long had for her Person, and as a small Recompence for the Pleasure and Happiness I have had in her Conversation."[41]

When he received word of Halifax's death and this startling bequest in 1715, Flamsteed, who had broken with Newton after a bitter quarrel over the publication of his star catalogue, could not resist the temptation to exploit the opening. "I doubt not but you have heard that the Lord Halifax is dead of a violent fever," he wrote to Abraham Sharp on July 9. "If common fame be true, he died worth £150,000; out of which he gave Mrs. Barton, Sir I. Newton's niece, for her *excellent conversation,* a curious house, £5000, with lands, jewels, plate, money, and household furniture, to the value of £20,000 or more."[42] Flamsteed's meaning, unnecessarily underlined for the sake of emphasis, was unmistakable, for in the eighteenth century conversation meant companionship, not mere social discourse.

"The life of the Earl of Halifax," Johnson wrote, "was properly that of an artful and active statesman, employed in balancing parties, contriving expedients, and combating opposition, and exposed to the vicissitudes of advancement and degradation."[43] This last was also true of his checkered private life. Having resigned his fellowship at Trinity College after the enormous success of *Country Mouse and City Mouse,* a burlesque of Dryden's *The Hind and the Panther,* Montague signed the invitation to the Prince of Orange and later sat in the Convention Parliament with Newton. The young man's fortune was literally made in 1688, two months before his twenty-seventh birthday, when he married the wealthy Countess Dowager of Manchester, a cousin more than twice his age. The Countess died childless from this union a decade later, in 1698, by which time Montague, something of a genius in economic matters, had become England's most powerful political leader. The Duchess of Marlborough, who despised both Montague's Whig politics and his breeding, described him as a "frightful figure" and a lecher who "pretended to be a lover, and followed several beauties, who laughed at him for it." The lady's judgment may have been a bit too harsh. The painting of Halifax in the National Portrait Gallery depicts not a frightful figure but a rather ordinary obese one, with features inclining toward the effeminate: an unwrinkled forehead; large, dark eyes beneath delicately arched brows; thick, sharply curving, sensual lips; a large but delicate nose; a round, fleshy face and bulging neck. He had the disposition of the overweight, easily aroused and as easily calmed. As to being spurned and laughed at by beautiful women, virtually every description we possess of Catherine Barton places her among the fairest of the fair. On the other hand, Halifax, a powerful, freewheeling, and gregarious peer of the realm, toasted many a pretty damsel who parried his advances, while Catherine, the lowborn daughter of a country parson, would hardly have figured in the calculations of the noble Duchess of Marlborough.

Voltaire's charge that Newton owed his public office to the seductive charms of a bartered niece almost certainly found its origin in the poisoned pen of Mrs. Mary de la Rivière Manley, notorious author of a 1710 roman à clef aimed at further defaming the Whig leadership, already under attack from several quarters. She, a fervent Tory, titled her satirical work *Memoirs of Europe, Towards the Close of the Eighth Century* and attributed its authorship to Eginardus, secretary to and court favorite of the Emperor Charlemagne. Reprinted several times, the popular work contained in its later editions a printed key in which Bartica, "ever a proud Slut," was identified as Newton's niece and Julius Sergius as Lord Halifax, the man she wishes to marry. Sergius calls his mistress "A Traitress, an inconsistent proud Baggage," and worse—"yet I love her dearly, and have lavish'd Myriads upon her, besides getting her worthy ancient Parent [Isaac Newton] a good Post for Connivance."[44] The savagery of political conflict in the age of Queen Anne was perhaps the worst in modern English history. "We fight with the poison pen of the tongue," Defoe ruefully observed, "with words that speak like the piercing of a sword, with the gall of envie, the venom of slander, the foam of malice."[45] Mrs. Manley was eventually brought to trial for her antics, but to no avail. In the *Adventures of Rivella* (1714), an autobiographical account of her life in London society, she gleefully told of being acquitted: "her defense was with much Humility and Sorrow, for having offended, at the same Time denying that any Persons were concern'd with her, or that she had a farther design than writing for her own Amusement and Diversion in the country."[46]

As others writing of this troubling matter have observed, the charge that Newton profited financially from Catherine's attentions to his patron is not supportable, unless one chooses to construe Montague's modest bequest of £100 to him in such a light. When Halifax obtained the position of Warden for Newton in 1696, his niece, only sixteen, was still living in Northamptonshire with her mother. Her uncle was alone in Cambridge. Not only was Halifax then ignorant of her existence, but one wonders if Newton himself had yet laid eyes on the girl. The Trinity College records of his travels to and from Cambridge suggest that he had not. Catherine was almost certainly living with him when he became Master of the Mint in 1699, the position specifically referred to by Voltaire, although the French philosopher probably never distinguished between the two Mint offices in his thinking. Even if he did, the evidence against a connection between Newton's further rise and Catherine's charms would again seem conclusive. Montague's career, which up to that point had been one of uninterrupted success, took a sudden, sharp downward turn. His considerable vanity and overbearing conduct made him vulnerable to charges of corruption at the Treasury; in November 1699 he resigned the office of Chancellor of the Exchequer, a month before Neale's death and Newton's selection as his successor. Meanwhile, Montague had also lost his position as leader of the House, which further undercut his political influence. An anxious Newton, identified as he was with Montague, may well have looked upon their old friendship as a political liability rather than an asset during this period of crisis, both public and private. By the time Montague was cleared completely of the many spurious charges brought against him, Newton had been Master for nearly six years.

If Newton can be absolved of profiting from his niece's attractions, we are still left facing the same nagging questions that so perplexed his early biographers: What form did the relationship between Baron Halifax and Catherine Barton take? If theirs was a sexual liaison, how is it possible, assuming he knew, or at the very least suspected, for one so puritanically virtuous as Newton to countenance it? The issue is further complicated by the fact that little if any new information on the subject has come to light since early in the eighteenth century.

The rumor has long circulated that Catherine lived in Halifax's home for some unspecified period before he died. On September 18, 1710, Swift dined "with Mrs. Barton alone at her lodgings." Since he referred to her as his "near neighbor," we are led to believe that she was then living with her uncle, who had lately moved to St. Martin's Street, a short walk from Swift's rooms in Leicester Fields. The satirist also dined at the home of Halifax during that period, but never once does he mention seeing the two together.[47] Yet it is possible, indeed probable, that Swift, suspecting his letters would one day have a far wider audience than the long-suffering Stella, was seeking to protect his friends, whose honor had recently come under vicious attack by Mrs. Manley. (It was perhpas also for this reason that he said nothing whatsoever of Newton, whom he almost certainly met on occasion.) Despite their often bitter disagreement over politics, Swift cared for Halifax almost as much as he did for Catherine Barton: "I told him he was the only Whig in England I loved, or had any good opinion of."[48] Whatever Swift's intent, there can be little doubt that Catherine was living under the same roof as Halifax by 1713, when the codicil bequeathing her an estimated sum of £25,000 went into effect. As further proof of this she sent her uncle a rather pathetic little note after Halifax died: "I desire to know whether you would have me wait here...or come home.... Your Obedient Neece and Humble Servt. C. Barton."[49] She was welcomed back, and Newton took on the onerous task of protecting her legal interests in the huge estate.[50] This, and the fact that in 1706 Newton had permitted Halifax to purchase for Catherine an annuity in his name, hardly supports the observation of one prominent twentieth-century historian who, in an effort to account for Newton's uncharacteristic behavior, chose the ultimate of rationalizations: It "may have been simply due to his not noticing."[51]

The Victorians certainly believed that Newton had noticed what was going on, and they went to ingenious lengths in an effort to dispel the dark cloud that threatened their godlike hero's reputation. To them, Halifax's great generosity toward Catherine was evidence of something more than a mere platonic relationship, however profound, something more than the gift of an enchanted lover to his mistress, however dazzling she may have been. Invoking the moral authority of good Christians, they posthumously united the unlikely couple in the bonds of holy matrimony.

Augustus De Morgan, a graduate of Cambridge and the first professor of mathematics at the new University of London, performed this service by proxy in his little volume entitled *Newton: His Friend: And His Niece*. On May 23, 1715, four days after Halifax's death from an acute attack of pleurisy, Newton wrote a brief letter to Sir John Newton, a distant relative, apologizing for not keeping a previous appointment. "The concern I am in for the loss of my Lord Halifax & the circumstances in wch I stand related to his family will not suffer me to go abroad till his funeral is over."[52] De Morgan accepted this single sentence as proof that Halifax died a relative of Newton, "meaning by *proved* what is meant by that word in the jury-box."[53] One would be hard-pressed to go to trial on such flimsy evidence today lacking, as the defense would, a high-minded Victorian jury. Assuming for a moment a morganatic marriage between Catherine and Halifax, kept secret owing to the disparity in their social standing, why has a thorough search of the legal registers turned up no record of it? And if there was good reason for secrecy during Halifax's life, why conceal the marriage after his death, thus exposing the "widow" to endless censure? Finally, when Catherine Barton gave legal notice of her intended marriage to Conduitt in 1717, she swore that she was a spinster.[54]

Alas, catty though he was, Flamsteed seems to have placed the correct interpretation on the language in Halifax's will: His are not the words of a husband but of a grateful lover, tenderly addressing his mistress. An anonymous biography commissioned by Halifax's nephew and major heir sought to salvage Catherine's honor by casting her as the superintendent of his domestic affairs, "a Woman of Strict Honour and Virtue."[55] How surprised her uncle, who had broken with Vigani simply for telling a lewd joke, would have been had he heard some of the salacious stories related to Swift by the virtuous Mrs. Barton: "she told for certain that Lady S_____ was with child when she was last in England, and pretended a tympany, and saw every body; then disappeared for three weeks, her tympany was gone. . . . No wonder she married when she was so ill at containing." On another occasion she entertained Swift with the story of an old gentlewoman who had died two months earlier. She had specified in her will that the sixteen coffin-bearers—eight men and eight maids—should have two guineas apiece, while the parson should have ten. There was one other stipulation: Each must swear to his or her virginity. "The poor woman still lies unburied, and must do so till the general resurrection." Swift also sat with Catherine in October 1711 after she received word of her brother's death in Canada: "I made her merry enough," he wrote, "and we were three hours disputing upon Whig and Tory. She grieved for her brother only for form, and he was a sad dog."[56]

The answer to this centuries-old riddle would appear to be that Newton was drawn to Catherine by the very qualities that so beguiled others: her grace, vivacity, wit, and great natural beauty. His only surviving letter to her is filled with tender concern for her welfare; he prays that her lovely face will bear no permanent marks. Quite simply, Newton loved Catherine like a daughter and came to look upon her as one. To be sure, his toleration of her relationship with Halifax constituted a clear breach of his Puritan scruples. Yet, considering the stark and mostly loveless character of his earlier existence, one can understand and perhaps forgive an aging bachelor—even one so unforgiving as Newton—for choosing companionship and requited affection over binding morality. The affair between his friend and his niece doubtless blossomed slowly, for Halifax would have done nothing to anger or alienate the great man whom he so deeply admired. The first codicil was drafted in 1706, three years after Halifax toasted her charms in public, the second seven years after the first, when Newton was seventy and presumably contemplating his own mortality. By the time the sexual issue came into play the possibility of shock was greatly diminished. Besides, who among Newton's acquaintances was going to broach the subject of a public scandal to the face of the man Bishop Burnet described as "the *whitest* soul he ever knew"?[57] As for attacks of the type delivered by Mrs. Manley, these could easily be rationalized and dismissed for what they were, the vitriol of shameless scandalmongers. Equally important is the fact that Newton nurtured an abiding affection for Halifax, who, in choosing not to abandon Catherine, had proved that his interest in her was as much familial as carnal. As the letter Newton wrote after the Baron's death suggests, he also looked upon the fallen peer as a member of his family; he remained at Catherine's side to console her in her hour of grief, which was no less his own. When the Abbé Alari, tutor of Louis XV, visited Newton in July 1725, he wrote that the old man kept a portrait of Lord Halifax in his chamber, hardly the conduct of a betrayed or outraged guardian.[58] He obviously knew Catherine was under the protection of Halifax and acceded to this relationship, advising her whenever necessary on the business side of the affair. Because the liaison did not bear directly on his own material fortunes, one can reasonably conclude that in this instance Newton's moral judgment was suspended in deference to the feelings of those for whom he cared above all others, not least of all himself. The brusque manner and the hostile intellectual aloofness that made him so difficult to know yielded, as they had for Fatio, to the inner stirrings of the heart.

IV

The beginning of the Mint's new fiscal year was only three days away when Thomas Neale died on December 23, 1699, and the Treasury was anxious to have a new administrator in place as soon as possible. Because Neale's demise was hardly unexpected, it is likely that his successor had been agreed upon for some time. The Mint accounts indicate that Isaac Newton became Master on December 25, his fifty-seventh birthday. Lest his appointment seem a routine matter, it is well to remember that only once before—during the reign of Elizabeth I—had a Warden risen to the post of Master. After Newton no Master was ever again chosen from among the officers of the Mint. Since Montague was not at the time a political force to be reckoned with, the selection of Newton must be credited to his meritorious service as Warden, but most especially to his established reputation as the leading natural philosopher of the age.

As Craig so astutely observed, "A Master of the Mint was hung about with parchments."[59] A year was required to complete the cumbersome, largely medieval paperwork associated with the appointment. Because Parliament had recently seen fit to end the life tenure previously enjoyed by the Master, Newton was forced to go through these same formalities on the respective accessions of Anne and of George I. He seems not to have minded, except for a major change in a long-standing policy whereby the incoming Master was required to post a £2,000 bond as insurance against theft and other forms of delinquency. Henry Slingsby, Neale's predecessor, had been suspended from his post in July 1680 for financial irregularities and was forced to resign in disgrace six years later. Neale, who had ceded most of his duties to Newton, incurred such heavy debts that only his huge income of £22,000, generated in large part by the great recoinage, saved him from bankruptcy. A wary Treasury asked that Newton post a bond of £15,000, the same amount it had requested of the financially troubled Neale. As much a point of honor to him as of inconvenience, Newton balked at so great a figure. He pored over old records and was able to show that, Neale excepted, all his other predecessors back to the rule of Richard III had been asked for only £2,000. "[I]f your Lordships please to accept the usual security," he wrote on June 21, 1700, "I am ready to give it that it may be entred in the Indenture now to be sealed."[60] A reduction to the lower figure was conceded ten days later after little if any debate. It was not until December 23, 1700, the first anniversary of Neale's death, that the Indenture or contract between the new Master and the crown was finally sealed.

Having watched with a disapproving eye while his feckless predecessor made and squandered a veritable fortune, Newton, who had done most of Neale's work for the fixed sum of £400 a year, eagerly looked forward to the income that was now to be his. With the recoinage at an end, he could not hope to earn what Neale had during the past three years, but the potential for the accumulation of considerable wealth was still there. In addition to an annual salary of £500, the Master received $3\frac{1}{4}d.$ per troy pound of coined silver and 1s. 10d. per troy pound of gold. From 1703 until at least 1717 he received £150 per year for overseeing the storage and sale of tin. When this source of revenue dried up, the loss was largely offset by income derived from the coining of copper money, which began in 1718. His yearly expenses, which included special meals for officials at the Mint, his deputy's salary, and certain prescribed gifts, came to roughly £180. But since he, too, received substantial gratuities ranging from barrels of wine to fine books, these expenditures were recouped for the most part. Thus his average annual income, excluding the profits from his lands and various investments, came to about £1,650, though it could oscillate wildly from year to year. In 1703, for example, when minting was at a virtual standstill, he earned some £663, only £13 of which came from royalties, while in 1715, a year of exceptional activity, the figure rose to £4,250.[61]

Newton's most enduring achievement as Master was the conversion of the Mint to standards of high accuracy and precision, something he had pointed toward ever since first setting foot in the Tower. Considerable strides had already been made in this direction, but the Mint was still turning out coins of the same denomination that varied in weight by as much as two or three grains of precious metal. So long as they remained within the remedy—the variation in weight and fineness allowed by the Indenture—the coins were issued. And since they were weighed in pound lots only, significant differences continued to be commonplace. When a customer obtained a packet of new guineas he simply culled out the heavy ones and sent them back to the Mint for recoinage at government expense. On average, he got back 53 guineas for every 52 he returned. These heavy coins were aptly dubbed "Come-again Guineas." According to Newton's own estimates, at least one-eighth and possibly one-fifth of all newly issued gold coins were returned to the Mint in this manner.[62] The practice, while quite legal, served to enrich the goldsmiths and moneyers at the taxpayers' expense. Not only did they reap the benefits of the remedy, they received handsome royalties for their part in the perpetual remelting and reminting of virtually uncirculated coin. Newton had been in office little more than a year before he denied the goldsmiths the full advantage of the remedy and so improved the sizing of gold coins that the culler soon went the way of the clipper. The outraged goldsmiths, who practiced culling themselves, naturally protested Newton's actions, but to little avail. They must have been bewildered as well as angered, for why would the new Master wreck a nearly perfect scheme that served to line his own pockets with royalties as well as theirs?

Newton had left Cambridge in 1696 like a man pursued, and there is nothing to indicate that he ever seriously contemplated a permanent return. Still, he had refused to relinquish his sinecure. Should trouble come, the Lucasian chair and his Trinity College fellowship provided insurance in an era of precious little security for the common man, no matter how uncommonly talented. Besides, the revenues they continued to provide enabled him to cope better with London's high cost of living. He served as Master for a year before deciding to act in December 1701. Deeply reassured by an income that approached £3,500, he resigned his professorship in favor of Whiston and gave up his fellowship about the same time. He soon had occasion for second thoughts, however, for the shaky peace secured by the Treaty of Ryswick did not hold, plunging the country back into the depths of armed conflict and taking his income down with it for the next five years.

On November 26, less than three weeks before he resigned his chair, Newton was once again elected to Parliament by the University. As in 1689 he came in second in the balloting. The other successful candidate, also a Trinity man, was Henry Boyle, nephew of Robert Boyle, afterward Lord Carleton. The voting slips reveal that Newton himself voted for his colleague. However, Bentley, who was now Master of Trinity, "had the satisfaction of assisting in the return of his illustrious friend."[63] The records indicate that Newton had no more to say during this session than during the Convention Parliament. While he may have taken part in committee work, he remained a silent spectator during the debates. His most important vote was cast in favor of Halifax and the other Whig leaders on a Tory motion seeking their impeachment. The Whigs nevertheless lost in the House of Commons, but the charges were later dismissed in the House of Lords, to which Halifax had recently moved, sparing him the politician's ultimate humiliation.

The life of this Parliament was brief. A sickly William III breathed his last on March 8, 1702, and Princess Anne ascended the English throne. Parliament was prorogued in May and dissolved less than two months later. Newton was invited to stand for reelection by the University's Vice Chancellor, but he was less than enthusiastic about the prospect. "I would have it understood that I do not refuse to serve you," he replied, "... but by reason of my business here I desist from soliciting, and without that, I see no reason to expect being chosen. And now I have served you in this Parliament, other gentlemen may expect their turn in the next."[64] The authorities bowed to his wishes and did not enter Newton's name into candidacy.

Other honors and offers of honors were gathering thick around him. Lord Pembroke, a close friend of Locke, pressed Newton to accept the mastership of St. Catherine's Hospital as a means of augmenting his income from the Mint, an offer he gracefully declined.[65] Bentley became the new Master of Trinity College in 1700 only after Newton refused the appointment, which would have required that he take holy orders. The offer was made by none other than the redoubtable Archbishop of Canterbury Thomas Tenison, forcing Newton to adopt some evasive terpsichorean tactics. "Why will not you?" the Archbishop implored, "you know more than all of us put together. Why then said S[r] I[saac], I shall be able to do you the more service by not being in orders."[66] Chary of rousing Tenison's suspicions by a second refusal, he accepted an appointment that same year as one of nine original trustees of the Golden Square Tabernacle, a chapel personally endowed by the Archbishop. He remained in office for twenty-two years and was elected Treasurer at least once, in 1701.[67] His service on the commission appointed to oversee the completion of St. Paul's Cathedral proved less long-lived. According to his niece, Newton, ever the Puritan, entered into an acrimonious exchange with Archbishop Wake over whether pictures should be hung in the sanctuary and never attended another commission meeting.[68]

V

Anne officially became Queen of England the moment William expired early in March 1702, although the coronation itself did not take place until April 23. The new sovereign disliked the idea of political parties but initially leaned toward the Tories because of their traditional support of both the crown and the Church of England. Marlborough, Halifax's old political foe, received the long-delayed garter on March 13, and on the following day he was appointed captain-general of the English Army at home. Instead of following precedent by choosing the Archbishop of Canterbury to preach her coronation sermon, Anne turned to Archbishop Sharpe of York, a leader of the high-church party. And at the prorogation of Parliament in May the Queen declared in no uncertain terms that she would favor those who exhibited "the truest zeal" in support of the established church. These words were underscored by deeds: The membership of the Whig leadership or Junto, comprising Somers, Halifax, Wharton, Russell, and Spencer, lost office to a man, making the new ministry almost exclusively Tory.

These and similar developments gave Newton cause for concern, even though he had been authorized by warrant to continue acting under his Indenture of 1700 until a new one could be drawn.[69] The times dictated that he maintain a low political profile, and he acted accordingly by requesting that another be chosen to run for his seat in Parliament. The Queen further gratified the new majority by dismissing Bishop Lloyd from the office of almoner for allegedly seeking to influence his clergy against the Tory can-

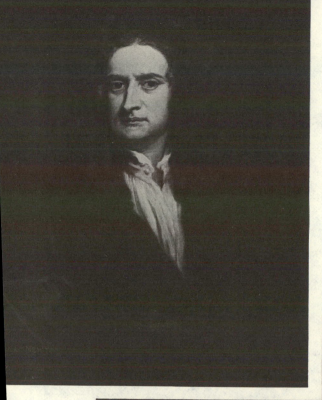

Newton at age fifty-nine
(1702) by Sir Godfrey Kneller.
*(By permission of the National
Portrait Gallery, London.)*

Newton at age sixty (1703) by
Charles Jervas. This portrait
was presented by Newton to
the Royal Society in 1717.
*(By permission of the Royal
Society.)*

Newton at age sixty-seven (1710) by Sir James Thornhill. *(By permission of the Master and Fellows of Trinity College, Cambridge.)*

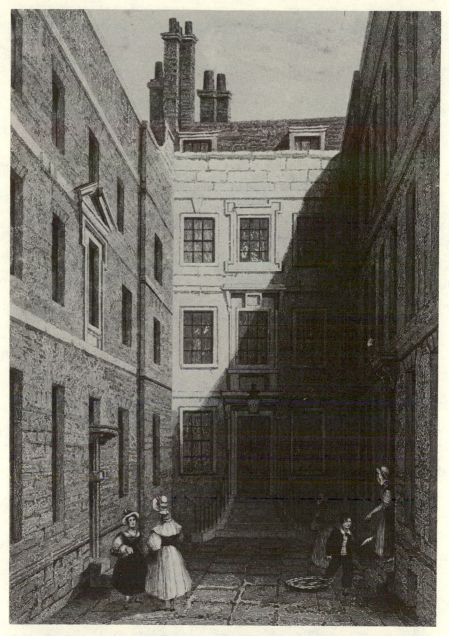

Crane Court, the first permanent home of the Royal Society. Purchased at Newton's insistence in 1710 and occupied in 1711. *(By permission of the Royal Society.)*

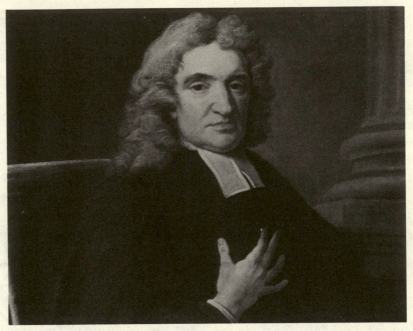

John Flamsteed, Astronomer Royal, (c. 1712) by Thomas Gibson. *(By permission of the Royal Society.)*

Gottfried Wilhelm Leibniz by Bernard Francke. *(By permission of the Herzog Anton Ulrich-Museum, Braunschweig. Photo source: Museumsfoto B.P. Keiser.)*

Newton at age eighty-two (1725) by John Vanderbank. *(By permission of the Master and Fellows of Trinity College, Cambridge.)*

Louis François Roubiliac's superb statue of Newton in Trinity Chapel with its apt quotation from Virgil. *(By permission of the Master and Fellows of Trinity College, Cambridge.)*

Newton at age eighty-three (1726) by John Vanderbank. The artist's identity has been
questioned. It may be the work of Michael Dahl. *(By permission of the Master and
Fellows of Trinity College, Cambridge.)*

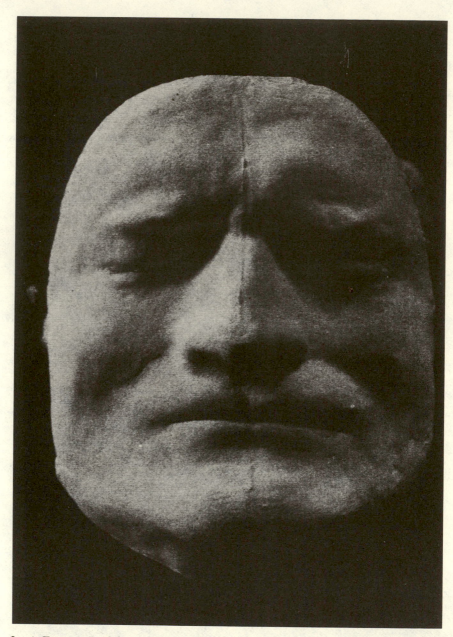

Louis François Roubiliac's death mask of Newton (March 1727). *(By permission of the Master and Fellows of Trinity College, Cambridge.)*

didate in the Worcestershire elections. It was just conceivable that she might be inclined to do the same to a politically repugnant Master of the Mint, whose tenure was entirely dependent upon the Queen's sufferance. The prospect, however remote, of being thrown back upon his own devices and the good graces of political friends whose careers were in eclipse was enough to make even Newton shudder.

No sooner had Anne ascended the throne than her ministers presented the Master with a welcome opportunity to serve his Queen. The Mint was ordered to design and produce a coronation medal in gold and silver for distribution among the attending dignitaries. Eager to please, Newton personally took charge of the design, a classical motif in which Anne was portrayed as the goddess Pallas Athena destroying a great Egyptian army symbolized by many-headed giants. According to Roman myth, the goddess entered the fray in relief of Jupiter, who had wearied of the carnage. Since William's coronation medal had depicted the King as Jupiter, Newton's design was a clever extension of the resurrected epic. The Mint struck 300 medals of gold and 1,200 of silver at a total cost to the public of £2,485. These were delivered to the Earl of Bradford, Treasurer of the Queen's Household, on April 22, the day before the coronation.[70] What Anne thought of the finished product is not known, but seven months later, on January 14, 1703, Newton's indenture with the Queen was officially sealed.

In the autumn of 1702 Newton made the brief journey northward to Essex for a reunion with his ailing friend John Locke. The two had worked together during the recoinage when Locke served as a Commissioner and briefly as President of the Board of Trade. Their correspondence then was largely concerned with stabilizing the relation between gold and silver. The philosopher had resigned his position in 1700 over the protests of King William, arguing that he was simply too old and too ill to be of further service to the crown. Unable to endure the foul and frequently poisonous air of London any longer, he left the city forever to pass his last days at the estate of the Mashams, his devoted friends. Newton was shown Locke's recently completed commentary on St. Paul's first two Epistles to the Corinthians, with which he seem pleased, though he did not have the time to examine it in detail. He asked Locke to send him a copy for his more careful perusal, and the promised documents arrived just before Christmas. Having heard nothing by the following March, Locke wrote to Newton in hopes of eliciting a reply. Receiving no answer, Locke wrote a second letter on April 30. This he enclosed in a message to his cousin Peter King, commissioning the young man to wait upon the Master of the Mint for the purpose of delivering his letter in person. Locke had become nervous and wary during the long delay: Reflecting back on the unpleasant developments of September 1693, he gave King careful instructions on just how to approach the great man. In so doing, Locke inadvertently provided posterity with one of the few convincing sketches we have of Newton's character:

The reason why I desire you to deliver it to him yourself is, that I would fain discover the reason of his so long silence. I have several reasons to think him truly my friend, but he is a nice [touchy] man to deal with, a little too apt to raise in himself suspicions where there is no ground; therefore, when you talk to him of my papers, and of his opinion of them, pray do it with all the tenderness in the world, and discover, if you can, why he kept them so long, and was so silent. But this you must do without asking why he did so, or discovering in the least that you are desirous to know. . . . Mr Newton is really a valuable man, not only for his wonderful skill in mathematics, but in divinity too, and his great knowledge in the Scriptures, wherein I know few his equals. And therefore pray manage the whole matter so as not only to preserve me in his good opinion, but to increase me in it; and be sure to press him to nothing; but what he is forward in himself to do.[71]

Shortly thereafter King informed his cousin that he had visited Newton. Since others were present, they did not discuss the theological papers openly, but Newton whispered that he had read about half of them. Embarrassed by this personal reminder of his neglect of a true friend, Newton immediately completed the task. He wrote Locke an apologetic letter on May 15. While he differed with certain of the philosopher's interpretations, he concluded on a positive note: "I think your paraphrase & comentary on these two Epistles is done with very great care and judgement."[72]

It is quite possible that Newton's decision to visit Oates in 1702 was predicated on an ulterior motive. Abigail Hill, cousin of Queen Anne's favorites the Marlboroughs, was engaged to the son of Sir Francis and Lady Masham. Newton had apparently turned to them for aid in his unsuccessful quest for a preferment in January 1691, and it is not unreasonable to suppose that he had set out in the hope that they might lobby on his behalf now. The fact that he neglected Locke's commentaries, which arrived only six weeks before his second Indenture was sealed, would also tend to support this view. Having achieved his objective of retaining the Mastership, Newton evidently decided that the theological papers could wait. In any case, Newton and his friend were never to meet again. Well aware that he was dying, Locke, in a final letter written to Peter King on October 25, 1704, asked his cousin to deliver a sealed packet and a brass ruler to Isaac Newton. What the packet may have contained is anyone's guess, and so is the symbolism, if any, of the ruler. Three days later, at three o'clock in the afternoon, Locke lifted his hands to his face and died. He was buried, as he had instructed, unadorned in a plain wooden coffin not covered with cloth. "That cost," Locke wrote near the end, "will be better laid out in covering the poor."[73]

Chapter Seventeen

Sir Isaac

Dark Care sits enthroned behind the Knight.

Horace

I

A stillness alien to Newton's experience at the Mint settled over the institution now that the great recoinage was complete. The giant presses, which had hammered precious metal into coin at a peak rate of many thousands of pounds a day, stood idle for months on end. Not a single coin was struck from May to December 1703, and during the following six years minting was limited to a total of only seventy-five days.[1] Much of the additional machinery installed for the recoinage both in London and in the country mints was sold off, and temporary workers were released by the hundreds. Even those belonging to the once powerful and prestigious Company of Moneyers sank to the level of casual laborers outside the Mint.

Thus far Newton had made a point of keeping his distance from the Royal Society, but that was about to change. Not only had the day-to-day demands of managing the Mint subsided, but nature, as if by design, simultaneously intervened and removed a final obstacle that Newton, try as he might, could never have overcome on his own. Robert Hooke, who in his last years had been reduced to a shriveled, pain-racked husk, died on March 3, 1703, at the age of sixty-eight. Hooke had been present at the creation and had served the Society for forty difficult but rewarding years, first as Curator of Experiments and later as Fellow, Council member, and Secretary. In 1700 he was still attending meetings and taking part in discussions. However, during his final year Hooke was bedridden and nearly blind, living

435

what his biographer and editor Richard Waller termed "a dying Life." Yet his mind remained clear to the last.

Hooke's failure to draw up a will may have cost the Society dearly, for the bachelor, who kept his money locked in an iron chest, had spoken of bequeathing sufficient funds with which to erect badly needed quarters, complete with meeting rooms, a well-equipped laboratory, and large library. A melancholy Waller termed the plan "an airy Phantom" that vanished into nothing, as had so many of Hooke's other grand designs. But while he was no Newton, Hooke possessed a restive, indefatigable brilliance whose Baconian bent enabled him to make wide-ranging contributions to seventeenth-century science. Despite his absolute refusal to admit it, Newton owed much to Hooke, because it was his insight on circular motion that set Newton on the track of universal gravitation. The Fellows of the Royal Society hardly needed to be reminded that the frequently irascible and petulant savant represented the pivot of their scientific enterprise. They did their duty and paid tribute by attending his funeral as a body. As to whether Newton himself went, there is no evidence. That he shed not a tear and experienced no lasting regret is certain.

There is no reason to suppose that the Master of the Mint quietly kept himself posted on Hooke's declining health. Nevertheless, his death opened the way for Newton's takeover of the Royal Society, a viselike domination that lasted for almost a quarter-century. The historic first chapter of this episodic tale was written on St. Andrew's Day, November 30, 1703, when the Fellows assembled for the annual election of Council members and officers. The record of the proceedings themselves, dry as dust, contains no hint of the behind-the-scenes maneuvering that brought Newton's name to the fore. According to Conduitt, Dr. Hans Sloane, the Society's venerable but crusty Secretary, approached Sir Christopher Wren with the request that he run for the presidency, which had fallen vacant on the resignation of Lord John Somers, leader of the Whig Junto. Wren is said to have declined and recommended Newton instead, a choice not unpalatable to the no-nonsense Dr. Sloane.[2] Before he could run for office, however, Newton, in compliance with the Charter, had to hold a seat on the twenty-one member Council, as did every other officer and candidate for office. When the preliminary ballots were counted, he received a total of twenty-two votes, placing him in about the middle of those selected. The election of officers followed: Newton was chosen as President with twenty-four votes, a total exceeded only by Waller, who garnered twenty-six votes for his reelection, along with Sloane, as one of the Society's two secretaries.[3]

To say that the sixty-year-old Newton took over the reins of an institution *in extremis* is to exaggerate only slightly. The Royal Society was, quite simply, bankrupt—intellectually, spiritually, and financially. For those who had lived through the last two decades of the seventeenth century, they must have seemed far removed from the institution's early halcyon years and the promise that Bacon's New Atlantis lay waiting just beyond the horizon, to lavish its intellectual bounty on keen and unfettered minds. While the chair was occasionally occupied by men of science, many of the past presidents were aristocrats and politicians who not only failed to lead but looked upon attendance at the Council meetings and general sessions as an intolerable inconvenience. Lord Somers, Newton's immediate predecessor, attended all of three meetings during the five years of his presidency, one of them apparently called in his London residence; never did he sit with the Council. Halifax, who bore the title of President from 1695 to 1698, missed all but one of the ten Council meetings held during his incumbency. The Society suffered no less severely from a precipitous decline in membership. It had averaged about 200 Fellows for each of the five years between 1676 and 1680. Fifteen years later that number had dropped to 115, rebounding slightly on the eve of Newton's presidency.[4] Few members attended weekly meetings and fewer still the Council, which occasionally made its policy decisions at gatherings little larger than the foursome needed for a table of whist.

The prolonged crisis of leadership and the apathy to which it gave rise gradually reduced most general meetings to exercises in intellectual futility. Hooke's pace had slowed when his health failed during the 1690s; by 1700 his contributions in the form of stimulating experiments were neither so frequent nor so challenging as they had been a decade before. Part, but by no means all, of the slack was taken up by Sloane, who had already served as Secretary for ten years before Newton became President. The physician was no experimental scientist; his contribution lay mainly in a vast correspondence, both at home and abroad, requesting communication on subjects of interest to the Society. Sloane's zeal and force of character occasionally led to conflicts with Fellows who thought him too ambitious and dictatorial. Undeterred by their censure, he became one of Newton's staunchest allies, at least in the beginning, helping him to formulate and institute the basic reforms needed to put the ailing patient back on its feet.

Such discussions as had taken place at meetings were largely concerned with medicine and teratology, a reflection of the fact that physicians and surgeons outnumbered all other occupational groups combined,[5] as well as a measure of Sloane's considerable influence. Unfortunately, medical curiosities and tales of the bizarre proved much more appealing to the membership than did attempts to further professional understanding of the illnesses commonly encountered in daily practice. The Fellows doted on live specimens of crocodiles and opossums, were fascinated by lectures on the relative merits of various poisons employed by murderers, and debated the medicinal value of bovine urine. The pages of the *Philosophical Transactions* became so cluttered with intellectually questionable accounts of medical prodigies that in 1700 an anonymous pamphleteer published a lampoon titled *The Transactioneer, with some of his Philosophical Fancies, in two dialogues.* The "his" in the title referred not to the author himself but to Sloane, editor of the *Transactions,* a man of "much reputation, . . . though he has neither parts nor learning to supply it." The pseudojournal's satirical table of contents listed the following articles:

> Eggs in the cauda [tail] of a barnacle.
> Four sorts of lady's bugs.
> A buck in a snake's belly.
> A shower of whitings.
> A shower of butter to dress them.

On another day the Fellows might have had themselves a good laugh before dismissing tract for what it was—a harmless piece of mischief. But in this instance the author struck a highly sensitive nerve. The Council expended much effort to discover his identity for the purpose of initiating legal action, but its exertions were fruitless.[6]

II

The new President of the Royal Society, like Walt Whitman, yielded to the potent narcissism in his makeup by having his image reproduced ever more frequently as he grew older. He had sat for Kneller only a year earlier, but Newton's election to high office gave him yet another excuse to call upon the services of an artist. This time he turned to the portraitist Charles Jervas. In Jervas's rendering the savant sits stiffly in a high-backed armchair, his head crowned by the same flowing peruke worn for the Kneller portrait. The fingers of his left hand, grown chubby over the years, rest on one arm of the chair, while his right forefinger points to the poorly defined image of a book or bound manuscript lying on a nearby table, a prefiguing of an important event to come. The eyes, ever protuberant, gaze fixedly back at the viewer. This is Isaac Newton at the height of his prestige. No covering robe does he wear. Satin knee breeches, an elegantly tailored jacket, and a white surplice adorn this pillar of the establishment. It was almost certainly the image of himself Newton wished to project for posterity; after accumulating even more accolades and honors, he presented the portrait to the Royal Society in 1717. Seven years earlier the Society's portrait of Hooke, the only one known, was lost forever, along with certain of his scientific instruments, when the Society transferred from Gresham College to its new quarters in Crane Court. Newton decreed that the portraits of the other Fellows were to be cleaned and lettered in gold.[7] To suggest that he might have been somehow responsible for the Hooke portrait's disappearance is perhaps going too far; on the other hand, he had repeatedly struck Hooke's name from the *Principia* in a fit of passion reminiscent of the ancient pharoahs, while Hooke had confided in his diary the nocturnal dream that Newton had died.

Surprising as it may seem, Newton was not present at the Royal Society meeting of December 8, 1703, the first after his election as President. However, he presided at the next meeting and missed fewer than one out of four thereafter. He cast an even longer shadow over the Council. The governing body met a total of 177 times during Newton's tenure. He missed only ten of those sessions, and seven of his absences came after his eightieth birthday when declining health increasingly kept him confined indoors.[8] Just how far the Society as a whole had to go in this direction was brought home at the meeting of March 8, 1704, three months after Newton took control: "The number of Fellows not makeing up the Quorum there is no meeting."[9]

Besides setting an example of regular attendance, he brought with him a master plan for institutional reform, which he somewhat arrogantly titled a "Scheme for establishing the Royal Society," as if it did not as yet exist. The document typically went through seven drafts as Newton debated with himself over its final form. His central concern, as one would expect, was intellectual; the Society must renew its commitment to the revolutionary principles underlying its original foundation. "Natural Philosophy," the "Scheme" began, "consists in discovering the frame & operations of Nature, reducing them (as far as may be) to general Rules or Laws, establishing those Rules by observations & experiments, & thence deducing the causes & effects of things." He envisioned the appointment of as many as four paid demonstrators who would be obliged to attend all meetings and to present regularly scheduled experiments and lectures in mechanics and mathematics, optics and astronomy, zoology, chemistry, and botany. While he could never bring himself to mention Hooke by name, there is no question that Newton was thinking of just such a model—"a person who hath already invented something new, or made some considerable improvement in . . . philosophy, or is eminent for skill therein, if such a person can be found." Still, it was not the new President's intent to follow the example of the French Academy, which had resolved to limit its membership to learned professionals only. Instead, he sought a return to the days when intelligent laymen were challenged by associating with accomplished natural philosophers dedicated to the growth and widest possible dissemination of the new learning. The Society had heard enough of the wild tales spun by naive travelers in distant lands; of "authentik" personal accounts concerning the supernatural; of mad dogs and laughing apes; of prodigies having no foundation save rumor and hearsay. The institution must rededicate itself to its original purpose: The "reward will be an advantage to the Royal Society to have such men at their meetings, and tend to make their meetings numerous and useful, and their body famous and lasting."[10]

Newton soon located the first of the skilled demonstrators for whom he was searching. Indeed, he probably had him in mind even before completing the first draft of his plan for basic reform. Called "the elder" to distinguish him from a nephew of the same name, Francis Hauksbee made his debut before the Society at its meeting of December 15, 1703, the first to be chaired by Newton. "Mr Hawksbee Showed a New Invented Air Pump and in it, upon the Mercury descending a Light & c. He was thanked."[11] About Hauksbee's origins and background we know virtually nothing. He seems almost to have sprung full-blown from the mind of Newton. If Hauksbee was not Newton's creation, he was surely his creature.

Hauksbee's first experiment was calculated to make a lasting impression, and by his own account he must have accomplished his design. The room was darkened as all eyes were fixed on the demonstration table. Hauksbee released mercury into the evacuated receiver of his new air pump. On spilling over an inverted glass vessel it gave off "a Shower of Fire descending all round the Sides of the Glasses."[12] Hauksbee put in an appearance nearly every week thereafter, performing a wide range of experiments with the pump. He received his first fee the following February, when the Council voted him two guineas for his efforts. The members awarded him five more guineas in July but declined to formalize the association by appointing Hauksbee to Hooke's old post as Curator of Experiments, doubtless as much a blow to Newton as to the candidate himself. Hauksbee was asked to appear as a paid demonstrator in the future with the promise of being rewarded "according to the proportion of his Services."[13] This rather loose arrangement continued for the next decade. Hauksbee was elected a Fellow in November 1705 and performed experiments until the onset of a fatal illness early in 1713.

Aside from his striking performance of 1703, most of Hauksbee's early demonstrations were essentially repetitions of previous ones conducted by Hooke, Boyle, and especially Dr. Denis Papin, Curator of Experiments from 1684 to 1687. Hence there is some reason to think that Hauksbee may have once served as Papin's assistant. His most original work came after 1705, when he became fascinated with the phenomena of electricity and of capillarity in vacuums, territory not unfamiliar to Newton. Much to the President's delight, Hauksbee demonstrated beyond doubt that the law of attraction delineated in the *Principia* acts on minute bodies in a vacuum as well as on far larger ones. Capillary rise may be "handsomely accounted for" by its action. Newton incorporated this insight into the second English edition of the *Opticks* (1717), carrying it a crucial step farther through the process of quantification, an exercise Hauksbee humbly admitted was beyond his powers. In the second edition of the *Principia* (1713), in his discussion of the motion of bodies through a resisting medium, Newton also employed Hauksbee's results derived from dropping glass balls of different weights from the cupola of St. Paul's, an experiment Newton himself probably suggested. Finally, Hauksbee's demonstration that glass is a convenient and pliable medium for producing static electricity opened the way for such luminaries as Benjamin Franklin, first recipient of the Copley Medal, the highest prize awarded by the Royal Society.

Whether one can call them true collaborators seems relatively unimportant. Chances are they were not, for sharing as an equal was alien to Newton's nature. What we do know is that Hauksbee, something of an experimental genius but a scientific neophyte, came to a better understanding of the theoretical importance of his discoveries through contact with Newton. For his part the aging lion of English science depended upon Hauksbee's talented hands to test a number of his more intriguing conjectures, even as they supplied him with fresh insights into long-familiar fields.

An advocate of leadership by example, Newton broke with aristocratic tradition and took an active part in the Society's weekly demonstrations from the outset of his tenure. A paper written by John Lowthorp on the subject of burning glasses was read at the meeting of January 20, 1704, the first of many sessions moved to a Thursday afternoon in deference to the President's obligations at the Mint. Newton volunteered his thoughts on the matter and agreed to try his hand at fashioning such a glass.[14] The unique result—a powerful instrument comprising six concave glasses, each $11\frac{1}{2}$ inches in diameter, arranged round a seventh to form the segment of a sphere—was first shown in May. The Fellows were so taken with the invention that Newton demonstrated its awesome cutting and liquifying powers on several occasions thereafter. John Evelyn, who had been very ill and did not see it in operation until April 1705, wrote: "Sir Isaac Newtons ... burning Glasse which dos such wonders as that of the K. of France which cost so much, dos not come-neere, it penetrating Cast Iron of all Thicknesse, vitrifies Brick, mealts all sorts of mettals in a moment."[15] Meanwhile the Society was treated to a highly specialized display of animal anatomy. A Mr. Cowper showed the penis of the Society's recently deceased possum and was duly ordered to draw up an account of his findings. Diligent to a fault, Cowper left nothing to the imagination in his written report of Febraury 2. According to the minutes, the self-styled anatomist took particular note of the animal's "forked Penis, its Erection and Retraction into the body,"[16] though how Cowper managed his detailed analysis of the lifeless form remains something of a minor mystery.

In addition to his many other concerns, Newton had become chief officer of an institution threatened with the loss of its quarters. Ever since receiving a formal charter in 1662, the Royal Society had been meeting on the premises of Gresham College in Bishopsgate Street, the vast timber-and-brick mansion originally constructed as a private residence by the Elizabethan merchant and ambassador whose name it bore. Allotted no rooms of their own, the members assembled in the apartments of various professors over the years. Robert Hooke had served as their most recent host, letting his quarters to the Society for the nominal sum of £10 per annum from 1688 until his death in 1703. The passage of time and a lack of proper maintenance had exacted a heavy toll on the elegant structure in which Queen Elizabeth I and the opulent Sir Thomas Gresham once drank wine laced with the essence of rare pearls. In 1701 the trustees of the college, with the consent of all the professors except Hooke, introduced a bill into Parliament for razing the historic edifice and rebuilding at the same location. The trustees agreed to provide the Society with its own rooms, in return for which the Gresham professors would have free access to its excellent library and museum. The bill passed the Commons with no difficulty but was thrown out at the second reading in the House of Lords on the petition of Hooke. Exactly why he opposed the project is difficult to say, but it may have had something to do with his putative design to play the role of benefactor by personally underwriting the cost of constructing separate Society quarters. The tense situation reached crisis proportions immediately following Hooke's death. On March 24, 1703, the thwarted trustees ordered the Society to clear out its library and to hand over the keys to Hooke's chambers at once. Luckily Dr. John Woodward, Gresham's prickly but talented Professor of Physic, offered his desperate colleagues the use of his rooms, whereupon the disgruntled trustees backed off, if only for the time being.

These events had come about before Newton's incumbency, but the issue was far from settled when he assumed the chair a few months later. At that time a committee of six appointed by the Council was actively seeking alternate quarters. Newton apparently effected some sort of rapprochement with the Gresham trustees, who agreed to reintroduce the construction bill. In February 1704 Newton brought word to the Council that Lord Halifax had assured him that the parliamentary committee in charge of the measure fully intended to provide all the room required by the Society. Encouraged by this development, the Council returned their thanks to the trustees and delegated Sir Christopher Wren to draw up a design for the proposed accommodations. The resulting document, which Sir Christopher titled "Proposals for building a House for the Royal Society," was ambitious to say the least. The master architect envisioned an imposing 40-by-60-foot structure set on grounds large enough to accommodate the coaches of the members, "some of which are of very elegant quality."[17] Newton, Wren, and Sloane presented the plan to an astonished committee for Gresham College, along with the proposal that all costs should be assumed by the Mercers Company and the City of London, joint administrators of the present building and grounds. When these plans were rejected they naively returned with a design for a structure about half the size of the one in the original proposal, which they were willing to lease gratis for 999 years. The Gresham trustees assembled on December 13 and voted unanimously to reject the second plan as well. What is more, they voted to withdraw the construction bill from Parliament.[18]

A report of the trustees' action reached the Council on the very day it was taken. The curt minutes state that it had been resolved "not to grant the Society any room at all,"[19] which left the erroneous impression that the Fellows had received a second ultimatum to withdraw from Gresham forthwith. Just who had been responsible for grossly overestimating the trustees' willingness to do the Society's bidding is open to question, but most signs point to Halifax and his glowing report of Parliament's feelings on the matter. From that time until the Society's removal to Crane Court in 1711, the Council minutes contain frequent references to the labors of the committee charged with securing a new home. The Council turned to the crown in January 1705. It addressed a petition to Queen Anne for a parcel of land on which to build in the Royal Mews in Westminster, but to no positive effect. They next applied to the trustees of the Cotton Library for permission to meet in its apartments, only to be denied. Other possibilities also came to nothing—a house in Whitehall, ground for a building in the Savoy and later near St. James's Park, the offer of an estate by the Duke of Bedford. The real problem confronting the Royal Society was not so much a dearth of suitable accommodations but a lack of ready funds with which to strike out on its own. Time had proved Thomas Sprat, the Royal Society's first historian, all too correct: Public faith in experimental philosophy was not strong enough to move men and women of all conditions to bring in their bracelets and jewels "towards the carrying of it on."

III

Secretary Hans Sloane made the following entry in the Journal Book on February 16, 1704: "The President presented his book of Optics to the Society. Mr Halley was desired to peruse it and give an abstract of it to the Society. The Society gave the President their thanks for the Book and for his being pleased to Publish it."[20] Unlike the *Principia*, whose publication was preceded by considerable fanfare and much anticipation, the *Opticks* received no mention in contemporary accounts—not even in the Royal Society minutes—immediately prior to its coming off the press. It is the only one of Newton's major works that he prepared for publication strictly on his own. He did not dedicate this, the second pillar of his immense scientific legacy, to the Royal Society, as he had the first. The institution did receive mention on the title page, however, for Newton employed Samuel Smith and Benjamin Walford, printers to the Society.

In the "Advertisement," which doubles as a preface, Newton informs the reader that the volume is largely the product of research completed many years earlier, when he was a young man and at the height of his powers. "To avoid being engaged in Disputes about these matters, I have hitherto delayed the printing, and should still have delayed it, had not the Importunity of Friends prevailed upon me."[21] The veiled reference to Hooke was not easily mistaken by those conversant with the bitter controversy sparked by Newton's submission to the Society of his revolutionary paper on light a generation before. While these words indicate that he still harbored some serious reservations concerning the wisdom of publishing at this time, Hooke's recent death had removed the one impediment that had caused him to remain intransigent.

It was also partly owing to the encouragement of friends that Newton had printed and bound with the *Opticks* two mathematical papers, written long before but lately revised: "A Treatise on the Quadrature of Curves" (*De quadratura*) and "Enumeration of Lines of the Third Order." Yet, as with the publication of the *Opticks* itself, there was more behind his actions than the mere entreaties of such admirers as Halley and Gregory. "[S]ome Years ago," Newton wrote, "I lent out a Manuscript containing . . . Theorems [on squaring curves], and having since met with some Things copied out of it, I have on this Occasion made it publick."[22] The biographer L. T. More took this for a "covert insinuation" that Leibniz had seen the manuscript of *De quadratura*, to which Newton had given Gregory and Halley access, during a visit to London and that the mathematician-philosopher had obtained from it certain hints that led him to the discovery of the calculus.[23] In point of fact Newton attributed the plagiarism of this particular manuscript not to Leibniz but to George Cheyne, a young Scottish physician who had decided to try his luck in London. Cheyne brought with him from Edinburgh a manuscript on fluxions, which he hoped Newton would evaluate with a view to publication. An introduction was arranged through Dr. John Arbuthnot, physician to Queen Anne, and Newton, not wishing to hurt anyone's feelings, including those of Cheyne's doting mentor Archibald Pitcairne, pronounced the manuscript "not intolerable" after briefly perusing its contents.[24] According to Conduitt, Newton later offered Cheyne a bag of money to finance its publication. When the young man refused his offer, "Sr I. would see him no more."[25] Cheyne went on to publish his book, *The Inverse Method of Fluxions* (1703), without Newton's financial assistance or, what is more, his imprimatur. Sparks flew when Newton discovered elements of his own work in the treatise, a result, no doubt, of Gregory's having shared notes on *De quadratura* with his countryman. How Newton settled matters with Gregory is not known, but the mathematician was moved to write the following in a memorandum of March 1, 1704: "Mr. Newton was provoked by Dr. Cheyns book to publish his Quadratures, and with it, his Light & Colours, & c."[26] Cheyne, who was active in Royal Society affairs before Newton became President, faded quickly from the scene.

"My Design in this Book is not to explain the Properties of Light by Hypothesis," Newton began, "but to propose and prove them by Reason and Experiments."[27] Thus did the gifted theoretician, whose genius had manifested itself in the *Principia*, lay claim to equal stature as an experimental scientist, a rare combination in the history of modern thought. Whereas the *Principia* was a mathematical treatise, involving intricate geometrical relationships and only a handful of significant experiments, the

Opticks overflowed with detailed accounts of the phenomena of reflection and refraction, the spectral decomposition of white light, the colors of the rainbow, the manner of the eye's operation, the formation of images by lenses, the construction of the reflecting telescope, and much more. Nor would the author be confined to subjects related only to the behavior and analysis of light. The range of topics touched upon by Newton included questions of gravitation, metabolism and digestion, sensation, the circulation of blood, the Creation and the Great Flood, moral philosophy, the inductive method, and the vivid images haunting the dreams of madmen. Written in English as opposed to classical Latin, the *Opticks* contained little mathematics, making its contents accessible to a far wider audience than the abstruse *Principia* had been. Like so many others, John Locke had been awed but also cowed and frequently baffled by the earlier work. By contrast, the philosopher read "with pleasure" a copy of the *Opticks* sent to him by Newton only months before he died, "acquainting himself with every thing in them."[28] Locke's delight proved an augury of things to come, for while the book became a natural focal point of experimental research in the infant eighteenth-century sciences of chemistry, electricity, magnetism, biology, and geology, it also exerted a profound influence over the poet, essayist, and philosopher. John Machin, another of Newton's widening circle of young disciples, spoke for many when he told Conduitt that the *Opticks* contained more philosophy than the *Principia*.[29]

Until the early 1690s Newton had called upon ether and its varying densities to explain such optical phenomena as refraction and the periodicity of thin films. When the ether evaporated in the wake of his pendulum experiments conducted in a vacuum and failed to rematerialize in the alchemy, Newton was bereft of a medium capable of accounting for the behavior of light, not to mention universal gravitation itself. Prior to this discovery, he had planned four books for the *Opticks* instead of the three ultimately published in 1704. In Book IV, which he withheld from the press, optical phenomena in support of the concept of forces at a distance were to be meshed with the laws of the *Principia*, giving birth to a harmonious new philosophy of nature—the grand synthesis of his dreams:

How the great bodies of ye Earth Sun Moon & Planets gravitate towards one another what are ye laws & quantities of their gravitating forces at all distances from them & how all ye motions of those bodies are regulated by their gravities I showed in my Mathematical Principles of Philosophy to the satisfaction of my readers. And if Nature be most simple & fully consonant to her self she observes the same method in regulating the motions of smaller bodies [including the corpuscles of light] wch she doth in regulating those of the greater.[30]

It was not to be. Shipwrecked on the rocks of his own experimental acuity and estranged from Fatio, Newton faltered and broke in the summer of 1693. When David Gregory visited him the following May, he saw three books of optics rather than four. Still, he was sufficiently moved to observe that "if it were printed it would rival the *Principia Mathematica*."[31]

Having already traced Newton's research on light and colors in considerable detail elsewhere, there is little to be gained by going over the ground a second time. Let Gregory's expression of wonder on seeing the manuscript in 1694 serve as my benediction as well. What Gregory in all probability did not see at the time was a set of sixteen queries with which Newton concluded the first edition of the *Opticks*. Their purpose, he submitted, was to advance the research of others, for "I . . . cannot now think of taking these things into farther Consideration."[32] This seems a half-truth at best; the queries are couched in the form of rhetorical questions, the affirmative answers to which Newton scarcely doubted. Having despaired of the theory of ether, he nevertheless abandoned any pretentions to strict neutrality with regard to his overarching hypothesis, namely, that an attractive force emanating from solid bodies influences light at a distance. Therefore, light too must be composed of tiny corpuscles or atoms, like every other form of matter.

> *Query* 1. Do not Bodies act upon Light at a distance, and by their action bend its Rays; and is not this action . . . strongest at the least distance?

> *Qu.* 3. Are not the Rays of Light in passing by the edges and sides of Bodies, bent several times backward and forwards, with a motion like that of an Eel?

> *Qu.* 4. Do not the Rays of Light which fall upon Bodies, and are reflected or refracted, begin to bend before they arrive at the Bodies; and are they not reflected, refracted, and inflected [diffracted], by one and the same Principle, acting variously in various Circumstances?

> *Qu.* 5. Do not Bodies and Light act mutually upon one another; that is to say, Bodies upon Light in emitting, reflecting, refracting and inflecting it, and Light upon Bodies for heating them, and putting their parts into a vibrating motion wherein heat consists?

Not all of the queries deal with forces. In query 11, for example, Newton asks: "Do not great Bodies conserve their heat the longest . . . ?" "And are not the Sun and fix'd Stars great Earths vehemently hot, . . . and whose parts are kept from fuming away, not only by their fixity, but also by the vast weight and density of the atmosphere incumbent upon them . . . ?" The final four queries deal with vision, which, he suggests, results when vibrations propagated by rays of light on the retina are conducted to the optic nerve, and from there to the brain. In query 13 various sorts of rays are associated with the several colors, "much after the manner that the Vibrations of the Air, according to their several bignesses excite Sensations of several sounds."[33] That he did not add more to the queries at this time is a measure of the lingering paranoia resulting from Hooke's assault on the "Hypothesis of Light," which Newton had submitted to the Royal Society in 1675. As a young man he had dreamed of the day when he would be welcomed into the world of the virtuosi, the day when he would unveil before an awed constituency of like spirits the entire range of his masterly skills. What was it Newton had called his discovery of the properties of light in January 1672, to which Hooke also took exception, "the oddest if not the most considerable detection wch hath hitherto beene made in the operation of Nature"? The dream had gradually been transformed into a thirty-year nightmare. Even the presidency of the Royal Society and the certain immortality assured him by the *Principia* could not make it go away.

IV

George of Denmark, the likable Prince Consort of Queen Anne, was a good-natured, devout Protestant to whom his spouse remained blindly devoted throughout their married life. In the general estimate, the Prince's intellectual capacity scored rather lower than his social graces, however. Charles II had hit the mark fairly well with his famous line, "I have tried him drunk and I have tried him sober and there is nothing in him."[34] Still, George, who cultivated a fervent if shallow interest in natural philosophy, was warmly welcomed as a new member of the Royal Society at its anniversary meeting of November 30, 1704. This was a significant triumph for Newton, who, along with Sloane, personally waited on the Prince when he entered his name in the official register. The occasion proved rewarding from Newton's point of view for another reason as well: He was reelected President for the coming year.[35]

In April 1705 Queen Anne and her whole court, including the Prince, embarked on a journey to the royal residence at Newmarket. The party then made the short trip to Cambridge on the morning of the sixteenth, where it was met by the mayor and other city officials on the broad green still known as Christ's Pieces. Anne was later escorted to Regent Walk to be received by Lord Somerset, Chancellor of the University, and addressed by Dr. Ayloffe, the Public Orator. William Stukeley, who was in his third year at Christ's College, retained vivid memories of the occasion. "The whole University," he wrote, "lined both sides of the way from Emmanuel college, where the Queen enter'd Town, to the public Schools."[36] After a customary meeting with the congregation of the Senate, the *Regia Comitia*, Anne conferred twenty-three doctorates by royal mandate. She then summoned court in a completely refurbished Trinity Lodge, currently occupied by Richard Bentley, and knighted three men: John Ellis, the Vice Chancellor of the University and Master of Caius; James Montague, brother of Halifax and University Counsel; and Isaac Newton, Master of the Mint, President of the Royal Society, and natural philosopher extraordinaire. A sumptuous dinner was later given for the Queen in the newly redecorated Trinity Hall, where the most illustrious guest after Anne herself had once waited on tables to help meet educational expenses resented by his wealthy but sparing mother. After attending evening services at the beautiful chapel of King's College, the royal party took its leave and returned to Newmarket. It was later learned that Her Majesty had been so short of funds that she found it necessary to borrow £500 from Trinity for the purpose of underwriting this extravagant entertainment, whose total cost was estimated at more than £1,000.[37] As for Sir Isaac, who remained in Cambridge on other important business, this was a moment to be savored. Shunned by all but a small handful of students as a professor, he suddenly found himself the object of their rapt attention. "We [students] always took care on Sundays to place ourselves before him," Stukeley later remembered, "as he sat with heads of the colleges; we gaz'd on him, never enough satisfy'd, as on someone divine."[38] The immense compensatory skills Newton had developed to make up for the rejection of his youth and early manhood were what now brought him the very recognition and acceptance he so deeply craved.

Conferring honors on the academic community, however, was not the primary reason for the royal visit to Cambridge. The Queen had made her final speech to Parliament on March 14, after which it was prorogued until its dissolution on April 5 and the issuance of writs for new elections on April 23. The Tories, whom she had strongly favored at the outset of her reign, were badly split over her policy of pursuing the war with France and retaining Marlborough as commander of the army, notwithstanding his stunning victory at Blenheim the previous August. If Anne were to govern effectively, she could no longer exclude the Whigs from royal favor. Her recent expedition to the north, where she campaigned on behalf of opposition candidates at Cambridge and dined with Lord Russell, a member of the Junto, at his nearby country home, indicated that she both recognized and accepted this hard political fact. The Queen was now irrevocably committed to a policy of "moderation" between the extremes of the two parties, a course she was to follow for the balance of her reign. As she told Sidney Godolphin, Lord Treasurer, in May 1705: "All I wish is to be kept out of the power of both" parties.[39]

Halifax, who after years of successfully fending off charges of corruption and malfeasance stemming from his government service under William III, was not going to allow this golden opportunity for a political resurgence to pass him by. He persuaded Newton to seek his old seat in Parliament and outlined plans for the Cambridge elections in a letter to his friend on March 17, 1705. "Mr Godolphin will go down to Cambridge next weeke and if the Queen goes to Newmarket, and from thence to Cambridge, she will give you great Assistance the Torys say she makes that tour on purpose to turn Mr Ansley out." The Godolphin referred to by Halifax was Francis Godolphin, son of the Lord Treasurer and a fellow Whig, who was to stand for election with Newton. Arthur Annesley, the future Earl of

Anglesey, was the Tory incumbent who had taken Newton's place in 1702 when he decided not to seek reelection. Newton had already made one trip to Cambridge to sample the political waters, which he found cool at best. Writing to an unidentified Cambridge friend (probably Bentley) in March, he expressed the intent to oppose neither Mr. Godolphin nor Mr. Patrick, "they being my friends." However, certain others were anxious that he make the effort, and "I beg the favour of you & the rest of my friends to reserve a vote for me till I either write to you again or make you a visit; wch will be in a short time."[40]

Halifax was to have his way after all, the result of a decision almost surely sweetened by the promise of a knighthood for his friend after Halifax discreetly consulted with advisers to the crown. Newton more than went through the motions by visiting Cambridge three times in the month of April; it was during the middle visit that the yeoman's son knelt and took the oath of fealty to his monarch, sword blade resting on his proud shoulders. Halifax reaped spoils as well in the form of a knighthood for his brother James and an honorary doctorate for himself. Still, as Flamsteed wrote to his friend and confidant Abraham Sharp on April 24, even the Queen's demonstrative support was no guarantee of political success: "Mr. Newton is knighted; stands for parliament at Cambridge; and is going down thither, this day or tomorrow, in order to his election. 'Tis something doubtful whether he will succeed or no, by reason he put in too late."[41] Hardly too late. Newton had begun his campaign nearly two months before Parliament was dissolved. Yet Flamsteed was not informed of this. Newton's inferior must be given no reason to think him fallible on any score. By early May Halifax, who had pulled every string at his command, acknowledged that his friend had no real chance of winning back his former seat. He wrote to Newton, who was in Cambridge electioneering: "I am sorry you mention nothing of the Election, it does not look well but I hope you will still keep your Resolution of not being disturbed at the event, since there has been no fault of your's in the Managemt." He could tell Newton some stories of chicanery at court, "but complaining is to no purpose and now the Die is cast, and upon the whole Wee shall have a good Parliament."[42]

When the ballots were counted on May 17, both Whig candidates went down to defeat. Annesley received 182 votes; Dixie Windsor, 170; Godolphin, 162; and Newton, a dismal 117.[43] Sir James Mansfield later told a friend that he knew an old man at Cambridge who, recalling this election, said that all the residents voted for Newton but were outnumbered by the large number of nonresident voters.[44] There may be some truth in this story, for many more votes were cast in the 1705 election than in the others in which Newton was involved. To make matters worse, the extreme Tory election cry was "the Church in danger." On polling day "a hundred or more young students, encouraged in hollowing like schoolboys and porters," taunted the Whig candidates, "crying, No Fanatic, No occasional conformity."[45] Few displays could have unsettled Newton more. Luckily the students, but more importantly the authorities, had no true idea what an "occasional" conformist he was. Newton never sought parliamentary honors again, which, judging from the record, was no great loss to either himself or to the nation.

V

Relations between Newton and Flamsteed had reached their apparent nadir in January 1699, when Newton unleashed a scathing assault on the Astronomer Royal for his mention of Newton's name in connection with a forthcoming publication by John Wallis. Flamsteed tried to heal the rift a few months later by making overtures to Newton through Christopher Wren. He invited both men to dine with him at Greenwich and afterward to inspect his growing catalog of lunar and stellar observations at their leisure. "There are none about town but yourself, and the Master of the Mint (Mr. Newton), that thoroughly understand how they conduce to the improvement of navigation, geography, and natural philosophy. You are both my friends; both zealous for the honour of the King and nation; . . . and equally desirous that they should be published with all convenient expedition."[46] Neither accepted the invitation, and six weeks later, on May 10, 1700, Flamsteed saw it as his Christian duty to warn John Low- thorp, a friend and colleague in the Royal Society, "to be careful of your behavior towards Mr. Newton." The Astronomer Royal went on to explain that he had recently visited Newton at his home in London and was re- buffed once more: "When I urged . . . again that he would come down [to Greenwich], he asked me a little quick, 'what for?'" Flamsteed proceeded to lecture Newton on the value of his astronomical endeavors—"that my work was like the building of St. Paul's; I had hewed the materials out of the rock, brought them together, and formed them, but that hands and time were to be allowed to perfect the building and cover it." Flamsteed added, "This, with some uneasiness, was allowed." He thought Newton "a good man at the bottom," but suspicious and too easily influenced by calumnies. To cure his friend of that, Flamsteed left him a little gift in the form of a moral lesson. Having arrived before Newton was up, he passed the time, as any good Protestant cleric might, by reading the Bible. This gave Flamsteed an idea: He borrowed a piece of paper and wrote down a couplet he remembered from a satire:

> A bantering spirit has our men possessed,
> And wisdom is become a standing jest.

Halley was the real culprit, to Flamsteed's way of thinking, and Newton had to be made to see the light, whether he wanted to or not. Flamsteed added these instructions: "Read Jeremiah, ch. ix. to the 10th verse," an invective against false prophets and liars. "I do not know whether he has seen it," Flamsteed further confided to Lowthorp, "but I think he cannot take it amiss if he has; and if he reflects a little on it, he will find I have . . . showed him the way of the world much better than politics or a play could do."[47] Had Newton but known what the morning would bring, he might well have chosen to stay in bed.

Things had scarcely improved during the ensuing four years. In February 1702 Flamsteed informed Sharp that he was no longer attending meetings of the Royal Society because it was packed with a company of "ingenious young gentlemen" who were serving the interests of his enemies. One or another of them occasionally visited Greenwich to spy on the progress of the star catalog, but they understood little: "I receive them and use them as Scipio did Hannibal's spies, show them what they desire, dismiss them smiling." Flamsteed was convinced that this and other alleged stratagems were hatched by Newton, whose name is cast in a negative light in nearly every other letter drafted by the Astronomer Royal during this period. Only Halley and Gregory, Newton's "darlings," fare worse. When Flamsteed obtained a copy of Gregory's recently published *Astronomiae Physicae & Geometricae Elementa* in July 1702, he accused the mathematician of attempting to "kill me" with false words. Six months later, in February 1703, Flamsteed made a prediction regarding Newton's place in history: "'Tis known very well whence he had his materials and principles; he makes but a small figure now, and will probably make a less in a little time, though, had he been wise, he might have made a much bigger than he did formerly."[48]

For all of Flamsteed's bluster, and despite the deep pain he suffered from Newton's roughshod treatment, he stood forever in awe of the man. Flamsteed had been a young astronomer of twenty-five when the first paper on light and colors was submitted to the Royal Society early in 1672. "I could not at first yield to this theory," he wrote in 1707, when his breach with Newton was approaching its widest point, "but, upon trial, found all the experiments succeeded as he related them; which kept me silent and in suspense."[49] But as he had grown older and come to know Newton personally, Flamsteed had also grown bolder. If he did not possess the true eye of the lion tamer he was still one of the very few willing to challenge Newton to his face, something not even the brash Hooke would do. It was partly the old story of being pushed too far and of fighting back with any weapon at hand. Flamsteed simply would not allow himself to become grist for any man's insatiable ego, Newton's included. Besides, he possessed something that Newton had wanted desperately for years, an exceedingly rare circumstance and one Flamsteed did not scruple to exploit to the full for want of treatment as an equal.

Late by four years, Newton came at last to dine at Greenwich in April

1704. Flamsteed apprised his guest of some alleged faults in the fourth book of the *Principia*, "which instead of thanking me for, he resented ill," and asked "why I did not hold my tongue." When the conversation then turned to the recently issued *Opticks*, a copy of which Newton had dispatched to Greenwich, things went no smoother. "I thanked him for his book: he said then he hoped I approved it. I told him truly no."[50] Newton dropped the subject and revealed the real purpose of his visit, which had nothing to do with Flamsteed's evaluation of his scientific endeavors. He was interested in the Astronomer Royal's progress on the projected star catalog. According to Flamsteed's account, a friend of his had recently acquainted Prince George with the nature of the Astronomer's work and the need of financial assistance to bring it to fruition. Newton, just elected President of the Royal Society, had gotten wind of the overture via "a great courtier"— namely Halifax—and went down to Greenwich in hopes of thrusting himself into the picture. As has been noted elsewhere, there is only one credible explanation for his conduct: Plagued by the intolerable prospect of failure, he needed Flamsteed's observations in order to complete his own lunar theory, which would crown a second edition of the *Principia*.[51] Flamsteed showed Newton his tables, observations, and stellar maps; his visitor, who was under the impression that the work would soon be ready for the press, offered to recommend it privately to the Prince. "I know his temper; that he would be my friend no further than to serve his own ends," Flamsteed wrote later in a cutting memorandum, "and that he was spiteful, and swayed by those that were worse than himself. This made me refuse him. [H]e never intended me any good by it, but to get me under him, that I might be obliged to cry up as E. H. [Halley] has done hitherto." They parted on superficially amicable terms, Newton promising to speak to the Prince on Flamsteed's behalf anyway. "Do all the good in your power," he told his host on leaving. "[A] short expression of very good advice," Flamsteed later observed, "and [one which] had been the rule of my life from my infancy; though I do not know that it ever has been his."[52] Small wonder that the Victorians brought pressure to bear on Francis Baily in their unsuccessful attempt to suppress his publication of Flamsteed's prolonged critique of their tarnished Augustan hero.

It appeared for a time that Newton had been checkmated. Prince George visited Greenwich at some point during the next few weeks or months and, after being shown the body of Flamsteed's work, agreed to finance the printing and engraving.[53] Meanwhile, Flamsteed accused Newton's "flatterers" of "crying up" his book, asking "why I did not print." The gibe struck home, for the Astronomer Royal had always been extremely touchy on the subject of publication. Indeed, it was Flamsteed's public response to this very question five years earlier that had so infuriated Newton, whose name had been invoked without prior consultation. On November 8 Flamsteed drafted an "Estimate" showing the extent of the work he planned to publish, but in so doing he grossly underestimated the time required for its completion. The charts of the constellations could be drawn and engraved at once, while the first two parts, including the catalog of the fixed stars, were virtually ready for the printer, or so he stated. By the time these were set and run off, the third and final part "may in the mean time be fitted for the Press."[54] In self-defense Flamsteed gave a copy or two of the "Estimate" to James Hodgson, a nephew-in-law and former assistant, who was elected a Fellow of the Royal Society on the day Newton became President. Hodgson was instructed to circulate the document among those of Flamsteed's friends possessed of "unjust suspicions" concerning his intent to publish. The younger astronomer made the crucial mistake of taking the "Estimate" to the Society meeting of November 15, 1704, during which it was inadvertently passed to the Secretary, Hans Sloane.[55] Thinking it a matter of official business, Sloan read the paper to the assembly, whose members thanked Flamsteed and promised to encourage the project "as farr as they Can."[56]

Newton must have experienced some difficulty suppressing his jubilation. Pure chance now promised to deliver into his hands what years of intimidation, flattery, threats, and guile had not. He moved quickly to exploit the boon by making certain at the next meeting that the President was asked to "recommend" and "encourage" Flamsteed's design.[57] At the Society's annual gathering a week later the Prince Consort became a Fellow. Newton and Sloane met with George on December 7 and presented him with a copy of Flamsteed's prospectus. On December 11 the man whom Trevelyan described as a kindly but negligible mortal ordered his private secretary to draft Newton an official letter of acceptance. Flamsteed's intended *Historia Coelestis Britannica (British History of the Heavens)* would not suffer delay for lack of funds. His Royal Highness was also pleased to command Newton, Roberts, Wren, Gregory, and Arbuthnot to examine Flamsteed's papers and determine which among them were fit for the press.[58] Newton had served his own interests far better than even he could have imagined a few short weeks before. He was now head of a committee packed with loyalists (Francis Aston was shortly added to the group) and granted free access to the lifework of a man who had himself been denied membership. "With these persons," Flamsteed was soon to lament, "Sir Isaac Newton began to act his part, and carry on his designs."[59]

Prince George's letter reached Newton on December 17. Newton wrote to Flamsteed at Greenwich the following morning, informing him of the arrangements concerning the *Historia Coelestis* and requesting that he bring specimens of his papers to a dinner with the referees scheduled for the nineteenth. It was precious little notice, but Newton had disarmed Flamsteed with unexpected kindness in the past, and he resorted to this proven tactic again, signing himself "Your very loving Friend & humble servant."[60] Considering what his "loving friend" had been up to, Flamsteed's response was exceedingly enthusiastic; he overlooked the personal inconvenience of meeting the imminent deadline and dashed off to London to dine with the referees as scheduled. Unable to complete their business at one sitting, they arranged to meet again on December 27. Two days after the second meeting Newton, all charm and solicitude, visited Flamsteed at the Royal Observatory. He made the desired impression. "Mr. Newton is becoming exceeding kind of late," Flamsteed waxed in a letter to Sharp on December 30, "he was here to visit me yesterday; stayed from 12 till near 5 o'clock; dined with me; took a new view of my books and papers; and becomes solicitor with the Prince on their behalf." His inherent distrust of the man was not wholly allayed, however: "I shall be as cautious as I can that he do me no injury."[61]

It was Flamsteed's turn to wait on Newton in London a few days later. Newton requested copies of additional papers before making an official report to the Prince, and these Flamsteed cheerfully supplied on January 6, 1705. The referees submitted their findings to Flamsteed's royal benefactor on the twenty-third. The work was estimated to run to about 1,200 pages in folio. The expense of printing four hundred copies was put at £683, while £180 more was recommended for the employment of the calculators needed to complete work on some 1,400 still unrefined observations.[62]

Flamsteed voiced his first serious objection to what Newton was about at this juncture. The final figure of £863 had not taken into account the astronomer's plan to publish every significant star catalog from Ptolemy to Halley. Flamsteed envisioned a monumental work of no fewer than 1,450 pages, and he had previously heard that the Prince was willing to expend up to £1,200.[63] He rightly attributed this arbitrary change of plans to Newton, who was interested in one thing only—immediate access to the observations crucial to the completion of his lunar theory. This recurring bone of contention inevitably gave rise to others. Newton requested that Flamsteed draft a note of security for Prince George's money. "This I know was to oblige me to be his slave," Flamsteed fumed. "I answered that I had (God be thanked) some estate of my own, which I hoped to leave, for my wife's support, to her during her life ... that therefore I would not cumber my own estate with impress or security." He proposed only to vouch for the workmens' bills as they came due, leaving actual payment to the referees. Newton backed down, admitting that since Flamsteed refused to handle any funds, "There was no need of security or articles." The Astronomer Royal fared less well in a dispute with Newton over an unrelated finacial matter. Flamsteed somehow got it into his head that royal largess to the tune of £2,000 would be "an honorable recompense for my pains." He dropped this bombshell on Newton during a chance encounter at Garraway's on February 28. This particular source of discontent apparently sprang from Flamsteed's anger over the referees' insistance on employing

Awnsham Churchill, a respected publisher and bookseller, to handle the final preparation and printing of his catalog. Churchill's compensation was ultimately fixed at thirty-four shillings a sheet, an expense Flamsteed had largely hoped to avoid by employing a printer only. Having toiled on behalf of his country at the modest salary of £100 a year for three decades, Flamsteed was bitterly disappointed that he, the author, should not share in the Prince's bounty, while the publisher was to be handsomely paid. "'Tis a great dishonor to the Queen, his Royal Highness, and the nation," he reflected in a private memorandum, "that no reward is proposed for so long, difficult, and laborious work: and that the small one I might justly expect is cast upon those that have had no part in the labour and expense, nor hazarded their health, nor felt my severe pains of the stone and other distempers, caused by my night watches and day studies."[64] Flamsteed, of course, had a point, but Isaac Newton, who had also watched and studied by day and by night, and had suffered terribly for it, both physically and mentally, was hardly the one to turn to for sympathy.

On March 5 Flamsteed met with Newton and the other referees at Castle Tavern to examine the first specimen sheets supplied by the hated Churchill. Not surprisingly, Flamsteed found them to be "ill done." He visited London again two days later and, engaging the services of a Mr. Barber, had others printed "very well" at his own expense, an ultimately futile exercise. By March 21 word had reached him via Aston that all matters concerning the printing were agreed upon but the paper, a welcome harbinger of spring. Flamsteed remained justifiably cautious, however, ordering his servant to copy and file every one of Newton's letters, past and future. The astronomer's hopes were soon dampened by a delay in the printing, which resulted from Newton's unsuccessful political activities at Cambridge. Meanwhile Flamsteed had convinced himself that Churchill was plotting to deprive him of his due by surreptitiously printing extra copies of the *Historia Coelestis* to be sold off "before I can sell one of those the Prince allows me."[65] His fears, as it turned out, were not only unfounded but highly premature; the first copy of the ill-fated book did not come off the presses until early in 1712.

The articles of agreement were still unsigned, a matter Newton planned to resolve over dinner at his residence on June 11. In addition to the host himself, the party included Flamsteed, Gregory, Roberts, and Churchill. Flamsteed wrote that the referees were settled upon a sum of £1.14s. a page for the publisher, a preposterous figure to which he could not agree. The meeting thus broke up with this and other crucial matters unresolved. Flamsteed was dead set against the enrichment "at my cost" of a man he hardly knew and, as he informed Sharp the next day, prayed to God that Newton would soon see the light. Unfortunately, darkness prevailed a few days later when the Royal Society President visited Greenwich, bringing

Churchill and an associate with him—"not a word of any recompense for 30 years' pains and extraordinary expense, though occasion enough offered [Newton] to speak of it." Flamsteed again prayed for a miracle.[66]

Newton's sorely tested patience had nearly exhausted itself, as attested by the brief, none-too-polite notes he wrote as the project's first anniversary approached. "If you stick at any thing, pray give Sr Chr. Wren & me a meeting as soon as you can conveniently, that what you stick at may be removed."[67] Flamsteed, who considered Sir Christopher about the only referee from whom he could expect a halfway fair hearing, had already bombarded the poor man with complaints, both oral and written, throughout the summer. Not wishing to undermine his credibility by attacking Newton directly, he took aim on an old target. It was Halley who had set the Master of the Mint at a distance; Halley who had impugned his reputation by false suggestions; Halley the liar, Halley the sycophant, Halley the godless![68] Yet Flamsteed's lamentations and objurations notwithstanding, he feared that a complete break with Newton and the other referees could place a lifetime of labor in jeopardy of never being published. Finally, on November 17, 1705, after "much talk, little done," the wranging parties reached an accommodation of sorts. Churchill was to be the publisher after all, but at a rate of 34s. per sheet rather than the £1.14s. previously recommended by the referees. He would print just four hundred copies of the *Historia Coelestis*, which were to become the property of the author. After the printing either Flamsteed or his agent was empowered to "Breake the press, without any delay . . . or Molestation" from Churchill or his workmen. The article Flamsteed found most objectionable specified that the catalog of the fixed stars—the very heart of his work—would be published in Volume One rather than Volume Two, as he had planned, and that it would include only those observations made through 1689.[69] Flamsteed signed but only, as he subsequently claimed, because he was under duress. Writing to the loyal Sharp on November 20, he cried foul: "Sir Isaac Newton has, at last, forced me to enter into Articles for printing my works with a bookseller, very disadvantageous to myself." In another account contained in his autobiography, Flamsteed was even more blunt. The articles were read aloud but once, "and I was required to sign them immediately, else the work was at a stand." He could recall being asked to only one other meeting of the referees thereafter, "where nothing material was determined whilst I was present."[70] In point of fact, he attended two.

Having yielded to Newton's conditions, Flamsteed comforted himself with the thought that Volume One would proceed with all dispatch, for the articles called for a minimum printing of twenty sheets a month. The astronomer soon found himself "deceived." On January 17, 1706, he informed Sharp that the manuscript had been in the referees' hands seven weeks, but Churchill was still not settled on a printer. Two weeks later Sharp heard from his friend that it was hoped the first sheet would be ready in a fortnight. It was not to be. Flamsteed learned from Hodgson that Newton had compiled four or five pages of differences between his original observation notes and the manuscript submitted to the referees. Fearing the worst, the Astronomer Royal hastened to London, where he confronted Newton with the rumor. Newton admitted to the existence of the pages but said he did not believe Flamsteed had committed major errors. As it turned out, Gregory, not Newton, had raised some serious questions, and Flamsteed spent a long afternoon setting the two referees straight. He subsequently labeled the differences "all of the Doctor's making."[71]

Meanwhile Flamsteed was yet to honor an important part of the bargain, which he had always known would be impossible to keep. He had delivered all of the manuscript copy of Volume One but the star catalog itself. The reason for the delay is well known. Several years later he wrote a telling note in the margin across from the article requiring that the catalog be published, against his wishes, in Volume One: "This Article impossible for the observations to compleate it were not all gotten till five years after."[72] Newton, too, was well aware that the Astronomer Royal could not live up to this promise, which Newton himself had extorted from him by ultimatum. Now he compounded his tryanny by refusing to budge unless Flamsteed complied. His back to the wall, Flamsteed carried a copy of the star catalog, in an uncompleted form, to Christopher Wren on March 8, 1706, insisting, as he later wrote, that it remain sealed until it was ready for printing.[73] At least he had denied Newton the pleasure of taking the manuscript from his own hands.

Flamsteed next met with the referees and Churchill on March 23. They agreed to draw money from the Prince so that the actual printing could begin. Asked by Newton if things were not now to his content, Flamsteed replied, "it was strange that I should be so little taken notice of, who was the person mainly concerned: at which he seemed chagrined." Things went no better at a private meeting between Newton and Flamsteed on April 19. When Flamsteed requested funds to defray the legitimate cost of his calculators a "very grave" Newton lectured him on the delicate state of royal finances. Besides, he had just advanced Churchill £250. What he did not tell Flamsteed is that he had already drawn £125 more with which to reimburse the author as per the articles of agreement. Flamsteed raised the matter again in September, informing Newton that he was forced to dismiss both his amanuensis and calculators.[74] He received no reply. Newton, as Flamsteed well knew, could punish in ways both gross and subtle:

I have plowed sowed reaped brought in my Corne, with my own hired Servants & purchased Utensills. Sr I N. haveing been furnished from My Stores would have me thrash it all out my selfe & charitably bestow it on ye publick that he may have ye prayse of haveing procured it. I am very desirous to supply the

Publick with my Stores if it will but afford me what I have layd out in tillage and harvesting in Utensills & help. & afford me hands to work it up since the labor is both too hard & much for me. for an adequate recompense I doe not expect but I must stand upon a reasonable one since God has blest my labors with large fruites & not to doe it were not to acknowledge his goodness; & my Countries Ingratitude would be attributed, by Sr I N himselfe, to my *Stupidity*.[75]

At last, on May 16, the first sheet of the *Historia Coelestis* came off the press. It was June 3 before Flamsteed saw another. He complained bitterly of Churchill's dilatory ways but was told by Newton "that we must proceed slowly at first, and make more dispatch after." He accused Newton of arbitrarily starting and stopping the press without the slightest consultation with himself or the other referees. His combative spirit notwithstanding, Flamsteed had been a fragile physical specimen since childhood, and he was beginning to bow under the weight of his three score years. "I grow gouty," he complained to Sharp on June 11, "and pains of my feet hinder me from stirring much abroad; so I am confined in a manner to my business, have more time to prepare for death, and now can thank God that we are to die: for life, to one so weak as I am, is no very pleasant state at 60."[76]

A bequest in the will of Doctor Thomas Plume, the late Archdeacon of Rochester, added considerably to the animosity between Newton and Flamsteed at this juncture. It seems that Plume had read Huygens's *Cosmotheoros* on the recommendation of the Astronomer Royal and was sufficiently impressed to set aside £1,800 for the construction of an observatory and the establishment of a Professorship of Astronomy and Experimental Philosophy at Cambridge. Most prominent among the four trustees were Bentley and Whiston, both of whom owed considerable debts to Newton. Furthermore, they were directed by the will to frame the statutes and to choose the first Plumian Professor only after consulting with John Ellis, Newton's old friend and fellow knight; Flamsteed; and Newton himself. Flamsteed met with Bentley and Newton on March 15. The "Doctor would have had my hand to a paper for the election of Mr. Cotes to be professor: I refused till I saw him."[77] A Fellow of Trinity College, the handsome twenty-four-year-old Roger Cotes, who was destined to publish the second edition of the *Principia* in 1713, was a brilliant but ill-fated mathematician. His early death deeply saddened Newton, who looked upon him as a potential intellectual equal. Flamsteed, of course, had his own thoughts on the appointment, favoring his Greenwich assistant John Witty, "an expert calculator." Presented with a virtual *fait accompli*, Flamsteed bowed to the inevitable, something he had been forced to do on every previous occasion when his views on patronage clashed with Newton's. Dr. Bentley, he wrote to Newton, has determined "Plumes professor of Astronomy . . . without ever so much as letting me know yt hee was about such a business & I fear directly contrary to ye Archdeacons designe wherewith I am apt to thinke none of ye Trustees in Cambridge were so well acquainted as I am. I had not known of it but by accident."[78] How fitting that Flamsteed's sense of alienation should have been expressed in a letter reminding Newton that while three months had passed since the signing of the articles of agreement, he was yet to receive a copy of his own.

The printing went forward with discouraging slowness. On April 15, 1707, Newton, accompanied by Gregory, went down to Greenwich to present Flamsteed with another of his arrogant ultimatums. He would stop the press and pay the Astronomer Royal nothing unless he handed over a corrected copy of the star catalog together with the additional observations made with Flamsteed's great mural arc down to the end of 1705.[79] As always, it was the latter data that Newton truly wanted, suffering not the slightest compunction at plundering another's lifework to further his own. Flamsteed would have none of it and wrote that "we parted quietly, I cannot say very friendly." On May 19, in the midst of their standoff, Flamsteed wrote cryptically to Sharp of Newton, "an accident has lately happened, that has discovered his proud and insolent temper, and exposes him sufficiently. He has been told calmly of his faults, and could not contain himself when he heard of them. My affair was not forgot." He promised to tell Sharp more when they met in person, adding only that he thought the "accident" a manifestation of superhuman design. Whatever had occurred, it was Newton who finally backed down. Flamsteed announced in his autobiography that all the printed material intended for Volume One—with the exception of the catalog—was ready in December 1707, that is, ninety-seven pages in some eighty-nine weeks.[80]

With the presses now at a standstill for lack of additional manuscript, the unresolved question of the catalog loomed larger than ever. Both men remained steadfast on the order of arrangement. Three months passed before a meeting of those most directly concerned was finally called on March 20, 1708, at the Castle Tavern. The resulting settlement, although somewhat ambiguous, seems to have favored Flamsteed. He agreed to deliver into Newton's hands a lengthy manuscript of observations taken with his famous mural arc since 1689. He also relinquished a more advanced copy of his catalog on the fixed stars in return for the right to revise the one currently in Newton's possession. Newton agreed to issue Flamsteed the £125 due him for the past two years, which he did on April 11. It was also Flamsteed's understanding that his 175 pages of observations would be printed next, allowing him time to complete the star catalog as originally planned, and that it would be published in Volume Two after all.[81]

Months passed, and not a single sheet was printed. Newton blamed the delay on Flamsteed for failing to provide corrected copy of the proofs as specified in the March 20 agreement. The referees informed the Astronomer Royal on July 13 that if he did not take care another "Corrector" of their choosing would be employed.[82] Flamsteed was thunderstruck: He was nothing if not punctual, a Puritan trait documented with tedious regularity in his papers. So anxious had be been to return corrected pages "a second time" in 1706 that he took to horse, gout and all, in a rainstorm, only to find Churchill's gate "stock locked" on his arrival in London. Another time he visited the press while Newton was there and pointed out how careless the compositor was in setting the figures and lines. "He put his head a little nearer to the paper, but not near enough to see the fault, (for he was very near sighted,) and making a slighting motion with his hand said, 'Methinks they are well enough.'"[83] No, it was Newton who was guilty of perfidy. If he could make this false charge stick, the printing again could be under his absolute command.

Flamsteed presented an excellent, well-documented defense in a letter to Wren of July 19: "Sr Isaac Newton has 175 Sheets of the 2d Volume in his hands, that the Press may proceed with whilest I am Compleating the Catalogue. So there need be no stop on my Account as there never was, nor hereafter shall be."[84] The summer passed without further incident, but not so the autumn. Prince George died in October, and all work on the project came to an immediate halt. Newton, who had never truly succeeded in bending Flamsteed to his will, had the astronomer's name stricken from the list of Royal Society Fellows in 1709 for nonpayment of dues.[85] Flamsteed responded in a bitter letter to Sharp: "He . . . has been lately much talked of; but not much to his advantage. Our society is ruined by his close, politic, and cunning forecast."[86] Having already encouraged Newton to read Jeremiah of the Old Testament for enlightenment, Flamsteed might well have directed him to an even more appropriate passage in Matthew of the New: "Why beholdest thou the mote that is in thy brother's eye, but considerest not the beam that is in thine own?"

Chapter Eighteen

The Devil's Banter

As for those wingy mysteries in divinity, and airy subtleties in religion, which have unhinged the brains of better heads, they never stretched the pia mater of mine.

Sir Thomas Browne

I

While Newton can scarcely be said to have been swamped with business at the Mint, his responsibilities had increased within months of Queen Anne's accession. Lord Sidney Godolphin, his new superior at the Treasury, harbored designs on the Parliamentary seats of Cornwall, whose numbers were greater in the House of Commons than those allotted to metropolitan London. Godolphin persuaded the Queen to purchase the favor of those who held them by concluding a seven-year contract with the Convention of Tin Miners, whereby 1,600 tons of the metal were to be sold annually to Her Majesty at the inflated price of £70 per ton. To ensure a monopoly on tin mining and sales, it was further agreed that she would buy the yearly output (50 tons) of Devon at the same high price.[1] The contract took effect in the autumn of 1703, offering Newton an opportunity both to broaden his administrative authority and to ingratiate himself further with the new regime. On October 30 he informed Godolphin of his willingness to undertake the responsibility of overseeing the entire operation: "I ... am humbly of [the] opinion that I can do it wth two Clerks added to my own."[2] Moreover, the Master asked for nothing in the way of compensation beyond reimbursement for costs, although he was eventually to receive £150 per annum. Given these conditions, Godolphin could hardly have refused.

461

Newton had ample opportunity to think better of his impetuosity in the coming years, for the many unanticipated problems associated with the tin trade only served as additional irritants to a psyche already overburdened with *causae bellum*. The Master was immediately faced with the question of where to stockpile the incoming ingots, which came by sea to the Tower quay in ever increasing numbers. Luckily the Irish Mint, housed at the east end of Tower Mint, was no longer turning out copper coins. Its rooms were emptied of equipment on Newton's orders and set aside for the storage of tin. Despite the government's virtual monopoly on the product, sales remained sluggish from the outset. Efforts to negotiate contracts on the Continent did little to alleviate the surplus. When Newton engaged warships to cut the costs of transport to Europe, the great trading companies issued vehement protests. The unsold stocks swelled at an alarming rate, causing the Treasury to entertain a number of radical proposals. One John Williams pressed the government to raise the selling price from £76 to £100 or £120 per ton on the grounds that England possessed a corner on the market. When asked his feelings on the matter, Newton demurred. Writing to Godolphin on September 23, 1706, the Master noted that Williams had authored a similar proposal the previous year. At that time Newton had met with the principal traders to East India, "who were then of opinion that great quantity of Tin might be had in the East Indies, & if the price of her Majesties Tin were raised here, it would encourage the importation of Tin from thence, & thereby hinder the consumption of English Tin in Europe."[3] So far as he was concerned, nothing had happened to warrant a change of attitude. A more enterprising William Tyndale hit upon the idea of striking tin money from the surplus metal in denominations ranging from £10 down to sixpence. Newton again responded in the negative: In the first place, the available tin was being used as collateral for loans; in the second, it would require the Treasury to accept tin money for taxes, a dangerous precedent.[4]

The Master himself could think of only one viable alternative—"a cessation of digging Tin in Cornwall." Based on the rate of sales in 1705, England had enough tin in storage to last eight years. If some decisive action were not taken very soon the Queen might be expected to lose the staggering sum of £452,402 by 1713,[5] a debacle Newton wanted no part of, blameless though he might be. Fortunately for the Queen, the final outcome proved somewhat less disastrous than Newton had predicted. Sales of tin rose from a yearly average of 860 tons in 1705 to 1,520 tons in 1709, then fell to 1,260 tons by the time of Anne's death in 1714. Yet never in any given year were sales commensurate with production. When the original contract expired in 1710, a year of general elections, it was renewed for seven more; a wholly irresponsible clause was added, calling for an increase in production of 200 tons annually at war's end. Newton continued to fret, admonish, and calculate, but to no avail. The "Queen's Tin" lay unsold in the Irish Mint until well into the 1720s, while the Treasury begged patience on the part of her creditors. Newton put the final loss at more than £200,000, thus ranking the captive votes of Cornwall among the most costly in modern political history by almost any measure.

On April 10, 1706, two days before he declared his love for fair Catherine in writing, Lord Halifax was appointed one of the commissioners for negotiating a different kind of union—that of England and Scotland. The treaty drafted later in the year became law on May 1, 1707. Article Sixteen of the Act of Union provided that "the Coin shall be of the same Standard and Value throughout the United Kingdom as now in England, and a Mint shall be continued in Scotland under the same Rules as the Mint in England."[6] The very embodiment of efficiency, Newton, in consultation with Godolphin, had begun making preparations for the anticipated event well in advance.

The plan called for all silver coins previously minted in Scotland to be gathered by collectors and shipped to Edinburgh for melting and restamping in the current English denominations. (It was decided that "Northern coppers" and gold coin of Scottish design should be left to the mercies of time.) Because their minting procedures were not identical, Newton proposed that officers of the Edinburgh Mint come to London for training and that one of his men be sent north. The plan was not embraced by either the government or the Scottish officials, who chafed at the idea of leaving their homes and families for several months. In the end Newton persuaded three London moneyers and one of his clerks, Richard Morgan, to spend the next year in the northern capital, where they were to place the process of refining and record-keeping on a London footing. He also received the Lord Treasurer's permission to appoint David Gregory as his special representative in Edinburgh. The native Scot was to remain in the city for only three months at a salary of £250, plus a travel allowance of £50 more, an enviable dollop of Newtonian patronage.[7]

The London officials arrived in Edinburgh on August 1. A variety of machines and other materials that the Edinburgh Mint lacked had been shipped well in advance of their departure, but, Newton's foresight notwithstanding, the operation immediately ran into difficulty. When asked for a list of items they most required, the Scots had responded with a confusing letter whose terminology baffled Newton's clerks. The list turned out to be riddled with spelling and grammatical errors, which resulted in shipping delays and the misidentification of critical supplies. Some of the equipment was still at sea, while other machinery had been completely overlooked by the Scots, who had minted nothing since 1701. So out of practice were the moneyers and melters that the castings of the late summer and early autumn were not fit for minting. By November, however, when Gregory left Edinburgh for London, where he died the following year, the recoinage was moving ahead in earnest at the rate of £6,000 per week. Although the Tower employees stayed on until its completion in March 1709, their work was largely finished by the previous December. The Edinburgh Mint had coined some £320,372 of silver in thirteen months. It was destined to stop forever on August 4, 1710.[8]

The technical questions that beset Newton during Scotland's recoinage were matched by a flood of personal complaints from Mint officials. Broadly organized like its sister institution, the Edinburgh Mint had a master, warden, comptroller, and various subordinates. Their salaries were lower across the board, in several instances much lower, than in London, a vexing state of affairs to those who considered themselves professional (and political) equals. It was their hope, apparently nurtured by the palpable example of a handsomely paid Gregory, that Newton would become their champion in this matter, a fancy of which they were soon disabused. One Edinburgh post without parallel in the London Mint was the collector of bullion. When Newton learned that the incumbent, Daniel Stuart, had died, he immediately recommended to Godolphin that "the place . . . being irregular should cease," which it did.[9] On July 10, 1708, W. Boswell, the Edinburgh Comptroller, wrote to complain that his salary of £60 a year was too low and that a number of new clerks appointed during the recoinage should also have their wages increased. Newton, he continued, could expect to hear more on this matter in the near future from the Warden, William Drummond.[10] Sure enough, Drummond wrote to him only two days after Boswell, but to little discernible effect. Drummond wrote again on July 20 to inform Newton that he had delivered a copy of the Latin edition of the *Opticks* "in your name" to the prestigious Edinburgh College of Physicians. The members were flattered and wished Newton "a long life and perfect health." Both wishes were granted, but not so that of the Mint employees. Nor was the Master moved by the final request of a dying man. George Allardes, his counterpart at Edinburgh, wrote on August 9, 1709, that "my old indisposition of vomiting blood" had recurred. Should death come, Newton could pay him a great honor by appointing his seventeen-year-old son to the mastership—"we think his minority may be no scruple."[11] Newton thought otherwise; when Allardes expired a short time later, he chose John Montgomery to succeed the late Master and drew up an indenture that further extended his personal control over his Edinburgh subordinate. Yet to Allardes belonged a measure of retribution. He died, as had Newton's predecessor Thomas Neale, before he he could clear his accounts, leaving a hundred loose ends for Newton to unravel. Neither could the Master of the Mint change the fact that the drones of Edinburgh and their descendants continued drawing aggregate salaries of approximately £1,000 a year until a statute of 1817 finally abolished this lucrative anachronism.

II

Abraham de Moivre, one of the young mathematicians and natural philosophers closest to Newton during that period, later recalled how the
great man would wait for him in the evenings at a coffeehouse (probably
Slaughter's in St. Martin's Lane) while he completed the private
mathematical lessons by which he earned a modest living. The two would
then repair to Newton's home, where they passed many a pleasant hour in
philosophical conversation. Unfortunately the details of these long discussions were never entirely set down on paper. We may be quite certain,
however, that they centered on the *Opticks*, for Newton had entrusted his
young disciple with the task of seeing a new edition through the press. Emboldened by the enthusiasm of his countrymen for the experimental
treatise, he was eager to publish a second edition in Latin so that classically
educated European scholars might have ready access to his revolutionary
work. The translation itself was undertaken by the author's friend Dr.
Samuel Clarke. Newton was so pleased with the accuracy and elegance of
the finished product that he bestowed a gift of £100 on each of Clarke's five
children.[12] Whether de Moivre settled for the intangible but more enviable
reward of intellectual intimacy is not known.

The most striking difference between the English edition of 1704 and
its Latin counterpart of 1706 is the addition of seven new queries. These
carry the numbers 17 to 23 but were recast as queries 25 to 31 in all subsequent editions and shall be referred to by their later numbers here. As in the
first sixteen queries, the author is firmly convinced that major physical
phenomena can be accounted for only by the presence of forces between
all bodies, large or small. He had discovered nothing to support the Cartesian belief that a medium of dense ether fills the heavens, controlling all
motion and making nature sufficient unto itself. "Are not all Hypotheses erroneous," Newton asks in query 28, "in which light is supposed to consist in
Pression or Motion, propagated through a fluid Medium?" It is "inconceivable," as Huygens had argued, to account for the double refraction of
Iceland spar (calcite) by postulating two vibrating waves of light "without
[their] retarding, shattering, dispersing and confounding one anothers Motions." On the macrocosmic plane, "a great Objection [to the ether] arises
from the regular and very lasting Motions of the Planets and Comets in all
manner of Courses through the Heavens." In other words, the presence of
a fluid medium ought to retard the movement of celestial bodies through
space, yet there was no evidence of any such attrition, "the Motions of the
Planets and Comets being better explain'd without it." Not only had his
own research convinced him that the ether was nonexistent, or if present
extraordinarily rare and therefore inconsequential, but the ancient atomists, to whom he had been attracted as a young man in his early twenties,
also dismissed the conception of a universe filled with invisible matter.

"And for rejecting such a Medium, we have the Authority of those the oldest and most celebrated Philosophers of *Greece* and *Phoenicia*, who made a *Vacuum*, and Atoms, and the Gravity of Atoms, the first Principles of their Philosophy; tacitly attributing Gravity to some other Cause than dense Matter." Later philosophers (namely Descartes) had unwisely discarded the wisdom of the ancients, "feigning Hypotheses" in order to explain all things mechanically.[13]

Still, the *prisca sapientia* and its accompanying sense of *déjà vu*—the feeling of a return to some primordial knowledge—had not of itself disabused Newton of the conventional mechanistic view of nature. His theory of matter was forged in the crucibles of his private alchemical laboratory. Drawing upon "De natura acidorum," which dated from the early 1690s, he set forth one of the boldest hypotheses of modern science in query 31:

> Have not the small Particles of Bodies certain Powers, Virtues, or Forces, by which they act at a distance, not only upon the Rays of Light for reflecting, refracting, and inflecting [diffracting] them, but also upon one another for producing a great Part of the Phaenomena of Nature? For it's well known, that Bodies act one upon another by the Attractions of Gravity, Magnetism, and Electricity; and these Instances shew the Tenor and Course of Nature, and make it not improbable but that there may be more attractive Powers than these. For Nature is very consonant and conformable to her self.

He next presented a detailed discussion of the research that had led him to formulate the dynamic concept of force. As always, he was wary of postulating more than the evidence would allow and carefully modulated the dialogue between his two inner voices, the one boldly imaginative, the other deeply critical: "How these Attractions may be perform'd, I do not here consider. What I call Attraction may be perform'd by impulse, or by some other means unknown to me. I use that Word here to signify only in general any Force by which Bodies tend towards one another, whatsoever the Cause." Yet of this much he was certain: Gravity, magnetism, electricity, and certain related phenomena yet to be elucidated all result from the presence of active principles, that is, the constant exertion of forces of attraction and repulsion between every particle of matter and body in the universe. "These Principles I consider, not as occult Qualities, . . . but as general Laws of Nature, by which the Things themselves are form'd; their Truth appearing to us by Phaenomena, though their Causes be not yet discover'd." Having already formulated the principle of universal gravitation, he thought there were no more than two or three other fundamental laws of motion still to be revealed. When they were discovered, it would constitute "a very great step in Philosophy."[14]

Newton's early rejection of Cartesian mechanics was, as we have long since observed, a partial result of an ingrained aversion to the belief that the natural order is capable of functioning independently of God. He could never embrace a cosmology that dispensed with the Creator once the act of creation itself was accomplished. Neither could he bring himself to leave God out of the *Opticks*. Indeed, the depth of his belief is underscored by an eloquence reminiscent of the unknown author of Job, as well as of his own "De gravitatione".

> Whereas the main Business of natural Philosophy is to argue from Phaenomena without feigning Hypotheses, and to deduce Causes from Effects, till we come to the very first Cause, which certainly is not mechanical; and not only to unfold the Mechanism of the World, but chiefly to resolve these and such like Questions. What is there in places almost empty of Matter, and whence is it that the Sun and Planets gravitate towards one another, without dense Matter between them? Whence is it that Nature doth nothing in vain; and whence arises all that Order and Beauty which we see in the World? To what end are Comets, and whence is it that Planets move all one and the same way in Orbs concentrick, . . . and what hinders the fix'd Stars from falling upon one another? How came the Bodies of Animals to be contrived with so much Art, and for what ends were their several Parts? Was the Eye contrived without Skill in Opticks, and the Ear without Knowledge of Sounds? How do the Motions of the Body follow from the Will, and whence is the Instinct in Animals?[15]

Had Newton gone no farther in his rhetorical theological discourse, he probably would have been spared at least a portion of the derision that followed. But as David Gregory noted in another of his numerous memoranda, Newton, after much discussion and soul-searching, chose to set forth his belief that God is omnipresent in the literal sense. "And that as we are sensible of Objects when their Images are brought home within the brain, so God must be sensible of every thing, being intimately present with every thing: for he proposes that as God is present in space where there is no body, he is present in space where a body is also present."[16] Too late did Newton realize he had made a strategic blunder by proclaiming that infinite space is the sensorium of God. He called in as many copies of the second edition as he could lay his hands on, deleting the controversial passage with scissors and inserting a less sweeping one in its place. As luck would have it, one of the unaltered originals was already on its way to Leibniz, who, among other things, pronounced Newton guilty of being philosophically naive.

A hint of the Enlightenment was in the air.

III

The high praise with which the Edinburgh College of Physicians had greeted the Latin edition of the *Opticks* paled in comparison to the accolades it garnered for Newton in Rome. Henry Newton, the English Envoy Extraordinary to Florence, who claimed to have nothing in common with the author apart from the privilege of sharing his surname, wrote

Newton in December 1707 that all the experts on high literature in the Italian capital, including Giovanni Maria Lancisi, Professor of Anatomy and Physician in Ordinary to the Pope, "applaud the offspring of the super-human intellect of the Golden Knight Newton, whose fame extends throughout Europe."[17]

Newton had been far less pleased with the appearance of another of his works earlier in the year. Whiston, his disciple and successor at Cambridge, published Newton's Lucasian lectures on algebra under the title of *Arithmetica Universalis*. Dissatisfied with the proofs as they came to him in the summer of 1706, Newton complained to Gregory that he had opposed the publication from the beginning. However, he had reluctantly given his consent in return for political patronage when he had last stood for Parliament in 1705. He nevertheless refused to allow his name to appear anywhere in the volume, although the author's identity was an open secret even before the date of publication. As in so many other instances, the passage of time did nothing to assuage Newton's anger and sense of betrayal, a radical reversal of normal human conduct. He later informed John Conduitt that Whiston made so many mistakes that he was personally obliged to publish a second edition with the necessary corrections.[18] This Newton did in 1722, in his seventy-ninth year. It was his final go at mathematics and a feeble effort at best, resulting in changes of a minor nature only. As he had informed Rémond de Monmort in 1714, "since I left off the study of Mathematics [eighteen years ago] the disuse of thinking upon these things makes it difficult for me to take them into consideration." He expressed himself more bluntly to Whiston some years later: "no old man (excepting Dr. *Wallis*) love Mathematicks."[19] Yet, its technical errors aside, *Arithmetica Universalis* became one of the most popular textbooks of its kind and remained so throughout much of the eighteenth century. As Whiteside has pointed out, the irony in this is simply too great to be ignored: While Newton's algebra became famous and was republished time and again, his papers on the revolutionary calculus were unstudied by all but a very few until our own day.

The animus Newton bore for Whiston was anchored by additional threads, which also connected him to other headstrong disciples. These became increasingly taut in 1707, threatening to ensnare the master in his homespun web of heterodoxy. Whatever Newton's shortcomings as a teacher of mathematics and natural philosophy, he had succeeded in setting brilliant young minds on fire during private discussions concerning the origin and evolution of Christianity. Fatio de Duillier, in whom he had inculcated the Arian heresy, was the first to cause a public stir. Though they never again drew close after the rupture of 1693, their shared intellectual interests brought them into occasional contact. In 1704 Fatio, after having devised a method of improving clockwork by replacing certain metal parts with pierced rubies, presented two of his watches to Newton, who pronounced them very fine.[20] He also turned up at meetings of the Royal Society, arranging for the election of his brother John in 1706, an indication that he had not fallen totally from Newton's grace.[21] But shortly thereafter the man to whom Newton was once more tightly bound than to any other wandered into the abyss.

While Newton immersed himself in the recoinage of Scotland's currency, Fatio joined forces with the exiled leaders of the French Camisards: Durand Fage, Jean Cavalier, and Élie Marion. In 1702 these men and their peasant followers, whose traditional smock or *camise* gave them their name, openly rebelled against the brutal persecution of the Protestants brought on by the revocation of the Edict of Nantes. They defeated the troops sent against them by Louis XIV time and again, until the King thought the insurrection of sufficient gravity to require the presence of his great general, Marshal Villars. The revolt was finally extinguished with Huguenot blood, terminating in the virtual destruction of the province of Cévennes and the flight or banishment of those important figures left alive. They carried with them to England a wild, prophetic doctrine, which proclaimed the imminent coming of Judgment Day and the ability of their leaders to raise the dead. Fatio was first attracted to this bizarre circle in 1706; by the following year he had taken over the duties of secretary and record-keeper, transcribing verbatim the utterances of the inspired and fashioning his own interpretations of their ravings when they spoke in tongues. Pamphlet literature, both pro and con, abounded, although most men of common sense ridiculed the impassioned sect, harassing its members and delighting in such lampoons as Thomas D'Urfey's comedy, *The Modern Prophets*.

Some type of showdown with the authorities seemed inevitable. Diplomatic attempts were made to save Fatio, who had become an object of considerable attention, from punishment by securing his promise to leave the country. Having refused, he was brought to trial, along with his coreligionists, in November 1707. Convicted of spreading terror among Her Majesty's subjects, Marion, Jean Daudé, and Fatio received their sentences at the Court of Queen's Bench: to stand twice on the scaffold with a paper denoting their offenses. They also paid stiff fines and gave security for their good behavior during the coming year. The sentences were carried out at Charing Cross and the Royal Exchange on December 1 and 2.[22] There is no indication that Newton took any steps to aid his erstwhile friend, who continued to believe in the lost cause until his dying day. Only rarely after this debacle did Fatio return to the Royal Society. It was later written that "Sir Isaac Newton himself had strong inclinations to go and hear these prophets, and was restrained from it, with difficulty, by some of his friends, who feared he might be infected by them as Fatio had been."[23] This would appear to constitute exaggeration in the extreme. While the President of the Royal Society sympathized with the Huguenots for their sufferings, he had come too far and had paid too great a price to risk his reputation and position in pursuit of a cause whose failure was prefigured in the rational inter-pretation of Scripture. It was Leibniz, against whom Fatio had first raised the charge of plagiarism concerning the caluclus, who lamented the dis-graced zealot's fate. Writing to Thomas Burnet from Hanover in March 1708, he noted: "The Affair of the Cévennes Prophets has ended in an un-fortunate catastrophe and I am sorry for love of M. Fatio; I do not under-stand how so excellent a man in mathematics could have embarked on such an affair."[24] Leibniz might better have addressed such a question to Newton.

Following Fatio's public humiliation, the torch of heterodoxy soon passed to an equally excitable Whiston at Cambridge. The Lucasian Pro-fessor had come down to London in 1707, at the height of the stir created by the French prophets, to deliver the annual Boyle Lectures, an enterprise of more than passing interest to Newton. Whiston titled his lectures *The Accomplishment of Scripture Prophecies* and acquitted himself admirably, in Newton's eyes, by denouncing the Huguenot zealots: "If any person in this age, who pretend to a prophetic spirit do foretell events, whether of mercy or of judgment, which do not come to pass according[ly], we have the warrant of God himself for their rejection."[25] Whiston wrote even more bluntly of the prophetic phenomenon in his *Memoirs,* contending that its adherents were possessed by the supernatural: "I thought they were *evil* and not *good* Spirits that were the Authors of those Agitations and Im-pulses." Echoing the Newtonian line, he added, "They think the Scripture the Rule of Faith, contrary to all Antiquity. They reject the Use of Reason." Such pretense is *"the Devil's Banter."*[26]

Whiston had passed many an hour in Newton's company discussing such subjects as the prophecy of Daniel and the Book of Revelation. And while they had their differences—some of which were quite marked—the younger man concluded that his Cambridge predecessor did not often fail him in his exposition of the prophecies.[27] Another theological skein ran through their private conversations as well. Newton, as with Fatio, made Whiston privy to his belief that the Church had taken a critically wrong turn in the fourth century. "He had early and thoroughly discovered that the Old Christian Faith, concerning the Trinity in particular, was then changed; that what has been long called Arianism is no other than Old Christianity."[28] With this the mists suddenly parted for Whiston, and he became a convert to the Arian credo. Yet he also wrote that Newton was a "prodigiously fearful" man of "cautious and suspicious Temper." Whiston himself lacked even a modicum of Newton's ingrained reticence, a flaw of character that eventually led to his undoing, as it already had to Fatio's.

Whiston's first serious confession of doubt about the nature of Christ occurred in 1706, the year of his intensive discussions on prophecy with Newton. He began to converse and preach openly against trinitarianism in 1708, much to the chagrin of his closest friends. Word spread quickly, and

James Pierce, a preacher at Newberry, wrote to Whiston on July 10 that, after a visit to London, "I was forc'd to believe what was so much against my Inclination." Pierce quoted Scripture chapter and verse in an attempt to convince his friend that he should rejoin "the Defenders of this Blessed Trinity."[29] Though deeply moved by Pierce's affectionate concern, Whiston gave his answer to the world a month later when he presented his Arian *Essay upon the Apostolic Constitutions* to Dr. Lany, Vice Chancellor of Cambridge, for licensing. When Lany refused, Whiston published his heretical *Sermons and Essays Upon Several Subjects* in London, appending Novatian's *De Trinitate* to it for good measure. This act resulted in his being summoned before a commission composed of the Vice Chancellor and eleven of the Masters in October 1710, although Bentley, a sympathetic spirit, was conspicuously absent. Whiston was charged with promoting within the University doctrines contrary to the established creed of the Church of England. When given the opportunity to abjure his heresy and thus save his position, Whiston chose the martyr's path. The commission acted accordingly, stripping him of his professorship and banishing him from the University on October 30. Whiston chided those friends who lamented his loss by characterizing preferment as the "utter Ruin of Virtue and Religion! Poison, sweet Poison."[30] Once again it was Leibniz who exhibited a generous measure of charity to one of his rival's disgraced followers: "I pity the good Mr. Whiston who does himself harm with his excess of zeal."[31]

Newton's tension and embarrassment at having his hand-picked successor booted out of Cambridge for embracing a doctrine that he himself had instilled in Whiston were yet to run their full course. Whiston took his family to London, where he was forced to make a living by whatever means he could. Buoyed by his role as a self-styled martyr, he continued publishing morally superior attacks on the doctrine of the Trinity, commencing in 1711 with the *Historical Preface to Primitive Christianity Revived,* a defense of Arianism virtually identical to that found in Newton's private papers. Later that same year Whiston made the first of several appearances before the Court of Convocation to answer for his heretical views. He was convicted of contempt for arriving late at one of the hearings in July 1713, but he was never excommunicated or degraded. His apparent lack of concern for the seriousness of the proceedings was demonstrated one day while he waited at the door of St. Paul's for the spiritual court to convene. Instead of comporting himself in a manner befitting the circumstances, he blithely passed the time by handing out free copies of his *Proposals for Finding out the Longitude at Sea.*[32] "Nor do I remember," Whiston later wrote, "that during all the legal Proceedings against me, which lasted in all four or five years at *Cambridge* and *London,* I lost sleep more than two or three Hours one Night on that Account. This," he concluded, "affords a small Specimen of what support old Confessors and Martyrs might receive from their

Saviour, when they underwent such Miseries and Torments."[33] Newton didn't quite see it Whiston's way. When the latter was nominated for membership in the Royal Society in May 1716, no action was taken.[34] Whiston later heard that a deeply concerned Newton had closeted himself with some of the leading members and threatened to resign if Whiston was made a Fellow. As soon as he was informed of the President's uneasiness, Whiston told Dr. Samuel Clarke that "had I known his Mind, I would have done nothing that might bring the great Man's *grey Hairs with Sorrow to the Grave.*"[35]

It was only natural that Whiston should unburden himself to Clarke, for both men had acquired certain of the finer points of their theology at Newton's knee. Clarke received his B.A. from Caius College, Cambridge, in 1695, at the age of twenty, and became a close associate of Whiston shortly thereafter. The young theologian was given the privilege of delivering the Boyle Lectures in 1704 and again in 1705. Published later under the titles of A *Demonstration of the Being and Attributes of God* and A *Discourse concerning Natural and Revealed Religion*, the lectures, like those delivered by Bentley and Whiston, employed Newtonian natural philosophy to support established theology. So taken was Newton with Clarke's erudition, both scientific and theological, that he commissioned him to translate the *Opticks* into Latin.

It was during this period that Clarke began to evidence an uncommonly strong interest in the early Church. Like Newton and Whiston, Clarke also suspected that the currently accepted doctrine of the Trinity was not that of primitive Christianity. Newton sought to tighten his grip on the gifted theologian by offering him a position at the Mint, but Clarke, who had more ambitious plans for himself, politely but firmly refused this overture. He went on to take his Doctor of Divinity degree at Cambridge in 1710 and was nearly unmasked in the bargain. Whiston, who was still present but about to lose his chair, wrote that Dr. James, Regius Professor of Divinity, sifted through every part of Clarke's thesis with the "strictest nicety," pressing the candidate to the limit on point after point. The dogged inquisitor was probing for evidence of Clarke's rumored Arianism. Clarke replied "readily to the greatest difficulties proposed . . . so . . . that perhaps [there] never was such a conflict heard in the schools."[36] When it was finally over, Clarke and Whiston dined with Bentley to celebrate the outcome. The Master of Trinity, no stranger to Arianism himself, joked about the confrontation and wrote an irreverent jingle in Latin to commemorate it.[37]

Clarke experienced an even more harrowing escape in 1714. Known for his association with Whiston, who was then under investigation by the Court of Convocation, the theologian was brought before this same tribunal shortly after the publication of his *Scripture Doctrine of the Trinity*. Refusing to follow Whiston's example of self-sacrifice to principle, Clarke, whom many considered the most learned metaphysician of the day, suc-

cessfully held off his accusers with a pliant tongue and solicitious demeanor. Thus, like Newton, he clung to the "sweet poison" of preferment
while the morally superior Whiston lived a hand-to-mouth existence over
which he grew increasingly bitter through the years. Although Clarke's
failure of nerve could not help but rankle his friend, the deepest pain
Whiston endured was that inflicted by Newton. "But he then perceiving
that I could not do as his other darling Friends did, that is, learn of him,
without contradicting him, when I differed in Opinion from him, he could
not, in his old Age, bear such Contradiction; and so he was afraid of me the
last thirteen Years of his life."[38] There is doubtless some truth in what
Whiston wrote, but it is also true that this fear was reciprocated. Moreover,
among Newton's followers it was leavened by a genuine sense of awe. This
is attested by the fact that no member of his select Arian circle, including
William Whiston, dared write a word against their silver-maned god until he
had breathed his last. Even then few chose to do so, whatever they may
have privately thought of his flaws.

IV

The intellectual paralysis that had settled like an enervating blanket over
the Royal Society during the last decades of the seventeenth century was
paralleled by another crisis of the gravest order. Having spent the last of its
dwindling treasury on Francis Willoughby's *History of Fishes,* the Society
had been unable to underwrite the publication of the *Principia.* Even its
own *Philosophical Transactions* had been temporarily suspended for lack of
funds. Meanwhile, the Journal Book was peppered with melancholy financial reports listing Fellows who had not paid a farthing in dues for years.
Out of patience with the toothless strategy of cajolery, Newton and Hans
Sloane decided upon sterner measures. In December 1706 the Council approved the President's proposal that every candidate for membership must
pay the admission fee and sign a bond for his weekly dues before induction.
Fellows who had not previously signed a bond were also called upon to do
so. And no member who failed to comply or who was behind in his dues
could serve on the Council. It was subsequently agreed that these regulations should not apply to the professors of Gresham College or to foreign
members. At a meeting held a few months later the officers were also excused from the payment of a yearly subscription. To show that the Council
meant business Henry Hunt, the Society's clerk, was ordered to wait on the
members "whensoever the President Shall direct."

  It appears that the enactment of these desperate measures, modeled
after a precedent dating from the 1680s, came as a considerable shock. The
Council itself voted on them several times between December 11, 1706,
and March 5, 1707. A sweetener was added during the final deliberations:

Any member in arrears would be restored to good standing by the payment of one year's dues, no matter how much he might owe. The Council also took pity on their clerk, who, while quite innocent, was certain to bear the brunt of the defaulting memberships' outrage. It voted to pay Hunt, a former assistant of Hooke, half his meager annual salary of £40, which had been in arrears for six months.[39]

No records exist of the amount recovered by this action, but it was enough to see the Society through its straits. The institution also held some stock in the East India Company and the East Africa Company, and had invested a bequest from John Wilkins in fee-farm rents yielding a return of £24 a year, although these, like the dues of many Fellows, were not collected until long after the Council instituted legal action.[40] Still, Newton had achieved his primary objective of assuring that the organization he presided over was solvent. He now turned his attention to the realization of yet another ambition, the accomplishment of which was directly tied to the first.

While the President was preoccupied with finances, Dr. John Woodward, the hot-tempered Professor of Physics in whose Gresham College rooms the Society still met, had managed to alienate several of the members with his impolitic remarks. A complaint was lodged in the Council on May 15, 1706, and Woodward received notice that his abusive conduct would not go unpunished in the future: "if any Members of the Society Should hereafter cast reflections on the Society or any of the Fellows thereof the Statute concerning Ejection Shall be taken into Consideration."[41] Woodward managed to stay out of trouble with the Council until March 1710, when he again crossed swords with the formidable Sloane, who had never accepted Woodward's claim that he had played no part in authoring *The Transactioneer*, the satirical pamphlet that had made a laughingstock of the Secretary in 1700.

The confrontation took place during the regular meeting of March 8, a day when Newton happened to be absent from the chair. Sloane was reading a translation of an account originally addressed to the French Academy on the subject of bezoars, or gallstones, a hard intestinal mass found chiefly in ruminants and considered an antidote to poison. According to the Council minutes of March 29, Woodward interrupted the discourse by telling Sloane to "Speak Sense or English and we shall understand you. If you understand anatomy you would know better." It was Woodward's claim, substantiated by Walter Clavell and the author of an anonymous letter addressed to Newton in Woodward's defense, that he had only reacted to the provocation of Sloane. When Sloane asserted that gallstones are the cause of colic, Woodward countered with the observation that no man who understands medicine can hold such an opinion. "Dr Sloane averr'd that all Medical writers were of that Opinion; Dr Woodward replyed *none unless the Writer of the History of Jamaica*," a biting reference to Sloane's natural history of the island published in 1707. Bested in this exchange, Sloane "made Grimaces with a laughter" at Woodward, who only then uttered the offensive demand that Sloane speak sense or English, preferably both.[42]

With his future as a member of the Council on the line, Woodward requested that the body postpone any consideration of this unseemly matter until Dr. John Harris, a friend and one of the Society's two secretaries, could be present. Faced on March 29 with the largest turnout of Council members (sixteen) since Newton had assumed the presidency, the accused was clearly bidding for time. When this strategy failed, a motion was introduced asking that the President, who worked closely with Sloane, vacate the chair in the interests of fairness, but it too was voted down. That the physician and his allies were correct in their suspicion that a prejudiced Newton wanted to be shed of him as badly as did Sloane is confirmed by the existence of a "hit-list" written in the President's hand. It contains the names of Council members with an "X" by seven of those he wanted defeated at the next election because they had lined up behind Woodward.[43] Had it not been for the supporting testimony of Clavell, Woodward probably would have been dismissed on the spot.

The Council did not convene in April. At its next meeting, on May 3, Woodward's fate was all but sealed when Newton ordered that the Statute of Ejection be read, as it had been in 1706. The dénouement came three weeks later during the meeting of May 24. Newton had put out the word that attendance was imperative, and all but three of the twenty-one-member Council put in an appearance, probably the largest single gathering during his presidency. Sloane declared that he had meant no harm to Woodward by any of the gestures made on the previous occasion. The Council voted in the affirmative that "it is Sufficient Satisfaction." Its members then voted in favor of a motion "that D^r. Woodward declare that he is Sorry that he misunderstood D^r. Sloane and beg his pardon for the reflecting words he Spake." When the proud and combative Woodward refused to knuckle under, as Newton and Sloane doubtless had hoped he would, the Council performed as orchestrated by its conductor: "D^r. Woodward for creating disturbances by the Said reflecting words after a former admonition upon the Statute of Ejection and for restoring the peace of the Society be removed from the Council." Finally the Council voted to thank Dr. Sloane for his "pains and fidelity" in serving the Society.[44] Woodward refused to go quietly. He instituted legal action in the hope of obtaining a writ of mandamus calling for his reinstatement on the Council. The Court of Queen's Bench denied the petition, but Newton could not forgive or forget. Years later he informed Stukeley that he had taken leave of Woodward with the following words: "We allow you to have *natural* philosophy, but turn you out for want of *moral*."[45] Yet so pervasive was Newton's influence that even one so irascible as Woodward would later dedicate his brilliant treatise on the classification of fossils to him, the *"Vir illustris"* by whose example the work was born.

The righteous image Newton sought to project of himself through Stukeley cannot hide the fact that the explusion of Woodward was part of a far more ambitious plan to consolidate his hold on England's scientific establishment. Though hardly a model of objectivity, Flamsteed, who had recently been ousted from the Society for the nonpayment of dues, was exaggerating only slightly when he informed Sharp that "Sir I. Newton has put our Royal Society in great disorder by his partiality for . . . Dr. Sloane, upon a small and inconsiderable occasion: so that they have broke up some weeks before their time." Flamsteed added somewhat cryptically that "Dr. Harris has lost all reputation by actions not fit for me to tell you."[46] He seems to have been referring to the fact that Harris, whose election to the office of Secretary in 1709 was engineered by Woodward and strongly opposed by Sloane, had turned turtle under pressure from Newton and accepted a seat on the committee charged with contesting his old friend's legal battle for reinstatement.[47] The switch bought Harris precious little time, for his name was among those marked with an "X" by Newton, a technique reminiscent of his days as a pursuer of coiners and clippers. The Secretary lost his office and Council seat in November and was never again to serve in either capacity.

The anonymous author of the letter written in defense of John Woodward felt sufficiently close to Newton to speak of other machinations involving the Society. "Dr. Sloane, & his Junto of Non-Solvents, by which he is supported, are to impede & defeat you." Had not Newton himself complained of Sloane's practice of surprising him at Council meetings with matters unfit for serious consideration? "You had declar'd to more then one Friend how *little qualified he was for the Post of Secretary*, so, upon these occasions, you as freely declar'd him to be a *Tricking Fellow*: nay a *Villain, & Rascal*."[48] Considering later developments, it is likely that Newton did indeed harbor suspicions about Sloane and thought him something of a threat. It is also true, however, that these two headstrong personalities had largely put aside their differences in order to save the Royal Society from disaster, both intellectual and financial.

The German traveler Zacharius von Uffenbach chanced to visit London in 1710, at the very moment when the Society was in turmoil over the Woodward affair. Unaware of what Newton had already accomplished, Uffenbach made little effort to mask his disdain for the bizarre scene that greeted his arrival:

> 5 *July, Saturday* morning, we drove to *Gresham-college*. Miserable state of the royal society's apparatus. The guide, if asked for anything, generally said: 'a rogue had stolen it away;' or he shewed fragments of it, saying: 'it is broken.' The 'transactions' of the first six years . . . are worth all the rest together. . . . The society never meets in summer, and very little in autumn. The present secretary,

Dr. *Sloane*, is indeed a very learned man, but engrossed with his practice and his own large cabinet. The president, *Newton*, is an old man, and too much occupied as master of the mint, and with his own affairs, to trouble himself much about the society. For the rest, excepting Dr. *Woodward* and a couple more *Englishmen*, and the foreigners, there remain only apothecaries and the like, who scarce understand latin.[49]

Had Uffenbach visited London later in the year, his assessment of the President as a preoccupied old man would probably have altered. Woodward's departure from the Council supplied Newton with the long-sought lever needed to fulfill one of his most ambitious objectives as leader of the Royal Society.

Newton could hardly have forgotten that it was Woodward who had made his quarters at Gresham College available for Society business after the death of Robert Hooke in 1703. He would have had to be aware as well that Woodward's ouster from the Council might leave the Society with no choice but to acquire separate quarters of its own. While there is no suggestion that the Gresham Professor wanted the Society, of which he apparently remained a member, to clear out, Newton's fierce sense of independence and honor rendered the existing arrangement intolerable; having been the instrumentality of the doctor's downfall, he could remain beholden to him no longer. The regular summer recess was not yet over when Newton called a special meeting of the Council on September 8. He informed those assembled that the house of the late Dr. Edward Browne in Crane Court off Fleet Street was for sale and, "being in the middle of the Town out of noise . . . might be a proper place to be purchased by the Society for their meetings."[50] The Council appointed a special twelve-member committee, including Newton and Sir Christopher Wren, to pursue the matter and specified that any four of its number would constitute a quorum. Wren and his son were sent at once to examine the structure, which they pronounced both suitable and sound. Expressing the hope that "they like the price also," an ecstatic Newton on September 13 addressed a brief communication to Sloane calling for a Council meeting the following Saturday, an alteration of established procedure and a measure of his desire to act quickly.[51] But when only four others besides himself put in an appearance on September 16 he postponed action until the following Wednesday, the Council's regular meeting day.

The Royal Society's perennial quest for a building of its own finally bore fruit on September 20, 1710. With Newton in the chair and thirteen members of the Council also in attendance, the question was put whether they agreed to purchase Dr. Browne's house for £1,450. Twelve voted in favor of the measure and only one against. The committee appointed on September 8 was charged with completing the transaction and with fitting up the house for the use of the Society "with all Necessary Expedition."[52] When the Council next met on October 26, the President reported the sale a *fait accompli*. On hearing this, Walter Clavell, who had not attended the previous meeting and who had testified on behalf of Woodward against Sloane, got up and left. He was joined by Harris, Woodward's erstwhile friend, who realized too late that he had become nothing more than a pawn in Newton's end game of administrative chess. They, together with John Lowthorp, another of the Woodward faction and the one Council member who voted against the purchase of Browne's house on September 20, failed to win reelection to the Council at the anniversary meeting of November 30.[53]

For all intents and purposes the brief struggle was over, but the bitterness engendered by Newton's highhandedness remained. November 22 saw the publication of an anonymous thirty-two-page pamphlet titled *An Account of the Late Proceedings in the Council of the Royal Society. In Order to Remove from Gresham-College into Crane-Court, in Fleet-Street. In a Letter to a Friend.* There is no question that its author was either confused or misinformed as to certain of the matters, including some dates, cited in support of his claim that Newton had acted against the Society's best interests. Yet neither is the work a wholesale fabrication. It chronicles an extraordinary meeting, of which there is no discussion in the Council minutes, called by Newton for the purpose of hearing objections against the anticipated move. Those present were so stunned by what they heard that the President's remarks were greeted by profound silence. Hoping to stimulate discussion, he addressed particular members, asking if they had any reservations concerning the plan. When they reminded him that Gresham College had served them well and that there seemed to be little reason for moving, especially on such short notice, "The President was not prepar'd . . . to enter upon that Debate: But freely (tho' methinks not very civilly) reply'd, That he had good Reasons for their Removing, which he did not think proper to be given there." At this point Sloane, "who had engross'd the whole Management of the Society's Affairs into his own Hands, and despotically Directs the President, as well as every other Member," came to Newton's defense with the reply that the Gresham Trustees were anxious to be rid of the Society. This did not satisfy certain members, who asked that such an important decision be postponed at least until the anniversary meeting late in November. Newton, we are told, would not hear of it: "his Scruples were unmoveable. So that some of the Gentlemen with warmth enough ask'd him, To what purpose then had he called them thither?" He allegedly responded by summarily adjourning the meeting.[54] Amid such conflicting evidence, at least one thing seems absolutely clear. The effort mounted by those opposed to the venture never came close to thwarting Newton's grand design, because he had far more votes than were necessary secreted in the pocket of his waistcoat. Old he undeniably was, but preoccupied with other affairs to the point of abstraction—never.

V

The joint effort to improve the Society's financial health, which had been undertaken by Newton and Sloane in 1706, just barely made the purchase and refurbishing of 2 Crane Court possible. Even so, the institution was forced to shoulder new debts of an unprecedented magnitude. After making a down payment of £550, the Council secured a mortgage for the remaining £900 at 6 percent interest from a Mr. Collier. An additional sum of £310 went for repairs, and £400 more was expended on a repository gallery designed by Wren, exactly twice the architect's original estimate. The latter sum was borrowed from the amiable clerk Henry Hunt, who had made other loans to the Society and who held notes of £650 when he died in 1713. Newton's powers of persuasion again carried the day, however. He contributed £120 of his own to the cause and persuaded several others to follow suit. This reduced the debt to manageable proportions almost overnight, and the entire sum was fully paid off in less than six years. Indeed, so improved was the Society's general financial condition that Newton, radiating confidence at the Council meeting of November 13, 1710, called for and obtained the repeal of the statute requiring that those delinquent in paying their dues be stricken from the rolls. And this only three weeks after the decision to purchase the Crane Court property.[55]

The repairs and modifications took nearly ten months to complete. In the meantime Newton refused to have anything further to do with the errant Woodward. The Society temporarily assembled in the rooms of another Gresham Professor, Andrew Tooke, who had been appointed to Hooke's vacant post in 1704. Finally, on August 2, 1711, it was ordered that Hunt should officially take possession of the refurbished quarters, and Tooke was paid the half-year's rent voted him by the Council the previous October. Surprisingly, the best contemporary description of the building is contained in the anonymous pamphlet decrying its purchase, though it should be kept in mind that it was written before all repairs and alterations had been made.

> The approach to it, I confess, is very fair and handsome, through a long court: but then they have no other property in this than the street before it; and in a heavy rain a man can hardly escape being thoroughly wet before he can pass through it. The front of the house, towards the garden, is about 42 ft. long; but towards Crane Court not above 30 foot. Upon the ground-floor there is a little hall, and a direct passage from the stairs into the garden, about 4 or 5 foot wide; and on each side of it, a little room about 15 ft. long, and 16 ft. broad. The stairs are easie which carry you up to the next floor. Here there is a room fronting the Court, directly over the hall, and of the same bigness. And towards the garden is the Meeting-room, which is $25\frac{1}{2}$ foot long, and 16 foot broad. At the end of this room there is another (also fronting the garden) $12\frac{1}{2}$ foot long, and 16 broad. The three rooms upon the next [third] floor are of the same bigness with those I have

> last described . . . . [Y]ou will add the garrets, a platform of lead over them, and the usual cellars &c. below, of which they have more and better at Gresham College. The garden is but 42 foot long, and 27 broad, and the coach-house and stables are 40 foot long, and 20 foot broad.[56]

While not nearly so sumptuous, grandiose, or convenient as Wren's ambitious design of 1704—especially in the all-important matter of accommodating the members' private coaches—Crane Court, complete with a liveried doorman bearing the Society's arms in silver, provided its intellectual lord and master with an ideal setting from which to dominate all he surveyed.

Financially secure in its own quarters after enduring fifty years of a stepchild's hand-to-mouth existence, the membership now manifested an unprecedented willingness to follow Newton in whatever direction he might choose. Never one to question the privileges of rank, the high priest of modern science quickly secured the passage of the "Orders of the Council" in January 1711. Under these new rules only the President himself was permitted to sit at the head of the table, with the two secretaries toward the lower end, one on each side. This "sacred configuration" could be altered only at the discretion of the President in order to accommodate "Some very Honorable Stranger." No talking was permitted among members during meetings unless they first addressed themselves to the President. On a less magisterial note, all papers placed before the Society were to be "minuted" and, if written in a foreign language, transcribed into English before being read.[57]

Newton also established the custom, not codified in the "Orders," whereby the mace could be laid on the table only when the President himself occupied the chair, the stature and character of his substitute, the Vice President, notwithstanding. (Sloane, who succeeded Newton on March 29, 1727, considered this practice so offensive that he moved to abolish it as the very first item of business following his election.)[58] The following portrait left by Stukeley of Newton's conduct as President created the very effect his idol had intended:

> Whilst he presided in the Royal Society, he executed that office with singular prudence, with a grace and dignity—conscious of what was due to so noble an Institution—what was expected from his character. . . . Sir Isaac was very careful of giving any sort of discouragement to all attempts of improvement in natural knowledg. There was no whispering, talking, nor loud laughters. If discussions arose in any sort, he said they tended to find out truth, but ought not to arise to any personality. . . . Every thing was transacted with great attention and solemnity and decency. . . . Indeed his presence created a natural awe in the assembly; they appear'd truly as a venerable *consessus Naturae Consiliariorum*, without any levity or indecorum.[59]

A further measure of Newton's command is reflected in yet another unprecedented development. The average number of Council meetings had declined somewhat under his tenure, partly because they occasionally clashed with his duties at the Mint.[60] When this situation became aggravated owing to a flurry of Mint activity in 1711, the tradition-bound Council paid its leader the highest of compliments—"the weekly meetings of the Society shall be on Thursdays at four a Clock, with the design that the President may be present, he being obliged to attend the Mint on Wednesdays."[61] Such has been the custom ever since.

Year by year, decade by decade, Newton had taken great care to surround himself with deferential young mathematicians and astronomers whose loyalty was cemented more often than not by appointments to choice academic and administrative positions. Edward Paget had been the first so favored, acquiring the Mastership of the Mathematical School of Christ's Hospital on Newton's recommendation back in 1682. The first to teach Newtonianism in Scotland, David Gregory, who had embarked on a carefully orchestrated campaign of ingratiation, received Newton's imprimatur and became Savilian Professor of Astronomy at Oxford in 1691. Another protégé, William Whiston, was deeded the master's own Lucasian Professorship several years later, a virtual lifetime sinecure, had not the brash young cleric chosen to parade his offensive theological linen in public. Newton also wrote most of the rules for Cambridge's new Plumian Professorship, which went in 1706 to the brilliant Roger Cotes, future publisher of the *Principia*'s second edition. Yet none of his admirers proved more steadfast or more eager to serve him than the astronomer Edmond Halley. Having seen the *Principia* through the press in 1686, he had helped fend off the attacks of a deeply wounded Hooke and later, at Newton's bidding, put in two miserable years at Chester, by far the most troubled and corrupt of the temporary branch mints. Following a series of scientific voyages, Halley became Savilian Professor of Geometry in 1704 at Newton's command. (The greatest encomiums of his inaugural address were lavished upon his benefactor, whom Halley styled his "Numen.") He was yet to perform other vital services on behalf of the Royal Society President and to reap additional rewards, including an appointment as Astronomer Royal in 1720, the very thought of which caused Flamsteed to spit venom. There were many others, all of whom took an active part in disseminating Newtonian philosophy through preaching, teaching, writing, or laboring on behalf of the Royal Society: Colin Maclaurin, Henry Pemberton, James Stirling, Samuel Clarke, Hopton Haynes, John Keill, Abraham de Moivre, Richard Mead, Pierre Des Maizeaux, Francis Hauksbee, and John Theóphilus Desaguliers. Newton protected his disciples, advanced their careers, and, in return, demanded and received total obedience almost to a man. Any inkling of disloyalty was summarily punished by exile from court, lest it cast an unflattering light upon the exalted one at the center.

Those few who could not be brought to heel by the sheer force of Newton's personality were dealt with in less delicate but equally effectual ways, as illustrated by the case of Hans Sloane. Their relationship had been uneasy from the outset, with each giving up as little ground as possible for the greater good of the Society. Now that Newton had achieved his primary objective of absolute control, the strong-willed Sloane proved less a rival than a nettlesome inconvenience. It took several years to ease the stubborn Secretary out of the office he had held for two difficult decades. Finally, amid rumors that Newton had stepped up the pressure after first being rebuffed, Sloane agreed not to seek reelection at the anniversary meeting of 1713. His ear ever to the ground, Flamsteed heard that there were "high and furious debates" over the matter. Halley was chosen to replace Sloane as Secretary, and Dr. John Keill, another favorite, became a member of the Council. "Sir I. Newton," Flamsteed wrote, "sees now that he is understood."[62] John Chamberlayne, a Fellow who held several court offices, including that of Gentleman of the Privy Chamber, certainly understood him well enough. He wrote to the President a week before the 1713 elections, asking Newton to mark the list of candidates for office on his ballot save one: "I desire to choose Freely . . . whom I would make Perpetual Dictator of the Society, if that depended only on the vote of Your Most faithful Humble Serve."[63] Though a friend of the outgoing Secretary, Chamberlayne saw the writing on the wall and, like many others, had hastened to endear himself to Newton through an act of supplication accompanied by a few choice words that spoke volumes. A beaten Sloane, who had been appointed physician to Queen Anne in 1712, turned his attention to the famous collection of manuscripts and natural curiosities housed in his Chelsea residence, waiting patiently for the perpetual dictator to pass from the scene before returning to the center of Society activity nearly fourteen years later.

The grand monarchial style cultivated by the President was many-faceted. When Newton assumed office the number of foreign members averaged forty-two a year; that figure had slightly more than doubled by the time of his death in 1727.[64] He seemed especially fond of ambassadors, most of whom had little or no background in the new science. This consideration mattered little to one who loved to surround himself with visiting dignitaries, much like a king holding court. The guests were seated in "elbow-chairs" near the President and diverted by experiments carefully chosen for their spectacular effects. On one such occasion Signor Grimani, the Venetian Ambassador; Signor Gerardini, Envoy from the Grand Duke of Tuscany; and the Duke D'Armont, the French Ambassador, were shown two preparations "made of the veins and arteries of a human liver by injecting red wax into them, which was very curiously and beautifully performed." They also saw light produced by friction and were treated to a demonstration of the power of gravity in a vacuum. According to the minutes, the experiments were eagerly received and afforded much delight to the distinguished guests, all of whom were elected members.[65] Neither did the President neglect the most influential political and social figures of the day: the Lord Chief Justice Parker, Lord Viscount Dupplin, and Lord Harley were elected Fellows on March 20, 1712.[66] Barred from direct participation by the unwritten but universally acknowledged sex code, the Queen nonetheless gave ample evidence of royal favor as when, on February 7, 1713, she instructed her foreign ministers and governors to "contribute all they can, in their several stations, towards promoting the designs for which the Royal Society was first instituted." She was about to dispatch a new minister to the "Court of Mosco" and asked Newton to propose a set of instructions "whereby he may be usefull to you in those parts."[67] He did this and more. Prince Alexander Danilovich Menshikov, who as regent became the virtual dictator of Russia after Peter's death, was warmly embraced as a new society member the following year. When Anne died in 1714, Newton hastened to approach George I with the request that the Hanoverian King serve as patron and that his son, the future monarch, became a Fellow. And it was to Caroline, Queen Consort of George II, that Newton gave the first copy of his revised chronology. She in turn consulted him concerning the education of her many children. Total membership in the Society rose from approximately 173 in 1705 to some 234 in 1716.[68] The latter figure would have been larger by one had not the Council, presumably with the blessing of its President, voted to deny membership to a Mr. Williams, "a Black Native of Jamaica."[69] Whatever his other attributes, Williams was only the second candidate to be rejected since the Society's founding in 1662. Seven months earlier, in May 1716, William Whiston was devastated to see his name removed from consideration by his retributory patron.

For reasons that remain unclear, Newton sold the Jermyn Street home he had occupied for more than a decade and moved to Chelsea late in 1709. He may have been in search of cleaner, fresher air, for Chelsea was then no more than a bustling village on the Thames 3 miles west of London. The tiresome journey back and forth probably got the better of him, however, for he returned to the city in 1710. It is perhaps worth noting that Sloane also lived in Chelsea and that the anonymous letter denouncing the Secretary's supposed behind-the-back dealings, which Newton received in March, could conceivably have played a part in his decision to leave. The village, like the Council of the Royal Society, may have been too small to accommodate both. At any rate he settled at 35 St. Martin's Street in the parish of Westminster, holding the tenancy until his death, although he removed to Kensington in 1725 in the futile hope that the purer air of the village would restore his gradually deteriorating health. Leicester House, as this dwelling became known, was a three-story structure of conventional Jacobean design to which its new resident is believed to have added a special room as an observatory. First occupied in 1694, the stone edifice was only sixteen years old when Newton rented it for the handsome sum of

£100 a year. Nearby stood Orange Street Huguenot Chapel, a favorite gathering place of those who had fled the tyranny of Louis XIV. As we have seen, Newton's sympathies lay on the side of the French Protestants, a number of whom he befriended and brought into the Royal Society, including the impoverished mathematician Abraham de Moivre. Among his surviving manuscripts is a list of new household furnishings on which he expended some £27: a bell (presumably for summoning servants), glasses, a range, two bottle racks and three beer stands, fireplace equipment, a tapestry, a bedstead, wall hangings, and a wooden tub.[70] Of his plans for the last-named item a widowed neighbor is said to have reported: "Every morning . . . he takes his seat in front of a tub of soap-suds, and occupies himself for hours blowing soap-bubbles through a common clay-pipe intently watching them till they burst."[71] While this anecdote, like so many others, is almost certainly apocryphal, a Newton human enough to take delight in blowing bubbles in his bath is appealing. The highly stylized portrait of him commissioned by Richard Bentley in 1709 seems rather more convincing. Painted by James Thornhill, it depicts Newton in the classical Roman manner of an abstracted and timeless seer, an image reinforced by the knowledge that he was encircled by stalwart vindicators eager to shoo the pigeons from his path. Alas, Leicester House was demolished earlier in this century to make way for a new public library. The original woodwork and crimson fabric, of which Newton seemed inordinately fond, are now to be seen in a faithful reconstruction of the parlor in the library of Massachusetts's Babson Institute.

VI

John Flamsteed, by his own account, had no direct dealings with Newton after the death of Prince George in October 1708. "Being now not disturbed by him any more at present, I set myself to carry on such observations as I wanted and made good advances in it." Indeed, Abraham Sharp received a letter from his former employer in March 1709 stating that, "I have all things in good readiness for the edition of my works." Then, without warning, the euphoria of nearly having put the capstone on his lifework virtually dissolved in a single, sickening instant. "I was afresh disturbed," Flamsteed wrote in his autobiography, "by another piece of Sir Isaac Newton's ingenuity."[72]

He was referring to a warrant from the Queen delivered to the Royal Society on December 14, 1710, by her physician extraordinary, Dr. John Arbuthnot, who just happened to be a member of the Council. The document appointed the President and such others as the Council shall think fit to serve as "constant Visitors of Our said Royal Observatory." What this meant, in plain terms, was that Newton and a group of his hand-picked followers were empowered to oversee the course of observations at Greenwich, on which they must report each year, and to inspect the instruments for damage and other defects with a view to having them either repaired or replaced. A stunned Flamsteed received a copy of the Queen's warrant on the same day as the Council and was shortly to hear that it had wasted no time in making certain that Her Majesty's will would be done. "The President, M^r. Roberts, D^r. Mead, Sir Christopher Wren, M^r. Wren, and D^r. Sloane were Ordered a Committee to go to Greenwich...and to report their Opinion of the Observatory and the Instruments."[73] The very next morning Flamsteed hastened to London, where he lodged the strongest of protests before Henry St. John, the dissolute Secretary of State, over whose signature the offensive document had been drafted. Rightly arguing that his labors would only be impeded by the Visitors and that the observatory instruments belonged to him and not, as was mistakenly thought, to the crown, the Astronomer Royal asked that the warrant be rescinded. St. John would not hear of it and answered, as Flamsteed recalled, haughtily: "The Queen would be obeyed." The desperate astronomer then turned to Lord Rochester, Anne's uncle, showing him "what tricks and disingenuous usage were put upon me by Sir Isaac Newton." Though he was treated with greater kindness by the old aristocrat, the outcome proved the same. Having returned to Greenwich empty-handed, Flamsteed drafted a plaintive remonstrance to the Queen, laying the blame for many of his past and present woes at Newton's feet. Whether the document was actually sent to London is unknown.[74]

To argue that Flamsteed was mistaken in his renewed suspicions concerning the Royal Society President requires that credulity be stretched beyond the breaking point. Arbuthnot stood well within the Newtonian coterie on the Council, and the fact that he served as the royal messenger on this occasion was hardly a matter of chance. The Scottish physician had been credited with saving the life of Prince George when the consort suddenly became ill at Epsom in 1703, which won him the eternal gratitude of the Queen. He was described by one courtier some years later as "a very cunning man, and not much talk't of, but I believe what he says is as much heard [by the queen] as any that give advise now."[75] In short, the discreet and well-placed Arbuthnot was just the person Newton would have called upon for assistance in his latest scheme for bending Flamsteed to his will. Nor does one have to search far for a motive. Newton had recently decided that the time was at hand to publish a second edition of the *Principia*. Yet the anticipated perfection of his lunar theory remained as dependent as ever upon the acquisition of Flamsteed's observations. As a Visitor he could once more demand that the cornered Astronomer Royal give him what he wanted, the difference being that he was now backed by the full power of the crown.

Leaving nothing to chance, Newton was about to assault Flamsteed from another unexpected direction. At the Society meeting of February 21, 1711, it was voted that a letter be written by Sloane to the Astronomer Royal, "desiring him to furnish the deficient part of his Catalogue of the fixed Starrs now printing by Order of the Queen." A copy of Sloane's letter was read at the March 7 meeting, but some misgivings surfaced over its authorship. Sloane and Flamsteed nurtured an intense mutual dislike, and it was finally agreed that Arbuthnot should be the one to notify the astronomer of the Queen's desire.[76] This he did on March 14, concluding on a somewhat overbearing note of mock sincerity: "I am the more fully perswaded you will complye with so reasonable a request because of the regard you have for the Memory of the Prince, as well as for your own reputation, both which are interested somewhat in this performance."[77] Precisely how the Queen's order to resume the printing of the *Historia Coelestis* was obtained is open to conjecture, since the document itself has never been found. That it or some other form of proxy once existed is scarcely to be doubted, however, for no person, no matter how favored, would have dared invoke Anne's authority or name without her consent. Coming as it did shortly after the appointment of the Visitors, this development can reasonably be attributed to the manipulative hand of Newton. The Royal Society with himself as President had assumed the part of official referee.

Though Arbuthnot pleaded urgency, Flamsteed took his time before sending a reply, no doubt checking to be certain that the Queen's physician knew whereof he wrote. Having satisfied himself that Arbuthnot was truly doing Her Majesty's bidding, Flamsteed drafted a polite but circumspect answer on March 23. He had accomplished much since the presses fell silent in December 1707. All of the requisite observations for producing a work worthy of the British nation and the memory of Prince George were completed. Yet much remained to be done, particularly in the area of calculating new planetary tables based on his latest sightings. Arbuthnot was invited to dine with him at Greenwich for the purpose of discussing the particulars and of finding ways of keeping the work "free from such hindrance and delayes as have formerly retarded ye progress of it."[78] Flamsteed obviously expected to be permitted to enlarge his magnum opus as he saw fit, while Arbuthnot, who was acting as Newton's stalking horse, had been cautioned to accept nothing beyond the material previously agreed upon. When Arbuthnot handed Flamsteed's letter to Newton there was no stanching the venomous flow. Taking pen in hand on March 24, Newton accused Flamsteed not only of being dilatory and evasive but of lèse majesty as well:

> You are therefore desired either to send the rest of your catalogue to Dr Arbuthnot or at least to send him the observations wch are waiting to complete it that the press may proceed. And if instead thereof you propose any thing else or make any excuses or unnecessary delays it will be taken for an indirect refusal to comply wth her Majts order.[79]

Whether this letter was actually sent is not clear, nor is it really important. Its significance lies in the fact that it makes a mockery of the fiction that Newton played anything but the underhanded lead in the latest revival of this shabby Augustan drama.

Flamsteed sustained a series of additional blows in rapid succession. A polite but firm Arbuthnot, echoing the Newtonian line, wrote that he was permitted by his commission to publish only the observations given to the referees before Prince George's death. He would, however, do his best to obtain the Queen's promise that an appendix containing new material would be added to the work, but only after the delivery of Flamsteed's missing data from the catalog of the fixed stars.[80] Two days before, on March 25, an unidentified friend had informed Flamsteed that the catalog was already in press and that Halley, who had been selected to serve as editor, claimed to have discovered many errors and had publicly displayed some sheets of it in Child's Coffee House.[81] Flamsteed met with Arbuthnot at Garraway's in London on March 29 and asked him point blank whether the printing had begun. The physician assured him that *"not a sheet of it was printed."* Flamsteed knew this to be a lie since, in the next breath, he was offered £10 for every press fault he could find in it. Four days later he received the first printed sheets containing the constellations Aries and Taurus and proclaimed them "Fairly printed." He also noted that "the Doctor had told me *what he knew was not true."* Thus despite Arbuthnot's promise at their parting that he would have *"just, honorable, equitable and civil usage"* in the future, Flamsteed became the victim of what Baily accurately characterized as a gross deception.[82] Indeed, the astronomer would have been within his rights had he chosen to withdraw at that point.

Of the many injustices Flamsteed attributed to Newton, none vexed him more than the resumption of the printing without his knowledge, let alone his consent. It was his contention, reiterated many times and in various forums over the years, that Newton had agreed not to unseal the star catalog when he first received a copy of it in 1706. When the packet was opened two years later it was with Flamsteed's approval, so that the recently computed magnitudes of the stars could be inserted. Then, according to its author, it was resealed by mutual consent and left in Newton's hands. That his nemesis had recently revealed its contents to everyone from the high-riding Halley to the lowly printers without first seeking his permission "was one of the boldest things that ever was attempted." Flamsteed was even less temperate on the subject when, in a letter to Sharp on May 11, he labeled Newton's conduct part of a "villainous outrage" for which Halley was also partly to blame.[83] The fact that Newton justified himself by saying he had acted on the Queen's command hardly mattered, since that order could only have been issued at his or Arbuthnot's request, which amounted to the same thing. More to the point, even supposing that Flamsteed had unduly exaggerated the importance of secrecy, it would have cost Newton nothing to have kept the gouty astronomer abreast of the latest developments, something he had neglected to do ever since the project's inception in 1704.

Flamsteed's loathing for the jocular and discreetly irreverent Halley was keenly honed over the years; he had done all within his power to deny Newton's inquisitive confidant access to the fruits of his scientific labors. That the judgmental eyes of the Philistine were now privy to his lifework constituted a form of defilement he could never forgive, his Christian scruples notwithstanding. Each successive delivery of proof pages evoked ever more violent and long-lasting paroxyms of anger. He accused Halley of altering the names, positions, and magnitudes of the stars without "ever consulting me about it: which I would never consent to." Halley "was minding to spoil the work" and "has exposed him[self] to all the town, and they forbear not to say he is impudent f[ool] and k[nave]."[84] While the latter sentiment was largely the product of wishful thinking on Flamsteed's part, it in no way detracts from the poignancy or essential truth of what he wrote to an increasingly uncomfortable and apologetic Arbuthnot on April 19:

> [D]o not tease me with banter, by telling me yt these alterations are made to please me when you are sensible nothing can be more displeasing nor injurious, then to be told so.
>
> Make my case your own, & tell me Ingeniously, & sincerely were you in my circumstances, and had been at all my labour charge & trouble, would you like to have your Labours surreptitiously forced out of your hands, convey'd into the hands of your de[c]lared profligate Enemys, printed without your consent, and spoyled as mine are in ye impression? would you suffer your Enemyes to make themselves Judges, of what they really understand not? would you not withdraw your Copy out of their hands, trust no more in theirs and Publish your own Works rather . . . then see them spoyled and your self Laught at for his suffering.[85]

Having had his honor stained once too often, Flamsteed had no choice but to insist that a new copy of the star catalog be substituted for the mutilated, incomplete manuscript in his enemy's hands, and that it be printed at his own expense. If this demand was not acceptable, he could no longer cooperate with those in charge of the printing. Arbuthnot responded four days later, surely only after conferring with Newton, who must have bolstered his flagging spine. "I have no design to rob you of ye fruit of your labours but . . . I will not delay any longer but take ye same method to make out ye rest of the Catalogue yt you have done, which is to employ people to Calculate from the observations what is wanting; and why we should not succeed as well in this piece of Journey work I cannot imagine." If after perusing the finished work Flamsteed did not approve of what he saw, "you shall be free to print your own." Newton waited little more than a month before exacting further retribution. In a curt and imperious letter dated May 30, which was cosigned by his fellow Greenwich Visitors Mead and Sloane, he ordered the Astronomer Royal to observe the upcoming solar eclipse of July 4 and to send his data to the house of the Royal Society in Crane Court.[86] Anyone who might question the source of this command need only refer to the Society minutes of May 24, according to which Newton not only broached the matter but ruled on it *ex cathedra*.[87]

When Flamsteed next received news concerning the *Historia Coelestis* late in June, it came from Halley himself. He had delivered a corrected copy of the star catalog into the hands of Flamsteed's niece, Mrs. James Hodgson. It was his hope that Flamsteed would avail himself of this final opportunity to amend all of the mistakes he could find, for Halley was not so foolish as to believe that he had made no errors. Still, he minced no words on the subject: "Pray govern your Passion, and when you have seen & considered what I have done for you, you may perhaps think I deserve at your hands a much better treatment than you for a long time have been pleasd to bestow on Your quondam friend and not yet profligate Enemy (as you call me.)" Flamsteed was hardly overcome with remorse, the less so after perusing the catalog. "When I examined it, I found more faults in it, and greater, than I imagined the impudent editor could, or durst have committed." He resolved to print it as he had originally planned, and in his own good time.[88]

The official printing, as Arbuthnot had promised, went ahead over the vehement objections of the author, who had nothing further to do with Newton during the next several months. Having disregarded the command that he file a written report concerning the summer eclipse of the sun, Flamsteed was finally called before the Royal Observatory Visitors at Crane Court on October 26. Halley met Flamsteed as he entered and offered him a cup of coffee, perhaps in hopes of advising the man he still called friend on how best to deport himself. If such was the case this gesture produced the opposite effect. Assisted by his manservant, the gouty Flamsteed brushed by Halley and limped up the stairs to the meeting room, where he found Newton, Mead, and Sloane: "the two last, I well knew, were the assertors of the first, in all cases, right or wrong." An account of the ensuing scene, which Flamsteed recorded in both his diary and a letter to Sharp, is among the most dramatic we have of Newton's temper in action. He began by asking the Astronomer Royal exactly what instruments Greenwich contained and which ones required repairs. Flamsteed replied that he always made the necessary repairs himself and that all of the instruments were his personal property, not that of the Government. This nettled Newton, who responded, "As good have no Observatory as no instruments." Flamsteed then raised the issue of the star catalog, protesting that *"I was robbed of the fruits of my labors."*

> At this, the impetuous man grew outrageous; and said, 'We are, then, robbers of your labors?' I answered, I was sorry that they owed themselves to be so. After which, all he said was in a rage: he called me many hard names; *puppy* was the most innocent of them. I told him only that I had all imaginable deference and respect for Her Majesty's order, for the honor of the nation, & c.: but that it was

a dishonor to the nation, Her Majesty, and that Society (nay, to the President himself), to use me so.

Caught off guard by this rare face-to-face display of mettle, Newton raged all the more:

> At last, he charged me with great violence (and repeated it), not to remove any instruments out of the Observatory: for I had told him before that, if I was turned out . . . I would carry away the sextant with me. I only desired him to keep his temper, restrain his passion, and thanked him as often as he gave me ill names. . . . He told me, moreover, I had received £3600 of the Government.* I answered, what had he done for £500 a-year salary that he had, or to that purpose? Which put him to a stand: but, at length, he fell to give me his usual *good* words: said I was proud and insolent, and insulted him. . . . He said I had called him *Atheist*. I never did: but I know what other people have said of a paragraph in his *Optics*; which probably occasioned this suggestion. I thought it not worth my while to say anything in answer to this reproach. I hope he is none.

When it was finally over Sloane, who unlike Mead had remained silent throughout, helped Flamsteed down the stairs and was thanked for his civility. The astronomer again encountered Halley at the door. He had not been far off all the while, Flamsteed sarcastically observed in his autobiography, and doubtless had heard Sir Isaac Newton show his best.[89]

The strangeness in Newton's proportion had lost none of its vigor with age: The old lion was still a lion rampant. The perpetual dictator, as Chamberlayne called him, simply would not, indeed could not, let go. He dogged his self-made enemies with the same tenacity he showed in his scientific pursuits and criminal prosecutions. Nearly two years later, when he presented a copy of the second edition of the *Principia* to Queen Anne, revenge was still very much on his mind. At the next meeting of the Royal Society on July 23, 1713, he reported that her Majesty was "pleased of her own accord" to reaffirm her desire that the President and the rest of the Society take charge of Greenwich Observatory.[90] Newton might better have said of the Queen's own accord after some clever prompting by her devoted servant.

An official visit was scheduled for August 1, and Flamsteed noted that Newton arrived at three o'clock, about a half-hour before the others, Halley, Thorpe, Machin, Rowley, and Hodgson. He said little until everyone was present, then rose and dryly announced their purpose: to see what repairs were needed, and what new instruments. Flamsteed's remorseless gout kept him in his chambers, but he made certain that the Visitors were shown everything but his private library. After the tour of inspection, Newton commented on the poor condition of certain instruments, to which

---

*Flamsteed had earned £100 a year as Astronomer Royal over the preceding thirty-six years, hardly a munificent sum. He reckoned the cost of carrying out his duties at £300 per year, two-thirds of which came out of his own pocket.

Flamsteed replied that they were his own and constituted no obstacle to the fulfillment of his duties.[91] It was Newton's desire, subsequently confirmed in a letter addressed to the Ordnance, that it undertake the repairs and by so doing undermine Flamsteed's long-standing claim to immunity from outside control. In their reply of September 4, 1713, the Ordnance officers, who were housed on none too friendly terms near Newton in the Tower, pleaded ignorance of the issue presented. Newton proceeded to fill them in, but he was put off on the highly unwelcome grounds that the instruments were indeed Flamsteed's, and even if the Ordnance had authority over them, it had no funds with which to effect the repairs.[92] That is how matters stood when the Queen died in 1714, nullifying the Visitors' mandate and tightening Flamsteed's grip on the Royal Observatory. Newton's petulance continued to the last. On a visit to Greenwich he had promised to return a Greek Ptolemy and four volumes of Flamsteed's "Night-Notes," which had been in his possession for the past six years. He kept them four more years, Flamsteed wrote, "to no other purpose but to show his authority and *good* nature."[93] Nor were the volumes returned until their owner instituted legal action.

Meanwhile the presses had been rolling. Halley's one-volume edition of the *Historia Coelestis* appeared in the spring of 1712. The order of arrangement conformed to Newton's design, with the star catalog at the beginning. Halley's preface, or rather one should say Newton's, struck below the belt by asserting that Flamsteed had played the part of an obstructionist, doing all within his power to thwart the command of his great benefactor, Prince George. The Astronomer Royal labeled the preface a tissue of "lies and false suggestions," and he was not far wide of the mark. According to Flamsteed's diary, Halley faked an attack of conscience a few months after publication. He came down to Greenwich on June 18, his wife, children, and a neighboring clergyman in tow, and asked "if I wanted preferment (a snare!): said he would burn his copy of my catalogue, if I would print my own." Flamsteed, who was indeed in the process of printing his own edition, thought this a clever way of sidetracking him. He turned down Halley's offer, "so he went away not much wiser than he came."[94] Though he never said as much, Flamsteed surely believed that Halley was on Newton's errand, especially if the offer of a preferment had been made as be reported. Whatever the truth of the matter, there is no mistaking the motive behind certain other steps taken by Newton at the time. He saw to it that Halley received £150 for his labors, a sum greater than that which Flamsteed was to have been paid for additional work.[95] His vengeance unsatiated, he also did to Flamsteed what he had once done to that *persona non grata*, Robert Hooke. Newton combed the soon to be published manuscript of the *Principia's* second edition, striking Flamsteed's name from every section but that devoted to the comet of 1680–81, a total of fifteen deletions in all.[96] It was only natural under the circumstances that Flamsteed should have pronounced the new edition worse than the old, "save in the moon; and there he is fuller, but not so positive, and seems to refer much to be determined by observations." The Astronomer Royal thought the volume worth only about seven or eight shillings, considerably less than the eighteen he had paid for a personal copy.[97]

The final sheet of Flamsteed's corrected and enlarged edition of the catalog was printed on December 5, 1712, several months after the appearance of Halley's mutilated volume. He turned next to the observations obtained with the aid of the great mural arc, only part of which had been included in the previously published edition. Unfortunately for Flamsteed, much of the data remained in Newton's hands, and he refused to give them back. The impasse caused Flamsteed a great deal of additional labor and expense, and he was not yet finished with his calculations when the Queen, who was but forty-nine, passed away on August 1, 1714.

The Tory government fell after George I succeeded to the throne, but Newton's fellow Whig and court patron, the Earl of Halifax, also died the following May. This provided Flamsteed with the opening he had long awaited but had despaired of ever seeing materialize. He informed Sharp that he now had his own "prime officers" at court "that will not suffer me to be used as I have been formerly."[98] Most prominent among them was Charles Paulet, the Duke of Bolton, who held the important post of Lord Chamberlain in the new regime. Flamsteed learned from a mutual friend that Bolton would be willing to obtain for him the undistributed copies of the *Historia Coelestis*, which, by virtue of the contract he had signed with the original referees, were legally his to do with as he saw fit. The ecstatic Flamsteed, who credited this development to Providence, passed the word, and Bolton issued a warrant addressed to Newton on November 25, 1715.[99] The three hundred remaining copies were delivered to the Astronomer Royal the following March. He had them transported to Greenwich before performing a primitive dual right of sacrifice and revenge. Flamsteed first separated out Halley's edition of the star catalog and the abridged observations obtained with the mural arc, and then, "some few days after, I made a *Sacrifice of them to Heavenly Truth*." In other words he reduced them to ashes by fire, the ancient instrument of purification. He pledged to do the same should the "Author of Truth" see fit to deliver any of the previously distributed volumes into his hands.[100]

Flamsteed devoted his remaining three years to the printing of his magnum opus. It was not yet finished when the end came on December 31, 1719. His two former assistants, Joseph Crosthwait and Abraham Sharp, completed what their teacher and friend had instituted so many years before. The *Historia Coelestis Britannica* at last appeared in three volumes in 1725, just as Flamsteed had always dreamed it would. The Astronomer

Royal had written to Sharp in October 1713 that "Sir I. Newton continues his designs upon me,... but, I thank God for it, hitherto without success."[101] Although he was already sinking slowly into death, Flamsteed, while deeply marked by the lion, never did bend to Newton's will. Of the many others who had incurred his terrible wrath, precious few could claim as much.

Chapter Nineteen

"Second Inventors Count for Nothing"

The lion and the unicorn
Were fighting for the crown;
The lion beat the unicorn
All around the town.

Anonymous

I

Fatio de Duillier once dreamed of bringing forth a new edition of the *Principia*. By March 1690 he had already compiled a partial list of errata and addenda; when he sailed for Holland in April, he carried a copy in his trunk, which he entrusted to the care of Christiaan Huygens. Often pretentious in the extreme, Fatio was at least correct in his assertion that his new-found idol intended to publish a second edition of his masterpiece. Once stimulated by Halley, Newton had composed the *Principia* in hot blood, completing the work in the remarkably brief period of eighteen months. Both he and Halley knew that it contained a sizable number of errors and that several major problems were only partially resolved, notably the resistance of fluids, the trajectories of comets, and the theory of the moon. But the emotionally charged break between Fatio and Newton in 1693 doomed the eager mathematician's hopes of ever becoming his hero's editor. The resulting wounds proved so deep, in fact, that Newton never again permitted any young lion to approach so close to him. This did not keep others from trying.

Fatio's aspirations eventually became those of David Gregory, the Scottish astronomer and mathematician of whom we have already heard a good deal. Like Fatio, Gregory had been among the first to embark on an in-depth study of the *Principia*, examining every axiom, corollary, and scholium in detail. The results of these extensive labors were embodied in Gregory's "Notae in Newton's *Principia Mathematica Philosophiae Naturalis*," an unpublished work of 213 folio pages that still exists in the form of four manuscript copies.[1] The numerous memoranda Gregory drafted are also rife with references and allusions to a second edition. He visited Newton at Cambridge on May 4, 1694, and extensive corrections under thirty-eight headings were shown to him. Shortly thereafter he wrote that Newton planned to publish a small treatise on the squaring of curves after first publishing the *Principia*.[2] Yet a decade and more later found a wearying Gregory making similar notes. He thought it likely that the new edition would go to press in the winter of 1706–7, but it did not. The following April Newton told him that new experiments on the resistance of fluids were in order and that publication would be delayed for two years. To make matters worse, the perennial row with Flamsteed seemed light years from a productive resolution, and Newton had always claimed that he could not proceed without the Astronomer Royal's observations. A year later, apparently taken by surprise, Gregory wrote in his diary on March 25, 1708, that Newton "has begun to reprint his Principia Philosophiae at Cambridge."[3] This was just five days after Newton had negotiated a tenuous second agreement with Flamsteed on the *Historia Coelestis*. For reasons never satisfactorily explained Gregory too had been denied the editorship he so strongly coveted. Equally distressing to him was the fact that Newton had chosen to disregard his painstakingly crafted "Notae."[4]

Strange as it may seem, Newton's choice for publisher of the second edition was a classical scholar and theologian rather than a skilled natural philosopher or advanced student of mathematics like himself. Perhaps one should say that Newton's publisher chose him, for the little circumstantial evidence available suggests that Richard Bentley, Master of Trinity College since 1700, first approached Newton on the subject. John Conduitt later noted that when Newton was asked why Bentley had been allowed to print the *Principia* and pocket the proceeds, he replied that Bentley "was covetous & loved mony & therefore I lett him that he might get money."[5]

The first edition had been limited to a few hundred volumes at most and had long since sold out. While few who purchased the work were sufficiently learned to read it from cover to cover, nearly every person associated with natural philosophy, no matter how tangentially, felt compelled to acquire a copy of his own. In 1708 one William Browne, an affluent student, gave two gunieas for a used copy; another young scholar, unable to pay so high a price, performed an academic labor of love by copying out the entire treatise in longhand.[6] Near the end of the preface to the long-awaited second edition Roger Cotes underscored the economic motive behind the publication in remarkably blunt terms: "Since copies of the previous edition were very scarce and held at high prices, he [Bentley] persuaded by frequent entreaties and almost by chiding, the splendid man [Newton] to grant him permission for the appearance of this new edition."[7] Bentley's shameless avarice did indeed seem boundless, and it is partly for this reason that his most recent biographer chose this epigram for the title page: "His sins were scarlet, but his books were read."[8] Indeed, the fact that Bentley's books were read by many may have played a more significant role in Newton's choice of publishers than he let on. Bentley, a tireless worker, was generally acknowledged to be the most accomplished classical scholar of his day.

Following Gregory's diary entry of March 1708, the next extant reference to a second edition came from the pen of Bentley himself. Trinity's Master wrote to inform Newton on June 10 that the printing was about to begin. He also enclosed samples of pages 1 through 8 for Newton's approbation and noted the purchase of one hundred reams of fine Genoese paper, a deal facilitated by the good offices of Sir Theodore Janssen, Director of the doomed but still thriving South Sea Company. Bentley was more than a little apprehensive, and for good reason. Newton had not yet placed a final copy of the revisions in his hands: "By this I hope you have made some progress towards finishing your great work, wch is now expected here with great impatience, & the prospect of it has already lower'd ye price of ye former Edition above half of what it once was." He went on to outline his plans for producing a book whose aesthetic qualities would do justice to the surpassing grandeur of its contents. "Pray look on *Hugenius de Oscillatione,* wch is a book very masterly printed, & you'd see that is done like this." In a not so subtle attempt to convince Newton of his own indispensability in this regard, Bentley added, "Our English compositors are ignorant & print Latin Books as they are used to do English ones; if they are not set right by one used to the beauties of ye best printing abroad."[9] Bentley displayed considerably more enthusiasm for their precarious arrangement than did Newton, who was either unwilling or, more likely, unable to provide his publisher with a complete text at the time. Thus no further signs of movement surfaced for nearly a year.

By the spring of 1709, when Newton finally signaled Bentley his willingness to proceed, a third party, highly skilled in mathematics and natural philosophy, had entered the picture. Perhaps Newton had entertained serious doubts about leaving Bentley in absolute control of the editing, because he himself would then be forced to shoulder the entire responsibility for reading and correcting all of the more technical proofs, an unwelcome prospect to an aging savant who by his own admission had never truly loved mathematics. Perhaps Bentley had come to a similar conclusion on his own. Though unrivaled as a student of the classics, he was in well over his head when it came to passing judgment on a treatise grounded in analytical geometry. These misgivings could explain why Roger Cotes, Plumian Professor of Astronomy and Experimental Philosophy, was tapped to edit the second edition under Bentley's watchful eye.

Trinity's Master had paved the way for Cotes's election to the Plumian Chair in 1707 as part of his larger design to rid Trinity of its academic dead-wood. Cotes was but twenty-five at the time of his appointment, a year younger than Newton had been when he became Lucasian Professor in 1669. Working closely with the ill-starred Whiston, Cotes established a thriving school of physical science at Trinity. Unlike Newton, he was popular among the more intellectually active students, impressing nearly all who knew him with his brilliance. Whiston wrote that when it came to mathematics, "I was but a Child to Mr. *Cotes*"; he further observed that "I esteem mine [experiments] far inferior to his." [10] In addition to his unusual intellectual gifts, which fully qualified him for dealing with the myriad complexities of the *Principia*, Cotes possessed the equally important traits of nearly inexhaustible patience and modesty. Like Halley, he proved himself capable of both stirring Newton into action and simultaneously correcting his work without unduly antagonizing or offending him, a ballet choreographed on eggshells.

In March 1709 Newton gave Bentley a partially corrected copy of the *Principia*, which the latter in turn had delivered to Cotes as per Newton's instructions. Bentley then visited Newton in London on May 31, after which he informed Cotes by letter that Newton would be happy to see him for the purpose of turning over that part of the book corrected for the press. [11] Cotes did exactly as Bentley suggested and traveled to the capital a few weeks later, only to be disappointed. Newton put him off with the promise that the corrected materials would be ready in a fortnight, and Cotes returned to Cambridge empty-handed. The delay of two weeks stretched into a month, subjecting Cotes to a lesser but still poignant form of the mental anguish suffered by Halley after his visit to Cambridge in 1684. Unable to endure Newton's silence any longer, Cotes took pen in hand on August 18. "You was pleased to promise me... You would send down the [*Principia*] in about a Fourtnights time. I hope You will pardon me for this uneasiness from which I cannot free my self & for giveing You this Trouble to let You know it." [12] Cotes added that he had gone over the second corollary to Proposition XCI of Book I. He found it "to be true," but he had also discovered two mathematical errors, which he hoped Newton would see fit to correct in the new edition. Another month passed and still there was no word from London. A frustrated Cotes finally repaired to the country on personal business. It was just as well, for Newton waited a total of seven weeks before framing a reply.

As Halley had long since learned, any delay on Newton's part was almost certain to be more than compensated for by the product of his labors. On

October 11 Newton wrote that Whiston would soon deliver "the greatest part" of the book. He also thanked Cotes for his pains in ferreting out and correcting the two errors in Book I. However, he did not expect Cotes to peruse the text with the eye of a perfectionist: "I would not have you be at the trouble of examining all the Demonstrations in the Principia. Its impossible," he went on, "to print the book wthout some faults & if you print by the copy sent you, correcting only such faults as occurr in reading over the sheets to correct them as they are printed off, you will have labour more then it's fit to give you." Such was hardly Cotes's conception of his role; he would be betraying the great trust placed in him if every step of every calculation was not rechecked for accuracy, if every word of every line was not examined for clarity and sense. While he might not have cared for Bentley's pompous manner of expression, Cotes soon gave every sign that he agreed with what the Master of Trinity wrote to Newton on October 20: "You need not be so shy of giving Mr Cotes too much trouble: he has more esteem for you, & obligations to you, than to think yt trouble too grevious: but however he does it at my Orders, to whom he owes more than yt. And so . . . we will take care yt no little slip in a Calculation shall pass this fine Edition."[13] Cotes made Bentley's promise good. His rigorous but tasteful editorship gradually aroused Newton's partially dormant creative powers, causing him to produce a more ambitious and polished work than anything he had envisioned since those halcyon days with Fatio.

Book I required only minor revisions. Now that Bentley had it in hand he pressed forward at full speed, fearing, it seems, that his elusive quarry might yet slip the net at the last moment. Sticking to his previously announced plan of modeling the second edition after Huygens's *Horologium oscillatorium*, he informed Newton in late October that a full fifty pages were set and ready for printing. The only snag concerned one Livebody, a London printer engaged by Newton to make the many woodcuts. In his letter to Cotes Newton had ended with the suggestion that Livebody come to Cambridge so that he could work directly with the compositors. Bentley consulted with the master printer, Cornelius Crownfield, who agreed at first but later, after making inquiries concerning Livebody's character, demurred on the grounds that the man had a reputation worthy of his name—"he . . . being a mere sot, & having plaid such pranks yt no body will take him into any Printhouse in London or Oxford; & so he [Crownfield] fears he'll debauch all his Men."[14] As frequently happens with those who adhere to an unbending moral code, Newton displayed a proclivity toward placing his trust in degenerates. About the same time he engaged the nephew of Whiston in his household. Not long thereafter he discovered that £3,000 in bank notes were missing, and the finger of suspicion pointed squarely in the young man's direction, not least because he subsequently purchased an estate of that value. One hundred guineas were stolen, probably by this same person, who Newton believed had secretly taken the key to his desk from a pair of his breeches.[15] Yet he refused to prosecute, doubtless because he—the supreme strategist in the war against coiners and clippers and the officer in charge of the Queen's coinage—feared looking the fool. Even now one can almost hear the sly jokes and doggerel that would have resulted.

So far as can be determined, nothing further passed between Newton and Cotes during the next six months. When the editor at last wrote to Newton on April 15, 1710, he reported that everything sent to Cambridge had been printed: "I must now beg of You to think of finishing the remaining part, as soon as You can with convenience." The progress made up to that point was impressive. Some 224 pages, nearly half of the ultimately completed work, were in the bank. This included all of Book I and the first seven propositions of Book II. Even Newton would have been shocked had anyone told him that the final pages would not come off the press for three more years. He gently admonished Cotes again in June: "You need not give your self the trouble of examining all the calculations of the Scholium. Such errors as do not depend upon wrong reasoning can be of no great consequence & may be corrected by the Reader." Yet Newton's very words belie mounting evidence of a piquant change of attitude. The persistent Cotes had gradually succeeded in engaging him in a technical correspondence destined to touch upon nearly every major subject dealt with in Books II and III, the most sustained and wide-ranging scientific exchange of Newton's life. Aloofness had given way to mild interest, which in turn yielded to impassioned commitment. He came to the realization that Bentley had served his interests well by not allowing him to go back on his promise to publish and that he had found in Roger Cotes a disciple worthy of his fullest attention. "I am wth my humble services to your Master," Newton wrote on July 1, "& many thanks to your self for your trouble in correcting this edition." [16] Still, the long delay that preceded the appearance of the second edition resulted from many factors other than the difficulties of correcting the text. Newton changed residence twice within the year, and his duties at the Mint had become more demanding owing to the necessity of minting new coin to underwrite a costly Continental war. The quarrel with Flamsteed was nearing its worst juncture, and he was simultaneously involved in the purchase, financing, and refurbishing of Crane Court. As if these burdens were not sufficiently onerous, the smoldering controversy with Leibniz over the calculus suddenly flared into the open, illuminating the scientific horizon not so much by the light of reason as by the heat of egocentric battle.

II

Aristotle defined the tragic life as one that evokes both pity and fear. When applied to Isaac Newton such a definition qualifies him as a tragic figure of towering dimensions. Dutiful toward superiors but often overbearing when dealing with subordinates or rivals, he was sometimes vicious as well, not content just to defeat but eager to destroy. Some have argued that Newton became proud, dictatorial, and even devious only in old age, after being corrupted by power and fame, that his early and middle years contained but the faintest hint of what was to come. Yet the signs of youth and early manhood, if carefully read and interpreted, point to a different conclusion. The political and professional offices Newton held, in combination with an unassailable reputation, merely enabled him to expand the scale on which his Jovian fits of passion were played out after 1700. The combative temper and vindictiveness, whose roots were anchored in an obsessive paranoia over which Newton, try as he might, could exercise only the most tenuous control through self-imposed isolation, had become manifest early on—perhaps even in the schoolyard and classroom at Grantham, where he had beaten and bested his bullying rival. Whatever the biographer may make of the emotional legacy of Newton's childhood, there is no denying the youth's hatred for his stepfather and sense of betrayal at his mother's hands; after all, it was Newton himself who had confessed to wishing them the most horrible of punishments—death by fire in their own home.

Later Newton fought ruthlessly with his scientific detractor Robert Hooke for almost thirty years, carefully nurturing the seeds of dislike until they blossomed into an exquisite hatred. Hooke's death did nothing to disturb the continuity of battle, for there were others to take his place. One biographer has characterized the events surrounding the publication of Flamsteed's *Historia Coelestis* as the most unpleasant episode in Newton's life.[17] Newton harried the neurotic Astronomer Royal for years, employing virtually every dirty trick imaginable in an attempt to gain control of Flamsteed's lifework. When all else failed he had Flamsteed expelled from the Royal Society and then pirated his observations, an act of retribution whose ugliness he compounded by turning the astronomer's manuscript over to Halley for publication.

Assuming the classical fighting posture of a rampant lion, Newton had bared his fangs and claws to several lesser lights along the way. Ignance Gaston Pardies, the French Jesuit who raised certain legitimate questions concerning the hypothesis of light in 1672, was bowled over by the force of Newton's blows and was thus involuntarily retired from the field. Pardies's less able theological brother, the octogenarian Francis Linus, suffered a similar fate three years later. Guillaume Cassegrain, who claimed priority in the discovery of the reflecting telescope and whose design is widely employed in the great observatories of the world today, was dismissed by Newton as little more than an uninformed opportunist. And when the venerable Christiaan Huygens began to entertain serious second thoughts with regard to the optical discoveries, Oldenburg, acting as intermediary, felt compelled to preface Newton's acid reply with a letter of apology. All this and more occurred before Newton reached his mid-thirties. Hence his later conflicts with many, including Flamsteed, Whiston, Sloane, Woodward, Lord Lucas, and even the admiring Stukeley, not to mention his wish that Locke, like his parents, would die, were part of a long-standing pattern of highly ritualized behavior. However, it was his quarrel with Leibniz that ultimately revealed to the world his more ruthless side.

As a physical specimen Gottfried Wilhelm Leibniz must have seemed quite unprepossessing to his contemporaries. He was somewhat stooped at the waist and of medium height, and his broad-shouldered torso was propelled about by bandy legs and diminutive feet. His oval countenance, unusually high forehead, and large, elongated nose did nothing to endear him to women, and, like Newton, he died a confirmed bachelor, having taken only one true mistress—his work. Yet inside this undistinguished shell resided a universal genius destined to influence such diverse fields as theology, logic, history, mathematics, law, mechanics, linguistics, geology, and, of course, philosophy, a mind whose irresistible currents dominated the intellectual life of Germany in the late seventeenth and early eighteenth centuries. An indefatigable worker whose mountainous manuscripts have not yet all been critically examined, let alone published, he exchanged letters with more than six hundred correspondents before his death in 1716. Yet Leibniz seems to have been as comfortable plying the hazardous roads of Europe in winter, or roaming the even more treacherous mines of the Harz Mountains, as he was occupying the same chair for days on end, engrossed in some particularly compelling problem.

Gottfried Leibniz was born in Leipzig on July 1, 1646. His parents, Friedrich and Katherina, were pious Lutherans whose ancestors had resided in the city for several generations. The father held various posts at the local university, including a professorship of moral philosophy, but his direct influence on the boy was short-lived. Gottfried was a student of six at the Nicolai School when Friedrich died in 1652. Fortunately, a perceptive relative recognized the child's gifts for what they were and recommended that he be allowed free access to the library left by his father, a development reminiscent of Hannah's decision to give Isaac the library of her deceased second husband. Gottfried learned arithmetic and a smattering of geometry at school, but he was largely self-taught from the books at home. By the time he completed his secondary studies at age fourteen, the omnivorous reading that was to be his lifelong habit had carried him through many of the works of the classical, scholastic, and even patristic writers. In 1661 he undertook the study of law at the University of Leipzig, at the time when Newton, almost four years his senior, was matriculating in Cambridge. Though he was attracted more strongly to the study of logic than to mathematics at this early stage of his career, it was during this period that the young German first came into intellectual contact with the giants who ushered in the modern age: Bacon, Kepler, Galileo, and Descartes. In 1663 Leibniz studied briefly at the University of Jena, where, under the guidance of Erhard Weigel, he quickly mastered the essentials of Euclidian geometry, an indispensable tool of the logician. Three years later, while Newton drafted the first papers on the calculus that would one day lock them in mortal combat, Leibniz wrote his *De Arte Combinatoria* ("On the Art of Combination"), a brilliant theoretical paper whose model for logical analysis has been characterized as the ancestor of certain modern computers. That same year, 1666, Leibniz completed his legal studies but was denied a doctorate of law because of his age (not quite twenty). Embittered, he left the city of his birth to seek his fortune elsewhere. He traveled to Altdorf, the university town of the free city of Nuremberg, where he found the atmosphere rather more congenial. His dissertation, *De Casibus Perplexis* ("On Perplexing Cases"), not only earned him a doctor's degree there but led to the offer of a professorship, which Leibniz respectfully declined.

The gifted jurist impressed more than the university authorities during his brief stay in Altdorf. There he encountered Johann Christian Boyneburg, one of Germany's most respected statesmen. Through Boyneburg Leibniz was introduced to the powerful Elector Johann Philipp von Shönborn, who took him into his service at Mainz. In 1672 the Elector dispatched Leibniz and Boyneburg to Paris on an ambitious diplomatic mission. In an effort to forestall French aggression in the Rhineland, the two envoys had hatched a clever scheme. They hoped to persuade Louis XIV and his ministers that France's true interests lay not in the conquest of its Dutch and German neighbors but in the colonization of north Africa. Egypt, they were prepared to argue, was ripe for the taking, and the French position could be consolidated by the construction of a great canal across the isthmus of Suez. Leibniz and Boyneburg were never granted an audience with the Sun King, but their far-seeing plan was dusted off and put to use more than a century later by another ruler of the French, a diminuitive general on a white horse.

The failed diplomatic mission and the death of Boyneburg in December 1672 were setbacks from which Leibniz soon recovered. In Paris he had immersed himself in the intellectual and scientific milieu that no German city could then offer. He met and became close to Huygens, who was reaching the pinnacle of his fame. Antoine Arnauld, the Jansenist theologian, introduced Leibniz to his co-religionist Etienne Périer, a nephew of Pascal, who in the summer of 1674 entrusted Leibniz with his uncle's unpublished papers. Meanwhile Leibniz had initiated contact with Oldenburg, the Secretary of the Royal Society, and in January 1673 he crossed the Channel with Melchior Friedrich Schönborn, nephew of the Elector, to spend a few exciting weeks in London.

Though the young German was a relative newcomer to physics and mathematics, the name of Leibniz was already known to the more active members of the Royal Society. Besides his having corresponded with

Oldenburg since 1670, his early meditations on optics, space, and motion, which were published under the general title *Hypothesis Physica Nova* (1671), had been the subject of a lively debate at one of the Society's meetings. Not surprisingly, a cantankerous Hooke had rejected the little treatise out of hand, while the more affable Wallis found its contents most appealing. In search of financial support as well as the approbation of his peers, Leibniz carried with him to London the unfinished model of an automatic calculating machine he had begun constructing in 1672. He demonstrated the device at the Society meeting of February 1, and Hooke, true to form, claimed to have designed something similar but better. We well know what Newton thought of "second inventors," but Leibniz's reaction to Hooke's claim provides a rare and penetrating insight into his own feelings on the subject. Writing to Oldenburg after returning to Paris in March, he minced no words:

> [A]s the substance of the invention is mine, or drawn from my own [resources, and] as whatever Hooke [may have accomplished] I myself have done as much, I hope that famous man, seeing the strength of my case, will leave the development and perfection . . . to me and that if he has any advice to give on that score he will liberally impart it to me, particularly through you. If he does this I shall praise his good spirit in public; if he does not, he will do something unworthy of his own estimate of himself, unworthy of his nation, and unworthy of the Royal Society.[18]

What Leibniz failed to mention is that the calculating machine, which he had so ingeniously fashioned with his own hands, was a more accomplished version of an earlier model developed by Pascal. The Fellows present at the February demonstration, coincidentally, were also treated to a long discourse by Hooke, who attacked Newton's model of the reflecting telescope and claimed priority for a superior design of his own.

Besides making the acquaintance of Hooke and Oldenburg, Leibniz was wined and dined by the illustrious Robert Boyle at the Pall Mall home of the chemist's sister, Lady Ranelagh. John Pell, whose reputation as an English mathematician at the time was second only to that of Wallis, joined in the evening's festivities, much to Leibniz's chagrin. When the topic of mathematics came up, the guest of honor made the mistake of reporting that he had recently made certain discoveries concerning the interpolation of series. He was stung to the quick when Pell acidly observed that another mathematician, Gabriel Mouton, had already published in this area. The next day Leibniz hastened to the library of Oldenburg and found that Pell had spoken the truth. The victim of his own incipient mathematical genius and naiveté, he sought to acquit himself of any suspicion that he had plagiarized by depositing an explanation of the incident with the Royal Society.[19] Leibniz was taken at his word—as well he should have been—and nominated for membership in the Royal Society before he returned to Paris early in March. Forty years later, while drafting the *Commericum epistolicum*, Newton cited this incident as evidence that Leibniz had always been a thief.

Despite its uncomfortable moments, the sojourn in London more than lived up to Leibniz's high expectations. He had cemented his relationship with an admiring Oldenburg while making invaluable new scientific contacts. Not least, he soon heard from London that he had gained acceptance into the English virtuosi's most exclusive professional club. Unfortunately, the fates proved less kind when it came to the question of his patronage. Unlike Descartes and Huygens, Leibniz, a man of common birth and exceedingly modest means, was destined to remain a lifelong pawn in the hands of the mighty. Having recently lost Boyneburg, a congenial friend and his principle advocate at the court of Mainz, he learned that the Elector too had died during his absence abroad; his services were no longer required. Leibniz loved Paris for its indefinable unity of atmosphere, which has fascinated intellectuals for centuries. Better, he thought, to remain in the City of Light at the edge of poverty for as long as possible than to hover sycophantically in relative comfort near the court of some vacuous, pleasure-loving prince. Keenly aware that his days in the French capital were numbered without an income-producing appointment, Leibniz labored feverishly to make his mark before time and diminished circumstances forced his return to the land of his birth.

One person Leibniz did not meet during his visit to London was John Collins, Oldenburg's unofficial mathematical adviser. Though Collins's grasp of higher mathematics always remained limited, he performed a valuable service by maintaining a thriving correspondence with such figures as Barrow, James Gregory, Wallis, and Newton. In reply to several questions from Leibniz, Oldenburg had Collins draw up the first in a series of mathematical reports on the state of the art in England. This the Royal Society Secretary translated into Latin before sending it on to Paris in April 1673. Newton's name, which probably first came to Leibniz's attention while he was in London, appeared a total of four times, once in regard to an especially intriguing reference to a general method of squaring curves and solving other knotty geometrical problems through the application of infinite series. "Mr Newton hath invented [it] before Mercator publish't his Logarithmotechnia."[20] Collins, his wary eye turned toward Cambridge, knew better than to elaborate, however. From what he could tell of Newton's exceedingly negative attitude toward correspondence, let alone publication, he had probably divulged too much already. Yet he must also have felt that he had done his young countryman an important service by establishing his literary priority. Collins realized that Leibniz was in regular contact with Huygens, and he wanted the Europeans to know that they must look to England for continued inspiration and guidance in these matters.

Newton remembered his early years at Cambridge with particular fondness, wistfully calling them "the prime of my age for invention." Leibniz could well have said the same of the time he spent in Paris, at least with respect to mathematics. He rapidly evolved from a mere neophyte in 1673 to a mature master in 1675, the year of his own *annus mirabilis*. After more than two and a half centuries of generally unproductive debate, there is no longer any question that Leibniz—acting independently of Newton—invented the differential calculus during the waning months of 1675, almost exactly a decade after his future rival achieved the fundamental insights of his fluxional method during the plague years of 1665–66.[21] Moreover, whereas Newton later (1691) introduced the "pricked" or "dotted" fluxional notation, Leibniz began with the more elegant yet practical symbolism still in use today, though both methods yield the same results. The fact that Collins nervously continued to feed Leibniz information on various mathematical subjects via Oldenburg later convinced Newton that he had been stabbed in the back. Yet, as before, his advances were communicated in such a general, nontechnical manner as to be useless as significant guideposts in Leibniz's independent journey of discovery. While nearly ten years behind his peer, Leibniz had indeed acted alone.

Meanwhile Collins, who was little impressed by the German at the beginning of their correspondence, gradually came to suspect that something profound was afoot. After 1674 Leibniz's letters were increasingly filled with brilliant, if often incomprehensible, mathematical surprises. Pleased that he should be a party to this intellectual explosion, yet mistrustful of foreign specialists working in the same fields as his countrymen, Collins beseeched Newton to publish before another could claim the harvest he alone so richly deserved. Yet even if Huygens was looking over Leibniz's shoulder, Collins dared not risk telling the Cambridge mathematician that he was freely using his name and descriptions of his revolutionary methods to impress a still relatively unknown German correspondent of whom Newton almost certainly knew little, if anything at all. In 1676, when it became clear that Newton was farther from publishing than ever, owing to the controversy surrounding his theory of colors, Collins and Oldenburg finally sought to draw him out by adopting a radical new strategy. Both men requested that he enter into a correspondence with Leibniz, who had recently raised some complex questions pertaining to series that only Newton himself was qualified to answer.

It was with the greatest reluctance that Newton took pen in hand in June. Though he could never have guessed it then, the resulting eleven-page letter, and a second drafted a few months later, in October, were destined to become prime exhibits in the priority dispute of the early 1700s. At that time Newton, who possessed the natural instincts and cunning of a first-class trial lawyer, labeled his exhibits the *Epistolia prior* (first letter) and *Epistolia posterior* (later letter), designations we too shall apply in our treatment of these documents. He began the first on a cordial note by complimenting Leibniz for his modesty concerning interpolation by infinite series: "I have no doubt that he has discovered not only a method for reducing any quantities whatever to such series, as he asserts, but also various shortened forms, perhaps like our own, if not even better."[22] The "our own" Newton referred to included nothing less than the binomial theorem, which he had discovered in 1664 or early 1665 and which James Gregory, another second inventor, came upon independently about 1668. Newton now wrote the theorem out for the first time in any of his correspondence and demonstrated its application by example, thus leaving no doubt that he, not Leibniz, had been the first to formulate and use it. He next turned to the solution of those problems concerning series about which Leibniz had specifically requested information. Further demonstrations followed on the squaring of curves, the determination of areas, and the computation of volumes. He refrained, however, from making any direct reference to his discovery of the calculus. Yet neither could he pass up the opportunity to let his eager correspondent know that the master was holding something of great moment in reserve:

> From all this it is to be seen how much the limits of analysis are enlarged by such infinite equations; in fact by their help analysis reaches, I might almost say, to all problems, the numerical problems of Diophantus and the like excepted. Yet the result is not altogether universal unless rendered so by certain further methods of developing infinite series.... But how to proceed in those cases there is now no time to explain, nor time to report some other things which I have devised.[23]

Newton's letter reached Oldenburg in mid-June. It was read at the Royal Society meeting of the fifteenth and translated into Latin before the cautious Secretary finally placed it in the hands of a private courier a month later. Leibniz did not receive it in Paris until August 26, and he hastened to draft a reply the following day. Yet when the *Epistolia prior* was later printed by John Wallis in Volume Three of his own *Mathematical Works*, either a printer's error or, more likely, deceit on Newton's part made it appear as though the document had been in Leibniz's possession for a considerable period before he framed a reply, the implication being—Newton was quick to point out—that Leibniz had milked its contents for precious clues to the calculus. What is more, Collins had begun putting together an account of James Gregory's mathematical discoveries, which later bore the title *Historiola*. It also contained allusions to the fluxional treatise *De Analysi*, which Newton had forwarded to Collins in 1669, after Barrow had made his existence known to the London virtuoso. Newton later erroneously convinced himself that a so-called abridgment of the *Historiola* had been sent to Paris in July and that this, together with his own letters, had supplied Leibniz with the critical mathematical mass with which to effect a breakthrough. In point of fact, the *Historiola* was never sent to Leibniz, though he inspected its contents during a second visit to London in October 1676, nearly a year after his independent discovery of the calculus. But for the time being relations between the two giants developed on a fairly high plane. "Newton's discoveries are worthy of his genius," Leibniz wrote in reply via Oldenburg on August 27, "which is abundantly made manifest by his optical experiments and by his catadioptical tube [reflecting telescope]. His method of obtaining the roots of equations and the areas of figures by means of infinite series is quite different from mine, so that one may wonder at the diversity of paths by which one can reach the same conclusion."[24] Leibniz plied Newton with more questions and selectively revealed additional elements of his own mathematical progress in the hope of drawing this strange and reclusive creature into the light, a strategy reminiscent of his even more clever use of challenge problems three decades later.

When Gilles Personne de Roberval died in 1675, Leibniz hoped to succeed him as professor of mathematics at the Collège de France. He also sought a lucrative membership in the prestigous Académie de Sciences on the strength of an improved model of the calculating machine he displayed at one of the Académie meetings that same year. Leibniz was rejected in both instances, dashing his dream of remaining in Paris for the rest of his days. In 1676 the grim realities of economics forced him to accept a position at the Court of Johann Friedrich, Duke of Brunswick-Lüneburg. Newton was hard at work on the *Epistolia posterior* when Leibniz paid a brief visit to London on his way to Hanover in the autumn. While the record of his activities in the English capital is sketchy at best, we know that he met with both Oldenburg and Collins. It was through the offices of the latter that Leibniz learned of Newton's great mathematical secret.

John Collins's cautious skepticism regarding Leibniz's mathematical ability had yielded to a sense of profound admiration for the apparent genius of his foreign visitor. Throwing caution to the wind, he allowed Leibniz what amounted to free access to the many papers contained in his burgeoning files. The German spent almost all of his limited time copying from two sets of documents in particular, Newton's *De Analysi* and Gregory's *Historiola*. In both cases Leibniz confined his excerpts to the expansion of series, the very subject on which he sought further enlightenment in his correspondence with Newton. The fact that he took no notes whatsoever on the calculus found in the concluding pages of *De Analysi* may at first glance seem strange indeed. Yet the simplest explanation is also the most convincing: Since Leibniz had already mastered the method on his own, Newton had nothing to teach him on the subject.[25] He may even have prided himself on the superiority of his notational system when compared with the cumbersome method employed by the mysterious Englishman.

When Collins wrote to the mathematician Thomas Strode on October

24, five days after Leibniz's departure, he claimed that the German had "outtopt our Mathematicks." He also alluded to a quotation from Virgil: "Rome stands out above all other cities as the cypress soars above the drooping undergrowth."[26] One can well imagine how Newton would have reacted had Collins chosen to share these sentiments with him. Collins, of course, did not. He was doubtless also painfully aware that he had violated Newton's confidence by making *De Analysi* available to Leibniz. He did not inform Newton of this gross indiscretion either. The first inventor of the calculus was to learn of it many years later, well after Collins was safely in his grave. Leibniz, for selfish reasons, breathed not a word.

Leibniz had just left Holland en route to Hanover when Newton completed and sent the *Epistolia posterior* to Oldenburg. The nineteen-page letter was translated and held in London for lack of a reliable means of conveyance until the following May and reached Leibniz in Germany only in June 1677, eight months after it was drafted. Newton began by presenting a detailed history of how he had discovered and expanded upon the method of series. Much to his later regret, he also heaped praise upon his worthy correspondent, which Leibniz delighted in quoting once they were joined in battle: "Leibniz's method for obtaining convergent series is certainly very elegant, and it would have sufficiently revealed the genius of its author, even if he had written nothing else. But what he has scattered elsewhere throughout his letter," Newton continued, "is most worthy of his reputation—it leads us also to hope for very great things from him." Newton proceeded to elaborate in such detail on the expansion of infinite series that his letter, or perhaps one should say little treatise, merited printing just as it stood. The thought must have crossed his mind as well, for two days later he wrote a follow-up letter to Oldenburg containing a familiar request: "Pray let none of my mathematical papers be printed wthout my special license."[27] The recent controversy springing from the publication of his theory of colors had cut deep.

Yet twice Newton seemed close to revealing elements of his fluxional method; twice he drew back, little suspecting that Leibniz already knew what he had chosen to conceal. He temporized and resorted instead to the use of an anagram, a popular seventeenth-century device for staking a scientific or mathematical claim while keeping its full potential and location a secret. "I cannot proceed with the explanation of [the fluxions] now, I have preferred to conceal it thus: 6accdae13eff7i3l9n404qrr4s8t12vx." The cipher represents a transcription of the letters in a Latin phrase which translates: "given any equation involving any number of fluent quantities, to find the fluxions, and vice versa."[28] Newton wrote the solution in the Waste Book and only revealed it to Leibniz and a still unsuspecting scholarly world via the *Principia* eleven years later, when professional courtesy had taken a back seat to rising malice. No matter; having studied the contents of *De Analysi* even as Newton was putting the finishing touches on the *Epistolia posterior*, Leibniz had no difficulty guessing what Newton believed his code so cleverly concealed.

Thrilled to learn that he had not been forgotten in London after all, Leibniz immediately wrote an answer to Oldenburg. As before, he praised Newton in the stately manner of the day and promised that he would respond to the *Epistolia posterior* in the near future "with the care and attentions which it deserves no less than it requires." Yet once he had begun, Leibniz could not contain himself. He filled ten pages with diagrams and calculations, including the essence of his differential calculus. He wrote again one month later, on July 12, begging Oldenburg to keep him informed of all important developments, ". . . would you from time to time see to the copying out for me of any other notable things that may be brought to your notice, not only in geometry but in other fields, especially if they are not of a kind to seem likely almost at once to figure in the [*Philosophical*] *Transactions*?"[29] Hardly the request of a plagiarist, as certain of Newton's followers later implied, this was a genuine *cri de coeur*. Leibniz had already grown impatient with the intellectual narrowness of Hanover and had tasted the loneliness that was to be his share in the dreary years to come. His largely unappreciated genius continued to evolve amid dashed hopes and blighted ambitions. The reclusive manner for which he became known sprang not from some inexplicable compulsion raging deep inside the inner man but from a dearth of human warmth and understanding.

It soon became clear to Leibniz that he would receive little additional sustenance from England. Oldenburg wrote in August that he should not expect a timely reply to his latest letters, for Newton was preoccupied with other matters. What Oldenburg did not tell Leibniz is that Newton had been uncompromising on the subject of further communication in his covering letter to the *Epistolia posterior*: "I hope this will so far satisfy M. Leibnitz that it will not be necessary for me to write any more. . . . For having other things in my head, it proves an unwelcome interruption to me to be at this time put upon considering these things."[30] It is possible, though not very probable, that Oldenburg might have changed Newton's mind, but he never got the opportunity to try. Having left London for Kent on his regular summer holiday in late August, Oldenburg came down with a severe fever, to which he succumbed within a matter of days. Meanwhile copies of Leibniz's letters reached Newton at Cambridge. Historians of science would give much to know precisely what thoughts crossed his troubled mind when he first laid eyes on their contents. We know only that in spite of what he saw, Newton remained as adamantly opposed to the prospect of publication as ever, nor did he choose to participate in an open dialog with Leibniz on the calculus, selfish decisions that inevitably led to needlessly tragic consequences. For his part Leibniz, in deciding not to confess to his secretly acquired knowledge of Newton's priority, must also bear no little share of the responsibility for later developments. Had he not censured Hooke for behaving in a similar way? By 1677 both parties, whether they were fully conscious of it or not, had contributed major ingredients to a Faustian brew, which simmered unobtrusively for the next several years before turning into a volcano of acrimony.

III

Nothing further was to be heard of the calculus until June 1684, when Newton received an interesting letter from David Gregory, together with an even more interesting little package. Gregory, who had recently been appointed to the mathematical chair at Edinburgh once held by his late uncle James, was nervously writing to Newton for the first time. Having inherited his uncle's papers, David had just published a fifty-page work titled *Exercitatio geometrica de dimensione figurarum* (*Geometrical Exercise on the Measurement of Figures*). It contained the essence of the gifted James's method of series, certain aspects of which Newton had long known about as a result of the dutiful Collins's correspondence. "But, as he read on, a growing feeling of the *déjà vu* must have come over him," Whiteside writes, "for Gregory devoted the remainder of his tract to elaborating...two of the three methods of reducing quantities to series...which Newton himself had set down in 'Reg. III' of his *De Analysi* fifteen years before." Gregory also implied that, depending upon what John Wallis might choose to include in his forthcoming *Treatise of Algebra*, he was considering the idea of publishing an even more advanced sequel on infinite series. Would Newton oblige him with a timely reply, "which I assure you I will justly value more then that of all the rest of ye world."?[31]

Newton could hardly summon up the bugbear of plagiarism in this particular instance. While James Gregory had neither corresponded with him directly nor been privy to the full scale of the Newtonian mathematical miracle of 1665, thanks to Collins Gregory knew enough to refrain from publishing until after the hard-bitten Lucasian Professor went into print himself. Unfortunately the gracious Scot was still waiting for Newton to act when death claimed him at the early age of thirty-seven. Neither could Newton rightfully vent his spleen on Gregory's nephew, for David had taken great care to laud his accomplishments in the *Exercitatio*. As with Leibniz seven years before, Newton found the agonizing consequences of his refusal to publish recurring like a bad dream. Above all, such challenges posed a direct threat to his most cherished belief, namely, that he shared a highly privileged relationship with his Creator. So certain had Newton been of his special place in the scheme of things that when Collins pleaded with him to publish in advance of Leibniz in 1676, he replied, "I could wish I could retract what has been done [meaning the theory of colors], but by that, I have learnt what's to my convenience, wch is to let what I write ly by till I am out of ye way."[32] He went on to assert that neither Leibniz nor anyone else could ever rival his mathematical achievement, no matter how long it might be denied to the world at large. Since posterity would have to assume the burden of deciphering his great works, time itself must stand still for Isaac Newton, the most accomplished mathematician since the ancients. Yet within a year he realized that the calculus was no longer his alone, though he ultimately rationalized this contradiction away by branding Leibniz a thief. Now, seven years later, it seemed within the realm of possibility that Gregory or Wallis, both of whom were close on his heels, might find their own separate ways to the most profound advance in mathematics since the age of Archimedes. Perhaps others whose identities were not yet known to him were closing in as well.

His very intellectual and psychological foundations thus threatened, Newton resolved to act with all dispatch. Choosing not to answer Gregory's letter in the conventional manner, he immediately began drafting a projected treatise of six chapters, the "Matheoseos universalis specimina" ("Specimens of a Universal System of Mathematics"). In so doing he was following a pattern established after the publication of Nicholas Mercator's *Logarithmotechnia* in 1668. At first he seemed bent on heading off further publication on the subject of infinite series by Gregory. It soon became apparent, however, that he also had Leibniz very much in mind. The thought of what the German had "done" to him had festered over the past several years. The time to even accounts was finally at hand. He would publish the entire correspondence of 1676–77, along with a full treatment of infinite series and the resulting fluxions. The latter material would serve as the centerpiece of the treatise, and Newton appropriately included it in Chapter Four. He began the chapter by transcribing the anagram inserted into the *Epistolia posterior* and concluded it by comparing his fluxional method with the differential calculus as it had been communicated in Leibniz's letter of June 1677.[33] But then, as the manuscript was nearing completion, he abandoned the project almost as suddenly as he had begun it. He had done this before, unable to face the moment of truth in the form of publication. Whether he would have carried through in this instance is something we will never know, for he was permanently distracted by the arrival of an unexpected visitor. Edmond Halley came up to Cambridge from London in the hope of finding out whether Newton possessed any inkling of what type of curve would result if a planet were attracted to the sun with a force reciprocal to the square of the distance between them.

On July 16, just weeks before Halley paid his fateful call, Otto Menke, editor of the *Acta Eruditorum (Transactions of the Learned)*, informed Leibniz by letter of rumors to the effect that Newton was being credited with a powerful method of squaring the circle, which was about to be published in England. Leibniz drafted a temperate reply: He possessed letters from Newton and Oldenburg "in which they do not dispute my quadrature with me." In point of fact Mercator, a German, had preceded both Newton and himself, or so Leibniz believed. "Mr Newton developed it further, but I arrived at it by another way. [O]ne man makes one contribution, another man another."[34] Enclosed with this letter was a paper on Leibniz's differential calculus, which Menke published in the October issue of the *Acta*. Newton's delaying tactics had cost him dearly once again, for Leibniz, his junior in invention by nearly a decade, was now the first in print. What is more, the article contained no hint of Leibniz's private admission that Newton had been laboring to profound effect in the same vineyard. Indeed, he gave no credit to anyone for the long series of contributing discoveries that had prepared the ground for his own stunning leap of genius. Though he was no plagiarist, Leibniz was, at the very least, being less than straightforward by concealing his knowledge of another's general statement of the same mathematical idea. Having taken no notes on the fluxions from *De Analysi*, he indulged in some gross rationalizing of his own.

It is unlikely that word of Leibniz's publication reached Newton before the spring, possibly even the summer, of 1685. By then Newton was deep in the throes of writing the *Principia*. Nothing but a mental breakdown or a serious bodily ailment could have diverted him from the Promethean task at hand, for he was literally stealing fire from the gods. Still, he became aware somehow that Leibniz had publicly laid claim to a portion of his own mathematical fire and decided that it would be wise to forestall the German until he could be dealt with more directly. Yet he also realized that a formal scientific treatise like the *Principia* was no place to begin a potentially sordid priority dispute. Instead he inserted a scholium on the fluxions into Book II, Proposition VII, of his masterpiece. Because of the great haste in which the *Principia* was drafted, the scholium is not without ambiguity. This is not to say that Newton intended it so, for he was anything but a conscious practitioner of the double-entendre. The scholium began with a reference to his correspondence with Leibniz ten years earlier, "when I signified that I was in the knowledge of a method of determining maxima and minima, of drawing tangents, and the like." He proceeded to translate the anagram from the *Epistolia posterior* and added, "that most distinguished man [Leibniz] wrote back that he had also fallen upon a method of the same kind, and communicated his method, which hardly differed from mine, except in his forms of words and symbols." As the mathematician Florian Cajori has so astutely observed, Newton's use of the words "hardly differed from mine" was something of an overstatement,[35] from my point of view a calculated one. His polite manner aside, Newton was not making a case for what A. R. Hall has aptly termed the "phenomenon of convergence," the independent solution of the same problem by two or more highly creative individuals.[36] His citing of the correspondence, the translation of the anagram, and, most important of all, the equation of Leibniz's method with his own add up at least to a conscious, though uncharacteristically subdued, assertion of priority, if not to the earliest printed hint of a questioning of Leibniz's conduct. Indeed, had it been otherwise Newton would have felt no need to broach the subject at all. The scholarly world might not now understand the true nature of his complaint, but Leibniz surely must. This scholium was modified slightly in the second edition; in the third, published long after his rival had died, the name of Leibniz disappeared altogether, going the way of numerous earlier references to the pariahs Flamsteed and Hooke.

After a lapse of two years Leibniz published a second and somewhat longer calculus paper in 1686, proudly noting that his advancements had won "no slight approval from certain learned men and are gradually indeed introduced into general use,"[37] words hardly calculated to warm the cockles of Newton's heart. As both an advocate and inventor of a universal language, Leibniz had instinctively realized that the correct choice of symbols would be equally crucial to the international propagation of the calculus. Accurate thinking, an essential element in the advance toward scientific truth, must be subject to the least ambiguity possible. Leibniz never tired of pointing out that the uniqueness of his method vis-à-vis Newton's lay in the formalism of his notational system, and in this he was indisputably correct. What he had sought, and from the still meager but mounting evidence had achieved, was a rigorously logical form of open communication. Newton, on the other hand, had chosen to hide his light under a bushel, not only by keeping his great discovery a secret from all but a special few but by wrapping it in esoteric swaddling clothes, much like the alchemists of old. When all is said and done, perhaps no other single example so clearly reflects the inherent differences of temperament and purpose dividing these two supremely gifted thinkers. Because of it, the next generation of English mathematicians, feeling honor bound to proceed by Newton's unquestioned symbolism, lost its position of leadership to the Europeans, who cheerfully marched to the German's drum from the outset, as all other students of the calculus have done ever since.

IV

In hindsight, it might have been better for everyone concered had not the seemingly inevitable collision between these titans been postponed by the *Principia*. Newton's capacity for anger tended to multiply geometrically over time, while Leibniz, who was deeply heartened by his singular identification with the calculus, gradually yielded to the illusion that his only equal would lay no claim to a share of the glory, let alone demand all of it. Yet unbeknownst to him, Newton had already taken other ominous steps beyond the drafting of a brief scholium in the *Principia*. The young Scottish mathematician John Craig, a former pupil of David Gregory at Edinburgh, had recently settled in Cambridge. Years later Newton testified that Craig was given access to his papers on the fluxions in 1685: "Mr Craige is a witness that in those days I looked upon the method as mine."[38] Knowing Newton as we do, we can be quite certain that what he vouchsafed to Craig he also showed to others. That same year John Wallis, who was then considered the doyen of English mathematicians, received Newton's permission to summarize the two letters to Leibniz in his *Algebra*, although Wallis made no mention of the fluxions. In an expanded version of the work issued in 1693, the Oxford professor explained his earlier silence by stating that he had omitted many things "worthy of note" because Newton himself planned to publish them. Be that as it may, by granting Wallis even so restricted a license, Newton was giving further notice of his refusal to roll over and play dead.

For the most part Newton nursed his grudge in silence until the early 1690s, when David Gregory again entered the picture. Having copied notes Craig had taken from Newton's work on quadratures, the Scot deduced the theory of binomial expansion in 1688. This he published as his own without a word of thanks to Newton, an incautious act of potentially explosive consequence. Escaping unscathed, Gregory prepared a general paper on quadratures, which he hoped to publish in the *Philosophical Transactions* of 1691. This time he sought Newton's imprimatur: "I know that ye have such a series long agoe I entreat ye'l tell me so much of the historie of it as ye think fitt I should know and publish in this paper."[39] It is apparent from what followed that Gregory had escaped censure only because his act of plagiarism reminded Newton of an even greater one, for by then he was fairly well satisfied in his own mind that the calculus had not been reinvented by Leibniz but purloined. What began as a reply to Gregory was quickly transformed into a full-blown fluxional treatise, the magisterial *De quadratura*. It opened with a discussion of Newton's correspondence with Leibniz and, like the aborted "Specimina" of 1684, contained a translation of the anagram inserted into the *Epistolia posterior*, presumably for the benefit of those who had missed it in the *Principia*. When Gregory first laid eyes on the manuscript three years later, he was dumbfounded: "The tract on Quadratures—he develops that matter astonishingly and beyond what can readily be believed."[40] Others in Newton's circle, including Fatio, Paget, and Halley, urged him to publish it on the spot, but to no avail. Falling into his old pattern, Newton not only refused to publish it but set the paper aside unfinished. An emasculated version of the original eventually appeared as an appendix to the *Opticks*, which itself was delayed until Hooke could do the author no further harm.

If Newton, for all his suppressed anger, remained gun-shy, Fatio, like an anxious retriever, thrust himself into the hunt for justice on behalf of his godlike friend. He wrote to Huygens on December 18, 1691, mincing few words:

It seems to me from everything that I have been able to see so far, among which I include papers written many years ago, that Mr. Newton is beyond question the first Author of the differential calculus and that he knew it well or better than Mr. Lebniz yet knows it before the latter had even the idea of it, which idea itself came to him, it seems, only on the occasion of what Mr. Newton wrote to him on the subject. (Please Sir look at page 235 of Mr. Newton's book). Furthermore, I cannot be sufficiently surprised that Mr. Leibniz indicates nothing about this in the Leipsig Acta.[41]

If Fatio had stopped short of charging Leibniz with plagiarism, he clearly implied it. He himself was certainly no friend of Leibniz and no stranger to hyperbole, but it is difficult to believe that his remarks were not a reflection of Newton's most private thoughts in this instance. Having been granted freer access to both the man and his papers than anyone else, the young Swiss could see for himself that what Newton claimed was true. He wrote to Huygens again two months later, pressing the same issue no less vigorously than before:

The letters that Mr. Newton wrote to Mr. Leibniz 15 or 16 years ago speak much more positively than the place that I cited to you from the Principles which nevertheless is clear enough especially when the letters explicate it. I have no doubt that they would do some injury to Mr. Leibniz if they were printed, since it was only a considerable time after them that he gave the Rules of his Differential Calculus to the Public, and that without rendering to Mr. Newton the justice he owed him. And the way in which he presented it is so far removed from what Mr. Newton has on the subject that in comparing these things I cannot prevent myself from feeling very strongly that their difference is like that of a perfected original and a botched and very imperfect copy.[42]

Huygens was well aware that Fatio had fallen under Newton's spell, nor was he himself ignorant of the darker side of Newton's nature. Thus, while he passed on certain of Fatio's mathematical intelligence to Leibniz, Huygens wrote nothing to his German friend of the priority issue, let alone the question of plagiarism.

Nevertheless, Leibniz had recently experienced a growing uneasiness brought on by word that an aging John Wallis was about to publish a multivolume edition of his complete mathematical works. Rumor had it that Newton had given Wallis permission to include detailed materials relating to the fluxions. Although he had not corresponded with Newton for seventeen years, Leibniz could endure the suspense no longer. He drafted a warm if somewhat disingenuous letter in March 1693, the first sentence of which happens to be the most revealing: "How great I think the debt owed to you, by our knowledge of mathematics and of all nature, I have acknowledged in public also when occasion offered." He then made a veiled reference to the calculus, which by then was well known to mathematicians on both sides of the Channel: "But to put the last touches I am still looking for something big for you." Truth can best be unearthed "by the friendly collaboration of you eminent specialists in this field."[43]

While Leibniz waited to see if Newton would take the not so subtle hint and reveal his plans, events of a wholly unexpected nature intervened. A parting of the ways with Fatio and the strain of overwork plunged Newton into the depths of mental depression. He did not answer Leibniz until October and excused himself by claiming to have mislaid the German's letter. Newton was not merely cordial but lavish in his praise of Leibniz: "I value your friendship very highly and have for many years back considered you as one of the leading geometers of this century, as I have also acknowledged on every occasion that offered." As we know, there is even greater reason at this juncture to doubt Newton's sincerity than that of Leibniz, but such were the ways of seventeenth-century gentlemen: What one granted to another in private was easily retracted in a published work, and vice versa. Indeed, Leibniz was also informed that he could expect to read about the fluxions in Wallis's forthcoming edition of the *Algebra*. For now, Newton would only favor his distinguished correspondent with a translation of the famous anagram, which told Leibniz nothing of which he was not already aware from having read the *Principia*. "[I]f there is anything that you think deserves censure," Newton audaciously observed, "please let me know of it by letter, since I value friends more highly than mathematical discoveries."[44]

Though Volume II of Wallis's *Mathematical Works,* an expanded version of the *Algebra* (1685), came off the press in 1693, nearly a year passed before a copy reached the anxious Leibniz in Hanover. It must have been readily apparent to him that Wallis, in drafting the first printed essay on the calculus of fluxions, had worked from copies of the letters Newton had sent to Germany through Oldenburg. But there was additional material comparing Newton's fluxions and his own differentials that he had not seen before. This, too, was largely the product of Newton's pen, and it had probably come into Wallis's hands in the form of two letters, since lost, written in 1692.[45] While it was argued that the fluxions were more easily grasped and applied than the differential method, a questionable proposition to say the least, the brief essay contained nothing of a contentious nature on the subject of priority. Wallis simply noted that Newton had long been in possession of the fluxions, a powerful mathematical tool of his own invention. Leibniz later remarked to Huygens that he had been somewhat disappointed with the book's lack of detail, but he also seems to have been considerably relieved.[46]

Again, what transpired behind the scenes was an entirely different matter, and ultimately of greater significance. In May 1694 Gregory visited Newton at Cambridge. Like Fatio, he was allowed to rummage almost at will through the great man's papers, a pattern of behavior consistent with

Newton's hope that his young followers would both record and later publish what they saw, thus becoming unimpeachable witnesses to his achievements before the bar of history. Gregory did indeed sketch out a treatise in which he compared the respective methods of Newton and Leibniz. Though never published, it was subsequently drawn upon by other mathematicians. In July Gregory recorded the substance of his recent conversations with Newton, which included a portentous question: "Whence the differential calculus of Leibniz?"[47] Wallis was not the relatively impartial figure he may have seemed in print either. The venerable mathematician passed up few opportunities to advance the English cause over the Europeans, and he wrote Newton an unsettling letter in April 1695, hoping to convince him that he should publish more: "I had intimation from Holland... that... your Notions (of *Fluxions*) pass there with great applause, by the name of *Leibniz's Calculus Differentalis*." Wallis also chided himself for not having printed Newton's *Epistolae* to Leibniz verbatim. It was too late to include them in the second volume of his projected three-volume *Mathematical Works,* but he altered its preface to make certain that Newton did not take a back seat to Leibniz on the question of priority.[48]

Newton's supporters were not the only ones to take up the priority issue. Having read Wallis's analysis of the fluxions, Johann Bernoulli wrote to his friend Leibniz in August 1696 to inquire whether Newton had not stolen from him. Leibniz was understandably unwilling to level a charge of plagiarism against his rival, yet his reply was nothing like an unequivocal denial: "I could easily believe that [Newton] possessed some very remarkable knowledge at that time [1676] which, in his usual way, he had greatly polished up in the subsequent period." Even more revealing is the fact that Leibniz had recently entered into a partnership with Bernoulli in hopes of demonstrating the superiority of the calculus vis-à-vis the fluxions. According to Catherine Barton, a vexed Newton dashed that dream by solving the Europeans' challenge problems before turning in at four o'clock of a winter's morning. Even so, Leibniz lamely and incautiously suggested in print that Newton's success derived from a sufficiently deep penetration "into the mysteries of our differential calculus."[49]

V

The year 1699 was by all accounts the point of no return in the calculus controversy. Leibniz published his suggestive comments in the *Acta,* and an angry Fatio de Duillier was not about to let them go unchallenged. He chose for his forum a mathematical tract bearing the daunting title, *Lineae brevissimi descensus investigatio geometrica duplex (A Two-fold Geometrical Investigation of the Line of Briefest Descent).* Nursing an old grudge, which dated back to 1687 when Leibniz had rejected the claim that he had formulated his own version of the calculus, Fatio brought a sweeping indictment against the German philosopher-mathematician:

> But I now recognize, based on the factual evidence, that Newton is the first inventor of this calculus, and the earliest by many years; whether Leibniz, the second inventor, may have borrowed anything from him, I should rather leave to the judgment of those who have seen the letters of Newton and his other manuscripts. Neither the silence of the more modest Newton, nor the unremitting exertions of Leibniz to claim on every occasion the invention of the calculus for himself, will deceive anyone who examines these records as I have.[50]

Now that the issue of plagiarism had at last been raised in print, the question is whether Newton was a party to Fatio's defamatory assault. The answer would seem to be a highly qualified no. Since they had not been on intimate terms for nearly six years, there is little reason to suppose that Newton would have asked his erstwhile friend to take up his cause, especially in so blunt a fashion as this. On the other hand, Fatio may have viewed this as a golden opportunity to regain the state of grace he had once occupied in Newton's eyes. In any case his comments sharply echo those imparted in the accusatorial letters to Huygens, both of which were drafted before he parted company with Newton. While Fatio was clearly guilty of gross indiscretion for trumpeting Newton's cause in so crass a manner, he had simply elevated the latter's private musings to the level of public consciousness. Leibniz, who was understandably aroused, wrote a sharp letter of protest to the Royal Society, for Fatio had cited his credentials as a Fellow on the tract's title page. Denied immediate satisfaction from this quarter, Leibniz authored a scarcely complimentary but anonymous review for the *Acta* and added a signed letter of protest for good measure. When Fatio asked permission to respond, he was told that it would not be possible because the journal sought to avoid private disputes. Touché!

Leibniz, well aware of the Newton–Fatio connection, doubtless had his suspicions about Newton aroused by this incident. He was soon presented with further cause for concern. That same year the third volume of Wallis's *Mathematical Works* came off the press. This in itself did not come as a surprise to Leibniz, since Wallis had requested permission to publish his early correspondence with Newton, which Leibniz cheerfully granted. The *Epistolae* were also published for the first time. What Leibniz did not know, but was about to discover, is that Newton was orchestrating events from behind the scenes. He had seen to it that documents on the fluxions that preceded the correspondence of 1676 were introduced into evidence. If Leibniz was not made to look the fool, neither did he come away as a man of unblemished honor. His previous neglect to name Newton as a fellow inventor reflected badly on his reputation, placing him on the defensive from the outset, a disadvantage from which he never truly recovered. Meanwhile

Newton began making notes to himself concerning the chronology of his disputed mathematical discovery, a chilling sign for anyone who dared to cross his path.

If the testimony of David Gregory can be taken at face value, Leibniz was now on the alert for an opportunity to square accounts with both Fatio and Newton. It will be recalled that when Fatio visited Huygens in 1690, he gave the Dutch savant a partial list of errata gleaned from the *Principia*. In 1701 the mathematician Johann Groening published the material as an appendix to his treatise *Historia cycloeidis*. "These were communicated . . . by M. Hugens to M. Leibnitz," Gregory charged, "& after M. Fatio's falling out with M. Leibnitz, malitiously published by the contrivance of M. Leibnitz."[51] An ardent if less than adroit Newtonian answered this volley in the summer of 1703. George Cheyne, who had previously sought to explain the circulation and heating of the blood by extending Newton's physics to the organic realm, published *On the Inverse Method of Fluxions,* in which he gave Newton sole credit for the discovery and perfection of the calculus. Leibniz, who took Cheyne for the mediocre mathematician he was, drafted an indignant rebuttal in a letter to Bernoulli after the latter had protested that Cheyne would make "Newton's apes" of all other mathematicians. "Certainly I have encountered no indication that the differential calculus or an equivalent to it was known to him [Newton] before it was known to me."[52] Leibniz knew better, of course, since he had perused *De Analysi* in 1676, the fluxions paper Newton had written for the benefit of Collins in 1669. What is more, Leibniz had set a trap for himself by admitting, in his reply to the *Epistolia Posterior,* that Newton had independently mastered a method not unlike his own.

While Newton may have welcomed Cheyne's defense of his priority, he was provoked by the slipshod, indeed plagiarized, presentation of his great analytical tool, not to mention his deep suspicion of Leibniz. At long last he decided to do what he should have done at least thirty years before— publish from the private reservoir of his advanced mathematical papers. When the *Opticks* finally appeared in 1704, it contained revised versions of *De quadratura* and *Enumeratio*. Lest there by any doubt concerning the historical antecedents of these works, he let it be known for the first time in print that "I gradually fell upon the Method of Fluxions in the years 1665 and 1666, which I have here employed in the Quadrature of Curves."[53] Five years earlier, in 1699, he had appropriated a blank page from his old mathematical notebook on which to trace his invention of the fluxions back to the plague years spent at Cambridge and Woolsthorpe.

Leibniz, currently in Berlin, received a copy of the *Opticks* late in 1704. The following January another of his anonymous reviews appeared in the *Acta*. Whether he was guilty of careless phraseology or the more serious offense of intentional provocation has been much debated ever since. In any event, his choice of words more than hinted that Newton's fluxions were merely the calculus by another name and that, like a master counterfeiter, Newton had disguised his false currency by devising a clever method of notation.

> Instead of the Leibnizian differences, then, Mr. Newton employs, and has always employed, *fluxions, which are almost the same as the increments of the fluents generated in the least equal parts of time.* He has made elegant use of these both in his *Mathematical Principles of Nature* and in other publications since, just as Honoré Fabri in his *Synopsis Geometrica* substituted the progress of motions for the method of Cavalieri.[54]

Leibniz fooled no one, least of all Newton, by attempting to hide his identity. Indeed, his penchant for anonymity did nothing to dampen his adversary's rising paranoia. Leibniz later protested that he never intended to imply that Newton was a plagiarist: "That is the malicious interpretation of a man who was looking for a quarrel."[55] So it was, but Leibniz could hardly escape responsibility for having compared Newton with Honoré Fabri, a notorious thief of the intellectual capital accumulated by others. Newton later claimed not to have set eyes on the aspersion until 1711, when John Keill brought it to his attention. Whether he was being truthful at that time is open to serious question. In any case, he chose to interpret Leibniz's remarks in the most defamatory sense imaginable.

Aside from having cooperated with Wallis in publishing certain documents relating to the development of his fluxions, Newton had been willing to leave the defense of his reputation largely in the hands of others. Indeed, it has been conjectured that there might have been no quarrel, at least of such major dimensions, had he not been goaded by his overzealous followers into taking reprisals against Leibniz. Such a view is scarcely consistent with his even harsher treatment of Flamsteed, who had done far less to kindle Newton's monumental ire than his German rival. We are dealing in both instances with classic examples of misplaced aggression. Having been unable to resolve the intractable problem of lunar dynamics, Newton, in one of his more impressive acts of psychological legerdemain, convinced himself that a jealous and dilatory Flamsteed meant to rob him of his due. In the matter of the calculus, Leibniz became the dark symbol of Newton's own failure to publish his findings before another could lay claim to them. Unable to judge himself accordingly, he chose instead to argue the case on the grounds of personal honor. What is more, the Newton–Leibniz dispute was played out against a shamelessly violent intellectual backdrop. Scholarly warriors of the Augustan Age rarely asked for quarter and just as rarely gave any. The compassion and respect displayed by Charles Darwin for his fellow naturalist Alfred Russel Wallace—the second and lesser known inventor of modern evolutionary theory—would have baffled men of the seventeenth century, who rejected rival claimants in the belief that scientific discovery was a uniquely individual achievement. Such considerations take on even deeper overtones when one focuses on special men like Newton, Leibniz, and Hooke. Each rose from a limited social and intellectual background to a position of relative prominence in an age when birth, rather than natural ability and hard work, still determined the paths trodden by most. So committed were they to their labors that none of the three married or fathered a child, their precious seed dying with them. For parvenus such as these the opinions of one's peers mattered less than the impression they made upon ministers, aristocrats, and princes. A charge of thievery was therefore a serious threat to one's social and political standing and could not be dismissed without squaring accounts. Newton never shrank from using any weapon or stratagem at his command, including the sneak attack under the cover of his disciples' names. While it is true that these men gladly wielded broadsword, battleax, and halberd in defense of their hero, it is no less true that from this point onward Newton directed their every stroke.

Enter John Keill, a rising Scottish astronomer-mathematician who had studied under Gregory at Edinburgh during the late 1680s and early 1690s. Keill followed his mentor to Oxford after Gregory became Savilian Professor of Astronomy. There he was appointed lecturer at Hart Hall and attracted considerable attention by offering the first course of lectures in any English university on the new experimental philosophy. A confirmed Newtonian, Keill instructed his auditors in the laws of motion, the principles of hydrostatics, and the theory of light and colors. And, like Newton, he refuted Descartes's concept of vortices and defended the literal interpretation of the Mosaic account of creation. Disappointed at not being elevated to Gregory's Oxford chair in 1708 upon the death of his former professor, Keill came to Newton's attention that same year and immediately projected himself into the thick of the calculus controversy. He had recently completed a paper entitled *The Laws of Centripetal Force*, which took the form of a letter to Halley, now Savilian Professor of Geometry at Oxford. Halley approved of what he saw and arranged for its publication in the *Philosophical Transactions* of September and October 1708. Near the end of his paper Keill included the following gratuitous passage, apparently as a riposte to Leibniz's implication that Newton was a plagiarist:

> All of these [laws] follow from the now highly celebrated arithmetic of fluxions which Mr. Newton, without any doubt, first invented, as anyone who reads his letters published by Wallis can readily determine; yet the same arithmetic, under a different name and method of notation, was afterwards published by Mr. Leibniz in the *Acta Eruditorum*.[56]

Newton later protested that he had nothing to do with the preparation or the printing of the offensive passage, asserting before the Royal Society that what Keill wrote had upset him until the mathematician produced a copy of Leibniz's defamatory remarks of 1705. Perhaps Newton told the truth; a letter from Keill dated April 3, 1711, would seem to indicate as much. On the other hand, Keill's paper had been presented to the Royal Society prior to its publication. On that day Newton was in his accustomed place, presiding from the chair.[57]

Of greater importance to Newton than the grand polemics of Keill was the recent acquisition of certain private papers by one William Jones, a young autodidact who taught mathematics in London before entering the service of Thomas Parker, the future Lord Chancellor. In 1708, the year of Keill's counterblast, Jones came into possession of Collins's mathematical papers, including a copy of De Analysi, which dated from 1669. Although Newton did not yet suspect that Leibniz had read the little treatise, here was the documentary evidence (separate from his own papers) that he required to vindicate his priority claim. Hence through pure chance Jones was about to succeed where others far closer to Newton had failed. He received Newton's permission to publish De Analysi, together with other important papers bearing on the calculus controversy. The little volume came off the press in January 1711, just weeks before the Royal Society received a bitter letter from Leibniz protesting Keill's offensive behavior. Without mentioning Leibniz by name, Jones, an attentive Newton at his shoulder, presented detailed evidence that the calculus dated as far back as 1665. Roger Cotes, who was still waiting for Newton's final revisions of the Principia, received a copy of the unexpected treatise in February. He concluded that the vindicating preface, which bore all the subtle marks of Newton at his stealthiest, would defeat future attempts to deprive the master of the justice due him.[58] The day was not far distant when Cotes, like Jones, would place himself entirely at Newton's disposal for the purpose of attacking Leibniz and his followers in the preface to the Principia's second edition.

The Transactions containing Keill's offensive remarks were not published until 1710, two years after they were drafted. Several more months passed before a copy reached Leibniz in Berlin. It was February 21, 1711, when he wrote a letter of remonstrance to Sloane, the Secretary of the Royal Society. Leibniz began by recounting an earlier treachery, which he had also protested: "Some time ago Nicholas Fatio de Duillier attacked me in a published paper for having attributed to myself another's discovery. I taught him to know better in the Acta Eruditorum." Why Keill had chosen to resurrect the ugly charge he did not know: "However, although I do not take Mr. Keill to be a slanderer (for I think he is to be blamed rather for hastiness of judgment than for malice) yet I cannot but take that accusation which is injurious to myself as a slander." Leibniz asked the Royal Society, of which he had been a Fellow since 1673, to extract a public apology from Keill, so that "a curb will be put on other persons who might at some time give voice to other similar" charges.[59] Fully aware of whose blessing he must receive for absolution, Leibniz characterized Newton as "a truly excellent person."

Following the consideration of such pressing matters as the President's claim that he had once observed a large worm in a dog's kidney, those present at the Royal Society meeting of March 22 heard a part of Leibniz's single-page letter read. There is no record of the ensuing discussion—if, indeed, one took place—but Sloane was ordered to draft a reply.[60] Newton apparently had second thoughts about pursuing this course, however, for no letter from Sloane to Leibniz has ever surfaced. Meanwhile Keill took the offensive by reminding Newton of the unfair unsigned review Leibniz had authored in 1705. His passion rekindled at the thought of having been compared to the plagiarist Fabri, Newton vented his emotions on paper:

> I have more reason to complain of the collectors of ye mathematical papers in those Acta then Mr Leibnitz hath to complain of Mr. Keil. For the collectors of those papers everywhere insinuate . . . that ye method of fluxions is the differential method of Mr Leibnitz & do it in such a manner as if he was the true author & I had taken it from him.[61]

On April 5, a day or two after Keill effectively appealed to Newton's baser instincts, the Society took up Leibniz's letter of protest for the second time. Keill, who was present on this occasion, repeated the charge that Leibniz had treated Newton ill in the pages of the *Acta*. Then the President himself took the floor. Building on Keill's moralistic prologue, he proceeded to recount the history of "his Invention." By the time it was over, the tables had been completely turned on Leibniz. Rather than being pressed to amend his abrasive remarks of the past, Keill was asked to draw up an account of the dispute, set it in a "Just Light," and "Vindicate himself from a perticular reflection in a Letter from M^r. Leibnitz to Doctor Sloane."[62] When the Fellows next assembled one week later, Keill, who had again come down to London from Oxford, remained at Newton's disposal. After the minutes of the previous week were read, Newton inserted a provision that a December 1672 letter to Collins, like those in Wallis's *Mathematical Works*, should also be taken into account in Keill's reply. Nor was Keill merely to vindicate himself; he was to draw up a paper asserting the President's rights in this matter.[63]

Keill drafted his apologia during the next six weeks. While we possess little documentary evidence of Newton's direct involvement in the project, there is much of a circumstantial nature to suggest that the resulting polemic was largely shaped by his hand.[64] To begin with, the letter contains intimate knowledge of Newton's mathematical history, including quotations from certain material (the 1672 letter on the quadrature of curves, for example) not previously made public. Furthermore, the evidence selected and the way in which it is presented are strikingly similar to the lines of argument followed by Newton on other occasions. Then, too, there is the matter of style, the most telling point of all. Taken as a whole, the document bears the indelible stamp of the master. It frames the priority issue as a historical question to be answered by the introduction of irrefutable empirical evidence, a technique Newton employed in everything from his religious and alchemical studies to the defense of his fiefdom at the Mint. It is also worth noting that he possessed a copy of Keill's letter, which contains a number of minor alterations in his hand. All things considered, he had almost certainly furnished Keill with the raw materials and directed him in their refinement and use, effectively turning the Oxford don into his willing mouthpiece. Bernoulli can hardly be said to have missed the mark when he latter referred to Keill as "Newton's toady." Apparently it never occurred to him that he himself was serving Leibniz in the same obsequious capacity.

Flexing his administrative muscle at the Royal Society meeting of May 24, Newton struck out at the two men he then despised the most. He first ordered that Flamsteed, whom he had recently attempted to hamstring by going over the Astronomer Royal's head to Queen Anne, observe the forthcoming eclipse of the sun and report his findings to the Society. This action was followed by the reading of Keill's reply to Leibniz, after which Sloane was asked to compose a covering letter before dispatching a copy to Berlin. (As it turned out, Newton himself wrote this letter, or so it would seem, since the surviving draft is in his hand.) His only concession was to postpone publication of Keill's reply until after Leibniz had had an opportunity to respond.[65]

Having enjoyed the honor of being the inventor of the calculus for twenty-seven years, at least on the Continent, Leibniz could not have been pleased with the condescending tone affected by the upstart Keill. It had not been his wish to disparage anyone, Keill disingenuously observed:

> I suggested only this, that Mr. Newton was the first discoverer of the Arithmetic of Fluxions or Differential Calculus; however, as he had in two letters written to Oldenburg (which the latter transmitted to Leibniz) given pretty plain indications to that man of most perceptive intelligence, whence Leibniz derived the principles of that calculus or at least could have derived them; But as that illustrious man did not need for his reasoning the form of speaking and notation which Newton had used, he imposed his own.

This limp disclaimer was completely undermined by what followed. Keill reiterated in lavish detail the charge that Leibniz had published nothing before he received Newton's letters from Oldenburg. The insinuation could not have been plainer: Leibniz was neither the first inventor of the calculus nor the second, but a clever thief who had finally been unmasked. Surely Keill's smug final remarks rankled the most:

I have detracted not a jot from Leibniz that was not Newton's; and there can be no doubt that candid judges of these things will with one voice allow that I said what I did say out of no slanderous spirit nor haste of judgement, which I have demonstrated to you with so much argument, clearer than the noonday sun.[66]

Leibniz's reply reached the Royal Society early in 1712 and was read at the meeting of January 31. The polite façade that had characterized his previous letter was abandoned. "What Mr John Keill wrote to you recently attacks my sincerity more openly than before; no fair-minded or sensible person will think it right that I, at my age and with such a full testimony of life, should state an apologetic case for it, appearing like a suitor before a court of law." While Keill might be a learned man, he was hardly experienced enough to judge the question of discovery with any degree of authority, much less objectivity. Score one for Leibniz. Yet unbeknownst to Keill, or Newton for that matter, the young mathematician had struck an even deeper nerve. In chiding the German for the manner in which the calculus had been announced in the *Acta Eruditorum*, he had reminded Leibniz of his own dishonesty in not giving Newton credit for having developed a similar method. The haunting memory of *De Analysi* surfaced once more, causing him to compound his original deceit further: "I find nothing [in the *Acta*]," Leibniz wrote, "that detracts anything from anyone." Having endured enough of Keill's "empty and unjust braying," he must now silence the toady-cum-jackass. In a tactical blunder he was destined to rue until his final hour, Leibniz appealed to the Royal Society for justice. Even Newton must disapprove of this libelous assault: "I am confident that he will freely give evidence of his opinion on this [issue]."[67] How ironic that these words, when reproduced in the *Correspondence* two and a half centuries later, should appear opposite those of Flamsteed in a letter to Abraham Sharp: "I have had another contest with ye PR.R.S. [President of the Royal Society] who formed a plot to make my instruments theirs."

Because of one of Newton's rare absences on February 7, the minutes indicate that no reply to Leibniz was forthcoming. The President's pen was anything but silent, however. He drafted a lengthy rejoinder, which was apparently meant for Leibniz after delivery before the Royal Society at some future date. He could agree that a quarrel with Keill was indeed beneath the older man's dignity, but he could not accept the claim that Leibniz was the first inventor of the calculus. Moreover, the papers Leibniz had published in the *Acta* "call my candor in question."[68] This letter was never sent. The ink had barely dried before Newton hit upon a devilishly clever scheme. Leibniz had asked for justice. Very well, let him have it. The Royal Society would appoint a special committee to settle the nagging priority issue— winner take all.

The committee, consisting of Arbuthnot, Hill, Halley, Jones, Machin, and William Burnet, was selected at the meeting of March 6, 1712. The name of Francis Robartes was added on the twentieth, followed one week later by the addition of Frederick Bonet, the London minister of the King of Prussia. Aston, Taylor, and de Moivre, the Huguenot emigré, joined its ranks on April 17, exactly seven days before the final report was delivered. Newton, who had hand-picked every member, later boasted that the committee "was numerous and skilful and composed of Gentlemen of several nations, and the Society are satisfied in their Fidelity ... without adding, omitting or altering any thing in favor of either Party." While the committee was empowered only to examine the letters and other papers pertaining to the accusation of libel brought by Leibniz against Keill, Newton subsequently denied that this was so. Its mission, he maintained, centered on the priority issue itself.[69]

So absolute was the President's control over subsequent developments that the committee's detailed report was ready on April 24, barely a month and a half after the appointment of its first six members. The names of those who served appear nowhere in the document, nor did they become publicly known until the nineteenth century, when the Royal Society's minutes were published by Turnor and De Morgan. Had it been otherwise, Newton's sanctimonious mask of impartiality would have dissolved like gossamer in the morning sun. For it was he who single-handedly pursued the "investigation," compiled the documentary evidence, which included the papers of Collins so fortuitously obtained by Jones, and drafted the final report, which survives in his hand. Every move Leibniz had made, from his London visits to his choice of notation for the calculus, was cast in a suspicious light. Newton, his anger penetrating spleen and marrow, again proved himself a master at rewriting history in a manner consistent with his own conspiracy theory. The details of the report, almost all of which are familiar to us by now, need not be reiterated at this juncture. The conclusion speaks for itself:

> For which Reasons we Reckon M^r. Newton the first Inventor and are of Opinion that M^r. Keill in asserting the same has been noways Injurious to M^r. Leibnitz and wee Submitt to the Judgment of the Society whether the Extract of the Letters and Papers now presented Together with what is Extant to the same Purpose in Doctor Wallis's third Volume may not deserve to be made Publick.[70]

The Fellows approved of the report to a man (*nemine contradicente*), for all intents and purposes branding Leibniz a plagiarist. They also agreed that the report, together with the documentary evidence, should be published "with all convenient speed" and named Halley, Jones, and Machin to the publication committee. The collection, bearing the title *Commercium epistolicum*, appeared nine months later, in January 1713. Relatively few copies of the book were printed, and these were distributed principally as gifts to those institutions and individuals who counted most—the universities and leading mathematicians. Newton was particularly anxious that the Europeans, who venerated Leibniz as a living saint among mathematicians, should know "the facts" of the matter. Using Keill as his shield, he also arranged for one T. Johnson, a Scottish bookseller living in The Hague, to take twenty-five copies at 3s. apiece. Keill, who had recently obtained the coveted Savilian Professorship of Astronomy at Oxford and a seat on the Royal Society Council, was more eager to please than ever.

A few years later, when Newton launched another vicious attack on Leibniz in his anonymous "Account of the Commercium Epistolicum," he wrote of his battered but unyielding foe: "But no Man is a witness to his own Cause."[71] This callously hypocritical remark, by which Newton subconsciously passed judgment on himself, reminds one of something Pepys wrote in his incomparable record of life in Augustan England: "My God what an age that makes a man act like a knave." As is true of all human beings, Isaac Newton cannot be separated from his times, and these, as we know, were anything but polite or forgiving, especially to the ambitious son of an obscure yeoman. But if Pepys's words ring of a universal truth, they also seem somehow inadequate when applied to the subject of this biographical study. It is doubtful that Newton would have acted very differently, no matter when he might have lived. Outwardly, he presented an impregnable defense. The vulnerability of his youth had hardened into an impenetrable shell. He prided himself on not displaying emotion, preferring to vent his inner conflicts on the written page. The world was a hostile home at best, desolate and without love, something no amount of success, whether material or intellectual, could change. Made for supremacy, he became the perfect example of the scientist whose conviction of his own genius renders him ruthless and uncompromising toward others. Whether he represented anything or anybody beyond himself is at best debatable. "The first inventor is the inventor & whether the second inventor found it by himself or not is a question of no moment," he wrote in one of several drafts of the preface to the *Commercium epistolicum.*[72] And after Leibniz died in 1716, he could not let the poor man rest, even as it took him years to bury Robert Hooke. The margins of his papers contain self-justificatory epithets of such a powerful order that he had no choice but to write them down; no choice, that is, if he was to know any peace of mind: "Second inventors," God's Privy Councillor thundered, "count for nothing!" A friend of Whiston later reported that he heard Newton pleasantly tell Samuel Clarke: "He had broke Leibnitz's Heart with his Reply to him."[73]

VI

Even as Newton was showing himself at his worst by waging simultaneous battles against Leibniz, Flamsteed, and Sloane, his creative juices were flowing freely for the last time, showing him at his intellectual best. Thanks to the able and persistent Roger Cotes, the ambivalence that had originally characterized his approach to a second edition of the *Principia* had yielded to genuine enthusiasm. It will be recalled that Cotes had edited and printed approximately half of the work by April 1710. Greater haste was possible in the beginning because Book I contained little new material. Book II, on motions in a fluid medium, proved rather less tractable, however, and Newton hit his first real snag when Cotes questioned the results of certain experiments performed in Cambridge many years before. Leaving as little as possible to chance, Newton devised a new series of ingenious demonstrations, which enabled him to measure the friction and velocity of water as it flowed through the orifice at the bottom of a large tank, a principle known as the *vena contracta*. The water, he was able to inform Cotes, does not flow out of the hole in a stream equal to its diameter, but contracts into a narrower vein. With the assistance of Francis Hauksbee, he extended his theater of operations to the great dome of St. Paul's Cathedral, a perilous 220 feet above the marble floor. A small hinged platform was erected on which Hauksbee placed balls of various sizes and densities. The latch on the platform was then tripped by means of a string extending to the floor, which also engaged a pendulum that timed the falling objects. Newton may not have possessed the physical resources needed for the demanding climb to the top of London's most impressive edifice, but the septuagenarian was still able to summon up his marvelous deductive powers at will.

Cotes was not the only one who stirred Newton by bringing to his attention significant errors in the second Book. Johann Bernoulli, a member of Leibniz's inner circle, had discovered that Newton's calculations designed to measure the force of gravity at any point on the trajectory of a falling body in a resisting medium yielded only two-thirds the numerical value they should have. Having employed Leibnizian differentials in his computations, Bernoulli interpreted his discovery as proof that Newton's alternative method of derivation was inferior and that he could not have been in full command of the calculus when writing the *Principia* in 1686. Knowing full well that a second edition was in the offing, Bernoulli selfishly decided to wait until Newton repeated his mistake before exposing it in print. As luck would have it, Johann's nephew Nikolaus, a promising young mathematician, paid a visit to London in September 1712. He was introduced to Newton by de Moivre and promptly informed him of Johann's disturbing discovery. Nikolaus also told Newton that he could not find the point at which his argument in the *Principia* went astray, and neither could his uncle. "[T]wo or three days later when I had gone to [Newton's] house," de Moivre wrote the elder Bernoulli, "he told me that the objection was valid, and that he had corrected the result . . . and it proved agreeable to the computation made by your nephew."[74] A grateful Newton begged de Moivre to bring Nikolaus back to his home so he could thank him in person, an unusual display of generosity for someone who hated contradiction, but one consistent with his even greater fear of embarrassing himself before his peers. Where others had succeeded only in part, Newton had triumphed once more, finding and righting the source of his mistake. Only after he repaid the Bernoullis by securing their election to the Royal Society did Newton learn of Johann's treachery. Indeed, Bernoulli had already prepared a paper for publication containing some barbed comments on Newton the mathematician. Fortunately, Newton took the important step of having Cotes reprint the page containing his blunder. This was pasted to the stub of the excised original, a galling reminder to Bernoulli of just how close he had come to avenging an earlier defeat arising from the challenge problems. Perhaps he was also reminded of something he had written on that occasion but, to his chagrin, had apparently forgotten—"from the claws of the Lion."

As Newton was wont to admit, nothing in the *Principia* so taxed his patience and natural gifts as the theories of the moon, the tides, and the comets, all of which are contained in Book III, or "System of the World." Armed with the new data he had coerced from Flamsteed, he hoped to bring a mathematical precision to the understanding of these vexing phenomena far beyond that expressed in the *Principia*'s first edition. The riddle of the tides was compounded by the joint gravitational action on the earth's oceans of the sun and the moon, while lunar motion was rendered intractable by the simultaneous pull of the sun and the earth. The tides and lunar theory were dealt with in a series of two dozen letters exchanged with Cotes between February 7 and September 15, 1712. Cometary theory, the last of the major celestial phenomena to receive consideration, was taken up in a further exchange of letters thereafter. While revisions were made, Newton, for all his magnificent powers, was again left with approximate values rather than the absolutes he always dreamed about. Thus the suffering he inflicted on the beleagured Astronomer Royal counted for next to nothing.

The Cartesian Christiaan Huygens, whose admiration for Newton's mathematical and deductive skills was considerable, had nevertheless labeled his principle of universal gravitation a "manifest absurdity." Leibniz, a dyed-in-the-wool Cartesian himself, had more recently taken up the hue and cry, which was echoed throughout Europe by his supporters. The burden of the mechanists' charge was this: The doctrine of action at a distance was tantamount to the Aristotelian belief in occult qualities, for by dispensing with the ether hypothesis Newton supposed that forces could be exerted between bodies without the operation of an intervening medium. The Cartesians, who refused to abandon their mechanism of causation, were content to explain the attraction of bodies by means of vortices in a dense and fluid ether, which the *Principia* rendered obsolete. As the calculus controversy reached the boiling point, the philosophical battle was also joined. To Leibniz, the idea of resorting to unknown, mysterious forces flew in the face of reason, and he rarely passed up an opportunity to score off his rival on this issue: "The fundamental principle of reasoning is, *nothing is without* cause. . . . This principle disposes of all inexplicable occult qualities and other figments." As for Newton's attempt to extricate himself from this dilemma by attributing universal attraction to a higher agent, something he had done in the second edition of *Opticks*, "But if he posits that the effects [of gravity] depend not on an occult quality but on the will of God or a hidden divine law, thereby he provides us with a cause, but a supernatural or miraculous one."[75] When Leibniz, in his *Essais de Théodicée*, assaulted Newton's philosophic position in 1710, Newton's xenophobic disciples rallied to the cause: "I have nothing of news to send you," Jones wrote Cotes on October 25, 1711, "only the Germans and French have in a violent manner attack'd the Philosophy of Sr. Is: Newton, and seem resolv'd to stand by Cartes; Mr Keil, as a person concern'd, has undertaken to answere & defend some things, as Dr Freind, & Dr Mead, does (in their way) the rest."[76]

Newton had always known that his only hope of eventually silencing the Cartesians lay in the mathematical precision he brought to scientific enquiry. While he never denied that a physical cause of gravity might be discovered one day, he could not allow the search for such a cause to obscure the fact that the universe moves according to fundamental laws of force, no matter how those forces are accounted for physically. Whereas Descartes had titled his great work *Principia Philosophiae* (1644), Newton emphasized his break with the archmechanist by calling his masterpiece *Principia Mathematica*, in honor of the universe of precision and quantitative order. With so much at stake, it was essential that the second edition offer even more compelling demonstrations of experimental and numerical accuracy than had the first. Leibniz was clearly on Newton's mind when he drafted an extended version of his rules of reasoning in Book III: "We are certainly not to relinquish the evidence of experiments for the sake of dreams and vain fictions of our own devising; nor are we to recede from the analogy of Nature, which is wont to be simple, and always consonant to itself." The second edition's celebrated General Scholium began on an equally blunt note: "The hypothesis of vortices is pressed with many difficulties."[77] As a means of repeatedly underscoring this point, Newton resorted to the clever manipulation of his sacred experimental data, feigning a level of mathematical precision that was unattainable by the scientific standards of his time.

So deft was Newton at the rearrangement of his numbers that his secret did not become widely acknowledged until the rather recent publication of a revealing article in the journal *Science*. A comparison of Newton's manuscript data with the *Principia*'s second edition yielded graphic evidence of significant numerical manipulation in three important areas: the acceleration of gravity, the precession of the equinoxes, and the velocity of sound. In the first two instances Newton purported to have achieved a precision of about 1 part in 3,000, while in the latter instance he settled for the more modest but still insupportable claim of 1 part in 1,000.[78] Small wonder that the mechanists, for all their protestation of occult qualities, were eventually "swept into oblivion." The concept of force acting at a distance had been so "precisely" delineated mathematically that by the end of the eighteenth century no self-respecting scientist could deny its existence.

In point of fact, Newton took relatively little risk by doctoring his figures, as he was well aware. Unlike the individual who intentionally fakes his experimental data to confirm a promising hypothesis, he knew full well that the fundamental insights gathered from a lifetime of meticulous experimentation were sound. He had simply, if dishonestly, elevated these demonstrations to a higher plane of accuracy than the figures warranted. When viewed in this light, such pretense obviously had one overriding purpose: to serve as a grand polemic against the mechanists in general and against Leibniz in particular. Writing to Newton on February 11, 1712, concerning the problem of calculating the rate of the moon's descent, Cotes captured the spirit of the enterprise: "In the Scholium to the IVth Proposition I think the length of the Pendulum should not be put 3 feet & 82/5 lines; for the descent will then be 15 feet 1 inch 11/3 line. I have considered how to make that Scholium appear to the best advantage as to the numbers & I propose to alter it thus." This was but one of several instances in which the young mathematician proved himself almost as adroit as his master at doctoring the figures. Two months later Newton praised him for further manipulating the lunar theory, and Cotes replied with reassuring words: "I am very glad to see the whole so perfectly well & fairly stated."[79]

His labors nearing an end, Newton wrote an intriguing letter to Cotes on January 6, 1713. "I shall send you in a few days a Scholium of about a quarter of a sheet to be added to the end of the book: and some are perswading me to add an Appendix concerning the attraction of the small particles of bodies."[80] This, of course, was the General Scholium concluding the *Principia*, which went through several drafts and grew considerably in length before Cotes received a copy early in March. In contrast the appendix, which was to have occupied six printed pages, had been reduced to a single curious paragraph grafted onto the end of the General Scholium. As noted above, Newton begins this, the most famous of all his writings, with a declaration that the mechanistic hypothesis of vortices is pressed with many difficulties. He then proceeds to discuss the motions of the planets and reiterates his findings that these motions are mathematically incompatible with Cartesian theory. Neither can the movements of the comets, which are governed by the same laws of force as the planets, be accounted for by the hypothesis of vortices. Finally, the celestial bodies suffer no resistance as they trace their paths through the celestial spaces, "the planets and comets will constantly pursue their revolutions in orbits given in kind and position," according to the law of universal gravitation. Yet Newton warns the reader that these bodies could not have established their regular positions by means of mathematical laws alone. It is God who put them in their respective places and who sustains their motion by virtue of His presence: "This most beautiful system of the sun, planets, and comets, could only proceed by the counsel and domination of an intelligent and powerful Being." And if, as Newton believed, the far distant stars are the centers of other solar systems, these too are the product of God's wise counsel and domination. "He is omnipresent not *virtually* only, but also *substantially*; for virtue cannot subsist without substance. In him are all things contained and moved; yet neither affects the other: God suffers nothing from the motion of bodies; bodies find no resistance from God." Still in agreement with the Cambridge Platonists whose acquaintance he had made as a young man, Newton clung to his belief that God acts as the sensorium of the universe. "He endures forever, and is everywhere, he constitutes duration and space." We know him only by his "excellent contrivances of things, and final causes"; we admire him for his perfection—"a god without dominion, providence, and final causes, is nothing else but Fate and Nature," the false god of the Cartesians.[81]

Having explained the phenomena of the heavens by the force of universal gravitation, Newton claimed not to have postulated the true cause of this force, a rather questionable assertion given his lengthy dissertation on the manifold attributes of the Almighty:

> But hitherto I have not been able to discover the cause of those properties of gravity from phenomenon, and I frame no hypotheses; for whatever is not deduced from the phenomena is to be called an hypothesis; and hypotheses, whether metaphysical or physical, whether of occult qualities or mechanical, have no place in experimental philosophy. . . . And to us it is enough that gravity does really exist, and act according to the laws which we have explained, and abundantly serves to account for all the motions of the celestial bodies, and of our sea.[82]

From a purely scientific standpoint Newton would have been better off if he had suppressed his theological bias and simply observed that hypotheses which seem incapable of verification by experimentation and mathematical analysis are to be viewed with great suspicion. Such a declaration would have left the Cartesians off balance and would have at least partially closed the opening he had left for them in the *Opticks*. But such was not Newton's nature: A Puritan and true believer, he never looked upon any material explanation of gravity as a serious contender with the power of God. In the words of Ecclesiastes, "the eye is not satisfied with seeing, nor the ear with hearing."

Having paid homage—both eloquent and earnest—to the manifold powers of his God, Newton chose to add a cryptic final paragraph to the General Scholium "concerning a most subtle spirit which pervades and lies hid in all gross bodies." This spirit possesses an electric force, which may account for various phenomena, such as the propagation of heat; the emission, refraction, and reflection of light; and muscular action in animals. Although he stops short of specifying it as the cause of gravity, the fact that his elastic medium explains the attractions and repulsions of diverse kinds of particles seems to suggest as much. No doubt Hauksbee's recent electrostatic experiments had a great deal to do with the revival of Newton's belief in an ethereal medium, but, like Descartes's vortices, it would not yield to an exact quantitative analysis. "These are things that cannot be explained in a few words," he wrote in conclusion, "nor are we furnished with that sufficiency of experiments which is required to an accurate determination and demonstration of the laws by which this electric and elastic spirit operates."[83] That he even bothered to raise the subject at this juncture is further evidence of his need to refute Leibniz's allegations of occultism, particularly with regard to the theory of universal gravitation. If the concept of an elastic spirit did not appeal to the mechanists, neither could they say that the *Principia*'s author was without a physical explanation of force, however dubious. The doctrine of action at a distance had been relieved of the stigma of purely supernatural influences, or so Newton falsely hoped.

Cotes, who was about to take holy orders, was less than satisfied with this equivocal apologia. He had wished for an uncompromising indictment of the materialism of Leibniz; what he received instead was an apparent softening of Newton's stance against the irreligious Cartesian mechanical philosophy. Newton disappointed Cotes on another count as well, for he had nothing to say about his priority in the calculus controversy. Rather than raise these sensitive issues with the volatile author, Cotes wrote to Bentley, who was then in London. Would it not be a good idea if the new preface (which Cotes had been charged with drafting) included a censure of Leibniz for pirating the calculus and usurping certain physical concepts set forth in the first edition of the *Principia*? "This I say upon Supposition that I write the Preface my self." It would be preferable, however, if Newton, or Newton and Bentley together, were to write it: "You may depend upon it that I will own it, & defend it as well as I can, if hereafter there be occasion."[84] This was yet another vivid example of Newton's mesmerizing hold over those who served his interests.

Bentley replied two days later, on March 12, 1713. He had discussed the matter with Newton, who was content to leave the task in Cotes's capable hands. The young mathematician should feel at liberty to write what he would about the calculus controversy and such other matters as he deemed fit. Newton had asked only that he spare the name of Leibniz and abstain from using epithets or words of reproach, "(not that its untrue) yt its rude & uncivil." What Newton had not said is that, having recently pilloried Leibniz in the *Commercium epistolicum,* he wished to maintain his public image as the very model of professional decorum. As usual, Cotes took the hint: "I think it will be proper," he informed Newton on March 18, "... to add something more particularly concerning the manner of Philosophizing [you] made use of & wherein it differs from that of De Cartes & others." Newton agreed but, acting the part of the fox as well as the lion, also warned Cotes that "I must not see it. for I find that I shall be examined about it."[85] Left to his own devices, Cotes produced the lucid exposition and defense of Newtonian dynamics he had promised. And while the name of Leibniz was nowhere to be seen, few among the learned missed the biting references to Newton's archrival: "Those who assume hypotheses as first principles of their speculations ... may indeed form an ingenious romance," Cotes chided, "but a romance it will still be."[86]

On June 30 Bentley sent welcome word to Newton: "At last Your book is happily brought forth; and I thank you anew yt you did me the honour to be its conveyer to ye world." He could expect to receive six copies in the near future and had only to ask for "what more you shall want." Bentley had already sent two hundred copies to France and Holland out of a printing of slightly more than seven hundred. According to his private accounts, the cost of producing the work came to a few shillings over £117. It proved a lucrative undertaking for Trinity's avaricious Master. By November 1715 all but seventy-one copies had been sold, leaving Bentley with a profit of £198, a substantial gift to him from Newton, largely earned by dint of Cotes's impressive labors.[87] So far as is known, Bentley shared none of these proceeds with the gifted editor, nor did Newton so much as thank him for his efforts. Cotes did receive a dozen copies of the *Principia,* two of which he gave to fellow Newtonians Samuel Clarke and William Whiston. Newton and Cotes exchanged a few letters during the three years preceding Cotes's untimely death in June 1716 at the age of thirty-three. Only then, after it was too late, was Newton sufficiently moved to express his regard for the talented man who had served him so patiently and well: "If He had lived," Newton told Cotes's cousin Robert Smith, "we might have known something."[88] Whether he truly meant what he said is impossible to tell.

Chapter Twenty

"They Could Not
Get Me to Yield"

*He who has a thousand friends has not a
  friend to spare,
And he who has one enemy will meet him
  everywhere.*

Ali ibn-Abi-Talib

I

On returning to Basel from Paris in the spring of 1713, Nikolaus Bernoulli carried a copy of the *Commercium epistolicum* to his anxious uncle Johann. Its contents were even more damaging than the elder Bernoulli had been led to believe by his friend and correspondent the Abbé Jean-Paul Bignon. He dashed off a letter containing the stunning news to an even more anxious Leibniz—"you are at once accused before a tribunal consisting, as it seems, of the participants and witnesses themselves, as if charged with plagiary, then documents against you are produced, sentence is passed; you lose the case, you are condemned." Bernoulli, a powerful advocate of Leibniz and a dangerous enemy of Newton—the more so because of his carefully guarded anonymity—proceeded to point out that it was Newton who had committed plagiary. After all, what had he really accomplished by 1676, or even 1687? Contrary to what Newton had written, he had employed the pricked or dotted fluxional symbolism only in the 1690s, long after Leibniz's papers on the calculus appeared in the *Acta*. If, as the Englishman maintained, he was truly master of the fluxions by the mid-1680s, how is it that "you can find no least word or single mark of this kind even in the *Principia Philosophiae Naturalis*" where he had "so many occasions" for using it? Finally, there was the matter of Newton's error in Proposition X, Book II, which Bernoulli himself had discovered, only to have it revealed to the *Principia*'s author by the loose-tongued Nikolaus. Still smarting from Newton's brilliant riposte, Bernoulli had recently published a paper in the *Acta* in which he drew attention to the place where sheets had been cut out of the second edition and new ones substituted in their stead. "At any rate it is clear that the true way of differentiating differentials was not known to Newton until long after it was familiar to us." Bernoulli, who has been aptly described as a lion by night and a jackal by day, gave Leibniz permission to make use of his remarks but begged him not to reveal their source, "for I am reluctant to be involved in these disputes or to appear ungrateful to Newton who has heaped many testimonies of his goodwill upon me."[1] The pusillanimous Bernoulli had good reason to be concerned. Newton had dispatched presentation copies of the *Principia* to several prominent European mathematicians, including the Abbé Bignon, Varignon, and Fontenelle. Even Leibniz, toward whom Newton was still cultivating a public image of moderation, had been so favored. Bernoulli, however, was left out, and he fretted over this slight as, indeed, Newton had hoped he would. Contrary to Newton's polite excuse that his supply of presentation copies was limited, he had gotten wind of Bernoulli's critical paper in the *Acta*. As usual, he took his revenge, which, considering Bernoulli's false neutrality, seems justified in this instance.

Leibniz had succeeded fairly well so far in controlling his passion. Bernoulli's letter, which caught up with him in Vienna, effected a drastic change of emotions. "I have not yet seen the little English book directed against me," he fumed; "those idiotic arguments which (as I gather from your letter) they have brought forward deserve to be lashed by satirical wit." Newton no more knew the calculus than Apollonius knew the algebraic method of Francois Viète and Descartes. "He knew fluxions, but not the calculus of fluxions which (as you rightly judge) he put together at a later stage after our own was already published. Thus I have myself done him more than justice, and this is the price I pay for my kindness." It seemed obvious to Leibniz that the Royal Society, in taking Newton's part, was furthering what amounted to a national conspiracy: "For many years now the English have been so swollen with vanity, even the distinguished men among them, that they have taken the opportunity of snatching German things and claiming them as their own." Boyle stole from Glauber and Guericke, Sir Paul Neile from Hendrik von Heuraet. Now Newton would lay parental claim to yet another German brainchild, the calculus. Leibniz begged Bernoulli not to be seduced by the blandishments of a plagiarist. He agreed to keep Bernoulli's name out of the fray for the time being, but he very much needed his colleague's support. "I expect from your honesty and sense of justice that you will as soon as possible make it evident to our friends that in your opinion Newton's calculus was posterior to ours, and say this publicly when opportunity serves."[2] Bernoulli suddenly found that he was pinioned between titans, a fate he could blame on no one but himself.

Leibniz was so provoked by the latest intelligence from Bernoulli that he decided to retaliate even before obtaining his own copy of the *Commercium epistolicum*. On July 29, 1713, a curious document known to history as the *Charta Volans* or, as Newton dubbed it, the "Flying Paper" suddenly appeared. The little work carried no clues regarding its origin; the names of the author and printer and even the place of publication were scrupulously omitted. As it turned out, the author's very concern for anonymity proved to be the most important clue to his identity. Leibniz had become well known in scholarly quarters for writing unsigned reviews that lauded his own accomplishments as an admirer would. Hence few were deceived as to the author of the *Charta Volans*, least of all Newton, who, as the *Commercium* demonstrates, was far more adept than Leibniz at playing the game of academic hide-and-seek.

Leibniz, who speaks of himself in the third person, begins by stating that he is now living in Vienna and has not yet seen the little volume *Commercium epistolicum*. However "a leading mathematician," whose name he is pledged not to reveal, has assured him that the case presented by the Royal Society is without factual foundation. Quoting extensively from Bernoulli's letter of June 7, he argues that Newton should content himself with the considerable honor of having advanced the understanding of series, since he could not pretend to have grasped the differential calculus until long after it was familiar to others. Whereas Newton was used to lowering his head and charging with all the fury of an enraged bull, Leibniz preferred the maddening barbs of the light-footed banderillero. Few words that he spoke or wrote of Newton could have stung more keenly than those he now chose:

> ... Newton took to himself the honour due to another of the analytical discovery or differential calculus first discovered by Leibniz ... he was too much influenced by flatterers ignorant of the earlier course of events and by a desire for renown; having undeservedly obtained a partial share in this, through the kindness of a stranger, he longed to have deserved the whole—a sign of a mind neither fair nor honest. Of this Hooke too has complained, in relation to the hypothesis of the planets, and Flamsteed because of the use of his observations.[3]

When and how Leibniz learned of Newton's bitter quarrels with Hooke and Flamsteed is uncertain, but he had doubtless waited a long time before choosing this particular moment to strike. By so doing he had gone beyond his promise to Bernoulli that the arguments of his opponent must be "lashed by satirical wit."

Details of the *Commercium epistolicum* had also reached Leibniz via his friend Christian Wolf, Professor of Mathematics at the University of Halle. The calculus controversy was a standard topic of discussion in subsequent letters exchanged between the two men, and it was Wolf who undertook to publish and circulate the *Charta Volans* for his German colleague. After Leibniz read Keill's "Lettre des Londres," an extension of Newton's version of the priority dispute published in the May and June 1713 issue of the recently launched Dutch *Journal Literaire de la Haye,* he wrote another anonymous defense ("Remarks") of his own cause and sent it to Wolf. The latter forwarded "Remarks," together with a French version of the *Charta,* to the editors of the new journal, who, having struck an unexpected bonanza, gladly published the juicy material in the issue for November and December.

Newton first laid eyes on the publication in April 1714. What he read stung to the quick. "And now it is made publick," he angrily wrote to Keill, "I think it requires an Answer. It is very reflecting upon the Committee of the Royal Society, & endeavours to derogate from the credit of some of the Letters published in the Commercium Epistolicum as if they were spurious." He had a suggestion to make: "If you . . . consider of what Answer you think proper, I will . . . send you my thoughts upon the Subject, that you may compare them with your own sentiments & then draw up such an answer as you think proper." Keill need not set his name to it. As before, the man whom Bernoulli and Leibniz variously dubbed the ape, toady, and jackass of Isaac Newton was only too willing to do his master's bidding. "I think I never saw any thing writ with so much impudence falshood and slander," Keill replied on cue, "I am of opinion that they must be immediately answered." The two exchanged eight more letters during the next month. On May 25 Keill sent Newton his final comments together with a short letter reiterating his complete subservience: "I leave my whole paper to You and Dr Halley to change or take away what you please, I only desire that it may be done quickly and sent over" to Holland. The finished product, which bore the Newtonian imprint from beginning to end, was published in the *Journal Literaire* for July and August 1714. Keill had diverged from his idol on only one point: He insisted on signing the article. Its appearance coincided with the delivery to Newton of a slaughtered deer from his half-sister Hannah Barton, an appropriate occasion for a feast.[4]

II

A year had passed since the second edition of the *Principia* was published in June 1713, time enough for the learned world to render at least a partial verdict. Perhaps the greatest compliment to the new edition was paid to it by the enterprising booksellers of Amsterdam, although Newton and especially Bentley could scarcely have approved of their means. The two hundred copies Bentley had shipped to France and Holland the previous summer had apparently sold out in a matter of months, leaving the needs of Continental scholars unsatisfied. In the summer of 1714 the first Amsterdam reprint, an unauthorized but scrupulously faithful reproduction of the second edition, issued from the presses. A major undertaking, it required a completely new setting of type and the preparation of new woodcuts for the diagrams. How many copies of the pirated edition were sold has never been established, but subsequent developments would strongly suggest that the booksellers more than recouped their considerable investment. A second Amsterdam reprint, also requiring new woodcuts and type, was issued in 1723.

This time there was no laudatory review of the *Principia* in the *Philosophical Transactions* à la Edmond Halley, and the *Bibliothèque Universelle*, which had carried Locke's unsigned and largely uninformed tribute to the first edition in March 1688, had since gone out of existence. As before the *Journal des Sçavans*, the official publication of the *Académie Royal des Sciences*, carried an anonymous review. However, the decidedly pro-Cartesian stance of the first review now gave way to a wholly objective treatment of Books I and II, a measure of the inroads made by Newtonian physics during the previous quarter-century. Not even in his concise summary of the author's views on absolute time and space in Book III did the reviewer level the slightest criticism. Nor would he condemn Newton's rejection of vortices, the sacred cow of the Cartesians, citing instead the evidence offered against their existence. As regards the new General Scholium, the reviewer closed without making so much as a single comment on it.[5]

The anonymous review published by the editors of the *Acta Eruditorum* proved an altogether different animal. Here praise was intermingled with cunning allusions to Newton's limitations and unacknowledged debts to others, namely, Bernoulli and Leibniz. After a painstaking comparison of the first and second editions, the reviewer broached the subject of the calculus: "[E]ven in this new edition [Newton] does not deny that the illustrious Leibnitz had communicated its foundations to him, although [Newton] zealously concealed a certain technique of his own." The writer then repeated the story, published in the *Acta* in 1713, of how Bernoulli had discovered the mathematical boner in Proposition X, Book II, which Newton then corrected at the last minute after consulting with Bernoulli's nephew Nikolaus, "and the sheets cut out and the new ones substituted in their place certainly declare that the errors of the first edition already had crept into the second." The reviewer focused at length on the General Scholium, repeating both Newton's admission that he could not deduce the cause of gravity from physical phenomena and his assertion that he would feign no hypotheses. Yet Newton had also thrown the Cartesians a bone in the form of an electric and elastic spirit, which, he hinted, just might account for such unexplained phenomena as attraction at a distance. The reviewer, obviously a Cartesian himself, deftly turned the argument against Newton in a concluding sentence calculated to rankle: "[I]t is to be feared that most people may assign a greater value to hypotheses than to the author's most subtle matter of the Cartesians."[6]

Whether, as Newton had good reason to suspect, Leibniz, rather than one of the German's followers, was the unnamed reviewer in the *Acta* cannot be proven absolutely. That the review was full of the same old innuendoes and half-truths was what really mattered.

John Chamberlayne, a court official under the dying Queen Anne and the Royal Society Fellow who had once voiced a desire to see Newton made "Perpetual Dictator" of the institution, aspired to the role of peacemaker between the aging giants. "[I]t would be very Glorious to me, as well as Advantageous to the common Wealth of learning," he wrote to Leibniz on February 27, 1714, "if I could bring such an Affair to a happy end." Leibniz's reply, which he drafted in May, was somewhat less than encouraging. He raked Keill over the coals for the umpteenth time and was scarcely more charitable toward Newton, condemning him for having turned the Royal Society into a one-sided tribunal and then publishing a book "printed expresly for discrediting me." "The pretended judgment, & this affront [was] done without cause to one of the most ancient members of the Society." Chamberlayne had his permission to redress "the evil" if he could, but Leibniz was anything but optimistic.[7]

Chamberlayne was aware that he was swimming against the tide, but powerful undercurrents about which he knew nothing doomed from the outset his meager chances of effecting a reconciliation. He carried Leibniz's prickly missive to a Master of the Mint engaged in yet another struggle for supremacy—this one against his Warden, Craven Peyton—and to a President of the Royal Society who was busily preparing multiple drafts of "Keill's" acid reply to Leibniz's "Remarks" of 1713. Newton responded by translating the Leibniz letter into English for the purpose of laying it before the Royal Society. Chamberlayne, who by now wished that he had never yielded to the tempting dream of glory, tried to dissuade Newton without offending him. "I am not sure it wil be agreable to the writer," he cautioned. Prudence also dictated that future political developments be given serious consideration, for in Leibniz Newton was dealing "with a Gentleman that is in the Highest Esteem of the Court of Hanover," the domain of the soon to be crowned George I.[8]

Newton would not be swayed. On June 30 Chamberlayne reluctantly drafted a long-postponed letter to Leibniz. "I am sorry to tel you Sr that my Negotiations have not met with the desired success, & that our Society had been prevail'd upon to vote that what you writ was insufficient & that it was not for them to concern themselves any further in that Affair." He later sent Leibniz a transcript of the May 20 meeting at which the Fellows voted against taking up the German's numerous complaints, not that Leibniz, whose letter was addressed to Chamberlayne as opposed to the Royal Society, had asked them to. Chamberlayne's last known written communication to Newton, the perpetual dictator, is dated October 28, 1714. Citing a previous engagement, he declined an invitation to meet with the Royal Society President at Crane Court—"besides I must frankly own to you, that I don't make a Cypher [of myself] anywhere." As far as Leibniz was concerned, Chamberlayne, who deeply regretted his failure as a self-appointed peacemaker, twice quoted the rather lame and melancholy words of an unidentified classical poet: *"Non nostrum est Tantas componere Lites"* ("It is not our role to resolve such great disputes.")[9]

When Queen Anne died on July 31, 1714, George I, Elector of Hanover, great-grandson of James I, and long-time patron of Gottfried Wilhelm Leibniz, was instantly proclaimed King. To the new fifty-four-year-old monarch England always remained a foreign country for which he had no love and of whose language, sentiments, and thought he was profoundly ignorant. Poorly educated, boorish, and attracted to blowzy foreign women, he made no appeal to the admiration or the fancy of his new subjects. Yet his very weaknesses proved an unexpected boon to the evolution of English democracy, since a king of more brilliant parts might have been an impediment to the development of constitutional government. Supported by the Whigs and undisguisedly partial to them, George left all questions of domestic policy in the hands of his ministers: Stanhope, Sunderland, Townshend, and, later, the skilled Robert Walpole, who was falsely convicted of corruption by the Tory administration in 1712 and spent some months in the Tower near Newton's office. Leibniz despised his uncouth patron and took advantage of every pretext to absent himself from Hanover. Indeed, he was serving as Imperial Privy Councillor in Vienna when word of the succession reached him. He immediately returned to Hanover but found, much to his disappointment, that George had left for England three days earlier. All he could do was petition the new King for a post as Court Historian in London. While Leibniz waited for a reply, Newton took charge of minting the official coronation medal, a simple design picturing George seated on the throne and Britannia placing a crown on his head. Newton later told de Moivre that when the Court of Hanover came to London Leibniz's friends "endeavoured to reconcile us in order to bring him over ... but they could not get me to yield."[10] Whether Newton was referring to Chamberlayne's unsuccessful overtures or to some other attempt at peacemaking is uncertain. Be that as it may, George absolutely refused to consider Leibniz's request to be brought to London until he finished his monumental history of the house of Brunswick-Hanover. Plagued by gout, Leibniz spent the last two years of his life under virtual house arrest, sadly neglected by the nobleman he had served for sixteen years, while his rival enjoyed ready access to the new and far distant court, a bitter irony to say the least.

Generations of historians have searched the documents associated with the Newton–Leibniz controversy in vain for the slightest signs of a moral calculus to match the mathematical prize that both claimed as their exclusive property. This ethical vacuum was especially pronounced on the part of Newton, a man who hated with so much force one can only conclude that he hated something deep within himself. Within weeks of the Hanoverian succession he began composing what has been characterized as the most shameful of all the many hundreds of documents on the calculus controversy, "An Account of the Book entitled Commercium Epistolicum." The consummate product of years of rationalization and convenient twisting of facts, it was destined to fill all but three pages of the *Philosophical Transactions* for January and February 1715. Like the author of the work he was purporting to review, Newton dared not sign his name to this volatile polemic. He laboriously recounted every scintilla of evidence damaging to his foe and favorable to himself, no matter how tortured or circumstantial. The "ancient Letters," meaning those written before 1677, were quoted as if they were holy writ; not only did they prove that Leibniz's mastery of the calculus came much later than his own, but these letters had been used by the German to discover the very method of fluxions Newton had invented, or so he claimed. A cunning thief, Leibniz had patiently bided his time before publishing his purloined goods, waiting until the older generation of English mathematicians, who could testify against him, had died off. "And therefore it lies upon Mr. *Leibniz* to prove that he found out this Method long before the Recipt of Mr. *Newton's* Letters. And if he cannot prove this, the Question, Who was the first Inventor of the method, is decided." Newton was still smarting from Leibniz's recent charge (leveled in the *Charta Volans* and elsewhere) that since the *Principia* contained nothing of the calculus, its author must have been ignorant of the revolutionary method at the time he wrote the work. Newton now countered with the assertion that he had in fact employed the new mathematics to devise "most of the Propositions in [the] *Principia Philosophicae*" but had chosen to retain the method of demonstration preferred by the ancients, namely, the mathematics of geometry. By falsely implying that he had composed not one manuscript but two, Newton inadvertently launched future generations of scholars on a fruitless quest for a lost masterpiece that in truth had never existed.[11]

The more Newton wrote, the more his rampant temper betrayed him. Perverse, misleading, and sometimes illogical, he resorted to the defamatory epithet with increasing frequency. "[S]econd Inventors have no Right," he railed, only to repeat the same charge in the very same words two pages later. "It has been said the Royal Society gave judgment against Mr *Leibniz* without hearing both Parties," he continued. "But this is a Mistake.

They have not yet given judgment in the matter."[12] Any person who had read the *Commercium epistolicum* knew better, of course. And when, as we have already noted, he wrote that "no man is a witness to his own cause," it appears obvious that his range of vision was much too narrow to include himself. If the thought had ever occurred to Newton that this declaration was equally applicable to his own self-righteous conduct, it would only have multiplied his rage against his enemy.[13] Newton concluded with a defense of his philosophy of nature, which had recently been attacked in the *Acta Eruditorum* by the anonymous reviewer of the *Principia*'s new edition. Even if Leibniz had not actually authored the review, as Newton believed he had, the ultimate responsibility for its contents was his rival's nonetheless. Leibniz had foolishly raised the mechanical philosophy to a level of certainty that no true believer could tolerate in good conscience:

> But must the constant and universal Laws of Nature, if derived from the Power of God or the Action of a Cause not yet known to us, be called Miracles and occult Qualities, that is to say, *Wonders* and *Absurdities*? Must all the Arguments for a God taken from the Phaenomena of Nature be exploded by *new hard Names*? And must Experimental Philosphy be exploded as *miraculous* and *absurd* because it asserts nothing more than can be proved by Experiments, and we cannot yet prove by Experiments that all the Phaenomena in Nature can be solved by meer Mechanical Causes? Certainly these things deserve to be better considered.[14]

True to himself, Newton reaffirmed his belief that the ultimate explanation of reality must be both scientific and sacramental.

The "Account" has been characterized as a "sermon for the converted," and so it was.[15] At the same time, however, Newton was most anxious that it reach the widest audience possible. De Moivre translated the paper into French for publication in the *Journal Literaire,* and a review of it was also sent to the *Nouvelles Literaires.* Not satisfied with these measures, Newton had the French version published as a separate pamphlet and circulated among natural philosophers and mathematicians throughout Europe. In 1722 he republished the work in Latin, after which it became familiar to Newton scholars as the *Recensio,* or *Review.* Leibniz, whose declining health and advancing age were beginning to exact a heavy toll, dismissed this latest polemic in a few lines written for the *Nouvelles Literaires,* sarcastically referring to it as nothing more than twice-cooked cabbage.

In the spring of 1715 Newton's attention was temporarily diverted from the calculus controversy by a sad event that bore directly on the fortunes, both emotional and material, of Catherine Barton, the beautiful niece who remained the one bright and calming star in his otherwise harried existence. Halifax suffered an acute attack of pleurisy early in May and succumbed a few days later at the age of fifty-four. Whether Catherine was by his side at the end is not known, but it seems virtually certain that she had been a member of Halifax's household for some time. The lengthy codicil added to his will on February 1, 1713, now came into play. Nullifying a former bequest to his companion and mistress, it granted Catherine the sum of £5,000, a life interest in the rangership and lodge of Bushy Park, and additional proprietary interests with an aggregate value of some £25,000. The grieving beauty eventually returned to live in her uncle's house off Leicester Fields, serving as an old man's comfort and buffer against the riptides of contention and the winds of adversity that swirled without respite through his hostile world. Halifax was not forgotten, as Newton kept a portrait of his departed friend and political benefactor in his chambers. Neither was he willing to forget the fortune left to his niece by the generous peer in appreciation of what a sneering Flamsteed termed "her *excellent conversation*." Newton spent much of the following year defending Catherine's legal interests against the encroachment of Halifax's nephew, the only other major heir, George Montague, but with little apparent success. Montague, who was created Viscount Sunbury and Earl of Halifax on June 14, 1715, received under the Great Seal all offices held by his uncle when he died, including the rangership of Bushy Park. And while the record is less clear on the subject, it appears unlikely also that Miss Barton inherited Apscourt Manor, Halifax's country estate.[16] In eighteenth-century England mistresses possessed no more rights than second inventors.

A few weeks before Halifax's passing a delegation of French scholars arrived in England to observe the solar eclipse of April 22. They were accompanied by an Italian cleric of noble birth, the Abbé Antonio-Schinella Conti, who, being interested in the work of Newton and his circle, decided to stay on. Conti had spent the previous two years in Paris, where he became acquainted with the leading figures of French science, including Fontenelle and Malebranche, whose philosophy he particularly admired. It was probably through Malebranche that Conti gained a written introduction to Leibniz, to whom he wrote on the eve of his departure for London. He pledged his continuing support in the conflict with Newton, an encouraging gesture to the German philosopher, who was languishing in virtual isolation back at the deserted court in Hanover. About to enter the lion's den, Conti had heard, and expected, the worst of Newton, and he wisely resorted to the strategy of ingratiation that had served him so well in Paris, a delicately balanced concoction of charm, flattery, and ungrudging deference. Conti, to his lasting surprise, found Newton not only tolerable but quite likable. What is more, Newton both liked and admired the urbane newcomer, at least in the beginning. On August 30, 1715, the Italian wrote an ecstatic, almost childlike, letter to his good friend Pierre Rémond de Montmort in Paris:

> I go to Newton's house three times a week and when I return to Paris I know that you will be pleased both with him and with me. You have no idea how learned he is in ancient history and how reasonable and accurate are the reflections he makes on the facts. He has read much and meditated a great deal on the

> Holy Scriptures, and he speaks about them with great wisdom and good sense, stripping the words of their allegorical meaning and reducing them to history.[17]

Whether the breathless Conti was actually visiting Newton as frequently as he claimed may perhaps be questioned, but the fact that he was elected a Fellow of the Royal Society in November under Newton's sponsorship speaks for itself, as does Newton's willingness to reveal a great secret to the Abbé. "I was myself when young a Cartesian," he confessed, but he hastened to add that the light had soon dawned, for Cartesian metaphysics ultimately proved to be nothing more than a "tissue of hypotheses."

Enjoying the best of both worlds, Conti was being courted no less ardently by Newton's embittered rival. An intimate friend and adviser of Caroline of Anspach, who, on the accession of George I, became Princess of Wales as a result of her marriage to the Crown Prince, Leibniz arranged for Conti to meet the future queen and the rest of the court. Leibniz also warned the Princess to take care lest her uncomplicated faith be compromised by the irreligious English. She must be particularly wary when confronting the atheism of Hobbes and the materialism of Newton, whose *Principia* had seemingly reduced the Creator of the universe to the role of a master mechanic, and a rather bumbling one at that. Caroline was soon expressing her appreciation of Conti's incisive wit and becoming an enthusiastic collaborator in his recently hatched grand design. Like Chamberlayne before him, Conti, who had very little hope of gaining lasting fame in his own right, dreamed of the glory that would be his if he only could heal the scandalous rift between Leibniz and Newton.

That dream took a serious blow at year's end. Leibniz sent Conti a letter in December 1715, to which he added a lengthy postscript clearly intended for Newton's eyes, for the German philosopher was done with Keill. (Amid unfounded rumors that Keill frequented taverns and bawdyhouses with students entrusted to his care, Leibniz, who possessed a puritanical streak almost the equal of his rival's, had written Christian Wolf in May that, "I wish to have no dealings with a man of that sort." Besides, "Keill writes like a bumpkin.") The postscript, or *Apostille* as it became known when first published in Joseph Raphson's *History of Fluxions*, began on a sarcastic note: "I am thrilled that you are in England; you will profit by it as it is true that there are able men there; but they wish to pass as almost the only inventors, and that is apparently what they will not succeed in doing."[18] Leibniz then reviewed the calculus controversy, restating his earlier assertion that Newton had not been master of "the Essence and the Algorithm of the infinitesimal method before me." He also decided that he could no longer keep Bernoulli's identity a secret and named the Swiss mathematician as one who could bear witness to his priority. Bernoulli, who was greatly upset by Leibniz's disclosure, should have known that it would happen one day. Indeed, as previously noted, he had been forewarned by Leibniz in June 1713 that his mask of secrecy must soon be discarded. Taking no chances with his faint-hearted ally, Leibniz had cleverly tightened his grip on Bernoulli by extensively quoting the "leading mathematician" in the *Charta Volans*. Newton now realized that Bernoulli and the leading mathematician were one and the same individual, as Leibniz knew he would. Whether at this juncture Newton was more angry with Leibniz or with the traitorous Bernoulli is anyone's guess.

The *Apostille* challenged Newton on other equally sensitive grounds. In January 1715 Bernoulli had written a pompous letter to Leibniz urging him to publish a *Commercium Literarium* as an antidote to the poisonous *Commercium epistolicum*. "For as the English try to prove everything by letters and narratives of events ... you should disclose yours ... left out by Newton's toadies or craftily suppressed." He also urged Leibniz to issue yet another mathematical challenge to Newton: "You would do well to publicize some [problems] where Newton would, as you know, find himself in difficulties."[19] After giving the matter much consideration, for this strategy had backfired on two previous occasions, Leibniz, who nurtured the dream of dispatching his opponent with a single crushing blow, finally decided to test the old maxim that the third time is a charm. His challenge, which otherwise became known as "feeling the pulse of the English," invited the solution to a complex problem involving intersecting curves. To his considerable dismay and embarrassment, Leibniz was careless in his statement of the problem, implying that he would be satisfied with a solution for one special case rather than the general method he truly had in mind. Word spread quickly among Newton's disciples in the mathematical community, who closed ranks to defend their hero's honor and that of their nation. Keill, Halley, Pemberton, and Brook Taylor all supplied answers. Even James Stirling, a promising student at Oxford, entered the fray. Newton's own attempted solution, which was published anonymously in the *Philosophical Transactions*, proved remarkably feeble, a graphic indication that his once magnificent powers were fast ebbing away. Frustrated by the English repulse, Leibniz was forced to turn to Bernoulli for aid in formulating a more suitable problem (which was eventually sent to England in another letter to Conti), evidence that the ailing German's days as a leading mathematician were also at an end. Having already submitted their answers before the revised challenge problem arrived, the English mathematicians collectively cried foul, an embarrassment for the already overmatched Leibniz. Newton's failure to solve the problem as originally stated never became known to his rival, and Bernoulli was again cited for his devious and cowardly ways. The two old lions found themselves fighting over a piece of meat that neither possessed the teeth to chew.

III

"The more I consider the Postscript of Mr Leibniz," Newton wrote in one of several drafts of a letter to Conti in February 1716, "the less I think it deserves an answer. For it is nothing but a piece of railery from beginning to end."[20] But the Venetian cleric, as was to be expected of an ambitious courtier whose greatest gift was a clever and facile tongue, was not about to sacrifice so juicy a morsel on the altar of discretion. Indeed, he had already written of the matter to his friends the Lady Mary Wortley Montague and Rémond de Montmort. More important, the Abbé had dined with the King of England, to whom he revealed everything he knew about the quarrel between Newton and Leibniz, including details of the latter's most recent communication. The events that resulted from this exchange contributed significantly to a revision of Newton's initially favorable opinion of the urbane Italian. "[T]he Postscript was shewed to the King," Newton ruefully noted in the draft of a letter to Pierre des Maizeaux in August 1718, "& I was pressed for an answer to be also shewed to his Majesty, & the same afterwards sent to Leibnitz."[21]

Whether Conti was quite the culprit Newton later made him out to be is debatable, since it would appear that Newton, who was no less eager than Leibniz to curry Hanoverian favor, saw this as a golden opportunity to strike a decisive blow of his own. As usual, his *modus operandi* was deceptively simple yet demonically clever. According to Conti, Newton prevailed upon him to help assemble a number of ambassadors and ministers at Crane Court for the purpose of discussing the priority dispute and subjecting the documents quoted in the *Commercium epistolicum* to a detailed examination. As the only true master of the literature, Newton planned to orchestrate the outcome in a repeat performance of that given in 1712, when his hand-picked committee of Royal Society Fellows had done his bidding without complaint. On this occasion, however, a dissonant chord was struck by the Baron von Kilmansegge, Master of the Horse to George I and husband of the King's mistress, the Countess von Kilmansegge, who was indelicately dubbed the "Elephant and Castle" for her immense bulk and girth. Leibniz had plied that human pachyderm with flattering missives from Hanover, successfully characterizing himself as the proud champion of the German cause against the mighty English. When the elephant trumpeted, the Master of the Horse had little choice but to listen, and he informed the distinguished assembly that Newton's efforts to settle the issue were not satisfactory. (Newton, too, had wooed the influential Baroness by arranging a private demonstration of his optical experiments in her behalf.)[22] The best way to end the quarrel would be for Newton to write directly to Leibniz. The others approved, and the King, who was told of this recommendation, lent his support as well. His strategy thwarted, Newton had no choice but to comply. Fortunately for him the news was considerably better on another front. When Princess Caroline visited the Royal Observatory in June, Flamsteed informed her that Sir Isaac Newton was a great rascal who had stolen two stars from him. The Princess could not reply, she informed Leibniz, for giggling to herself. And what did this royal personage, who has been likened to a heroine fluttering in the box seats, understand of the motives that drove men like Newton and Leibniz? Quite obviously, more than women of her day were supposed to: "But great men are like women, who never give up their lovers except with the utmost chagrin and mortal anger," she wrote to Leibniz on April 24. "And that, gentlemen, is where your opinions have got you."[23]

Newton's letter to Leibniz took final form within days of his embarrassing setback at Crane Court, the scene of so many personal triumphs in the past. Dated February 26, 1716, it was addressed to Conti, who self-servingly brandished it at court for a month before sending it to Hanover with a covering letter of his own. Blessed are the peacemakers. Still, such notoriety did not prove all bad from Newton's point of view. King George, not noted for his diplomatic skills, read the letter and virtually pre-empted anything Leibniz might write in reply by remarking that the explanations were lucid and the facts difficult to answer. Leibniz took a rather more jaundiced view of the legalistic-sounding missive, terming it the *Cartel de Défi*. In sum, the letter presented nothing new in the way of factual evidence and may be fairly characterized as a précis of the *Commercium epistolicum* and its more recent sibling, the "Account." It doubtless also contained the essence of Newton's remarks given before Baron Kilman-segge and the other dignitaries, which, in addition to his deep familiarity with the issues, would explain why he was able to draft the document in such haste. But while Newton could be made to write a letter to his hated rival, he could not be made to temper his remarks by any man, the King of England included:

> But as [Leibniz] has lately attaqued me with an accusation which amounts to plagiary: if he goes on to accuse me, it lies upon him by the laws of all nations to prove his accusation on pain of being accounted guilty of calumny. He hath hitherto written Letters to his correspondents full of affirmations complaints & reflexions without proving any thing. But he is the aggressor & it lies upon him to prove his charge.

This after Newton, without so much as batting an eye, had repeated what we now know to be an absolute falsehood, "that the Commercium Epistolicum contains the ancient Letters & Papers . . . collected & published by a numerous Committee of Gentlemen of severall nations appointed by the R. Society for that purpose." How dare Leibniz complain "of the Committee . . . as if they had acted partially in omitting what [was] made against me."[24] One wonders if Newton himself was truly capable any longer of separating fact from fiction, if dream and reality had not finally merged into a twilight fog of perpetual self-delusion.

Newton was not going to let Bernoulli go unpunished either. Though he did not mention the Swiss by name, now that Bernoulli's identity was known to him he reduced the "leading mathematician" to the rank of "pretended mathematician," a demotion calculated to exact the stiffest penalty of all—a loss of respect both in Newton's eyes and in those of the scientific community in general. Leibniz immediately informed Bernoulli of this development, presumably to convince him that the point of no return had at last been reached. "The letter [quoted by Leibniz in the *Charta Volans*] which he knows is yours, he says is written by a ... pretended mathematician ... as if he were ignorant of your work. He calls the whole paper in which your letter is inserted defamatory." Bernoulli replied on May 9. Trying not to show his anguish, he began on a positive note. "It is a good thing that Newton has at last entered the ring himself, in order to fight under his own name, and laid aside his mask," which was more than Bernoulli could say for himself. After raising a number of technical matters, none of which were very important, he broached the subject that most concerned him: "I wonder how Newton could have known that I was the author of that letter ... when actually no living soul knew that I had written it except you to whom it was written." Hoping against hope that Leibniz had not betrayed him, Bernoulli, like a fated mortal in a Sophoclean drama, grasped at every straw. Perhaps the expression "pretended mathematician" possessed "a meaning other than you think: for it could even be read as if Newton had believed that the letter itself was not genuine, but written as the work of some fake mathematician." Desperate, Bernoulli was later to deny explicitly to Newton the authorship of the disputed letter, but Newton, as he wrote in the appendix to Raphson's *History of Fluxions,* knew better: "The Letter was published in a clandestine, back-biting manner (as defamatory Papers use to be) ... and was dispersed above two Years before we were told that the Mathematician was *John Bernoulli*." He left the faint-hearted traitor to twist slowly in the wind until it suited his purpose to effect a tepid reconciliation well after Leibniz had died.[25]

As great controversies are wont to do, the quarrel between Newton and Leibniz spread over vast territories only marginally connected with the original *casus belli*. Prior to her arrival in England, Princess Caroline had been both a close friend and a disciple of Leibniz. They continued that relationship by corresponding on a regular basis after the Hanoverian succession, but Leibniz was deeply concerned as to the resolve of his distinguished pupil, and for very good reason. Another philosophical mentor was vying for the future queen's allegiance, meeting with her weekly and frequently bringing with him other men of equally dangerous persuasion. His name appeared in virtually all of her letters written after October 1715. "I have talked to-day to the Bishop of Lincoln about the translation of your *Theodicy*," she wrote on November 14; "he assures me there is no-one capable of doing it except Dr. Clarke, whose books I sent you." Clarke, the Princess added for Leibniz's information, was a close friend of Chevalier Newton. She wrote again two weeks later with the news that "Dr. Clarke is too opposed to your opinions to do it; ... he is too much of Sir Isaac Newton's opinion and I am myself engaged in a dispute with him."[26] She accused Clarke, Newton's hand-picked choice to deliver the Boyle Lectures of 1704 and 1705 and the translator of the *Opticks* into Latin, of "guild[ing] the pill" for his unwillingness to admit that Newton held the philosophical opinions Leibniz ascribed to him after having read his scientific works. Still, the Princess always looked forward to Clarke's return each week and, as Leibniz feared might happen, gradually became a convert to a number of the theologian's views, which were also those of Newton.

Meanwhile Leibniz wrote the Princess to express his deep concern about the alarming spread of Newtonian thought: "Natural religion itself, seems to decay (in England) very much. Many will have human souls to be material: others make God himself a corporeal being." What is worse, "Sir Isaac Newton says, that space is an organ, which God makes use of to perceive things by." Finally, Newton and his followers argue that "God Almighty wants to wind up his watch from time to time: otherwise [the universe] would cease to move." How odd that a supposedly perfect God should construct a machine of such imperfect design "that he is obliged to clean it now and then by extraordinary concourse, and even to mend it, as a clockmaker mends his work."[27] The calculating Caroline, who had joined Conti's conspiracy to effect a fruitful dialogue between the two sides, turned Leibniz's letter over to Clarke, and thus began one of the most celebrated of many famous eighteenth-century philosophical debates.

As Leibniz described it to Bernoulli in the summer of 1716, "Clarke sent the sermon [his formal reply] to me, I responded, he replied; I duplicated, he triplicated; I have just quadruplicated; that is, I have answered now his third letter."[28] In all, the exchange lasted through five rounds. Each successive letter (there were ten) was longer than the last; each was less focused and penetrating. Clarke began by refuting Leibniz's charge that God needs any sense organ whatsoever to command the universe, and further argued that the Creator's periodic interventions were simply part of the original divine plan. They ended the exchange a year later by debating such concepts as the nature of time and space, with Leibniz claiming that Newton's belief in the vacuum was both philosophically and scientifically absurd. Clarke countered by citing Guericke's experiments with the air pump and Torricelli's work with mercury barometers. Princess Caroline, amid repeated assertions of her hope that Newton and Leibniz might yet be reconciled, gave some indication of how the debate was going, at least from her unique perspective. "Mr. Clarke's knowledge and his clear way of reasoning have almost converted me to believing in the vacuum," she wrote to a seriously ailing and deeply disheartened Leibniz on April 24, 1716. Three weeks later she wrote of going to see experiments on colors and of having just seen another on vacuums, which again "has almost converted me. It is for you to lead me back into the right way, and I await the answer which you make to Mr. Clarke." And, as if Leibniz hadn't guessed, she added that none of Clarke's replies were written without Newton's advice.[29]

Indeed, Clarke was Newton's own minister at St. James's, Piccadilly, where they worked together on various charities. The clergyman had only to walk round the corner to consult with the master whenever he chose, which one senses was quite often.

How long this increasingly unproductive exchange might have continued had not Leibniz's health failed is impossible to say. Racked by gout from head to foot, the philosopher, who was almost constantly bedridden now, was seized with an especially acute attack of the arthritic disease in late October. He resorted to the pharmacopeia of a Jesuit at Ingolstadt, but it was too late. He died on November 4, having just reached the Biblical age of seventy. Conti, who had left England for Germany in the hope of meeting the great man face to face, arrived too late. He wrote to Newton from Hanover on November 29: "M. Leibniz is dead; the dispute is finished."[30] Conti, subsequent developments would soon reveal, was only half right.

Chapter Twenty-One

Infinity

The created world is but a small parenthesis in eternity.
Sir Thomas Browne

I

In the summer of 1717 John Conduitt, scion of a wealthy Hampshire family, became a familiar face at Leicester House, Newton's comfortable Westminster residence. While there is no doubt that Conduitt was fascinated and awed by the great man, he came not to speak with Newton of science, mathematics, religion, or even of the day-to-day operations of the Mint, matters that would one day dominate their frequently long and occasionally enlightening conversations. Preoccupied for the moment with less mundane concerns, Conduitt came to 35 St. Martin's Street for the purpose of wooing the still beautiful Catherine Barton, whose period of mourning for Halifax had come to an end. At thirty-eight years of age Catherine was nine years older than her suitor, a marked contrast to the deceased Halifax who had been her senior by eighteen years.

Very little is known of Conduitt's early life or of the circumstances that led to his acquaintance with Newton and his niece. Born and baptized in London in March 1688, he attended Westminster School and in June 1705 entered Trinity College as a Westminster scholar, a step that may have provided his first indirect link with the erstwhile Lucasian Professor. Conduitt left Cambridge before taking a degree and is thought to have traveled for some time on the Continent. When next heard of, in 1711, he was serving as judge advocate with the British Army in Portugal and as secretary to the Earl of Portmore, the commanding general during the War of the Spanish

Succession. The following year Conduitt was made captain in a regiment of dragoons, and in 1713 he became commissary to the British forces at Gibraltar, an appointment that seems to have added considerably to the proceeds from a generous inheritance. While in Spain he succeeded in identifying the site of Carteia, a lost city once occupied by the Romans. Word of this significant discovery was communicated to the Royal Society in December 1716 and again in March 1717. Three months later, when the twenty-nine-year-old venturer was back in England for good, he was invited to read his paper on Carteia before the Royal Society on June 20.[1] Newton, who was working on a chronology of ancient kingdoms, may have taken Conduitt home for a private discussion of the Mediterranean antiquities so recently viewed by his younger and sharper eyes. If indeed Conduitt and Catherine did not meet until that summer, theirs was a classic whirlwind romance. The Hampshire gentleman and the slightly disreputable London beauty were wed on August 26. Respectable at last, Catherine bore the couple's first and only child, a daughter, two years later. They christened her Catherine—Kitty for short—and when she married in 1740 it was to Viscount Lymington, eldest son of the first Earl of Portsmouth. Their son became the second Earl and inherited from his mother the papers of Sir Isaac Newton, the bulk of which were presented to the Cambridge University Library by members of a later generation of the family.

According to tradition, the Conduitts lived with Newton after their marriage. While there is little question that they stayed at Leicester House for extended periods when they were in London, Catherine's letters suggest that much of their time was spent at Cranburg Park, the ancestral home of her devoted husband.[2] She may have sensed that the hitherto warm relations with her uncle might cool if she attempted to bring an infant daughter into the household of an aging bachelor who still demanded the daily solitude of a monastic. (The Conduitts later purchased a home of their own in George Street, Hanover Square.) Conduitt, who represented Whitchurch in Parliament, probably visited Newton often when he was in the city; it could have been on such occasions, over quiet dinners, that he collected much of the information for his projected biography. The book was never written, but his undigested mass of papers and notes, hagiographical though they are, provide some rare glimpses of Newton in his more unguarded and expansive moments. When, after succeeding Newton as Master of the Mint, Conduitt died on May 23, 1737, he was buried, as he had arranged, near his hero in Westminster Abbey. There the three men who loved Miss Barton most (Halifax, too, was interred in the Abbey) waited for their fair Catherine to join them, which she did in 1739 at the age of sixty.

The marriage of Catherine took place at the end of the lull in the priority dispute following Leibniz's death. Most men in Newton's position would have exercised discretion and taken advantage of the opportunity to bury the hatchet along with the enemy, but Newton was nothing if not atypical. When Bernard le Bovier de Fontenelle's *Eulogy* of Leibniz appeared on behalf of the French Academy, Newton was disturbed because his rival was exonerated of plagiarism. ("I reccon that Mr Fontenell was not sufficiently informed," he wrote in an unsent letter to Varignon).[3] Taking pen in hand, he attempted again to set the record straight in a series of extensive notes, which, mercifully, never saw the light of day. Dissatisfied with the result, he composed several drafts of yet another history of the fluxions, but it too remained unpublished and forgotten.[4]

Newton's reaction to Bernoulli's latest polemic on the priority dispute was rather less circumspect. Published anonymously and titled "Letter Written on Behalf of the Eminent Mathematician, Mr. Johann Bernoulli, against a Certain Adversary from England," the essay appeared in the *Acta Eruditorum* for July 1716 but did not reach Newton's hands until the following spring. The adversary to whom Bernoulli referred in the title was none other than Newton's lackey John Keill, who immediately deduced the author's identity from an editorial oversight. Though the article had been written in the third person, an unnoticed "m"—*meam* for *eam*—gave Bernoulli away. It was not until a year later that the error was called to his attention. Deeply embarrassed, Bernoulli wrote to ask for a correction and even had his son spread the word that the author was someone other than himself, adding an element of farce to an increasingly virulent form of trench warfare. Bernoulli's denial proved futile, of course, and an outraged Keill wrote Newton on May 17, 1717, that "I was amazed at the impudence of Bernoulli[.] I believe there was never such apeice for falshood malice envie and ill nature published by a mathematician before."[5] Newton was inclined to agree. Bernoulli had cast aspersions not only on Keill's mathematical ability but on his own as well by reiterating an earlier claim that Newton had not known of the calculus when he wrote the *Principia*.

As usual, Keill let it be understood that he would entertain whatever suggestions Newton might care to make concerning the inevitable reply. When Keill's draft reached him, Newton went over it with a fine-toothed comb, rewriting several of the longer passages and modifying others. He accused Bernoulli of participating in a conspiracy with Leibniz to share the credit and the glory for the respective discoveries of the integral and differential calculus, "cit[ing] your self as a Witness for your self & for him." In the future, "whenever I meet with such anonymous papers wherein you are applauded or cited as a witness or your enemies abused I shall ... look upon them as written by your self or at least by your procurement."[6] For unknown reasons, Newton held on to Keill's paper for two years. Perhaps even he was growing weary of a war neither side could win; perhaps his friends urged restraint now that Leibniz was dead. Only when Bernoulli provoked him anew did he release the broadside for publication in the *Journal Literaire*.

A reluctant warrior under the best of circumstances, Bernoulli panicked on the death of his leader. Fearing that he might now become the sole target of Newton's wrath, the Swiss mathematician wrote a letter to Rémond de Monmort, their mutual friend in Paris, who sent it on to Brook Taylor in England for Newton's perusal. Bernoulli was not yet aware of the self-incriminating printing error in his recently published "Letter"; he thus affected a tone of righteous indignation at the thought that Newton should think him capable of duplicity. "I desire nothing so much as to live in good friendship with him, and to find the occasion to make him see how much I value his rare merit; in fact I never speak of him without much praise."[7] While it was true that he had helped Leibniz formulate the challenge problem of 1715, he had done so only with the greatest reluctance.

If ever a strong rebuke was in order, this was the time, but Newton stayed his pen once more. The strain of not knowing when or where Newton or one of his followers might strike proved too much for Bernoulli. Early in 1718 he wrote Newton a second letter of appeasement via Montmort.[8] It too was met with stony silence.

Meanwhile, Pierre Varignon, a distinguished professor of mathematics and a member of the *Académie Royale*, was working behind the scenes to effect the truce that neither Chamberlayne nor Conti, for all their good intentions, had succeeded in bringing about. Twice he sent copies of the *Opticks*, one in Latin, the other in English, to Bernoulli as though instructed to do so by its author, while to Newton he wrote that Bernoulli viewed the loss of his friendship as "a very unhappy accident." Having succeeded in convincing Bernoulli that all was not lost, Varignon persuaded him to write directly to Newton. The letter, dated June 24, 1719, carried a heading of such transparent design that its recipient must have scoffed in disdain: "Johann Bernoulli presents a grand salute to the most illustrious and incomparable Isaac Newton." As before, Bernoulli insulted Newton's intelligence by casting himself in the role of the innocent victim:

> Beyond all doubt they are mistaken who have reported me to you as the author of certain flying sheets [a reference to Bernoulli's anonymous letter in the *Charta Volans*] in which perhaps you were not treated with sufficient respect; but I earnestly entreat you, famous Sir, and implore you in the name of everything sacred to humanity, that you convince yourself thoroughly that whatever has been published anonymously in this way is falsely imputed to me. For it has not been my custom to issue anonymously what I neither wish nor desire to acknowledge as my own.[9]

One's sympathy for Newton would be much greater were it not for the fact that his own anonymous indictment of Leibniz via a packed committee of the Royal Society was easily as reprehensible as the aggregate of Bernoulli's sins.

The first draft of his reply called Bernoulli's veracity into question. After all, no less a witness than Leibniz himself had identified Bernoulli as the "eminent mathematician" quoted in the *Charta Volans*. But as Newton pursued his usual custom of composing multiple drafts, his combative spirit waned. Age, more than any man, was his mortal enemy now. Approaching seventy-seven, he would overlook Bernoulli's hypocrisy and make peace. "But since from the letter now received from M. Bernoulli, I understand that *he is not the author of it* [the letter in the *Charta Volans*], I readily welcome and court his friendship on which account I have written the enclosed letter to him." Hardly a stranger to hypocrisy himself, Newton freely mingled fact with fiction in his reconcilatory missive: "Now that I am old I take very little pleasure in mathematical studies, nor have I ever taken the trouble of spreading opinions through the world. . . . For I have always hated disputes."[10]

Having come this far, Newton was prepared to pay an additional price in order to keep the fragile truce with Bernoulli from crumbling. He attempted to stop the publication of Des Maizeaux's forthcoming *Collections* (*Recueil de diverses pièces sur la religion naturelle, l'historie, les mathematiques & c*), which contained unflattering references to Bernoulli's authorship of the anonymous letter quoted by Leibniz in the *Charta Volans*. In November 1719 he "begged" Des Maizeaux by letter to postpone publication, offering twelve guineas to the bookseller "if he pleases to defer publishing . . . till Lady Day next." De Moivre, who was quite close to Newton during this period, later informed Varignon that Newton had offered to buy out the work entirely, thus ensuring Des Maizeaux a handsome profit.[11]

Anticipating the worst, Newton, in a good faith gesture, alerted Bernoulli to the imminent publication, as did Varignon. Instead of thanking Newton for his concern, Bernoulli drafted a haughty reply in which he suggested that Newton could have dissuaded Des Maizeaux if he had truly wanted to. "The world of learning . . . will sharply regret that you are not able to undertake mathematical studies, now that you are an old man," he added gratuitously. Bernoulli had also heard via a friend that his name had been stricken from the rolls of the Royal Society. "Did you so flatteringly procure for me a fitting place in that illustrious Society, without my soliciting it, in order afterwards to expel me in so disgraceful a manner?"[12] Newton, who had rarely been more conciliatory, had done nothing of the kind. Still, Bernoulli, knowing of Flamsteed's expulsion at Newton's hands, may have thought that his unnamed friend, who had recently visited London, was telling the truth.

The stiffly polite draft of Newton's reply, which ended in mid-sentence, was never sent. What is more, he now gave the hot-blooded Keill permission to proceed with the publication of his long-delayed riposte to Bernoulli's anonymous and defamatory "Letter Written on Behalf of the Eminent Mathematician." A soured Conti, whose efforts at peacemaking had won him nothing but the enmity of both camps, wrote cynically from the sidelines that the false truce made him laugh.[13]

Despite ample forewarning Bernoulli greeted Des Maizeaux's *Collections* with the outrage of one betrayed. While he continued to deny that he was the unnamed mathematician quoted by Leibniz, Bernoulli protested that Newton had called him a pretended mathematician, a novice, and most insulting of all a "Chevelier errant." Only if Newton made a public retraction would the difficult Bernoulli be willing to forgive and forget. A distressed Varignon reluctantly asked de Moivre to approach Newton about writing such a letter. Mincing few words, Newton, now an octogenarian, declined on the grounds that he could not be certain how Bernoulli would use such a document. De Moivre observed that he was too fearful of Newton to press him further.[14] Varignon, the last of a triumvirate of failed peacemakers, died in December 1722, and de Moivre absolutely refused to have anything further to do with Bernoulli and his outrageous demands.

Newton had one last trump to play. He revised the *Commercium epistolicum,* to which he appended the anonymous letter from the *Charta Volans* along with a detailed rebuttal of it. Although he did not mention Bernoulli by name, few of his more perceptive readers could have mistaken the Swiss mathematician for anyone else. In February 1723 Bernoulli himself made one final approach to Newton, having received a copy of the *Opticks* recently published at Paris: "[W]hat you have most successfully discovered . . . concerning light and the system of colours, has in me a very great admirer; indeed, it is a discovery more enduring than any bronze [monument], and one to be prized by posterity more than it is now." Having said this, Bernoulli shamelessly broached the matter of Des Maizeaux's *Collections* once again: "If you should consider me worthy of a reply, I would very gladly learn from your pen what you decide to do, in order to defend your innocence."[15] Newton most certainly did not consider Bernoulli worthy, nor did he propose to do anything further. His unbroken silence marked the end of the calculus controversy, at least on the public level. Newton's private thoughts, as always, were another matter, however, for the bitterness spawned by the decades-long dispute had eaten into his very soul.

II

On that January day half a century past when he had written Henry Oldenburg of the *experimentum crucis,* calling it "the oddest if not the most considerable detection wch hath hitherto been made in the operations of Nature," Newton had put the scientific world on notice that this discovery destined him for immortality. Despite many virulent attacks on his optical research over the years, never for a moment did he pause to reconsider seriously what he had so boldly declared to the Royal Society Secretary when still an obscure Cambridge don of twenty-nine. With death seeming little more than an arm's length away and an intellectual legacy that had grown to astounding proportions, Newton decided to take final stock of his lifework by publishing new editions of the *Opticks* and the *Principia*.

Any reader diligent enough to subject the first English edition of the *Opticks* (1704) to a page-by-page comparison with the new English edition brought out by Newton in 1717 would have had almost nothing to show for his taxing efforts. Scarcely a line appearing in the main body of the treatise had been revised by its author—a testament to his experimental genius as a young man, but also evidence that the physical stamina and intellectual acumen necessary for such a sustained effort could no longer be summoned up at will. Rather, Newton chose to concentrate on developing a new set of Queries (numbered 17 to 24) designed to reinforce his methodological position vis-à-vis that of his recently deceased rival, Leibniz. As he explained to Fontenelle in a letter written in the autumn of 1719: "Here I cultivate the experimental philosophy as that which is worthy to be called philosophy, and I consider hypothetical philosophy not as knowledge but by means of queries."[16] Or, as he had stated in the famous General Scholium of the *Principia*'s second edition, "*Hypotheses non fingo.*"

It will be remembered, however, that Newton had also concluded the second edition of the *Principia*, which had been edited by Cotes, on an ambiguous note. After rebuking Leibniz and his Cartesian followers for advocating certain unproved mechanistic principles such as the theory of vortices, he himself had written of an electric spirit that pervades and lies hidden in all gross bodies, a spirit that might account not only for such phenomena as heat, light, and muscular action in animals but for universal gravitation itself. Once again he took up that theme, expanding his thoughts on the matter in the new *Opticks*. "Is not this Medium," he postulated in Query 21, "much rarer within the dense Bodies of the Sun, Stars, Planets and Comets than in the empty celestial Spaces between them? And in passing from them to great distances, doth it not grow denser and denser perpetually, and thereby cause the gravity of those great Bodies towards one another . . . ?" And just how rare is this ether? "And therefore the elastic force of this Medium, in proportion to its density, must be above 700000×700000 (that is, above 490000000000) times greater than the elastick force of Air is in proportion to its density."[17] Under precisely what circumstances, if any, one is compelled to ask, had such astronomical figures been previously employed in the eighteenth century? Indeed, one wonders whether Newton was really serious when citing a figure of 490 billion or merely, in his years of decline, offering the arch-mechanists an additional sop in the hope of silencing them.[18]

As his revisions suggest, Newton remained concerned about the acceptance of the *Opticks* on the Continent, especially in France, where, aside from Hooke's assault, they had met with the greatest criticism from the likes of Huygens, Lucas, Gascoines, and Pardies. The first Latin edition

(1706) had won him relatively few converts, and he made arrangements for the publication of a new Latin edition, which appeared in 1719. A year later Pierre Coste's French translation was published in Amsterdam, and Varignon oversaw the printing of a second French edition in Paris in 1721.

Meanwhile Newton had served his own cause in other ways, equally practical. When a distinguished company of French savants came to England to view a solar eclipse in 1715, they visited the Royal Society, where its President arranged for J. T. Desaguliers to repeat several of the more critical optical experiments. The results were published in the *Philosophical Transactions* for April, May, and June 1716, together with a less edifying doctor's "*Account of what appear'd on opening the* big-belly'd Woman *near* Haman *in* Shropshire, *who was supposed to have continued many Years with Child*." The French returned home to declare the truth of something several of their countrymen had long denied, namely, that what Newton claimed to have discovered about the composition and behavior of light was absolutely correct. In 1721, after the publication of the second Latin edition, Sebastien Truchet, a Carmelite Brother and honorary member of the *Académie Royale*, at last successfully repeated Newton's optical experiments. "To confess the truth," the sequestered cleric wrote poignantly to Newton, "I envy my letter the journey it makes to you; how much more glorious to me would appear that day on which I might be allowed to enjoy a most welcome meeting and conversation with you, the wisest of men." Unfortunately, "it is not given to me to enjoy so great a pleasure and reward," owing to the religious life, "which prudence thrust upon me when barely sixteen years old."[19] Varignon echoed Brother Sebastien's glowing testimonial: "I have read the *Opticks* with the greatest delight, and all the more so because your new system of colours is firmly established by the most beautiful experiments."[20] When Varignon asked Newton to provide a sketch for an engraving of the French edition he published in 1721, Newton appropriately chose to illustrate the *experimentum crucis* first performed so long ago. He also wrote a typically brief but incisive caption: "Light does not change color when it is refracted." The French, on the other hand, like so many other early critics of Newtonian natural philosophy, were left with no choice but to change theirs.

Newton received a most painful reminder of his mortality early in 1722 in the form of kidney stones. Seriously ill at first, he was gradually nursed back to health by Richard Mead, his personal physician. By July he had recovered to the point where he could write to Varignon, who himself had less than six months to live, that "I am getting well little by little."[21] The moral of this agonizing ordeal was easily discerned. If a third edition of the *Principia* was to be published during his lifetime, the Fates must be tempted no longer.

In one sense preparations for the new edition had been under way for several years. Newton, as he had done after the publication of the first edition, continued inserting marginal notes on the interleaved sheets of a specially bound copy of the second. Yet at the age of eighty he found the task of revising the work without major assistance out of the question. Once again he sought the services of a young man who could share his burden as Halley and Cotes had done so admirably in the past. His choice fell to Henry Pemberton, who was twenty-eight years old when he made Newton's acquaintance in 1722. Details of their relationship are sketchy at best; indeed, exactly why Newton selected Pemberton, whom he had known for only a few months, to be editor of the projected crowning edition of his *magnum opus* remains something of a mystery, which the unfortunate disappearance of Pemberton's manuscripts and most of his correspondence has done nothing to solve.

Pemberton had spent the past several years at Leyden studying medicine under the gifted Hermann Boerhaave. While there he borrowed a copy of the *Principia*, which, to his surprise, he did not have too much difficulty understanding. He also developed fundamental skills as a mathematician and, after taking his medical degree, went to Paris for further study. There he purchased and mastered a large store of ancient and modern mathematical works. On returning to London Pemberton sought an introduction to Newton via Keill, but he was rebuffed by the aged lion. Shortly thereafter the fortuitous publication of a mathematical paper by a certain Professor Poleni of Padua worked a radical change in Pemberton's fortunes. Poleni had attempted to demonstrate that the theory of universal gravitation was invalid and that Leibniz's views on force were correct after all. Pemberton wrote an informed critique of the Italian's paper, which Mead, who was visiting the ailing Newton on a daily basis, passed on to his illustrious patient. Pemberton soon received the introduction earlier denied to him and just as quickly became a familiar face at Leicester House. "In a little time after," Pemberton wrote in the preface to his highly popular A *View of Sir Isaac Newton's Philosophy,* "he engaged me to take care of the new edition he was about making of his *Principia*." Pemberton further noted that even though Newton's memory was failing, he remained perfectly capable of understanding his own writings, which was contrary to the frequently repeated rumor that senility had begun to nibble away at England's finest mind. Pemberton had also been cautioned to be wary of crossing Newton, who, it was said, nurtured an inordinately high opinion of himself. He was therefore surprised to discover that "Neither his extreme great age, nor his universal reputation had rendered him stiff in opinion, or in any degree elated."[22] Both Conduitt and Stukeley reported much the same throughout their association with him. This seeming contradiction of character was in fact more illusory than real. Conduitt, Pemberton, and Stukeley posed no threat to Newton. Because they approached him as respectful admirers, they were treated as such. Only when Stukeley, without first gaining the President's approval, unsuccessfully offered himself for the post of Royal Society Secretary upon the resignation of Halley in 1721, did he incur

Newton's displeasure. "Sir Isaac show'd a coolness toward me for 2 or 3 years," Stukeley later wrote.[23] On the other hand, Whiston, who had quarreled with Newton, exaggerated only slightly when he called his erstwhile idol the possessor of "the most fearful, cautious, and suspicious temper" he had ever known, a judgment Hooke and Flamsteed would have been only too willing to confirm.

Indications are that the printer began work on the third edition of the *Principia* during the winter of 1723. Progress was slow to say the least, averaging only about twenty pages completed a month. Since both men were residents of London, Pemberton could visit Newton whenever necessary. He also wrote Newton thirty-one letters, mostly brief, only seven of which were dated. The absence of any replies suggests that when Newton did respond it was probably in the form of notes jotted in the margins of the proof sheets. A comparison of Pemberton's written suggestions with the finished work reveals that some of his proposed revisions were accepted and many were not.[24]

Two factors in particular combined to assure that the third edition contained no changes of great moment. In the first place Pemberton's modest talents were simply no match for the unusual gifts possessed by his distinguished predecessors, Halley and Cotes. A conscientious journeyman, the *Principia*'s third editor was neither very creative nor an especially accomplished mathematician. Such immortality as Pemberton might gain would be as an interpreter of Newtonian science rather than as an original thinker. At the same time Newton, who was nearing eighty-one when the printing began, could no longer perform original work on anything approaching a sustained basis. On those relatively few occasions when Pemberton raised significant issues he tended to brush them aside, usually without comment, casting considerable doubt on his editor's recollection that the "Remarks I continually sent him by letters on his *Principia* were received with utmost goodness."[25] The same treatment was accorded to Brook Taylor, William Molyneaux, and James Stirling, all of whom suggested revisions of a substantive nature, but to no effect. Nevertheless Leibniz, dead for several years, was far from forgotten. Newton composed his fourth and final rule of reasoning in philosophy with his old nemesis very much in mind.

> In experimental philosophy we are to look upon propositions inferred by general induction from phenomena as accurately or very nearly true, notwithstanding any contrary hypotheses that may be imagined, till such time as other phenomena occur, by which they may either be made more accurate, or liable to exceptions.

"This rule," he added pointedly, "we must follow, that the argument of induction may not be evaded by hypotheses."[26]

That Newton was still capable of striking terror in the hearts of his old associates is best proved by an incident bearing directly on the process of revision. Having asked Halley, then sixty-eight, to compute the place of the comet of 1680–81 for him, Newton received a response bordering on panic. "I was astonisht to find my self capable of ... an intollerable blunder," Halley wrote on February 16, 1725, "for which I hope it will be easier for you to pardon me, than for me to pardon my self, who hereby run the risk of disobliging the person in the Universe I most esteem. I entreat therefore that you would not think of any other hand for this computas, and that you please to allow me the rest of this week to do it."[27] While Newton did not turn to anyone else for assistance, neither did he keep faith with the unwaveringly loyal Halley, since the Astronomer Royal's revised computations on the comet were omitted from the third edition.[28] Being human and therefore falling short of the perfection Newton demanded, even the most faithful of his disciples sometimes failed him.

The slow production finally culminated in the appearance of the new edition in March 1726, by which time Newton had already passed his eighty-third birthday. Of the 1,250 copies issued, 50 were printed on superfine royal paper, 200 on royal general, and the remaining 1,000 on demy. A "richly bound" copy in morocco leather was presented to the Royal Society in Newton's name. The author also sent six copies to the *Académie Royale* in care of its *Secrétaire Perpetual,* Fontenelle, the man who had eulogized Leibniz in biographical form and would do the same for Newton little more than a year hence. Only as an afterthought did Pemberton, whose role was essentially limited to questions of style, receive mention in the preface. Newton's brief first draft contained no word of this "man of the greatest skill," a reference inserted in the second and final draft, which was more than had been accorded to Cotes. He also made Pemberton a generous gift of 200 guineas for his efforts, but Pemberton informed Dr. James Wilson, his first and only biographer, that the recognition meant far more to him.[29] Apparently it was recognition that carried limited clout with Newton. A young Benjamin Franklin was introduced to Pemberton by a mutual friend at Bateson's Coffee House, a favorite meeting place of physicians. Franklin wrote in his autobiography that his new acquaintance "promised to give me an opportunity some time or other of seeing Sir Isaac Newton, of which I was extremely desirous; but this never happened."[30] Forced to settle for second best, Franklin became acquainted with Hans Sloane, who became President of the Royal Society at Newton's death. There is a story, never confirmed, that Franklin set some of the type for Pemberton's *A View of Sir Isaac Newton's Philosophy,* printed by Samuel Palmer, for whom the still undiscovered genius was working in 1728.

III

In May 1724 Newton received a one-paragraph letter from one Guillaume Cavelier, a member of the famous Paris family of booksellers. A short manuscript attributed to him on the subject of ancient chronology had recently come into Cavelier's possession, although the bookseller did not explain how. He very much wanted to publish the little work but was afraid it might contain errors that would upset such a great man. Thus he respectfully requested that Newton provide him with a corrected copy at his earliest convenience. Newton, hoping his silence would quash Cavelier's plan, chose not to reply. The bookseller appealed to him a second time in October and was again greeted with silence. In March 1725 Cavelier wrote a third and final time. If he did not hear from Newton in the near future, "I will take your silence for your consent." Faced with this highhanded ultimatum, Newton drafted the briefest possible reply consisting of four meager sentences in May. It was true that he had once written out a chronological index for a friend, but only on the condition that it should not be circulated. Whether the manuscript in Cavelier's hands was one and the same he did not know, nor did he care to know. "I intend not to meddle with that which hath been given under my name, nor to give any consent to the publishing of it." Cavelier, who proceeded with the printing anyway, subsequently twisted the facts to absolve himself of wrongdoing by maintaining that Newton's demurrer had arrived too late.[31]

The bookseller sent the reluctant author a copy of the *Abrégé de la Chronologie de M. le Chevalier Isaac Newton*, which he received on November 11, 1725. The perfidy behind the conspiracy to publish the confidential manuscript—for Cavelier had not acted alone—so angered Newton that he penned a short essay titled, "Remarks upon the Observations made upon a Chronological Index of Sir Isaac Newton." The degree of his agitation can be gauged by the fact that he composed no less than seven drafts before settling on the one published in the belated *Philosophical Transactions* for July and August 1725. There, as in conversations with a few privileged admirers, he recounted certain elements of the curious events that had led up to this gross breach of confidence.

It began innocently enough in 1716, while Newton was talking with Princess Caroline on the subject of how best to educate her children. For some reason their conversation turned to history, at which time Newton revealed that he had been working on a new system of ancient chronology ever since his professorial days at Cambridge. The Princess, who thanks to the adroit coaching of Samuel Clarke had recently deserted Leibniz's camp for Newton's, requested a copy of the work. Not daring to provide a verbatim draft of the heretical manuscript, Newton stalled for time by arguing that his papers were faulty and disordered. Meanwhile, he hastily drafted a sterilized abridgement of the original variously known as the "Abstract" or "Short Chronology." The Princess, as Newton had hoped, seemed well satisfied, but he soon ran into potential trouble on another score. The Abbé Conti, who had once figured rather prominently in the waning stages of the calculus controversy, asked for a copy of his own. When Newton hedged, Conti persuaded the Princess to intercede on his behalf. Faced with little choice in the matter, Newton provided a second copy of the text but exacted a promise from the French cleric that he would reveal its contents to no one.

The years passed, and Newton became preoccupied with other matters. For Conti, who had failed in his attempt to heal the rift between Newton and Leibniz, the little manuscript was tangible evidence that he had once communed with the great. He carried the document back to Paris in 1716, and while he did not turn it over to a printer, he gossiped freely about its contents and accorded a number of intimates the privilege of reading what he had promised to keep secret. Among them was the Jesuit Father Etienne Souciet, an authority on ancient chronology. Though scarcely heretical, the "Abstract" gave the learned cleric a bad turn, and he wanted to learn more about the theoretical basis of Newton's startling revision of long accepted historical dates. Souciet's questions were dispatched to John Keill in 1720, and the faithful protégé dutifully laid them before Newton. The latter would only say that the manuscript derived from a much longer work and that his revolutionary system of dating major events had an astronomical and, by implication, a scientific basis. But since Newton refused to put his comments in writing or to elaborate further, Souciet wisely chose to let the matter drop.

When the chronology was published posthumously in full by John Conduitt in 1728 as *The Chronology of Ancient Kingdoms Amended,* Newton's method of computation became clear. Basing his system on the principle of the precession of the equinoxes, he calculated that the Argonaut expedition had taken place around 936 B.C., fully four or more centuries later than the traditional record of Greek history allowed. Using that data as a benchmark, Newton lopped off even greater chunks of time from the chronologies of ancient Egypt, Assyria, Babylonia, and Media. All ancient empires, he observed, vainly extended their antiquity. While their chronologers had calculated the average reign of a king or emperor at something over thirty years, the correct figure is eighteen to twenty. Only one ancient people, the Hebrews, escaped Newton's sword. Their written record, the Old Testament, was to him the earliest left by any civilization. Those compiled by other peoples had to be amended to conform to it. In the religion of Moses and the prophets lay the seeds of primitive Christianity, which all nations must one day embrace or perish. What Newton did not say—dared not say—is that the Christianity he had in mind could take only one form, the Arianism of the earliest Fathers.[32]

In 1724 Conti's unconscionable brandishing of Newton's manuscript in the scholarly salons of Paris led to the almost inevitable contretemps. He entrusted the text to M. de Pouilly of the *Académie des Inscriptions et Belles-Lettres,* who in turn allowed Nicholas Fréret, a scholar of ancient history, to copy the "Abstract" and translate it into French. After drafting a refutation of the document, Fréret arranged for their joint publication by the bookseller Cavelier. Not wishing to own up publicly to his duplicity, Fréret simply signed himself the Observator.

The anonymous Fréret and the scheming Cavelier were both pilloried by Newton in the "Remarks," and justly so. He had encountered Fréret's parasitic type many times before: "the Observator . . . hath undertaken to translate and to confute a Paper which he did not understand . . . to get himself a little Credit." As for Cavelier, the cat's paw of this conspiracy, did he not have the basic intelligence to realize that no sane man would have consented to the printing of an unseen translation of his papers, made by an unidentified person, with a rebuttal attached? Not surprisingly, Newton aimed his sharpest barbs at Conti, the former intimate who had violated what to Newton constituted an almost sacred trust. The man was a false scientist, a veritable Judas who, while "pretend[ing] to be my friend," was actually assisting Leibniz "in engaging me in disputes." As if that weren't bad enough, Conti had been responsible for stirring up criticism of the optical experiments on the Continent, even though "they have all been tried in France with Success." Let the man reflect on his sins, and let this latest be his last.[33] So much for the image of magisterial calm sketched by Henry Pemberton.

The fires of hell burned fiercely before the mind's eye of Isaac Newton. He believed implicitly in punishment everlasting, believed in it more than in happiness, whether earthly or eternal. In his last days he must have shuddered when reflecting upon the narrow passages and treacherous currents he had negotiated while growing up in a world populated by idolators and false prophets. Among the very few to have seen the light, this secret Arian had long since made peace with his God. Yet the frenetic pace at which he continued to pursue his religious and ancillary chronological studies until the very end belies the natural uneasiness of one who believed his convictions were about to be subjected to the ultimate test. Zachery Pearce, Rector of St. Martins-in-the-Fields, drew a compelling portrait of the old man a few days before he died:

> I found him writing over his *Chronology of Ancient Kingdoms*, without the help of spectacles, at the greatest distance of the room from the windows, and with a parcel of books on the table, casting a shade upon the paper. . . . He read to me two or three sheets of what he had written . . . which had been mentioned in our conversation. I believe that he continued reading to me, and talking about what he had read, for near an hour, before the dinner was brought up.[34]

When Stukeley had visited Newton on Christmas Day 1725, the savant's eighty-third birthday, the scene was very much the same. Newton showed his much younger friend a drawing of the plan for Solomon's Temple, which became the centerpiece of the *Chronology*'s fifth chapter. "He had a good notion," Stukeley wrote, ". . . that the *Divine* lays his mysterious plan of future things in the scenes of the Jewish temple and service." On Sundays "he turn'd over the sacred volumes with great diligence, and full conviction of the divine spirit that dictated them."[35]

Perhaps Newton also experienced secret pangs of guilt for having masked the unorthodox aspects of his religion. Rather than martyr himself as Fatio and Whiston had done, he turned a deaf ear to the entreaties of his two most fervent protégés and cast his lot with the theological Establishment. First, last, and always, he remained the heretic *sotto voce*. The man who had once secretly agonized over surrendering his Trinity College fellowship in order to escape the mark of the Beast now served as a trustee of the Golden Square Tabernacle and as a member of the committee established by Parliament to erect fifty new churches in the city of London. Until recently he had also sat on the committee charged with overseeing the completion of St. Paul's Cathedral. Only on rare occasions did subtle hints of Newton's private theology surface, as in 1716, when Joseph Morland, a physician and Royal Society Fellow, wrote him a touching letter as Morland lay dying. "I have done and will do my best while I live to follow your advice to repent and I pray often as I am able that god will make me sincere & change my heart. Pray write me your opinion whether upon the whole I may dye with comfort." Lest Newton think badly of him for betraying a confidence, Morland added: "This can do you no harm written without your name."[36] If Morland's distinguished but inordinately cautious correspondent drafted a reply, which seems unlikely, he violated a long-standing practice by keeping no copy of it. Still, so celebrated a personage as Newton could not prevent rumors of his heterodoxy from circulating. In 1720 he was accompanied by Stukeley to a sitting for one of three portraits by Godfrey Kneller. Stukeley listened with delight while Sir Godfrey, "who was not famous for sentiments of religion," sought to draw the great man out. Newton "answered him with his usual modesty and caution."[37]

Such considerations aside, religious study remained as always the commanding passion of Newton's life. While much has recently been written about the theological pursuits of his declining years, he had altered few of his fundamental precepts significantly since the 1670s, when he first traced the historical roots of his antitrinitarian credo in copious detail. As before, he focused his attention on the prophets, centering almost exclusively on the Revelation of St. John and the Book of Daniel. So far as he had been able to determine, the prophetic statements uttered by these extraordinary men had proved to be factually true down to the last detail. Using their works as a foundation, he composed perhaps the most important of the later theological papers, the *Irenicum: or ecclesiastical polity tending to peace*. His thesis, which exists in at least seven drafts, is a more elaborate version of the one so tentatively hinted at in the pirated "Abstract." Tracing the history of the Church back to the earliest days of Judaism, Newton wrote that all nations were originally of one religion based on the moral precepts of Noah's sons. This religion was passed on to the great Hebrew patriarchs Abraham, Isaac, and Jacob. Moses later carried it to Israel. The Greeks learned it through the wanderings of Pythagoras, who passed it down to his disciples on returning to Hellas, and it also spread to Egypt, Syria, and Babylonia. The two great commandments of this primitive religion were profoundly simple: to love God and to love one's neighbor as oneself. Thus Newton viewed the Jews and their early patriarchs not as another secular kingdom but as the progenitors of the true and everlasting Church, an ancient people apart from all others. To their rules of love he added a third principle foretold in prophecy: that Jesus was the long-anticipated Savior. Christ's incarnation was a signal for the establishment of a second covenant. He and his disciples had preached a simple message: Repent and be freed of sin. "This was the religion of the sons of Noah established by Moses & Christ & [is] still in force," he wrote.[38]

Once again Newton had reiterated his position that all things necessary for salvation were present from the beginning in the primitive church. And once again he had reasserted his belief that Christ is a lesser and therefore a subservient being to God, different from the Father both in substance and in nature. The doctrine of the Trinity had been formulated many years after the age of prophecy and therefore lacked divine authority. Christ Jesus was a very special man to be sure, but a man nonetheless. Drawing upon the teachings of St. Paul as an argument against the complex institution that the Church had become, he referred to the fundamental truths of primitive Christianity as "milk for babes." Ever in search of a villain, Newton accused St. Jerome of having inserted Athanasius' heretical doctrine of the Trinity into the Vulgate when he translated the then existing Greek versions of Scripture into Latin during the fourth century. What is more, he dismissed the hated saint with the very same words he had applied to Leibniz in the *Commercium epistolicum:* "For no man is a witness in his own cause."[39] While *Irenicum* may have stood for peace, its author remained unalterably opposed to the slightest accommodation with the idolatrous Church of Rome or with any form of Christianity that embraced the Trinity. Stukeley's seemingly straightforward observation that "Sir Isaac was an intire Christian, upon fundamental principles" concealed a meaning that few of their contemporaries could have imagined.[40]

IV

Isaac Newton's last years were a time of natural decline, much like the autumn days of a ripened apple soon to fall from the bough. Yet there is little evidence that he mellowed as he grew older or resigned himself to a world he could no longer hope to control. It was not his way to accept the limitations of age, to substitute serenity for purpose. While those who visited him most frequently, especially Conduitt and Stukeley, tried to por-

tray him as a cheerful, rather outgoing old soul, the anecdotes they compiled were often more suggestive of a querulous and dissatisfied octogenarian who frequently vented his disappointment with life upon the
lesser mortals around him. In many respects Newton's second childhood
was as miserably unhappy as his first. Still, he clung to life with a tenacity of
purpose rarely equaled, doing everything within his power to prolong it.
Perhaps he experienced a certain preverse delight as old rivals and younger
disciples alike fell by the wayside while he carried on. Aided by a strong
following wind, he had outjumped the other boys at Grantham School on
the day Cromwell died; now he experienced the ultimate triumph of outliv-
ing them as well. Perhaps, too, the renown he had won made it seem worthwhile to him to endure. The shy and introverted lad from Lincolnshire had
become an object of veneration and international pilgrimage.

Entrée to the great man was never granted indiscrimininately. As has
been noted, the young Benjamin Franklin did not receive the introduction
he sought, and neither did the equally admiring François Marie Arouet,
who wrote under the pen name Voltaire. The closest the iconoclastic savant got to Newton was an acquaintance with his niece Catherine Barton
Conduitt, who regaled her visitor with stories of her uncle's sometimes
peculiar habits. Voltaire repaid her by spreading scandal about her being
Halifax's onetime mistress.

One who did enter the Newton home regularly was William Stukeley, a
physician and native of Lincolnshire. Befriended by Richard Mead, as
Henry Pemberton had been, Stukeley first met Newton in March 1718,
when he was admitted as a Fellow to the Royal Society on Mead's recommendation. "From that time," Stukeley wrote, "I was well receiv'd by him,
and enjoy'd a good deal of his familiarity and friendship." Stukeley attributed Newton's longevity to his great prudence and naturally good constitution. Though not tall in stature, Newton was well made with a deep but
pleasant voice and a large chest for a man of his height. The protuberant
eyes, which had rendered him nearsighted in youth, had since grown
somewhat flatter, as is normal with advancing age. In 1725 Stukeley
watched transfixed as Newton added up a sheet of figures without the aid
of spectacles, pen, or ink. On the occasion of their last meeting, about a
year later, Newton, who was suffering from a bladder disorder that
rendered him incontinent, told Stukeley that his breakfast consisted of
boiled orange peel, sweetened tea, and a little bread with butter. He drank
more water than in the past and took wine only in small quantities at dinner.[41] Conduitt reported that Newton observed no particular regimen but
was very temperate concerning his diet, yet he still became plump in his
later years. He ate little flesh and lived chiefly on broth, vegetables, and
fruit, "of which he always ate very heartily." The bladder problems, which
were aggravated by painful kidney stones, forced him to give up his carriage
for good and to travel about in a slower but more smooth-riding sedan chair,
his arms dangling out through the openings on either side. His eyes were

lively and piercing, and his silver hair remained thick and full, a "venerable sight" when his peruke was removed. "And to his last illness he had the bloom and colour of a young man, and never wore spectacles, nor lost more than one tooth to the day of his death."[42]

Francis Atterbury, Lord Bishop of Rochester, differed with at least a part of Conduitt's appraisal: "In the whole aire of his face and make, there was none of that penetrating sagacity which appears in his composures. He had something rather languid in his look and manner, which did not raise any great expectation in those who did not know him."[43] Thomas Hearne, the scholarly gossip, seconded Bishop Atterbury's impression, calling Newton a man of disagreeable conversation and no promising aspect.[44] Consulting the many portraits in search of definitive evidence with which to settle this issue, one is confronted with equally contradictory views. A late painting by Kneller, when his subject was about eighty, presents the viewer with a wigless, slightly smiling, vacant-eyed old man, flirting with the prospect of senility. Yet three years later John Vanderbank saw him as a distinguished, impeccably groomed gentleman, eyes lucid and filled with purpose. The truth, as usual, probably lay somewhere between these two extremes, as did Stukeley's subtle description of Newton's demeanor. "But his natural disposition was of a chearful turn, when not actually engag'd in thought. He could be very agreeable in company, and even sometime talkative."[45] Like many of us, Newton had his moods, and these obviously varied depending upon the occasion, the company, and the subject being discussed.

Despite the new relationships with young admirers, the visits of foreign dignitaries, and the marriage of his favorite niece, the sustained output of Newton's quill pen suggests that he continued to live much as he always had, taking his meals alone and passing most of his days in his private study, reading and writing. His duties at the Mint, which were no longer as onerous as they once had been, were lightened even further by Conduitt, his successor as Master, who shouldered an increasing share of the administrative burden. For more than a year before his death Newton rarely visited the institution, or so Conduitt reported. (One thing is certain: Newton's inflexible morality with respect to coiners and clippers had not softened with age. When queried by Lord Townshend as to whether one Edmund Metcalf, who had been convicted at the Derby Assizes of counterfeiting, should be hanged as scheduled, he replied, "I am humbly of opinion that its better to let him suffer, than to venture his going on to counterfeit the coin & teach others to do so until he can be convicted again, For these people very seldom leave off.")[46] Newton attended the regular meetings of the Royal Society when he felt up to it, but his excellent record became tarnished after 1724 as illness took its toll. He rarely dined out any longer and entertained little at home, although he occasionally confounded those who knew him by turning up in unexpected places.

One evening in 1721 a gathering for natives of Lincolnshire was held at the Ship Tavern, Temple Bar. Stukeley reported that when he went into the upstairs dining room, where the better sort of company had gathered, he heard that an old gentleman seen downstairs was thought to be Sir Isaac Newton. The disbelieving physician went down immediately and, to his amazement, found Newton seated alone. When word reached those on the second floor that the rumor was indeed true, they asked the two distinguished companions to join them. "I answered, the chief room was where Sir Isaac Newton was. Upon which the upper room was immediately left to ordinary company, and the better sort came to us." The conversation turned to the subject of opera, which was then coming into vogue. When asked his opinion of the art form, Newton, who was otherwise reportedly fond of music, called it too much of a good thing. "'I went to the last opera,' says he, 'The first act gave me the greatest pleasure, The second quite tired me: at the third I ran away.'" On another occasion, during a discussion with Stukeley of Lord Pembroke's famous collection of classical busts and statues, Newton referred to them as "My Lord's old fashioned babys."[47] According to one source, his opinion of poetry was no better. When asked his thoughts on the subject, he is said to have replied, "I'll tell you that of Barrow; he said that poetry was a kind of ingenious nonsense."[48] Whether this anecdote is true or not, Newton's library was distinguished by the almost total absence of works by Chaucer, Shakespeare, Spenser, Milton, or his contemporary, Dryden. The unusual evening at Temple Bar ended with Newton shelling out five guineas to cover the cost of drinks for his entranced countrymen.

Three days later Stukeley breakfasted with Halley and Newton. During the interim, the latter's jovial mood had reverted to one of pique, fed by grievances old and new. Inveighing against the dead still constituted one of Newton's favorite pastimes, and this time it was Flamsteed's turn to be execrated once again. Newton complained that Halley's predecessor at the Royal Observatory had provided him with only a handful of observations when he was struggling to complete his lunar theory. He owed the man no thanks for the limited success he had achieved on that score. Newton also boasted that he could now finish his work on the moon if he wanted to, "but he rather chose to leave it for others." On a different occasion he told Halley he was contemplating the prospect of having "another shake at the moon," while to Benjamin Smith, his nephew, he once spoke of having yet "another touch at metals."[49] Such talk, of course, was based on the self-deluding fancies of old age, to which Newton, as much as any man, was certainly entitled. Yet these ruminations are even more revealing for another reason: They tend to prove that Newton never believed his natural philosophy had been brought to a satisfactory end, a painful thought, which he carried with him to the grave. That same morning he proudly showed his guests the great telescopic lense once owned by Huygens. He had purchased it at considerable expense from a source in Italy and had had it transported to Wanstead for use by the astronomer Dr. James Pound. Newton complained that the officer at the Customs House made him pay £20 duty on the lens. Had he only told the man that it was plain glass, the tariff would have been a fraction of what he wound up paying. "Sir Isaac," Stukeley remarked, "was too conscientious to deal with them."[50]

Twenty pounds was an inconsequential sum to a man of Newton's considerable wealth. Indications are that he expended much greater amounts on less rewarding enterprises, the most costly being the infamous South Sea Company, which failed disastrously in 1720. Founded in 1711 by Robert Harley to reap the anticipated benefits of peace negotiations to end the War of the Spanish Succession, the new enterprise was granted extensive trading concessions with the islands of the South Seas and South America. In 1720 it proposed that it should assume responsibility for the entire national debt by offering its own stock in exchange for government bonds. After much debate the government approved the plan, touching off a wave of unprecedented speculation, which drove up the price of the company's stock from £128$\frac{1}{2}$ in January to £1,000 in August. In September 1720 the South Sea Bubble burst, resulting in the widespread failure of banks unable to collect their loans on the inflated stock. Thousands were ruined, including numerous government officials, and fraud was discovered in the South Sea Company. The Royal Society lost £600, which Newton offered to make good out of his own pocket. Stukeley, who was on the Council, wrote that the Society would not permit it, so generous had Newton been in the past.[51]

Exactly how much of his own funds Newton lost is impossible to determine. Catherine Conduitt is said to have placed the figure at no less than £20,000, but the record of her uncle's financial transactions is sketchy and inconclusive. He had begun to invest in the South Sea Company as early as 1713, when he held £2,500 of its stock. Indications are that he purchased several hundred additional shares over the years. He sold off some of his holdings in April 1720, before the debacle, but then purchased more stock in June as the fever engulfed a frenzied investment community. A letter of August 1722, written to the accountant of the company, stated that he then owned well over £21,000 of stock.[52] Thus it is difficult to believe that he did not suffer a serious loss of funds. Lord Radnor related that Newton, reflecting back on what had taken place, said that "he could not calculate the madness of the people."[53] Whether he secretly included himself among their number seems destined to remain a tantalizing mystery.

Newton had rather more urgent uses for his money. Like many public figures of his times, including Robert Hooke, he died intestate, a somewhat curious state of affairs for one of his fastidious nature. Conduitt attributed the absence of a will partly to Newton's largess: "He was generous and charitable without bounds, he used to say, that they who gave away nothing till they died, never gave."[54] Stukeley was scarcely less assertive on the subject: "Infinite instances might be given of the extensiveness of his charitys; not those of a little and low kind, a weak mind, a false pity; but what show'd the noble spirit, that gave...profusely."[55] Hyperbole aside, the record clearly supports the claim that Newton became a significant dispenser of charity in his last years.

In the almost total absence of other forms of relief, the ethical standards of the eighteenth century dictated that the prosperous individual assist his less fortunate connections. Not surprisingly, the list of Newton's beneficiaries was led by members from both sides of his family, none of whom, with the exception of Catherine (who got her start in her uncle's household) seems to have fared very well. Newton showed particular concern for the welfare of his widowed half-sister Hannah Barton and her three minor children. He made financial provisions for each of them in the form of a generous annuity, and when Robert, the only son and the brother of Catherine, became old enough, he actively promoted his career in the military. After Lieutenant Colonel Barton was killed during an unsuccessful assault on Quebec in 1711, Newton further saw to it that his widow received a government bounty of £30 in addition to a yearly pension.[56] Shortly before he died, Newton paid £4,000 for an estate at Boyden in Berkshire for the deceased Robert's three children. He found afterward that he had paid twice its actual value. The contract might have been set aside in court, Conduitt wrote, but "he said he would not for the sake of £2,000 go to West r Hall to prove he had been made a fool of."[57] Such had also been his reaction to being relieved of a large sum by Whiston's light-fingered nephew. It must be said that Newton seems to have managed public funds with greater prudence than his own. Kitty Conduitt, daughter of John and Catherine, also received an estate in Kensington from her great uncle valued at £4,000. Nor did Newton overlook his other half-sister, Mary Pilkington, who, like Hannah, was widowed before her several children were grown. Humphrey Newton's sad description of the Pilkingtons seems to have mirrored the family fortunes in general: "Mr Pilkinton...died in a mean condition (tho' formerly he had a plentiful estate) whose widow with 5 or 6 children Sir Is. maintained several years together." In the case of Mary, the daughter and namesake of his half-sister, Newton provided quarterly payments of £9 year after year.[58]

Stukeley further noted that Newton attended the weddings of his younger relatives when he found it convenient. "He would on those occasions lay aside gravity [one more indication that he was usually less "chearful" than portrayed], be free, pleasant, and unbended. He generally made a present of £100 to the females, and set up the men to trade and business."[59] Little wonder that the male offspring of those marriages were sometimes christened Isaac or Newton, which pleased the couples' elderly relative more than they knew. While few of his relations truly fathomed what their distinguished kinsman had accomplished, they obviously recognized a good thing when they saw it. With every reason to be jealous of Catherine's special relationship with the childless old man, they may have been startled to learn that no will existed leaving most of his fortune to her, and that his half-nieces and half-nephews would therefore share equally in his estate. For his part, the charitable acts, the social gatherings, and the occasional letters of thanks for the granting of some particular favor served as belated substitutes for the fellowship and love he had sacrificed on the Faustian altar of the mind.

Relations between Newton and his many relatives, generally harmonious, were not entirely without friction. Most vexing to him by far was his scandalously profligate half-nephew Benjamin Smith, who is said to have lived in London with his uncle for a time after Catherine's marriage. A wild youth who consorted with the very type of reprobate element Newton had so despised during his student days at Cambridge, Benjamin later took holy orders and became Rector of Linton-in-Cavern, where he referred to his small congregation as "baptized brutes." The parishioners returned his scorn with interest, causing Bishop Warburton to term Smith's ordination a "furious scandal." Previous to these unsavory developments, Newton had written to his nephew in the plainest and severest terms. These "vulgar" letters, as they were described by one source, were discovered after Smith's death by his successor to the rectory, the Reverend William Sheepshanks, who, living up to his name, destroyed them lest they adversely reflect on England's most shining genius.[60] At the same time Newton was upbraiding his half-brother's son, Conduitt pictured him as occasionally giving way to uncontrollable outbursts of sentimentality, much like a child losing its innocence: "He had such a meekness and sweetness of temper, that a melancholy story would often draw tears from him, and he was exceedingly shocked at any cruelty to man or beast; mercy to both being the topic he loved to dwell upon."[61]

V

The end came gradually and, though painfully, with dignity. As we have already seen, Newton became seriously ill during the spring of 1722. In August 1724 he passed what Conduitt described as a stone about the size of a pea, which came away in two pieces some days apart. From that time onward he suffered from incontinence, which forced him to curtail his activities, both official and social. Since his attendance at the Royal Society and the Mint could no longer be assured, he had Martin Folkes appointed his deputy at Crane Court, while Conduitt assumed most of his duties at the Tower. In January 1725 Newton contracted a violent cough and severe inflamation of the lungs. His recovery took weeks, after which he was persuaded to move to the village of Kensington, which the bad air and smoke of the city did not yet reach. Despite an attack of the gout, which his friends assured him was a guarantee of long life, he found the new surroundings to be of considerable benefit. When Conduitt sought to convince him that it would be better if he rode to church rather than walked, he countered, "Use legs and have legs." His nephew-in-law also noted that while Newton was always the worse for leaving the country air, "no methods that were used could keep him from coming sometimes to town." During one of his progressively rarer visits to London and the Mint he burned several boxes of papers, the unknown contents of which have troubled the sleep of Newton scholars ever since.[62]

On a Sunday night in March 1725, shortly after Newton had recovered from the debilitating lung ailment and the attack of gout, Conduitt visited him at his new lodgings. He found Newton to be of clearer head and stronger memory than "I have known them for some time. He then repeated to me . . . that it was his conjecture (he would affirm nothing) that there is a sort of revolution in the heavenly bodies." Vapors and light emitted by the sun gradually coalesce into bodies, which attract more matter from the planets until they turn into satellites, like the earth's moon. Growing even larger, these secondary objects eventually become planets in their own right, and then, increasing in size still more, turn into comets, which, after revolving ever nearer to the sun, fall into the star, replenishing its spent matter and completing the cosmic cycle of transmutation. "He seemed to doubt whether there were not intelligent beings superior to us, who superintended these revolutions of the heavenly bodies, by the direction of the Supreme Being." The inhabitants of the earth had not been long upon its surface, nor were they destined to remain. There were visible marks of past ruin everywhere, which could be attributed only to such cataclysms as the Great Flood. When asked if the earth would be repeopled after the next disaster, Newton replied that "that required the power of a creator." When Conduitt asked why he did not publish his ideas, the classic answer prevailed: "I do no deal in conjectures."[63] Stukeley wrote of a similar exchange. The solar system of which the earth is a part "is but a sort of picture of the Universe," Newton explained. "God always created new worlds, always creates new worlds, new systems, to multiply the infinitude of his beneficiarys, and extend all happiness beyond all compass and imagination."[64]

It was during one such wistful conversation with an unnamed visitor that Newton attempted a description of his feelings about a life as completely devoted to the pursuit of human understanding as any in history: "I do not know what I may appear to the world; but to myself I seem to have been only like a boy, playing on the sea-shore, and diverting myself, in now and then finding a smoother pebble or prettier shell than ordinary, whilst the great ocean of truth lay all undiscovered before me."[65] In one sense Newton's metaphor of the wonder-filled child on a beach seems peculiarly strange, if only because there seems to have been so precious little of the child in the man. But in another sense his choice of images could not have been more apt, for the most profound scientific discoveries, as he had come to realize, are made by adults who continue asking the questions of children. Still, the ultimate answers had eluded him, whatever his discoveries may have seemed to the world.

More secure in the embrace of nostalgia than ever before, Newton often allowed his thoughts to carry him back to the scenes of his early youth. In April 1726, when Stukeley told him that he was moving back to his native Grantham, Newton heartily approved of the idea, adding that he too had frequently thought of spending the last of his days in the town where he had attended grammar school. He asked his friend to inquire about the house to the east of the church, where Mrs. Vincent had once lived. If it was available and price was right, Stukeley was authorized to purchase it in his behalf. While nothing came of this passing fancy, one could almost picture the aged Newton, walking stick in hand, shuffling about the grounds where his mind first took flight after he had beaten his bullying rival into submission.[66] He remained equally fond of Colsterworth, the hamlet just across the River Witham from Woolsthorpe. He gave £12 to erect a gallery in the church on whose grounds his ancestors were buried and £3 more to repair its decaying floor. Upon learning that there was a surplus of funds, he asked that they be used to teach the young people of the parish to sing psalms. According to Thomas Mason, the Rector, he also spoke of founding and endowing a school in Woolsthorpe, but no provision was made to carry out the plan. Six weeks before Newton died, he wrote Mason a letter, perhaps his last, informing the cleric that some ore delivered to him in London on behalf of a parishioner contained no trace of malleable metal, one of several favors granted his countrymen during this period.[67]

On the last day of February 1727 Newton came into London to preside at the March 2 meeting of the Royal Society. Conduitt thought he had not seen him looking so fit in years and told him so. Newton replied that he was sensible of it himself, "and told me smiling, that he slept the Sunday before, from eleven at night to eight in the morning without waking." But by the time he returned to Kensington it was apparent that the rigors of travel and of making several social calls had been taxing in the extreme. Conduitt, who shortly thereafter received word that Newton's "old complaint" had returned in a violent form, contacted Doctors Mead and William Cheselden, and together they hurried to his bedside. The prognosis could not have been worse. Another stone had lodged itself in the bladder, and both physicians concluded that there was no hope of recovery. During the next several days the patient alternated between long sieges of excruciating pain and brief periods of tranquility: "[T]hough the drops of sweat ran down his face," a grieving Conduitt observed, "he never complained, or cried out, or shewed the least signs of peevishness or impatience, and during the short intervals from that violent torture, would smile, and talk with his usual cheerfulness." Stukeley painted an even more dramatic picture of the death scene: The pain "rose to such a height that the bed under him, and the very room, shook with his agonys to the wonder of those that were present. Such a struggle had his great soul to quit its earthly tabernacle!"[68] And it was clearly with his soul that Newton remained most concerned. Conduitt, grudgingly and with chagrin, admitted that he refused to receive the final rites of the church, the last important act of a dying man who, for more than half a century, had secretly viewed the trinitarian credo with abhorrence.[69] Isaac Newton would not yield to this final temptation of Satan and be lost.

On Wednesday, March 15, Newton seemed somewhat better, raising hopes that he might confound the experts once again. On Saturday morning he read the newspapers and talked with Mead at some length, but that evening at six o'clock, twilight yielded to night. Newton slipped into a coma and remained unconscious until he died between one and two o'clock in the morning of March 20, almost the exact hour of his Christmas birth more than eighty-four years earlier.[70]

On March 23 the following entry, simple yet eloquent, appeared in the Journal Book of the Royal Society: "The Chair being Vacant by the Death of Sir Isaac Newton there was no Meeting this Day."[71] Word of his death reached the Reverend Mason at Colsterworth that same morning. Newton's liquid assets, which totaled some £32,000, were to be divided equally among his eight nieces and nephews, but the estate at Woolsthorpe was now legally the property of the next surviving Newton. Conduitt, who served as executor, asked Mason to identify that individual. He turned out to be one John Newton, descendant of a brother to Newton's father. Mason accurately described him as "God knows a poor Representative of so great a man, but ys is a case yt often happens."[72] John Newton more than lived up to Mason's assessment of him. He gambled and drank his inheritance away, dying by accident when, after a round of drinking, he stumbled and fell with a pipe in his mouth, the broken stem lodging in his throat.

Newton's body lay in state in the Jerusalem Chamber of Westminster Abbey from March 28 until April 4, the day of the funeral. The pall was carried into the main sanctuary by the Lord High Chancellor, the Dukes of Montrose and Roxborough, and the Earls of Pembroke, Sussex, and Macclesfield, all members of the Royal Society. The Hon. Sir Michael Newton, a knight of the Bath, led the procession of mourners, which included several relatives and distinguished public figures. The service was performed by the Bishop of Rochester, assisted by the prebendary and the choir. Newton's final resting place, which had been denied to several men of noble birth, was chosen for its prominence in the nave of the great Abbey. When the service was over and the last of the mourners had departed, the body was lowered into its crypt. His life's journey completed, this rare mortal, who came into the world as a quart-sized babe, was now a part of that vast inexorable drift called infinity.

Bibliography and Abbreviations of Frequently Used Sources

Note: The reader who seeks a more exhaustive bibliographical treatment of Newtonian scholarship than the one contained below is referred to the difinitive work by Peter and Ruth Wallis, *Newton and Newtoniana, 1672–1975* (Folkstone, England, 1975).

Manuscript Collections and Edited Papers

D.T.W.	*The Mathematical Papers of Isaac Newton.* Ed. by D. T. Whiteside. 8 vols. Cambridge, England, 1967–80.
F.M.N.	Fitzwilliam Museum Notebook, Fitzwilliam Museum, Cambridge, England.
I.B.C. *Papers*	*Isaac Newton's Papers & Letters on Natural Philosophy.* 2d ed. Ed. by I. Bernard Cohen. Cambridge, Mass., 1978.
I.N. *Corres.*	*The Correspondence of Isaac Newton.* Ed. by H. W. Turnbull, J. F. Scott, A. R. Hall, and Laura Tilling. 7 vols. Cambridge, England, 1959–77.
J.E.	*Correspondence of Sir Isaac Newton and Professor Cotes.* Ed. by J. Edleston. London, 1850.
Keynes MS.	Keynes Manuscript Collection, King's College Library, Cambridge, England.
Mint MS.	Mint Manuscript, Public Record Office, London, England.

U.L.C. Add. MS.   University Library Cambridge, Additional Manuscript, Cambridge, England.

Yahuda MS.   Yahuda Manuscript Collection, Jewish National and University Library, Jerusalem.

Works by Newton

I.N. "Account"   Isaac Newton. "An Account of the Book entitled Commercium Epistolicum," *Philosophical Transactions*, 29 (1715): 173–224.

I.N. *Opticks*   *Opticks or A Treatise of the Reflections, Refractions, Inflections & Colours of Light.* Based on the 4th ed. New York, 1952.

I.N. *Principia*   *Mathematical Principles of Natural Philosophy.* Trans. and ed. by Andrew Motte in 1729. Rev. by Florian Cajori. Berkeley and Los Angeles, 1934.

Books

A.R.H.   A. Rupert Hall. *Philosophers at War: The Quarrel Between Newton and Leibniz.* Cambridge, England, 1980.

A.R.H., M.B.H.   *The Correspondence of Henry Oldenburg.* Ed. by A. Rupert Hall and Marie Boas Hall. 9 vols. Madison, Wis., 1965–73.

B.J.T.D.   Betty Jo Teeter Dobbs. *The Foundations of Newton's Alchemy.* Cambridge, England, 1975.

C.A.R.   Colin A. Ronan. *Edmond Halley: Genius in Eclipse.* New York, 1969.

C.H.C.   Charles Henry Cooper. *Annals of Cambridge.* 5 vols. Cambridge, England, 1842–1908.

C.R.W.   Charles R. Weld. *A History of the Royal Society.* 2 vols. London, 1848.

E.S.D.B.   *The Diary of John Evelyn.* Ed. by E. S. De Beer. 6 vols. Oxford, 1955.

E.T.   *Collections for the History of the Town and Soke of Grantham.* Ed. by Edmund Turnor. London, 1806.

F.B.   Francis Baily. *An Account of the Rev.[d] John Flamsteed, the First Astronomer Royal.* London, 1835.

F.M. *Portrait*   Frank Manuel. *A Portrait of Isaac Newton.* Cambridge, Mass., 1968.

F.M. *Religion*   Frank Manuel. *The Religion of Isaac Newton.* Oxford, 1974.

H.P.   Henry Pemberton. *A View of Sir Isaac Newton's Philosophy.* London, 1728.

H.W.R., W.A.   *The Diary of Robert Hooke, 1672–1680.* Ed. by Henry W. Robinson and Walter Adams. London, 1935.

I.B.C. *Intro.*   I. Bernard Cohen. *Introduction to Newton's 'Principia.'* Cambridge, Mass., 1971.

J.A.	John Aubrey. *Brief Lives*. Ed. by O. L. Dick. 3d ed. London, 1968.
J.B.M.	James Bass Mullinger. *The University of Cambridge*. 3 vols. Cambridge, England, 1873–1911.
J.C. *Mint*	John Craig. *The Mint*. Cambridge, England, 1953.
J.C. *Newton*	John Craig. *Newton at the Mint*. Cambridge, England, 1946.
J.E.B.M.	J. E. B. Mayor. *Cambridge Under Queen Anne*. Cambridge, England, 1870.
J.H.	John Herivel. *The Background to Newton's 'Principia.'* Oxford, 1965.
J.R.H.	John R. Harrison. *The Library of Isaac Newton*. Cambridge, England, 1978.
J.S.	Jonathan Swift. *Journal to Stella*. Ed. by Harold Williams. Oxford, 1948.
L.P.K.	Lord Peter King. *The Life and Letters of John Locke*. New ed. London, 1858.
L.T.M.	Louis Trenchard More. *Isaac Newton: A Biography*. New York, 1934.
M.C.J.	Margaret C. Jacob. *The Newtonians and the English Revolution: 1689–1720*. Ithaca, N.Y., 1976.
M.E.	Margaret 'Espinasse. *Robert Hooke*. Berkeley and Los Angeles, 1962.
P.H.O.	Percy H. Osmond. *Isaac Barrow: His Life and Times*. London, 1944.
R.L., W. M.	*The Diary of Samuel Pepys*. Ed. by Robert Latham and William Matthews. 11 vols. Berkeley and Los Angeles, 1970–83.
R.S.W. *Biog.*	Richard S. Westfall. *Never at Rest: A Biography of Isaac Newton*. Cambridge, England, and New York, 1980.
R.V.	Richard de Villamil. *Newton: The Man*. London, 1931.
S.D.B.	Sir David Brewster. *Memoirs of the Life, Writings, and Discoveries of Sir Isaac Newton*. 2 vols. Edinburgh, 1855.
S.H.L.	Sir Henry Lyons. *The Royal Society, 1660–1940*. Cambridge, England, 1944.
T.B.	Thomas Birch. *The History of the Royal Society of London*. 4 vols. London, 1756–57.
T.S.	Thomas Sprat. *History of the Royal Society*. Ed. by Jackson I. Cope and Harold Whitmore Jones. St. Louis, 1958.
W.G.H.	*David Gregory, Isaac Newton, and Their Circle*. Ed. by W. G. Hiscock. Oxford, 1937.
W.S.	William Stukeley. *Memoirs of Sir Isaac Newton's Life*. Ed. by A. Hastings White. London, 1936.
W. W.	William Whiston. *Memoirs of the Life and Writings of Mr. William Whiston*. London, 1749.

Journals

C.W.F.

Canon C. W. Foster. "Sir Isaac Newton's Family," *Reports and Papers of the Architectural Societies of the County of Lincoln, County of York, Archdeaconries of Northampton and Oakham, and County of Leicester*, XXXIX, pt. I (1928): 1–62.

R.S.W. "Writ."

Richard S. Westfall. "Short-Writing and the State of Newton's Conscience, 1662 (I)," *Notes and Records of the Royal Society of London*, 18 (June 1963): 10–16.

Others

C.M.R.S.   Council Minutes of the Royal Society, London, England.
C.S.P.D.   Calendar of State Papers, Domestic
C.T.B.   Calendar of Treasury Books
C.T.P.   Calendar of Treasury Papers
J.B.C.R.S.   Journal Book (Copy) of the Royal Society, London, England.
J.H.C.   Journals of the House of Commons

Notes

Chapter One:
Inside a Quart Pot

1. The parish register, showing the damaging effects of time, was kindly made available to me by Mr. P. G. Isaac, retired headmaster of Colsterworth School and warden of the village church.
2. C.W.F., pp. 13–15.
3. *Ibid.*, p. 15.
4. The source of this characterization is Thomas Maude, who mistakenly confused Newton's father with one John Newton, an heir-at-law of Sir Isaac. *Viator, a poem: or, A Journey from London to Scarborough by the Way of York* (London, 1782), pp. iv–v.
5. For the complete "Will and Inventory of Isaac Newton, 1642," see C.W.F., pp. 45–47.
6. *Ibid.*, p. 15.
7. Keynes MSS. 130 (10), p. 1; 130, p. 15.
8. Keynes MS. 130, pp. 9–10.
9. I.N. *Corres.*, I: 2.
10. Keynes MS. 130 (8). For a record of the college absences, see J.E., p. lxxxv.
11. For Hannah's will, see C.W.F., pp. 50–53.

12. C.W.F., p. 17.
13. The parish register of North Witham (1592–1756) is still kept in the local church.
14. Keynes MS. 130 (2), p. 11.
15. Smith's wealth is attested to by his will dated August 17, 1653, and proved February 8, 1654. See C.W.F., pp. 53–54.
16. Keynes MS. 130 (2), p. 11.
17. Keynes MS. 136, p. 6. Though Newton received nothing beyond the Sewstern property when his stepfather died, he did benefit financially, if indirectly. Hannah inherited a substantial amount of money and a considerable annual income, with which she purchased additional land. Both this land and part of the remaining money went to Newton when Hannah died in 1679.
18. Most biographers have written of this period as though James Ayscough the elder had already passed from the scene. In point of fact he was very much alive. James Ayscough the elder and his son James the younger witnessed Barnabas Smith's will in August 1653. See C.W.F., p. 54. Drawing upon private papers, Edmund Turnor, whose family purchased

Woolsthorpe in 1733, noted that "in 1650 James Ayscough, Gent. is stated to be guardian to Isaac Newton, lord of the manor, under age." E.T., p. 158.

19. R.S.W. "Writ.," p. 13.

20. Quoted in Christopher Hill, "Newton and His Society," in *The Annus Mirabilis of Sir Isaac Newton, 1666–1966*, ed. by Robert Palter (Cambridge, Mass., 1970), p. 26.

21. Robert Newton, Isaac's grandfather, was churchwarden of Colsterworth in 1594, 1595, 1613, 1614, 1626, and 1627. C.W.F., p. 13.

22. Hill, "Newton and His Society," p. 31.

23. Keynes MS. 130 (3). This belief still survives "in an attenuated form" in the Woolsthorpe–Colsterworth locality. F.M. *Portrait*, p. 28. Also see F.M. *Religion*, p. 17.

24. Quoted in F.M. *Portrait*, p. 34. Newton's Latin exercise book is in a private collection in Los Angeles.

25. Parts of this work, usually referred to as the Morgan notebook since it is now in the Pierpont Morgan Library, New York, were published earlier in this century. See David Eugene Smith, "Two Unpublished Documents of Sir Isaac Newton," in *Isaac Newton, 1642–1727*, ed. by W. J. Greenstreet (London, 1927), pp. 16–34.

26. Compare, for example, D.T.W., I: 1–2, with G. L. Huxley, "Two Newtonian Studies," *Harvard Library Bulletin*, 13 (1959): 349, 354–61.

27. Charles Henry Cooper, "Facts Respecting Henry Stokes, Newton's Schoolmaster," *Communications Made to the Cambridge Antiquarian Society*, II, no. 12 (1862): 161–63.

28. Keynes MS. 136, p. 4.

29. *Ibid.*

30. *Ibid.*, pp. 4–5.

31. Keynes MS. 130 (2), p. 21.

32. Keynes MS. 136, p. 5.

33. Keynes MS. 130 (3), p. 9.

34. Keynes MS. 136, p. 4.

35. *Ibid.*

36. *Ibid.*, p. 5.

37. L.T.M., p. 15.

38. Keynes MS. 136, p. 6.

39. See H. W. Robinson, "Note on Some Recently Discovered Geometrical Drawings in the Stonework of Woolsthorpe Manor House," *Notes and Records of the Royal Society of London*, 5 (1947): 35–36.

40. The link between Newton's early intellectual development and Bate's popular book was first established by E. N. da C. Andrade, "Newton's Early Notebook," *Nature*, CXXV (March 9, 1935): 360. For a more detailed treatment, see G. L. Huxley, "Two Newtonian Studies," pp. 348–54.

41. U.L.C. Add. MS. 3975, p. 178. This remedy, like most items in the notebook, was copied by Newton from the works of the natural philosopher Robert Boyle.

42. Keynes MS. 130 (2), p. 18.

43. L.T.M., p. 11.

44. R.S.W. "Writ.," pp. 13–14.

45. Keynes MS. 136, pp. 6–7. Rumors have persisted to the effect that Isaac's dilatory behavior was costly not only in time but in money. Among the Lincolnshire Archive papers of the Turnor family is an abstract of the fines levied at Court Leet and Court Baron in Colsterworth. On October 28, 1659, three farmers were each fined 3s. 4d. "for suffering . . . sheep and cattle to break ye stubbs." One of these was named "Isaac Newton." This same individual was further fined for other minor infractions. This probably refers to another Isaac Newton, a descendant of Newton's great uncle, who was baptized in 1593. This third Isaac reportedly lived to the astounding age of ninety-five and died at Colsterworth. (Stukeley briefly refers to him in Keynes MS. 136, p. 7.) This seems even more likely given the fact that "the wife of Isaac Newton" was fined 3s. 4d. at the same session of court. The woman could have been Newton's mother, but Hannah had long since taken the name Smith (Smyth), by which she appears on other contemporary documents. In 1660 a "Mrs Smyth" was fined 3s. 4d. for permitting her swine to run loose and 2s. more "for carting over Maurice Myles lands after it was sown." This was probably Isaac's mother, though her son may have been the real culprit. I am grateful to a Colsterworth resident, K. A. Baird, C.B.C., M.A., for bringing this information to my attention.

46. Keynes MS. 136, p. 6.

47. *Ibid.*, pp. 7–8.
48. *Ibid.*, pp. 6–7.
49. *Ibid.*, pp. 5–6. Stukeley's "Miss Storey" was, in fact, a Miss Catherine Storer, sister of Arthur and Eduard (Edward) Storer, all stepchildren of the apothecary Clark. According to the parish records of Boothby Pagnall, a village 3 miles from Woolsthorpe, Catherine Storer married one Francis Bakon, a Grantham attorney, in 1665, a union which produced four children. After Bakon's death, she married the widower John Vincent on July 7, 1685. Widowed a second time in 1715, Catherine died in November 1727, nine months after Newton. (The latter information is contained in the parish records of St. Wulfram's Church, Grantham.)
50. Keynes MS. 136, p. 6.
51. *Ibid.*, p. 3. The "Dr. Newton" mentioned by Stukeley is Humphrey Newton, Isaac's amanuensis at Cambridge. The shelves have disappeared. A partition now encloses the southeast corner of the bedroom. According to tradition, Newton himself constructed it, but the wood appears too new and the carpentry too refined for a youth, no matter how clever, working with simple tools.
52. *Ibid.*, p. 7.
53. J. W. N. Sullivan, *Isaac Newton, 1642–1727* (New York, 1938), p. 5.
54. Keynes MS. 130 (2), p. 33.

Chapter Two: Of Giants and Dwarfs

1. Keynes MS. 130 (7).
2. G. M. Trevelyan, *Trinity College: An Historical Sketch* (Cambridge, England, 1946), p. 10.
3. Charles Webster, *The Great Instauration: Science, Medicine, and Reform, 1626–1660* (London, 1975), p. 117.
4. Phyllis Allen, "Scientific Studies in the English Universities of the Seventeenth Century," *Journal of the History of Ideas*, X, no. 2 (April 1949): 219–21. For a more detailed treatment, see William T. Costello, *The Scholastic Curriculum at Early Seventeenth-Century Cambridge* (Cambridge, Mass., 1958), pp. 36–105.
5. Mark H. Curtis, *Oxford and Cambridge in Transition, 1558–1642: An Essay on Changing Relations Between the English Universities and English Society* (Oxford, 1959), pp. 96–97.
6. Allen, "Scientific Studies in English Universities of Seventeenth Century," p. 220.
7. Curtis, *Oxford and Cambridge in Transition*, p. 227.
8. Trevelyan, *Trinity College*, pp. 14–15.
9. The Wallis letter is dated January 29, 1697. *Peter Langtoft's Chronicle*, ed. by Thomas Hearne (Oxford, 1735) I: cxlvii–cxlviii.
10. Allen, "Scientific Studies in English Universities of Seventeenth Century," p. 223. For a discussion of the tutor's role see Curtis, *Oxford and Cambridge in Transition*, pp. 107–14, 119–22.
11. J.B.M., III:107.
12. *Ibid.*, III:235.
13. Webster, *The Great Instauration*, p. 132.
14. William Oughtred, *The Circles of Proportion* (London, 1633), appendix, "The Just Apologie," unpaged.
15. Allen, "Scientific Studies in English Universities of Seventeenth Century," p. 229.
16. See Mark H. Curtis, "The Alienated Intellectuals of Early Stuart England," *Past and Present*, no. 23 (1962), pp. 25–43.
17. Jan Amos Comenius, *A Reformation of Schooles Designed in Two Excellent Treatises* (London, 1642), pp. 20–21.
18. R. S. Westfall, *Science and Religion in 17th Century England* (New Haven and London, 1958), pp. 72–73. Also see R. Hooykass, *Religion and the Rise of Modern Science* (Edinburgh and London, 1972), and Eugene M. Klaaren, *Religious Origins of Modern Science* (Grand Rapids, Mich., 1977).
19. Galileo Galilei, *Letter to the Grand Duchess Christina*, in *Discoveries and Opinions of Galileo*, trans. by Stillman Drake (New York, 1957), pp. 182–83.
20. Richard Foster Jones, *Ancients and Moderns: A Study of the Rise of the Scientific Movement in Seventeenth-Century England*, 2d ed. (Berkeley and Los Angeles, 1965), p. 22.
21. Quoted in *Ibid.*, p. 123.
22. Quoted in *Ibid.*, p. 127.

Chapter Three: The Transit of Genius

1. Thomas Fuller, *History of the University of Cambridge published with his Church-history of Britain* (London, 1655), p. 122.
2. Trinity College, Cambridge MS. R4. 48ᶜ. The Book is so small that it is commonly referred to as the "Pocketbook." On the third page, in bold "Old Barley" script, is Newton's signature and the date: "Martii 19, 1659."
3. J.E., p. xli.
4. Augustus De Morgan, *Essays on the Life and Work of Newton*, ed. by Philip E. B. Jourdain (Chicago and London, 1914), pp. 8–9.
5. F.M.N., unpaged. According to Turnor, who once more had access to the legal documents, "Sir Isaac Newton . . . came into possession of the family estate here [Woolsthorpe], and at Sewstern, in 1663; for in 1650 James Ayscough, Gent. is stated to be guardian to Isaac Newton, lord of the manor under age." (E.T., p. 158.) Although most historians have written that Woolsthorpe did not become Newton's until his mother's death in 1679, her will supports Turnor's conclusion. Each parcel of land owned by Hannah Newton Smith is individually listed, and those in an around Woolsthorpe were all purchased *after* Smith's death. (C.W.F., p. 52.) The manor house and surrounding fields receive no mention in the document, nor should they have if Isaac took possession in 1663. Despite her son's ownership, however, Hannah managed the estate in his absence, exercising considerable control over its revenues.
6. S. Brodetsky, *Sir Isaac Newton: A Brief Account of His Life and Work* (London, 1927), pp. 14–15. A perusal of the *Alumni Cantabrigienses* suggests that the typical matriculant was between sixteen and seventeen, with the average age rising toward the end of the seventeenth century. The youngest students admitted in Newton's day were boys of thirteen, while the oldest, in their early twenties, were already young men. John A. Venn and J. A. Venn, *Alumni Cantabrigienses*, pt. I, 4 vols. (Cambridge, England, 1922–27).
7. W.S., pp. 50–51.
8. Venn and Venn, *Alumni Cantabrigienses*, pt. I, III:406.
9. Keynes MS. 130 (4), pp. 1–2.
10. Keynes MS. 137. For additional details see J.E., p. xliii. A possible clue to the identity of Newton's first "chamberfellow" survives. A 1662 entry in the Fitzwilliam Museum notebook has Newton confessing to the sin of "Using Wilfords towel to spare my own." This is undoubtedly Francis Wilford, who, according to the *Alumni Cantabrigienses*, was admitted pensioner at Trinity in June 1661, the same date as Newton. The reference is all the more suggestive since a towel is hardly an article one goes any distance to use.
11. D.T.W., II: ix.
12. Keynes MS. 137.
13. F.M. *Portrait*, pp. 409–10, fn. 21.
14. Shorthand was commonly used by university students of the period. Newton adopted the system pioneered by Thomas Shelton in his immodestly titled work: *Tachy graphy the most exact and compendious methode of short and swift writing that hath euer yet beene published by any . . . Approouved by both Universities* (London, 1641).
15. R.S.W. "Writ.," pp. 13–14.
16. Philip Greven, *The Protestant Temperament: Patterns of Child-Rearing, Religious Experience, and the Self in Early America* (New York, 1977), Chapters I–III.
17. U.L.C. Add. MS. 3996. Newton used several notebooks while at Cambridge, each devoted to a different subject: mathematics, theology, alchemy, philosophy, and so on. This work has been termed the "philosophical notebook." See Richard S. Westfall, "The Foundations of Newton's Philosophy of Nature," *British Journal for the History of Science*, I, no. 2 (1962): 171.
18. Walter Charleton, *Physiologia Epicuro-Gassendo-Charletoniana* (London, 1654), Chapter III, Section II.
19. U.L.C. Add. MS. 3996, ff. 88–89.

20. *Ibid.*, f. 102^v.
21. *Ibid.*, f. 101^v.
22. *Ibid.*, ff. 111–12.
23. See William T. Costello, *The Scholastic Curriculum at Early Seventeenth-Century Cambridge* (Cambridge, Mass., 1958), p. 103.
24. Before graduating B.A., Newton had begun to acquire a library of his own. At the time of his death it consisted of some 1,900 volumes and many unbound pamphlets. Of the bound works, 862 are in the collection of Trinity College Library (J.R.H.). For a discussion of the major scientific books in the Trinity College Library during Newton's student days, see I. B. Cohen, "Newton's Attribution of the First Two Laws of Motion to Galileo," in *Tratto da Atti del Simposio su Galileo Galilei nella storia e nella filosofia della scienza* (Firenze–Pisa, 1964), pp. xxvii–xxxi.
25. U.L.C. Add. MS. 3996, ff. 90^v, 114–16.
26. I.N. *Corres.* III:153.
27. U.L.C. Add. MS. 3996, ff. 109^r, 125. Newton headed this section: "Starred into the the Sun."
28. A. R. Hall, "Sir Isaac Newton's Note-book, 1661–1665," *Cambridge Historical Journal,* 9 (1948): 245.
29. Keynes MS. 130 (4), p. 9.
30. J.E., p. xlii.
31. Hall, "Sir Isaac Newton's Note-book," p. 246. Conduitt seems to have discovered the fault in his chronology, for he later moved the prism purchase up to 1663, which seems to be at least a year too early.
32. *Ibid.*, p. 249. For additional background on Newton's early experiments with light see J. A. Lohne, "Isaac Newton: The Rise of a Scientist, 1661–1671," *Notes and Records of the Royal Society of London,* 20 (1965): 125–39.
33. U.L.C. Add. MS. 3996, f. 128.
34. Keynes MS. 130 (10).
35. I.N. *Corres.*, III:284.
36. Roger North, *The Lives of the Norths,* ed. by Augustus Jessopp, 3 vols. (London, 1890), III:284.
37. D.T.W., I: 5.
38. U.L.C. Add. MS. 4004. This volume is often referred to as the *Waste Book.* The earliest dated entries are October, November, and December 1664.
39. U.L.C. Add. MS. 4007, ff. 706–7.
40. U.L.C. Add. MS. 4000, f. 14^v. For further scattered references made by Newton regarding his early mathematical studies and discoveries see D. T. Whiteside, "Newton's Marvellous Year: 1666 and All That," *Notes and Records of the Royal Society of London,* 21 (1966): 32–41.
41. H.P., Preface.
42. I.N. *Corres.*, II:375. Emphasis mine.
43. *Ibid.*, III:155–56.
44. Keynes MS. 136, p. 8. A slightly different version appears in W.S., pp. 53–54.
45. The Latin text of the original statutes of the Lucasian Professorship is reproduced in D.T.W., III: xx–xxvii.
46. J.E., p. xlv.
47. U.L.C. Add. MSS. 3968.5, f. 21^r; 3968.41, f. 86^v.
48. P.H.O., p. 114.
49. Keynes MS. 130 (10), f. 2^v.
50. Newton's copy of Barrow's *Euclid* (1655) is in the Trinity College Library collection (NQ. 16. 201). For details and photographs of its contents see H. Zeitlinger, "A Newton Bibliography" in *Isaac Newton, 1642–1727,* ed. by W. J. Greenstreet (London, 1927), pp. 168–70. At least one of Newton's marginal diagrams bears a strong resemblance to the etchings on the inner walls of Woolsthorpe.
51. Keynes MS. 130 (7).
52. R.S.W. "Writ.," p. 13.
53. Trinity College, Cambridge MS. R4. 48^c. Barnham Oliver and Francis Wilford were Trinity pensioners, while John Pollard, John Bigg, and John Andrew (Andrews) were sizars of the College. The Guy to whom Newton refers was probably Joseph Guy (Gye), a sizar of St. John's and former classmate of Newton at Grantham. Agatha and Gosh are unaccounted for in the *Alumni Cantabrigienses.*
54. W.S., p. 19.
55. Keynes MS. 136, p. 8. Each person standing for a degree deposited nine groats (small coins) in the hands of an academic officer. Only if he did well in the disputations was the money returned to him. See Francis Grose, *A Classical Dic-*

tionary of the Vulgar Tongue, ed. by Eric Partridge (London, 1963), p. 172.

56. J.E., p. xli.

Chapter Four: A Movable Feast

1. Revelation 6:8.
2. R.L., W.M., VI:93, 145.
3. Thomas Vincent, *God's Terrible Voice in the City* (London, 1667), p. 29.
4. J. F. D. Shrewsbury, *A History of the Bubonic Plague in the British Isles* (Cambridge, England, 1970), p. 5.
5. The *Directions* of the College of Physicians are reprinted in Daniel Defoe's fictional but factually based diary, *A Journal of the Plague Year* (New York, 1966), pp. 57–65.
6. J.B.M., III:618–20.
7. F.M.N.
8. J.E., p. xlii.
9. J.B.M. III:620.
10. F.M.N.
11. U.L.C. Add, MS. 3968.41, f. 85^r.
12. W.W., p. 39.
13. John Maynard Keynes, "Newton, the Man," in *Newton Tercentenary Celebrations* (Cambridge, England, 1947), p. 29.
14. D.T.W., I:3.
15. U.L.C. Add. MS. 3958.3, ff. 48–63. For a profound step-by-step analysis of these developments, see D.T.W., vol. I.
16. U.L.C. Add. MS. 4004, f. 14^v, 20^r.
17. I.N. *Corres.*, II:179–80.
18. W.S., pp. 19–20.
19. Voltaire, *Elémens de la Philosophie de Newton* (London, 1741), p. 289. An earlier edition of the work was published in 1738, but it strangely omits any reference to the falling "fruits" which in Voltaire's mind were synonymous with apples. See Charles Coulston Gillispie, "Fontenelle and Newton," in I.B.C. *Papers*, p. 434, fn. 21.
20. F.M. *Portrait*, p. 80.
21. H.P., Preface.
22. W.W., I:36–37.
23. J.H., p. 7.
24. *Ibid.* For Galileo's influence on Newton, see pp. 35–41.
25. *Ibid.*, pp. 42–54.
26. U.L.C. Add. MS. 4004, f. 10^v.
27. *Ibid.*, f. 11^v.
28. J.H., p. 9.
29. U.L.C. Add. MS. 3968.41, f. 85^r.
30. H.P., Preface.
31. U.L.C. Add. MSS. 3958.2, f. 45^r; 3958.5, ff. 87^r–88^r. Both have been reproduced and extensively annotated in the *Correspondence*, the first in III:46–54, the second in I:297–303. They also appear in J.H., pp. 182–98.
32. I.N. *Corres.*, I:301, fn. 1. Emphasis mine.
33. Herivel conjectures that the concept of universal gravitation may have been carried further by Newton in 1666 than the surviving evidence would indicate. J.H., pp. 71–74.
34. See, for example, U.L.C. Add MS. 3958.2, f. 45^r.
35. J.H. pp. 68–69.
36. U.L.C. Add. MS. 3958.2, f. 45^r
37. J. Picard, *La measure de la terre* (Paris, 1671).
38. I.N. *Corres.* I:356.
39. This famous quote first appeared in print in an article on Newton in the *Biographica Britannica* (London, 1760), 5:32–41.
40. U.L.C. Add. MS. 4007, f. 706^v.
41. U.L.C. Add. MS. 3958.5, f. 87.
42. I.N. *Corres.*, II:440.
43. R.L., W.M. VII:274.
44. *Ibid.*, pp. 267–68.
45. J. P. Malcom, *Londinium Redivivum* (London, 1802), IV:74.
46. R.L., W.M., VII:369.
47. Walter George Bell, *The Great Fire of London of 1666* (London, 1923), p. 18.
48. Vincent, *God's Terrible Voice in City*, p. 46.
49. Bell, *The Great Fire of London of 1666*, p. 177.
50. "A true and perfect inventary [sic] of all and singular the Goods, Chattels, and Credits of Sir Isaac Newton," is reproduced in R.V., pp. 50–61.
51. David Eugene Smith, "Two Unpublished Documents of Sir Isaac Newton," in *Isaac Newton, 1642–1727*, ed. by W.J. Greenstreet (London, 1927), pp. 19–23.
52. I.N. *Corres.*, I:92.
53. U.L.C. Add. MS. 3975, p. 15.
54. U.L.C. Add. MS. 3996, f. 122^v.
55. I.N. *Corres.*, I:96–97.
56. *The Works of the Honourable Robert*

Boyle, ed. by Thomas Birch, new ed., 6 vols. (London, 1772), I:738.

57. Richard S. Westfall, "The Development of Newton's Theory of Colors" *ISIS*, 53, pt. 3, no. 173 (1962): 342.

58. Copious notes on the *Micrographia* occur in U.L.C. Add. MS. 3958.1.

59. Robert Hooke, *Micrographia* (London, 1665), p. 64.

60. U.L.C. Add. MS. 3996, f. 122^r.

61. I.N. *Corres.*, I:92. Emphasis in original.

62. U.L.C. Add. MS. 400, ff. 26–33^v.

63. R.S.W. *Biog.*, pp. 161–62.

64. U.L.C. Add. MS. 3975, pp. 1–20. The essay is undated and because it contains few corrections may be a synthesis of other papers now lost. See R.S.W. *Biog.*, p. 163, fn. 61.

65. U.L.C. Add. MS. 3975, p. 2. In the paper to Oldenburg Newton cited somewhat different measurements, suggesting at least two experiments of a similar nature. (I.N. *Corres.*, I:93). At my request some years ago a Colsterworth resident, Kenneth Baird, kindly measured the distance from the wall to the inside of the small window at the southeast corner of Newton's Woolsthorpe bedroom. The interior distance is 21′ 4″, while it is almost exactly 22′ from the wall to the outside of the window, the very figure cited by Newton in his paper. It is therefore probable that a number of the optical experiments were undertaken in Lincolnshire during the plague years. Continuing his research after returning to Trinity, Newton would have sought to duplicate the conditions at Woolsthorpe as closely as possible, which would account for the slight discrepancy in his figures. The analysis of a spectrum cast on a distant wall by the steadily moving sun required the help of an assistant. One thinks of his younger brother or sisters while at Woolsthorpe and of John Wickins back at Cambridge. He later commented on the use of an assistant (I.N. *Corres.*, I:376–77).

66. I.N. *Corres.*, I:92.

67. *Ibid.*, I:93.

68. *Ibid.*, I:96. For a detailed discussion of this subject see Johs. Lohne, "Newton's 'Proof' of the Sine Law and his Mathe-matical Principles of Colors," *Archive for History of Exact Sciences*, 1, no. 4 (1961): 389–405.

69. Hooke, *Micrographia,* p. 54. Lohne has pointed out that Hooke did not believe the term to be original but thought he was simply borrowing from Bacon. Instead, Hooke confused Bacon's *Experimenta Lucifera* with his *Instantia Crucis*. See J. A. Lohne, "Experimentum Crucis," *Notes and Records of the Royal Society of London*, 23 (1968): 179.

70. U.L.C. Add. MS. 3975, p. 12.

71. U.L.C. Add. MS. 3996, f. 122^r.

72. U.L.C. Add. MS. 3975, pp. 12–13. Though perhaps only a coincidence, Newton recorded the purchase of three prisms at a shilling apiece in his accounts for 1667–68 (F.M.N.).

73. I.N. *Corres.*, I:183.

74. U.L.C. Add. MS. 1975, p. 10.

Chapter Five: A Kinde of Nothinge

1. C.W.F., p. 3.
2. J.E., p. xlii.
3. *Ibid.*, pp. xlii–xliii.
4. *Ibid.*, p. xliii.
5. F.M.N.
6. W.S., p. 56. Also see B.J.T.D., pp. 98–99.
7. F.M.N.
8. D.T.W., II:xi.
9. R.V., p. 38.
10. F.M.N.
11. W.W., p. 129.
12. U.L.C. Add. MS. 4005.5, f. 15^r.
13. F.M.N.
14. J.E., p. xliv.
15. *Ibid.*, pp. lxxxiii–lxxxiv.
16. R.S.W. *Biog.*, p. 181.
17. F.M.N.
18. I.N. *Corres.*, I:3.
19. Keynes MS. 130 (10).
20. F.M.N.
21. Keynes MS. 136, p. 7.
22. C.H.C., III:517. Somewhat different versions of this anecdote are to be found in J.E., pp. xlvi–xlvii, and in W.S. pp. 58–59. According to Stukeley, Newton also harbored an experimental interest in sound

which led to the fashioning of "speaking trumpets." Keynes MS. 136, p. 8.

23. Keynes MS. 130 (10).

24. Keynes MS. 136, p. 11.

25. U.L.C. Add. MS. 3975, p. 22.

26. Walter George Bell, *The Great Fire of London of 1666* (London, 1923), p. 275.

27. F.M.N.

28. I.N. *Corres.*, I:9–10. In composing this part of the letter, Newton drew upon an essay, "An Abridgement of a Manuscript of Sr Robert Southwell's concerning travelling." Written in an unknown hand, it was later found among Newton's papers. Keynes MS. 152.

29. I.N. *Corres.*, I:10.

30. U.L.C. Add. MS. 3975, pp. 227–57.

31. P.H.O., p. 3.

32. *Chambers's Encyclopedia* (Philadelphia and Edinburgh, 1871), I:715.

33. J.A., p. 20.

34. For the text of Barrow's sermons and theological writings, see John Tillotson, *The Works of the Learned Isaac Barrow, D.D.*, 3 vols. (London, 1700).

35. Hill's biographical sketch is prefixed to *Ibid.*, vol. I, unpaged.

36. *Ibid.*

37. P.H.O., pp. 30–31.

38. *Ibid.*, p. 47.

39. The Statutes and Charles II's Confirmation are reprinted in D.T.W., III:xx–xxvii.

40. U.L.C. Add. MS. 3970.10, f. 645^r.

41. D.T.W., III:xviii–xix. This neglect is somewhat mitigated by the fact that in October 1674, after holding the professorship five years, Newton submitted thirty-one "lectiones" on optics delivered between January 1670 and October 1672, and, about 1685, turned over ninety-seven lectures on algebra given between 1673 and 1683.

42. P.H.O., p. 92.

43. R.S.W. *Biog.*, p. 209.

44. J.E., pp. 242–43, fn. 2.

45. Keynes MS. 135.

46. Keynes MS. 130 (4), pp. 1–2. My suspicions are shared by Whiteside but for different reasons from those expressed here. See D.T.W., III:442, fn. 30.

47. F.M. *Portrait*, p. 97.

48. I.N. *Corres.*, I:13.

49. *Ibid.*, I:14.

50. *Ibid.*, I:14–15.

51. *Ibid.*, I:60.

52. J.E., p. xlv.

53. I. Bernard Cohen, *Franklin and Newton* (Philadelphia, 1956), p. 52.

54. L.T.M., p. 81.

55. Isaac Barrow, *Lectiones XVIII Cantabrigiae in Scholis publicis habitae; in quibus opticorum phaenomewn genvinae rationes investigantur, ac exponuntur* (London, 1669).

56. D.T.W., III:xiv, fn. 14.

57. R.S.W. *Biog.*, p. 207.

58. I.N. *Corres.*, I:252.

59. J.A., p. 20.

Chapter Six: Some Strangeness in the Proportion

1. J.E., p. lxxxv.

2. I.N. *Corres.*, I:53–54.

3. *Ibid.*, I:54.

4. F.M.N.

5. I.N. *Corres.*, I:16–17.

6. *Ibid.*, I:24.

7. *Ibid.*, I:27, 54.

8. *Ibid.*, I:20, 24.

9. *Ibid.*, I:30.

10. *Ibid.*, I:31.

11. *Ibid.*, I:32–33.

12. *Ibid.*, I:34–35.

13. *Ibid.*, I:43–44.

14. *Ibid.*, I:55.

15. U.L.C. Add. MS. 4002, p. 1.

16. U.L.C. Add. MS. 4005.5, ff. 14^r, 15^r.

17. *Ibid.*, f. 14^v.

18. J.A., p. 20.

19. Quoted in J.E., p. xlii.

20. U.L.C. Add. MS. 3975, p. 12.

21. U.L.C. Add. MS. 4002, p. 129. For a detailed discussion and analysis of the experiment, see J.A. Lohne, "Experimentum Crucis," *Notes and Records of the Royal Society of London*, 23 (1968): 169–99.

22. I.N. *Corres.*, I:94–95.

23. U.L.C. Add. MS. 4002, p. 22.

24. *Ibid.*, p. 47.

25. I.N. *Corres.*, I:65.

26. *Ibid.*, II:242.

27. *Ibid.*, I:66.

28. *Ibid.*, I:68.
29. *Ibid.*
30. J.E., p. lxxxv.
31. I.N. *Corres.*, I:54.
32. D.T.W., III:71.
33. U.L.C. Add. MS. 3968.41, p. 83.
34. I.N. *Corres.*, II:133.
35. J.R.H., p. 25.
36. I.N. *Corres.*, I:153.
37. *Ibid.*, I:95.
38. *Ibid.*, I:3-4.
39. *Ibid.*, I:53. Perhaps Collins simply misunderstood Newton on this point. The telescope delivered to the Royal Society had a magnifying capacity of about 40 power, the figure originally claimed by Newton.
40. T.B., III:1.
41. I.N. *Corres.*, I:88, 146.
42. *Ibid.*, I:89.
43. *Ibid.*, I:73.
44. T.B., III:1.
45. I.N. *Corres.*, I:79.
46. *Ibid.*, I:80.
47. *Ibid.*, I:82-83.

Chapter Seven: The Killing Ground

1. I.N. *Corres.*, I:84.
2. *Ibid.*, I:97-99.
3. *Ibid.*, I:102.
4. T.B., III:9.
5. I.N. *Corres.*, I:107.
6. *Ibid.*, I:108-9.
7. E.S.D.B., III:290. The seed of the East Indian tree *Strychnos nux-vomica* is one of the sources of strychnine. The poison itself was first produced in 1818.
8. R.L., W.M., VI:84-85.
9. Quoted in M.E., p. 52.
10. T.S., p. 223.
11. R.L., W.M., VII:373.
12. Thomas Kuhn, "Newton's Optical Papers," in I.B.C. *Papers*, p. 28.
13. T.B., III:9.
14. I.N. *Corres.*, I:198.
15. *Ibid.*, I:110-11.
16. M.E., p. 1.
17. *The Posthumous Works of Robert Hooke* . . . , ed. by Richard Waller (London, 1705), p. vii.
18. J.A., p. 165.
19. *The Posthumous Works of Robert Hooke* . . . , ed. by Waller, p. xxvi.
20. J.A., p. 165.
21. H.W.R., W.A., p. 235.
22. *Ibid.*, p. 3. The editors suggest that Hooke chose Pisces, the twelfth sign of the Zodiac, because of the legendary association of Venus and Cupid with the symbol.
23. F. M., *Portrait*, p. 135.
24. I. N. *Corres.*, I:116.
25. T. B., III:10.
26. I.N. *Corres.*, I:100.
27. *Ibid.*, I:4. This account, which refers to meetings of the Royal Society on January 18 and 25, 1672, was drafted by John Collins and appended to a copy he made of Newton's first written description of the reflecting telescope.
28. *Ibid.*, I:95, 111.
29. *Philosophical Transactions*, I (1665-66): 203.
30. Zev Bechler, "A Less Agreeable Matter: The Disagreeable Case of Newton and Achromatic Refraction," *The British Journal for the History of Science*, 8, no. 29 (1975): 101-26.
31. I.N. *Corres.*, I:116, 117, 135.
32. *Ibid.*, pp. 139, 137-39, 135, 136.
33. A.R.H., M.B.H., IX:7-8, 9.
34. I.N. *Corres.*, I:144.
35. A.R.H., M.B.H., IX:60, 63.
36. I.B.C. *Papers*, p. 106.
37. *Ibid.*, p. 109.
38. I.N. *Corres.*, I:151.
39. *Ibid.*, I:120-29, *passim*.
40. *Ibid.*, I:154.
41. *Ibid.*, I:147. This did not prevent Collins from including a list of mathematical works shortly to be published by Huygens, Slusius, and Hevelius, among others.
42. *Ibid.*
43. *Ibid.*, I:161.
44. *Ibid.*, I:122, 135, 137.
45. *Ibid.*, I:151.
46. *Ibid.*, I:155.
47. *Ibid.*, I:159-60.
48. *Ibid.*, I:160.
49. *Ibid.*, II:133.
50. *Ibid.*, I:193.
51. *Ibid.*, I:171-72.

52. *Ibid.*, I:173, 181–82.

53. H.W.R., W.A., p. 192.

54. T.B., III:52.

55. The decision, made at the February 15 meeting of the Royal Society, not to publish Hooke's paper applied only to the issue of the *Philosophical Transactions* that contained Newton's. Oldenburg was thus under no sanction to exclude Hooke's critique from any subsequent issue. T.B., III:10.

56. *Ibid.*, III:52–54.

57. Only what appears to be a partial copy of this paper survives. See I.N. *Corres.*, I: 198–203.

58. *Ibid.*, I:194, 210. Just who Mrs. Arundell was and why Newton spent a fortnight in her home have heretofore remained minor mysteries. However according to the *Alumni Cantabrigienses*, one Francis Arundell, son of Francis of Stoke Park, was admitted fellow-commoner at Trinity in June 1669. He matriculated in 1670 and was granted an M.A., by royal mandate (*Litterae Regiae*) in 1671. It was to Francis's widowed mother's home that Newton came to visit during the summer of 1672, though the nature of his relationship to Arundell, who became Sheriff of Northamptonshire in 1693, remains unknown. John A. Venn and J. A. Venn, *Alumni Cantabrigienses*, pt. I, 4 vols. (Cambridge, England, 1922–27), vol. I.

59. *Ibid.*, I:209.

Chapter Eight: "I Desire to Withdraw"

1. Keynes MS. 137.

2. Stanford University MS. 538. Newton, perhaps taking a cue from Barrow, also extolled the curative powers of opium; to what extent, if at all, he experimented with the drug is impossible to say (U.L.C. Add. MS. 3975, pp. 181–82).

3. M.E., p. 151.

4. W.S., p. 61.

5. John Ivory, *The Foundation of the University of Cambridge, with a Catalogue of the Principal Founders and Special Benefactours of all the Colledges, and totall number of Students, Magistrates and Officers therein being* (Cambridge, England, 1672).

6. C.H.C., III:560.

7. J.E., p. lxxxv.

8. This account, based on a contemporary manuscript, is quoted in C.H.C., III: 559–63.

9. H.W.R., W.A., p. 120.

10. Roger North, *The Lives of the Norths*, ed. by Augustus Jessopp, 3 vols. (London, 1890), II:326.

11. J.E., p. lxxxv. One can only assume, in the absence of contradictory evidence, that a ten-day absence in October 1675 was spent in Lincolnshire.

12. R.S.W. *Biog.*, p. 253. Further evidence of this is contained in the Junior Bursar's Book for the year ending at Michelmas, 1673: "for seiling Mr Newton's chamber," and "for mending the slating ... over Mr Wickins." Slating, of course, would pertain to a leaking roof. J.E., p. xliii.

13. J.E., p. xlvii.

14. B.J.T.D., p. 99.

15. Yahuda MS. 34, f. 2.

16. J.E., p. xliv.

17. A.R.H., M.B.H., IX:119, 249.

18. *Philosophical Transactions*, VIII (1673): 6086.

19. I.N. *Corres.*, I:262.

20. *Ibid.*, I:263.

21. *Ibid.*, I:282.

22. *Ibid.*, I:284.

23. *Ibid.*, I:294–95.

24. *Philosophical Transactions*, VIII (1673): 6108–11. The anonymous letter "N" is substituted for Huygens's name.

25. A.R.H., M.B.H., IX:571–72.

26. I.N. *Corres.*, I:264.

27. A.R.H., M.B.H., IX:676.

28. I.N. *Corres.*, I:283, 284, fn. 1, 290–95.

29. *Ibid.*, I:307.

30. *Ibid.*, I:309.

31. *Ibid.*, I:356.

32. *Ibid.*, I:318. For a discussion of Linus's previously distinguished career see Conner Reilly, S.J., "Francis Line, Peripatetic (1595–1675)," *OSIRIS*, 14 (1962): 222–53.

33. I.N. *Corres.*, I:328–29. Nevertheless, Linus's letter, together with a carefully edited version of Newton's "anonymous" reply, was published in the *Transactions* for January 25, 1674–75.

34. J.E., p. xlix.

35. Quoted in *Ibid.*, p. 1.

36. H.W.R., W.A., p. 148.

37. I.N. *Corres.*, I:334.

38. *Ibid.*, I:358.

39. *Ibid.*, I:318, 356–57.

40. *Ibid.*, I:358.

41. *Ibid.*, I:360.

42. *Ibid.*, I:359.

43. *Ibid.*, I:362–63.

44. *Ibid.*, I:361.

45. *Ibid.*, I:363–64.

46. *Ibid.*, I:366.

47. *Ibid.*, I:365.

48. H.W.R., W.A., p. 211. For Newton's clarifying letters to Oldenburg, see I.N. *Corres.*, I:404, 407.

49. The full text of the "Discourse" is contained in T.B., III:272–78, 280–95.

50. H.W.R., W.A., pp. 200, 205–6.

51. T.B., III:269.

52. F.M. *Portrait*, p. 141.

53. H.W.R., W.A., p. 188.

54. I.N. *Corres.*, I:405, 406.

55. *Ibid.*, I:408.

56. H.W.R., W.A., p. 213.

57. I.N. *Corres.*, I:412, 413.

58. *Ibid.*, I:416.

59. F.M., *Portrait*, p. 143.

60. I.N. *Corres.*, I:393–94.

61. *Ibid.*, I:394.

62. *Ibid.*, I:411.

63. *Ibid.*

64. T.B., III:313.

65. H.W.R., W.A., p. 228.

66. I.N. *Corres.*, II:6.

67. *Ibid.*, II:8–9.

68. *Ibid.*, II:76.

69. *Ibid.*, II:79.

70. For a more detailed discussion of this and related issues, see Johs. Lohne, "Newton's 'Proof' of the Sine Law and his Mathematical Principles of Colors," *Archive for History of Exact Sciences*, 1, No. 4 (1961): 389–405; L. Rosenfeld, "*La théorie des couleurs de Newton et ses adversaires*," *ISIS*, 9 (1927): 44–65; and

Richard S. Westfall, "Newton Defends His First Publication: The Newton-Lucas Correspondence," *ISIS*, 57, pt. 3, no. 189 (1966): 299–314.

71. I.N. *Corres.*, II:104–5.

72. *Ibid.*, III:83.

73. *Ibid.*, III:182–83.

74. *Ibid.*, III:184–85.

75. J.E., p. lxi.

76. Keynes MS. 130 (10), ff. 3ᵛ–4ᵛ.

77. Keynes MS. 136, p. 10.

78. I.N. *Corres.*, II:251.

79. *Ibid.*, II:240.

80. *Ibid.*, II:254, 257, 259, 263.

81. *Ibid.*, II:265.

82. *Ibid.*, II:269.

Chapter Nine: The Treasures of Darkness

1. I.N. *Corres.*, I:xvii; Richard S. Westfall, "Isaac Newton's Index Chemicus," *Ambix*, 23, pt. 3 (November 1975): 175.

2. Keynes MSS. 129 (B):11–12, and 130 (7). The "history" to which Conduitt refers is almost certainly *The Chronology of Ancient Kingdoms Amended*, which Conduitt himself edited and published in 1728.

3. S.D.B., v. II, pp. 373–75. "Fools and knaves" would have been more accurate, for Brewster cited the "contemptible" works of alchemical poetry from which Newton copiously borrowed: Thomas Norton's *Odinall of Alchimy* and Basil Valentine's *Mystery of the Microcosm*. So far as annotation was concerned, Brewster cited Eirenaeus Philaletes's *Secrets Reveal'd, or an Open Entrance to the Shut Palace of the King*.

4. John Maynard Keynes, "Newton the Man," in *Newton Tercentenary Celebrations* (Cambridge, England, 1947), pp. 27, 29.

5. Of the large number of works on the history of alchemy, I have drawn from two in particular for background: E.J. Holmyard, *Alchemy* (London, 1957), and John read, *Through Alchemy to Chemistry: A Procession of Ideas and Personalities* (New York, 1963).

6. Quoted in Bernard Jaffee, *Crucibles: The Story of Chemistry* (New York, 1957), p. 26.

7. W.S., p. 50.

8. U.L.C. Add. MS. 3996, f. 96^v.

9. U.L.C. Add. MS. 3975, pp. 61–66.

10. Bodleian Library, Oxford, MS. Don. b. 15, f. 3^r, hereinafter cited as MS. Don. b. (followed by the appropriate number). The brackets indicate material inserted for effaced writing in the damaged manuscript.

11. F.M.N.

12. Keynes MS. 130 (8).

13. R.S.W. *Biog.*, p. 285.

14. MS. Don. b. 15, ff. 4^r, 8.

15. B.J.T.D., p. 122. Any discussion of Newton's early alchemy, especially this one, would be much the poorer without frequent reference to Mrs. Dobbs's pioneering study.

16. Keynes MS. 29 contains some of the same material used by Newton in this letter, thus suggesting an experimental interest predating May 1669.

17. U.L.C. Add. MS. 3975, pp. 189–90.

18. E.S.D.B., III:336.

19. R.L., W.M., IX:415–16.

20. B.J.T.D., pp. 96–98.

21. U.L.C. Add. MS. 3996, f. 89^r.

22. B.J.T.D., p. 105.

23. Leon Edel, *Henry D. Thoreau* (Minneapolis, 1970), p. 8.

24. B.J.T.D., p. 88.

25. Richard S. Westfall, "The Role of Alchemy in Newton's Career," in *Reason, Experiment, and Mysticism in the Scientific Revolution,* ed. by M.L. Righini Bonelli and William R. Shea (New York, 1975), p. 203. With the exception of Index 3a, which is in the Yale Medical Library, the remainder of the *Index chemicus* is in Keynes MS. 30.

26. Richard S. Westfall, "The Influence of Alchemy on Newton," in *Science, Pseudo-Science and Society,* ed. by Marsha P. Hanen, Margaret J. Osler, and Robert G. Weyant (Waterloo, Ontario, 1980), p. 154.

27. J.R.H., p. 64. Harrison suggests that these few books call into question Dobbs's belief that Barrow had a deep interest in alchemy. It must be remembered, however, that Barrow sold his library in 1655 prior to going abroad. Thus he could have once owned many more books on alchemy than those listed in the inventory.

28. *Ibid.*, pp. 8, 58–59. This is not to say that every book with a publication date prior to 1696 was purchased before he left for London.

29. R.V., p. 110.

30. J.R.H., p. 125.

31. Keynes MS. 67.

32. R.S.W. *Biog.*, p. 286.

33. *Ibid.*, p. 288.

34. I.N. *Corres.*, I:305.

35. Keynes MS. 33, f. 5^r.

36. B.J.T.D., pp. 111–12.

37. J.A., p. 89.

38. *Ibid.*, p. 90.

39. Keynes MS. 29.

40. J. B. Craven, *Count Michael Maier* (London, 1910), p. 70. "Amicus Socrates, Amicus Plato, veritas magis amica," wrote Maier at the beginning of his book. "Amicus Plato amicus Aristotles magis amica veritas," wrote Newton as an afterthought across the top of the first page of the *Quaestiones* (see Chapter 4). Keynes MS. 32 contains 88 pages of Newton's notes from four works by Maier—about 50,000 words in all.

41. H.J. Sheppard, "The Mythological Tradition and Seventeenth Century Alchemy," in *Science, Medicine, and Society in the Renaissance,* ed. by Allen G. Debus (New York, 1972), I:53.

42. Frances A. Yates, *Giordano Bruno and the Hermetic Tradition* (London, 1964), p. 12.

43. Isaiah 45:3.

44. Keynes MS. 33, f. 5^v.

45. Keynes MS. 130, f. 5^v.

46. Keynes MS. 135.

47. J. E. McGuire and P. M. Rattansi, "Newton and the 'Pipes of Pan,'" *Notes and Records of the Royal Society of London,* 21 (1966): 126.

48. See, for example, E. N. da C. Andrade, *Sir Isaac Newton* (London, 1954), Chapter 8; R. J. Forbes, "Was Newton an Alchemist?" *Chymia,* 2 (1949): 28; and F.M. *Portrait,* pp. 160–90.

49. George E. Bates, Jr., "Seventeenth- and

Eighteenth-Century American Science: A Different Perspective," in *Eighteenth Century Studies*, 9, no. 2 (Winter 1975–76): 181.

50. R.S.W. *Biog.*, p. 294.

51. F.M. *Portrait*, p. 165.

52. Richard S. Westfall, "Isaac Newton in Cambridge: The Restoration University and Scientific Creativity," in *Culture and Politics from Puritanism to the Enlightenment*, ed. by Perez Zagorin (Berkeley and Los Angeles, 1980), p. 151.

53. Quoted in S.H.L., p. 41.

54. Keynes MS. 130 (10), f. 3^v.

55. Westfall, "The Role of Alchemy in Newton's Career," p. 196.

56. B.J.T.D., p. 175.

57. Yahuda MS. 259, no. 9.

58. *Theatrum Chemicum Britannicum*, ed. by Elias Ashmole (London, 1652), pp. 446–47.

59. U.L.C. Add. MS. 3975, p. 80.

60. *Ibid.*, p. 143.

61. Westfall, *The Role of Alchemy in Newton's Career*," p. 228.

62. B.J.T.D., p. 143.

63. Keynes MS. 55, f. 3^r.

64. MS. Don. b. 15, ff. 4^v–5^r.

65. U.L.C. Add. MS. 3975, p. 82.

66. I.N. Corres., I:82.

67. B.J.T.D., p. 150.

68. Keynes MS. 64, f. 4^r.

69. Keynes MS. 18. The Latin "*Clavis*" has been translated by Dobbs, whose rendering I have quoted (B.J.T.D., Appendix C).

70. B.J.T.D., p. 185.

71. J.R.H., pp. 65, 215, 243.

72. One student of Newton's alchemy has argued that he probably copied "*Clavis*" from a lost manuscript of Starkey. Karen Figala, "Newton as Alchemist," *History of Science*, XV (1977): 107. For an opposing view, see Westfall, "The Role of Alchemy in Newton's Career," p. 207, and B.J.T.D., pp. 175–78.

73. *The Works of the Honourable Robert Boyle*, ed. by Thomas Birch, new ed., 6 vols. (London, 1772), V:79–80.

74. See his letter to John Locke written on August 2, 1692. I.N. Corres., III:218.

75. *Conway Letters: The Correspondence of Anne, Viscountess Conway, Henry More,*

and their Friends, 1642–1684, ed. by Marjorie Hope Nicolson (New Haven, 1930), p. 479.

76. *Philosophical Transactions*, X (1675–76): 515–33.

77. I.N. *Corres.*, II:2.

78. *Ibid.*, I:364.

79. *Unpublished Scientific Papers of Isaac Newton*, ed. by A. Rupert Hall and Marie Boas Hall (Cambridge, England, 1962), p. 141.

80. Burndy Library, Norwalk, Connecticut, MS. 16, f. 1^r. Hereinafter cited as Burndy MS. (followed by the appropriate number).

81. James Boswell, *The Life of Johnson*, ed. by Christopher Hibbert (New York, 1979), p. 122.

82. Emile Mâle, *L'Art religieux de la fin du moyen age en France*, 2d ed. (Paris, 1922), pp. 303–4.

83. *Theatrum Chemicum Britannicum*, ed. by Ashmole, p. 279.

Chapter Ten: Heretic: Sotto Voce

1. J.E., p. lxxxv.

2. Keynes MS. 130 (8).

3. *Bishop's Transcripts of Colsterworth*, Lincoln Cathedral Archives, Lincoln, England.

4. C.W.F. p. 50.

5. Geoffrey Wolff, "Minor Lives," in *Telling Lives: The Biographer's Art*, ed. by Marc Pachter (Washington, D.C., 1979), p. 68.

6. C.W.F., p. 52.

7. J.E., p. lxxxv.

8. I.N. *Corres.*, II:300.

9. *Ibid.*, VII:373.

10. *Ibid.*, III:393. This letter has been attributed to the 1690s by the editors of the *Correspondence*. The opposite side contains mathematical calculations placed around 1680. D.T.W., IV:329.

11. I.N. *Corres.*, II:269–74, 502–4.

12. *Ibid.*, V:253–54.

13. J.E., p. lxxxv.

14. R.S.W. *Biog.*, p. 343, fn. 30.

15. Keynes MS. 137.

16. Keynes MS. 135.

17. S.D.B., I:471.
18. Keynes MS. 135.
19. Roger North, *The Lives of the Norths* ed. by Augustus Jessopp, 3 vols. (London, 1890), II:283–85.
20. *Ibid.*, II:277.
21. Quoted in J.E., p. lxiv.
22. North, *The Lives of the Norths*, II:328.
23. I.N. *Corres.* II:207.
24. R.S.W. *Biog.*, p. 336.
25. I.N. *Corres.*, VII:390.
26. *Ibid.*, II:415.
27. J.E., pp. xxvi, xxviii–xxix; J.R.H., pp. 154, 167.
28. W.S., p. 16.
29. J. R. H., pp. 59, 66.
30. R. S. W., "Writ.," p. 14.
31. U.L.C. Add. MS. 3996, ff. 128–29.
32. Roland N. Stromberg, *Religious Liberalism in Eighteenth-Century England* (Oxford, 1954), p. 39.
33. Keynes MS. 2, f. XIII^V.
34. *Ibid.*, f. XI.
35. R.S.W. *Biog.*, p. 311.
36. Keynes MS. 2, pp. 13–14, 49–50. "De Trinitate," or "On the Trinity" covers nine pages, making it the longest entry in the notebook, pp. 33–36, 79–82, 89.
37. Revelation 14:9–10.
38. Yahuda MS. 14, f. 25.
39. Keynes MS. 3, p. 14.
40. "The Dynamo and the Virgin," in *The Education of Henry Adams*, ed. by Ernest Samuels (Boston, 1973), pp. 379–90.
41. Keynes MS. 3, p. 14.
42. Yahuda MS. 15.3, f. 46^r.
43. Yahuda MS. 1.1., f. 2^r.
44. J. E. McGuire and P. M. Rattansi, "Newton and the 'Pipes of Pan,'" *Notes and Records of the Royal Society of London*, 21 (1966): 130.
45. Keynes MS. 33, f. 5^V.
46. Yahuda MS. 41, ff. 6–7.
47. Babson College Library MS. 434, f. 58.
48. Keynes MS. 135.
49. Yahuda MS. 18, f. 2^V.
50. F.M. *Religion*, p. 86.
51. Yahuda MS. 1.1, f. 1.
52. F.M. *Religion*, p. 88.
53. Many Biblical scholars now believe that St. John the apostle and St. John of Ephesus, reputed author of Revelation, were two different persons.

54. Yahuda MS. 1.1, f. 1.
55. *Ibid.*, ff. 12^r, 14^r.
56. Yahuda MS. 14, ff. 78–80.
57. Yahuda MS. 1.1, ff. 24–27.
58. Yahuda MSS. 4.1, 4.2.
59. M.C.J., p. 132.
60. R.S.W. *Biog.*, p. 321.
61. Yahuda MS. 1.1, f. 16^r.
62. Yahuda MS. 1.2, ff. 60–61.

Chapter Eleven: A Pitfall in Eden

1. Keynes MS. 135.
2. Yahuda MS. 2.4, f. 25^V.
3. Keynes MS. 130 (6), Book 2. The anecdote was related by John Conduitt.
4. I.N. *Corres.*, II:297.
5. *Ibid.*, p. 299, fn. 7.
6. J.A., p. 167.
7. From the *Lectiones Cutleriane*, quoted in *Ibid.*, p. 166.
8. I.N. *Corres.*, II:300–301.
9. *Ibid.*, II:436.
10. *Ibid.*, II:302.
11. *Ibid.*, II:304–5.
12. *Ibid.*, II:306.
13. *Ibid.*, II:307.
14. H.W.R., W.A., pp. 435–36.
15. I.N. *Corres.*, II:309.
16. *Ibid.*, II:312–13.
17. *Ibid.*, II:447.
18. U.L.C. Add. MS. 3965.1, ff. 1–3.
19. H.W.R., W.A., pp. 459–60.
20. I.N. *Corres.*, II:315.
21. Angus Armitage, *Edmond Halley* (London, 1966), p. 43.
22. U.L.C. Add. MS. 4004, f. 98^V.
23. I.N. *Corres.*, II:368.
24. U.L.C. Add. MS. 4004, f. 98^V.
25. I.N. *Corres.*, II:360.
26. L. Rosenfeld, "Newton and the Law of Gravitation," *Archive for the History of Exact Sciences*, 2 (1965):377.
27. I.N. *Corres.*, II:338, 346.
28. U.L.C. Add. MS. 4004, f. 105.
29. D.T.W., V:298–302 and Appendix 3.
30. C.A.R., p. 61.
31. I.N. *Corres.*, II:361.
32. T.B., 4:89–90.
33. J.R.H., p. 116.
34. U.L.C. Add. MS. 4004, f. 99. Indeed, Ellis

had shown Newton a copy of a letter written by Cassini, who at first supported the dual comet theory. Perhaps there was a later one detailing Cassini's change of mind (I.N. *Corres.*, II:342). The Bainbridge of whom Newton wrote was probably Thomas Bainbridge, Vice-Master of Trinity, 1700–1703.

35. U.L.C. Add. MS. 4004, ff. 103^r–103^v.
36. Robert Hooke, *Cometa*, reprinted in *Early Science in Oxford*, ed. by R. T. Gunther, 14 vols. (Oxford, 1920–25), VIII:223–24, 247.
37. U.L.C. Add. MS. 3965.14, f. 613.
38. I.N. *Corres.*, I:364.
39. I.N. *Principia*, pp. 325–26.
40. I.N. *Corres.*, II:288.

Chapter Twelve: The Most Perfect Mechanic of All

1. I.N. *Corres.*, II:433–35, 442.
2. *Ibid.*, II:442.
3. J.A., p. 121.
4. Quoted in C.A.R., p. 6.
5. U.L.C. Add. MS. 4004, f. 101^v.
6. U.L.C. Add. MS. 4007, ff. 707^r–707^v.
7. Quoted in R.S.W. *Biog.*, p. 452.
8. U.L.C. Add. MS. 4007, f. 707^r.
9. *Ibid.*
10. T.B., 4:347. While the minutes suggest that Halley had not seen *De Motu* before his second visit to Cambridge, most scholars agree that Paget had carried a copy to him in London. See, for example, I.B.C. *Intro.*, pp. 54–58.
11. On February 23, 1685, Newton wrote to Aston, now Secretary of the Royal Society, to thank him for "entring in your Register my Notions about Motion." I.N. *Corres.*, II:415.
12. J.E., p. lxxxv.
13. I.N. *Corres.*, II:413.
14. *Ibid.*, II:403, 412.
15. *Ibid.*, II:408.
16. *Ibid.*, II:413. As is obvious, Newton's general discovery that cometary orbits are conics (or ellipses when below a certain velocity) had not yet been applied to the movements of specific bodies.
17. *Ibid.*, II:407, 409, 413.
18. *Ibid.*, II:415.

19. J.E. p. lxxxv.
20. Antonia Fraser, *Royal Charles: Charles II and the Restoration* (New York, 1979), pp. 442–57.
21. C.H.C., III:606–11.
22. See I. Bernard Cohen, "Pemberton's Translation of Newton's *Principia*, With Notes on Motte's Translation," *ISIS*, 54, pt. 3 (September 1963): 319–51.
23. I.N. *Principia*, p. 397.
24. Keynes MS. 133, p. 10.
25. Keynes MS. 130 (6), Book 2.
26. I.N. "Account," p. 206.
27. R.S.W. *Biog.*, p. 424.
28. I.N. *Principia*, p. xvii.
29. *Ibid.*, p. xviii.
30. *Ibid.*, p. 13.
31. R.S.W. *Biog.* p. 411.
32. I.N. *Principia*, pp. 13–14.
33. *Ibid.*, pp. 193, 197, 202–3.
34. *Ibid.*, pp. 230–31.
35. *Ibid.*, pp. 387, 396.
36. T.B. 4:453.
37. I.B.C. *Intro.* pp. 122–24.
38. T.B. 4:479. Halley's paper, titled "Discourse Concerning Gravity, and its Properties," was subsequently published in the *Philosophical Transactions*, XVI (1686–87): 3–8.
39. T.B., 4:479–80.
40. C.A.R., p. 75.
41. T.B., 4:484.
42. I.N. *Corres.*, II:431–32.
43. *Ibid.*, pp. 433–34.
44. *Ibid.*, p. 434.
45. T.B., 4:486, 474, 545.
46. I.N. *Corres.*, II: pp. 435–37.
47. *Ibid.*, pp. 438–39.
48. *Ibid.*, pp. 441–43.
49. *Ibid.*, pp. 444–45.
50. *Ibid.*, pp. 446–47.
51. Keynes MS. 135.
52. *The Diary of Robert Hooke*, repr. in *Early Science in Oxford*, ed by R.T. Gunther (Oxford, 1920–25), X:98, 184.
53. I.N. *Corres.*, II:464.
54. T.B., 4:505, 516, 523.
55. I.N. *Corres.*, II:469.
56. *Ibid.*, II:472.
57. *Ibid.*, II:474.
58. I.N. *Principia*, p. 415.
59. *Ibid.*, p. 398.
60. *Ibid.*, pp. 424, 428.

61. E. N. da C. Andrade, *Sir Isaac Newton* (London, 1954), p. 86.
62. I.N. *Principia*, p. 435.
63. I.N. *Corres.*, II:419, 421, 437.
64. I.N. *Principia*, p. 498.
65. I.N. *Corres.*, III:234.
66. *Ibid.*, II:481–82.
67. *Ibid.*, III:233.

Chapter Thirteen: "Go Your Way, and Sin No More"

1. Keynes MS. 135.
2. From an examination of the watermarks on the manuscripts in the University Library, Cambridge, it seems clear that Newton had a decided preference for paper imported from Holland. I have traced the watermark to Jean Villedary, a French Huguenot, who emigrated to Holland around 1668. It depicts a lion rampant, fasces in one paw, sword in the other, striking at a hat impaled on a rod held by a mythological figure. Above the watermark is the inscription, *Pro Patria* (for the fatherland). See W. A. Churchill, *Watermarks in Paper* (Amsterdam, 1935), pp. 21–22, 71, xcix.
3. I.N. *Corres.*, II:483.
4. *Philosophical Transactions*, 16 (1686–87): 297.
5. I.N. *Corres.*, II:484.
6. *Ibid.*, II:501, III:7.
7. *Ibid.*, II:485.
8. "Éloge de Mr. Moivre," *Historie de l'Academie royale des sciences* (Paris, 1754), p. 262.
9. J.T. Desaguliers, *Experimental Philosophy* (London, 1734), vol. I, Preface.
10. I.B.C. *Intro.*, p. 146.
11. I.N. *Corres.*, III:71–76.
12. *Ibid.*, III:4–5.
13. *Ibid.*, III:40–42.
14. I.N. *Principia*, p. xviii.
15. *Unpublished Scientific Papers of Isaac Newton*, ed. by A. Rupert Hall and Marie Boas Hall (Cambridge, England, 1962), pp. 333, 341.
16. Keynes MS. 130 (5).
17. Maurice Ashley, *James II* (Minneapolis, 1977), p. 15.
18. C.H.C., III:615.
19. I.N. *Corres.*, II:467–68.
20. Quoted in C.H.C., III:614, 634.
21. Proverbs 28:1.
22. Keynes MS. 130 (10), ff. 3ᵛ–4.
23. Quoted in C.H.C., III:621.
24. *Ibid.*, p. 626.
25. Keynes MS. 116.
26. Quoted in C.H.C., III:631.
27. I.N. *Corres.*, II:502–3.
28. *Ibid.*, II:504.
29. J.E., p. lxxxi.
30. Ashley, *James II*, p. 229.
31. I.N. *Corres.*, III:8.
32. Keynes MS. 113.
33. J.E., p. lxxxi.
34. E.S.D.B., IV:633.
35. See the letter from William Herbert of the Middle Temple asking Newton to relay a personal message to Sacheverell. I.N. *Corres.*, III:66.
36. *Ibid.*, p. 10.
37. C.H.C., IV:2.
38. I.N. *Corres.*, III:12–13.
39. *Ibid.*, p. 24.
40. *Ibid.*, pp. 22–23. Newton also referred to "some indisposition" (p. 18), which kept him in his rooms for a few days in March. It is not clear whether or not this was the same illness.
41. *Ibid.*, p. 16.
42. J.E., p. lix.
43. *Ibid.*
44. C.H.C., IV:9–10.

Chapter Fourteen: Cul-de-Sac

1. E.S.D.B., IV:343.
2. Edward Ward, *A Step to Stir-Bitch-Fair: with Remarks upon the University of Cambridge* (London, 1700), pp. 12–13.
3. C.H.C., IV:12–13.
4. *Ibid.*, IV:12.
5. John Locke, *An Essay Concerning Human Understanding*, ed. by Peter H. Nidditch (Oxford, 1975), pp. 9–10.
6. I.N. *Corres.*, III:79.
7. L.P.K., p. 263.
8. I.N. *Corres.*, III:82.
9. *Ibid.*, III:195.
10. L.P.K., p. 263.
11. John Locke, *Epistola de Tolerantia, A*

Letter on Toleration, ed. by Raymond Kilban (Oxford, 1968), p. xxiv.

12. I.N. *Corres.*, III:147, 152.

13. *Ibid.*, III:185, 192.

14. *Ibid.*, p. 195. "Mr Paulin" was undoubtedly Robert Pawling, a friend with whom Locke stayed when he was in London.

15. *Ibid.*, p. 154.

16. J.E., p. lxxxv.

17. I.N. *Corres.*, III:193.

18. *Ibid.*, III:195.

19. *Ibid.*, III:215–16.

20. *Ibid.*, III:219.

21. William Seward, *Anecdotes of Distinguished Persons*, 5th ed. (London, 1804), II:178.

22. *Ibid.*, pp. 182, 190.

23. I.N. *Corres.*, III:45.

24. Quoted in Charles Andrew Domson, *Nicolas Fatio de Duillier and the Prophets of London:An Essay in the Historical Interaction of Natural Philosophy and Millennial Belief in the Age of Newton* (New York, 1981), pp. 32–33.

25. I.N. *Corres.*, III:390.

26. See, for example, Seward, *Anecdotes of Distinguished Persons*, II:173–95. Fatio, a foreigner, employed the *very same* superscript abbreviations as did Newton: y^e, w^{th}, w^{ch}, y^t, and so on.

27. I.N. *Corres.*, III:68–69.

28. *Ibid.*, III:69–70, 191.

29. *Ibid.*, III:69.

30. *Ibid.*, III:186–87.

31. *The Diary of Robert Hooke*, repr. in *Early Science in Oxford*, ed. by R.T. Gunther (Oxford, 1920–25), X:190–91.

32. I.N. *Corres.*, III:168.

33. J.E., p. lxxxv.

34. I.N. *Corres.*, III:170.

35. Christiaan Huygens, *Oeuvres complètes*, pub. by *Société hollandaise des sciences* (The Hague, 1888–1950), XI:271–72. Parts of this letter are also found in I.N. *Corres.*, III:186–87.

36. I.N. *Corres.*, III:230, VII:392.

37. F.M. *Portrait*, p. 199.

38. I.N. *Corres.*, III:231, VII:392.

39. *Ibid.*, III:232–33, VII:392.

40. *Ibid.*, III:241–43.

41. *Ibid.*, III:242, 245.

42. *Ibid.*, III:261, 263.

43. *Ibid.*, III:391.

44. *Ibid.*, III:266.

45. *Ibid.*, III:268–69.

46. J.E., pp. lxxxii, lxxxv, lxxxix–xc.

47. Huygens, *Oeuvres complètes*, XII:162.

48. Voltaire, *Letters concerning the English Nation* (London, 1733), pp. 116–17.

49. G. M. Trevelyan, *Trinity College: An Historical Sketch* (Cambridge, England, 1946), pp. 50–66.

50. I.N. *Corres.*, III:150–51, 155–56.

51. Eustace Budgell, *Memoirs of the Lives and Characters of the Illustrious Family of the Boyles* (London, 1737), Appendix, p. 25.

52. E.S.D.B., V:88.

53. *Letters and the Second Diary of Samuel Pepys*, ed. by R. G. Howarth (London and Toronto, 1933), p. 226.

54. Henry Guerlac and M. C. Jacob, "Bentley, Newton, and Providence (The Boyle Lectures Once More)," *Journal of the History of Ideas*, XXX (July 1969):307–18, and M.C.J., Chapter Four.

55. I.N. *Corres.*, III:233.

56. Guerlac and Jacob, "Bentley, Newton, and Providence," XXX:311–12.

57. I.N. *Corres.*, III:234.

58. *Ibid.*, III:274.

59. *Ibid.*, III:235–36.

60. *Ibid.*, III:240, 253–54.

61. *Ibid.*, III:164.

62. *Ibid.*, III:155.

63. *Ibid.*, III:166, 171, 181, 199.

64. *Ibid.*, III:203.

65. Quoted in C.A.R., p. 124.

66. Domson, *Nicolas Fatio de Duillier and Prophets of London*, p. 40.

67. Quoted in C.A.R., p. 122.

68. Keynes MS. 130 (7), p. 1.

69. I.N. *Corres.*, III:278.

70. *Ibid.*, III:279.

71. *Ibid.*, III:281–83.

72. *Ibid.*, III 280.

73. *Ibid.*, III:283–84, 286, 292.

74. *Ibid.*, III:369–70.

75. John Bartlett, *Bartlett's Familiar Quotations*, ed. by Emily Morison Beck, rev. ed. (New York, 1980), p. 313.

76. L.T.M., p. 389.

77. P. E. Spargo and C. A. Pounds, "Newton's 'Derangement of the Intellect':

New Light on an Old Problem," *Notes and Records of the Royal Society of London*, 34, no. 1 (July 1979): 28–29. Others have advanced a similar hypothesis: see L. W. Johnson and M. L. Wolbarsht, "Mercury Poisoning: A Probable Cause of Isaac Newton's Physical and Mental Ills," *Notes and Records of the Royal Society of London*, 34, no. 1 (July 1979): 1–9; R. Seitz and J. Y. Lettvin, "Mercury and Melancholy: The Decline of Isaac Newton," *Bulletin of the American Physical Society*, ser. II, 16 (1971): 1400.

78. Johnson and Wolbarsht, "Mercury Poisoning: A Probable Cause of Isaac Newton's Physical and Mental Ills," p. 2.
79. F.M. *Portrait*, pp. 219–20.
80. Domson, *Nicolas Fatio de Duillier and Prophets of London*, p. 63.
81. I.N. *Corres.*, III:282.
82. *Ibid.*, III:359–60.
83. Domson, *Nicolas Fatio de Duillier and Prophets of London*, p. 60.
84. This little known chapter in Newton's intellectual development was only recently delineated. See D.T.W., VII: 186–561.
85. I.N. *Corres.*, III:339.
86. Babson College MS. 420.
87. I.N. *Corres.*, III:293–303. Also see Florence N. David, "Mr Newton, Mr Pepys & Dyse: A Historical Note," *Annals of Science*, 13 (1957): 137–47.
88. I.N. *Corres.*, III:369, 372–73.
89. *Ibid.*, IV:131.
90 *Ibid.*, IV:84, 110.
91. *Ibid.*, IV:24.
92. Bertrand Russell, *The Autobiography of Bertrand Russell, 1872–1914*, (London, 1967), I:153.
93. D.T.W., VII:xxiii.
94. U.L.C. Add. MS. 3996, f. 109^r.
95. I.N. *Corres.*, IV:7.
96. *Ibid.*, IV:8.
97. Derek Howse, *Francis Place and the Early History of the Greenwich Observatory* (New York, 1975), pp. 18–20.
98. I.N. *Corres.*, IV:14, fn. 8; 80–81.
99. *The Diary of Robert Hooke*, repr. in *Early Science in Oxford*, ed. by Gunther, X:207; H.W.R., W.A., pp. 105, 330.
100. Quoted in C.A.R., p. 126.

101. I.N. *Corres.*, IV:13, 24.
102. *Ibid.*, IV:26, 35.
103. *Ibid.*, IV:45.
104. Keynes MS. 130 (6), Book 3.
105. I.N. *Corres.*, IV:54–55, 58.
106. *Ibid.*, IV:62.
107. *Ibid.*, IV:87, 106.
108. *Ibid.*, IV:113, 121.
109. *Ibid.*, IV:133.
110. *Ibid.*, IV:134.
111. *Ibid.*, IV:135.
112. *Ibid.*, IV:138.
113. *Ibid.*, IV:143.
114. *Ibid.*, IV:144.
115. *Ibid.*, IV:150.
116. *Ibid.*, IV:152.
117. Keynes MS. 130 (6), Book 3.
118. I.N. *Corres.*, IV:169, 192.
119. *Ibid.*, IV:188–89.
120. *Ibid.*, IV:188.
121. U.L.C. Add. MS. 3996.15, f. 370^v.
122. I.N. *Corres.*, IV:172.
123. *Ibid.*, IV:196–98.
124. *Ibid.*, IV:193–94.
125. *Ibid.*, IV:195.

Chapter Fifteen:
A Morality Play

1. *The Letters of John Dryden*, ed. by Charles E. Ward (Durham, N.C., 1942), pp. 77, 80.
2. Thomas Macaulay, *History of England* (New York, 1898), VI:96, fn. 1.
3. *Ibid.*, VI:94.
4. J.C. *Mint*, pp. 145–46.
5. A. E. Feavearyear, *The Pound Sterling: A History of English Money* (Oxford, 1931), p. 4.
6. J.C. *Mint*, p. 167.
7. My discussion of Mint operations was derived largely from J.C. *Newton* and J.C. *Mint*.
8. *Privy Council Register*, vol. 76, July 18, 1695.
9. Hopton Haynes, *Brief Memoires Relating to the Silver & Gold Coins of England: With an Account of the Corruption of the Hammerd Monys. And of the Reform by the Late Grand Coynage, At the Tower, & the Five Country Mints,. In the Years*

1696, 1697, 1698, & 1699 (British Library, Lansdowne MS. DCCCI), ff. 44, 71.

10. Dudley North, *Discourses Upon Trade* (London, 1691), p. 14.

11. G. Findlay Shirras and J. H. M. Craig, "Sir Isaac Newton and the Currency," *The Economic Journal*, LV (1945): 223.

12. *Ibid.*, LV: 222–27, and J.C. *Newton*, pp. 9–10.

13. Keynes MS. 130 (6), Book 2.

14. C.T.P., 10:1358.

15. I.N. *Corres.*, IV:201.

16. Mint MS. 19.1, ff. 9–10.

17. J.C. *Newton*, pp. 2–4, and Newton's "An Account of the Mint in the Tower of London [1697]," I.N. *Corres.*, IV:233–35.

18. W.S., p. 58.

19. J.H.C., 11:776.

20. Hall, who had outside business interests and held other government posts, left an estate valued at £42,000. Thus he died a somewhat wealthier man than Newton (Keynes MS. 148).

21. I.N. *Corres.*, IV:203.

22. E.S.D.B., V:228.

23. Narcissus Luttrell, *A Brief Historical Relation of State Affairs* (Oxford, 1857), IV: 86.

24. J.C. *Mint*, p. 191.

25. Feavearyear, *The Pound Sterling: A History of English Money*, p. 131.

26. I.N. *Corres.*, IV:204–5.

27. *Ibid.*, IV:205–6. This letter, written on or near June 16, also states that "the Warden is now at much greater expenses & pains then formerly in going constantly to the Mint," making it appear that he was not inhabiting the premises, even at this early date.

28. *Ibid.*, IV:206.

29. *Correspondence and Papers of Edmond Halley*, ed. by Eugene Fairfield MacPike (Oxford, 1932), p. 97.

30. I.N. *Corres.*, IV:254.

31. *Ibid.*, IV:230.

32. *Ibid.*, IV:254.

33. *Correspondence and Papers of Edmond Halley*, ed. by MacPike, p. 103.

34. I.N. *Corres.*, IV:246.

35. *Ibid.*, VII:400–401.

36. J.C. *Newton*, p. 15.

37. J.C. *Mint*, p. 193.

38. J.H.C., 11:701, 774.

39. J.C. *Mint*, p. 192.

40. I.N. *Corres.*, IV:241, fn. 2.

41. J.H.C., 11:776.

42. C.S.P.D. (1697), p. 219.

43. I.N. *Corres.*, IV:243, 240.

44. *Ibid.*, IV:244.

45. Newton invoked this and many other forgotten privileges in his 1697 memorandum, "An Account of Mint in Tower of London."

46. *Ibid.*, IV:243.

47. Mint MS., 19.3, ff. 388^v–89^r.

48. C.T.B., 13:59–60.

49. Mint MS., 19.1, f. 407.

50. I.N. *Corres.*, IV:258.

51. J.R.H., p. 72.

52. Keynes MS. 130 (7).

53. I.N. *Corres.*, IV:236, 216–17.

54. *Ibid.*, IV:208.

55. F.M. *Portrait*, p. 232.

56. I.N. *Corres.*, IV:209–10.

57. C.S.P.D. (1696), p. 362.

58. Mint MS. 19.1, f. 467.

59. I.N. *Corres.*, IV:217.

60. John Craig, "Isaac Newton and the Counterfeiters," *Notes and Records of the Royal Society of London*, XVIII (1963): 139.

61. Mint MS. 15.17, no. 27.

62. C.S.P.D. (1697), p. 439.

63. I.N. *Corres.*, IV:317.

64. F.M. *Portrait*, p. 232.

65. John Craig, "Isaac Newton—Crime Investigator," *Nature*, 182 (July 19, 1958): 150.

66. Mint MS. 15.17, nos. 215, 233.

67. Haynes, *Brief Memoires Relating to the Silver & Gold Coins of England . . . In the Years 1696, 1697, 1698, & 1699*, f. 68.

68. I.N. *Corres.*, IV:209.

69. Craig, "Isaac Newton and the Counterfeiters," pp. 138–39.

70. C.S.P.D. (1697), p. 117, and I.N. *Corres.*, VII:408.

71. R.S.W. *Biog.*, p. 570.

72. *Guzman Redivivus. A Short View of the Life of Will. Chaloner* (London, 1699), p. 8.

73. *Ibid.*, pp. 4–5.

74. I.N. *Corres.*, IV:261–62.

75. *Ibid.*, IV:231–32.

76. J.H.C., 11:777.

77. C.S.P.D. (1697), p. 357.

78. Mint MS. 15.17, nos. 129, 134.
79. I.N. *Corres.*, IV:305.
80. *Ibid.*, VII:407.
81. *Guzman Redivivus*, p. 9.
82. I.N. *Corres.*, IV:307.
83. *Guzman Redivivus*, p. 12.

Chapter Sixteen:
"Your Very Loving Unkle"

1. I.N. *Corres.*, IV:265.
2. E.S.D.B., V:284, fn. 5.
3. Robert K. Massie, *Peter the Great: His Life and World* (New York, 1980), p. 209.
4. E.S.D.B., V:283.
5. Keynes MS. 130 (7), f. 1.
6. I.N. *Corres.*, IV:225.
7. Keynes MSS. 130 (5), f. 1; 130 (6), Book 1.
8. I.N. *Corres.*, IV:220–28.
9. D.T.W., VIII:506.
10. *Ibid.*, VIII:73, fn. 1.
11. *Ibid.*, VIII:9.
12. I.N. *Corres.*, IV:267.
13. J.B.C.R.S., X:52.
14. I.N. *Corres.*, IV:296.
15. J.E., pp. xxxv, 299.
16. I.N. *Corres.*, IV:253.
17. *Ibid.*, IV:294. Emphasis Flamsteed's.
18. J.E., pp. xxxvi, lxix.
19. *Ibid.*, p. lxix, and R.S.W. *Biog.*, pp. 576–77.
20. James Boswell, *The Life of Johnson*, ed. by Christopher Hibbert (New York, 1979), p. 344.
21. James H. Monk, *The Life of Richard Bentley, D. D.* 2d ed. (London, 1833), I: 96.
22. I.N. *Corres.*, IV:250, 252.
23. *Ibid.*, IV:277.
24. *Ibid.*, IV:284, 291.
25. *Ibid.*, IV:292, fn. 2.
26. *Ibid.*, IV:292.
27. *Ibid.*, IV:295, fn. 8.
28. *Ibid.*, IV:289.
29. *Ibid.*, IV:296.
30. *Ibid.*, IV:303–4.
31. *Catalogue of the Newton Papers Sold by Order of the Viscount Lymington* (London, 1936), lot 208, p. 51.
32. C.W.F., pp. 51, 53.
33. I.N. *Corres.*, IV:349.

34. John Dryden, *Miscellany Poems*, 5th ed. (London: 1727), 5:61. Other Kit-Kat poems authored after Dryden's death are also contained in this work.
35. Samuel Johnson, *Lives of the English Poets* (Oxford, 1955), I:379.
36. Jonathan Swift, *Journal to Stella*, ed. by Harold Williams (Oxford, 1948), I:229–30.
37. Brook Taylor, *Contemplatio philosophica* (London, 1793), pp. 93–94.
38. F.M. *Portrait*, p. 252.
39. *The Works and Life of the Right Honourable Charles, Late Earl of Halifax* (London, 1715), Appendix, p. iv.
40. Voltaire, *Dictionnaire philosophique; Oeuvres complètes de Voltaire,* (Paris, 1787), 42:165.
41. *The Works and Life of the Right Honourable Charles, Late Earl of Halifax,* Appendix, pp. v–vi.
42. F.B., p. 314.
43. Johnson, *Lives of the English Poets*, I: 376.
44. Mary de la Rivière Manley, *Memoirs of Europe, Towards the Close of the Eighth Century. Written by Eginardus* (London, 1710), p. 294.
45. Quoted in E. S. Roscoe, *Robert Harley, Earl of Oxford, Prime Minister 1710–1714. A Study of Politics and Letters in the Age of Anne* (New York and London, 1902), p. 1.
46. Mary de la Rivière Manley, *The Adventures of Rivella* (London, 1714), p. 113.
47. Swift, *Journal to Stella*, ed. by Williams, I: 31, 36, 55, 105.
48. *Ibid.*, p. 39.
49. *Catalogue of Newton Papers Sold by Order of Viscount Lymington,* lot 176, p. 41.
50. I.N. *Corres.*, V:xlv–xlvi.
51. Christopher Hill, "Newton and His Society," in *The Annus Mirabilis of Sir Isaac Newton, 1666–1966,* ed. by Robert Palter (Cambridge, Mass., 1970), p. 40.
52. I.N. *Corres.*, VI:225.
53. Augustus De Morgan, *Newton: His Friend: And His Niece* (London, 1885), p. 49. Emphasis De Morgan's.
54. *The Marriage, Baptismal, and Burial Registers of the Collegiate Church or Abbey of St. Peter Westminster,* ed. by

Joseph Lemuel Chester (London, 1876), p. 354.

55. *The Works and Life of the Right Honourable Charles, Late Earl of Halifax*, p. 196.
56. Swift, *Journal to Stella*, ed. by Williams, I: 31, 230, and II:383.
57. Keynes MS. 130 (7), f. 1.
58. J.E., p. lxxviii.
59. J.C. *Newton*, p. 30.
60. I.N. *Corres.*, IV:348.
61. R.S.W. *Biog.*, pp. 606–7, and J.C. *Mint*, pp. 199–200.
62. J.C. *Mint*, p. 212.
63. Monk, *The Life of Richard Bentley, D. D.*, I:157.
64. I.N. *Corres.*, IV:382–83.
65. Keynes MS. 130 (6), Book 1.
66. *Ibid.*, Book 2. Conduitt seems to have confused some dates and names when recounting this anecdote. See R.S.W. *Biog.*, p. 589, fn. 131.
67. I.N. *Corres.*, IV:377–80.
68. Keynes MS. 130 (6), Book 2.
69. I.N. *Corres.*, IV:383–84.
70. *Ibid.*, pp. 384–88.
71. L.P.K., p. 263.
72. I.N. *Corres.*, IV:406.
73. Maurice Cranston, *John Locke: A Biography* (London and New York, 1957), p. 480.

Chapter Seventeen: Sir Isaac

1. R.S.W. *Biog.*, p. 626.
2. Keynes MS. 130 (6), Book 2.
3. J.B.C.R.S., X:53–54.
4. S.H.L., p. 343.
5. *Ibid.*, pp. 341–42.
6. C.R.W., I:353.
7. F.M. *Portrait*, p. 284.
8. R.S.W. *Biog.*, p. 630.
9. J.B.C.R.S., X:67.
10. U.L.C. Add. MS. 4005.2, f. 1.
11. J.B.C.R.S., X:55.
12. Francis Hauksbee, *Physico-Mechanical Experiments on Various Subjects*, 2d ed. (London, 1719), p. 9.
13. C.M.R.S., II:174.
14. J.B.C.R.S., X:58.
15. E.S.D.B., V:592.
16. J.B.C.R.S., X:58, 60.
17. The complete plan is more fully described in S.H.L., p. 132, and C.R.W., I: 363–64.
18. I.N. *Corres.*, V:62–63.
19. C.M.R.S., II:177.
20. J.B.C.R.S., X:63–64.
21. I.N. *Opticks*, p. cxxi.
22. *Ibid.*, p. cxxii.
23. L.T.M., p. 507.
24. D.T.W., VIII:16.
25. Keynes MS. 130 (7).
26. W.G.H., p. 15.
27. I.N. *Opticks*, p. 1.
28. Jean-Théóphile Desaguliers, *Course of Experimental Philosophy*, 3rd ed. (London, 1763), I:viii.
29. Keynes MS. 130 (5), pp. 1–2. For Newton's influence on literature see Marjorie Hope Nicolson, *Newton Demands the Muse: Newton's "Opticks" and the Eighteenth Century Poets* (Princeton, N.J., 1946); for his impact on science see I. Bernard Cohen, *Franklin and Newton* (Philadelphia, 1956).
30. U.L.C. Add. MS. 3970.3, f. 336.
31. I.N. *Corres.*, III:338.
32. I.N. *Opticks*, p. 338.
33. *Ibid.*, pp. 339–45.
34. Gilbert Burnet, *History of His Own Times* (Oxford, 1823), III:49.
35. C.M.R.S., II:175, and J.B.C.R.S., X:89.
36. W.S., p. 9.
37. James H. Monk, *The Life of Richard Bentley, D.D.*, 2d ed. (London, 1833), I:184; C.H.C., IV:72.
38. W.S., p. 10.
39. Quoted in Edward Gregg, *Queen Anne* (London and Boston, 1980), p. 135.
40. I.N. *Corres.*, IV:439, 441.
41. F.B., p. 239.
42. I.N. *Corres.*, IV:445.
43. *Ibid.*, IV:445, fn. 4.
44. S.D.B., II:218.
45. Quoted in J.E., p. lxxiv.
46. F.B., p. 173.
47. *Ibid.*, pp. 174–75.
48. *Ibid.*, pp. 199, 204, 212.
49. *Ibid.*, p. 29.
50. *Ibid.*, pp. 72, 216.
51. R.S.W. *Biog.*, p. 655.
52. F.B., pp. 66, 74.
53. *Ibid.*, p. 316.
54. I.N. *Corres.*, IV:421.
55. F.B., p. 75.

56. J.B.C.R.S., X:86.
57. *Ibid.*, X:87.
58. I.N. *Corres.*, IV:430.
59. F.B., p. 77.
60. I.N. *Corres.*, IV:432.
61. F.B., p. 232.
62. I.N. *Corres.*, IV:436.
63. *Ibid.*, IV:241; F.B., p. 66.
64. F.B., pp. 66, 219. 246.
65. *Ibid.*, pp. 78, 219, 238.
66. *Ibid.*, pp. 220, 79, 241.
67. I.N. *Corres.*, IV:447..
68. See, for example, F.B., p. 245.
69. I.N. *Corres.*, IV:452-58.
70. F.B., pp. 256, 81, 254-55.
71. *Ibid.*, pp. 257, 259.
72. I.N. *Corres.*, IV:458.
73. F.B., pp. 81-82, 223.
74. *Ibid.*, pp. 223-24; I.N. *Corres.*, IV:477.
75. I.N. *Corres.*, IV; 490-91. Emphasis Flamsteed's.
76. F.B., pp. 259-60.
77. *Ibid.*, p. 223.
78. I.N. *Corres.*, IV:473-74.
79. *Ibid.*, IV:487-88.
80. F.B., pp. 226, 264, 83.
81. I.N. *Corres.*, IV:513.
82. *Ibid.*, IV:524.
83. F.B., pp. 225, 83.
84. I.N. *Corres.*, IV:527.
85. C.M.R.S., II:210.
86. F.B., p. 272.

Chapter Eighteen: The Devil's Banter

1. J.C. *Newton*, p. 57; J.C. *Mint*, p. 208.
2. I.N. *Corres.*, IV:409-10.
3. *Ibid.*, IV:478.
4. J.C. *Newton*, p. 59.
5. I.N. *Corres.*, IV:466.
6. Statutes of the Realm, 6° Annae c. 11, p. 570.
7. J.C. *Newton*, p. 71; I.N. *Corres.*, IV: 494-95.
8. J.C. *Mint*, p. 210; J.C. *Newton*, p. 72; I.N. *Corres.*, IV:503-4.
9. I.N. *Corres.*, IV:525.
10. *Ibid.*, VII:463.
11. *Ibid.*, IV:522-23, 528, 541.
12. J.E., p. lxxiv; William Whiston, *Histor-ical Memoirs of the Life of Dr. Samuel Clarke* (London, 1730), p. 13.
13. I.N. *Opticks*, pp. 362-65, 368-69.
14. *Ibid.*, pp. 375-76, 401.
15. *Ibid.*, pp. 369-70.
16. W.G.H., p. 30.
17. I.N. *Corres.*, IV:507-8.
18. Keynes MS. 130 (5), sheet 1.
19. U.L.C. Add. MS. 3968.41, f. 99^v; W.W. pp. 315-16.
20. W.G.H., p. 17.
21. J.B.C.R.S., X:124-27; Charles Andrew Domson, *Nicolas Fatio de Duillier and the Prophets of London* (New York, 1981), p. 84.
22. A. Boyer, *The History of Queen Anne* (London, 1735), pp. 316-17.
23. Joseph Spence, *Anecdotes, Observations, and Characteristics, of Books and Men,* ed. by Samuel S. Singer (London, 1820), p. 72.
24. Quoted in F. M. *Portrait*, p. 210. For a different interpretation of Newton's relationship to the Camisards, see Margaret Jacob, "Newton and the French Prophets: New Evidence," *History of Science*, 16 (1978): 134-42.
25. Quoted in Maureen Farrell, *The Life and Work of William Whiston* (New York, 1981), p. 264.
26. W.W., pp. 137-38. Emphasis Whiston's.
27. *Ibid.*, p. 40.
28. William Whiston, *Collections of Authentick Records Belonging to the Old and New Testament* (London, 1928) II:1077.
29. W.W., pp. 141-42.
30. *Ibid.*, p. 156.
31. Quoted in F. M. *Portrait*, p. 210.
32. Farrell, *The Life and Work of William Whiston*, p. 273.
33. W.W., p. 152.
34. J.B.C.R.S., XI:124.
35. W.W., p. 293. Emphasis Whiston's.
36. *Biographia Britannica* (London, 1766), p. 1360. Clarke wrote his thesis on the proposition: "No article of Christian faith, delivered in Holy Scripture is disagreeable to right reason."
37. Whiston, *Historical Memoirs of the Life of Dr. Samuel Clarke* (London, 1730), p. 13.

38. W.W., p. 294.
39. C.M.R.S., II:185–91.
40. S.H.L., pp. 126–27, 129.
41. C.M.R.S., II:181.
42. Ibid., II:214, and I.N. Corres., IV:18.
43. Yahuda MS. 15.7, f. 180ᵛ and R.S.W. Biog., p. 674.
44. C.M.R.S., II:217–18.
45. W.S., p. 67.
46. F.B., p. 276.
47. C.M.R.S., II:219.
48. I.N. Corres., V:17, 19.
49. J.E.B.M., p. 365.
50. C.M.R.S., II:220.
51. I.N. Corres., V:61.
52. C.M.R.S., II:222.
53. Ibid., II:223, and J.B.C.R.S., X:249.
54. Quoted in I.N. Corres., V:77.
55. C.M.R.S., II:223–56.
56. Quoted in C.R.W., I:393–94.
57. C.M.R.S., II:233–34.
58. Ibid., II:385.
59. W.S., pp. 78–81.
60. S.H.L., p. 342.
61. C.M.R.S., II:236.
62. F.B., p. 306.
63. I.N. Corres., VI:43.
64. S.H.L., p. 343.
65. C.R.W., I:419, and J.B.C.R.S., X:401–3, 483.
66. J.B.C.R.S., X:373.
67. I.N. Corres., V:376–77.
68. S.H.L., p. 343.
69. C.M.R.S., II:299.
70. U.L.C., Add. MS. 3965.12, f. 209ᵛ.
71. H. D. Anthony, Sir Isaac Newton (London and New York, 1966), p. 197.
72. F.B., pp. 89, 270.
73. C.M.R.S., II:229–32.
74. F.B., pp. 92, 278–79.
75. Quoted in Edward Gregg, Queen Anne (London and Boston, 1980), p. 234.
76. J.B.C.R.S., X:260–62.
77. I.N. Corres., V:99.
78. Ibid., V:101.
79. Ibid., V:102.
80. Ibid., V:105.
81. Ibid., V:130, and F.B., p. 95.
82. F.B., pp. 94–95, 226. Emphasis Flamsteed's.
83. Ibid., pp. 93, 291.
84. Ibid., pp. 95, 291–92.
85. I.N. Corres., V:121.
86. Ibid., V:122, 131.
87. J.B.C.R.S., X:290.
88. I.N. Corres., V:165–66, and F.B., pp. 95–96.
89. F.B., pp. 228–29, 97. Emphasis Flamsteed's. C.f. pp. 294–95.
90. J.B.C.R.S., X:511.
91. F.B., p. 229.
92. I.N. Corres., VI:24, 69–70.
93. F.B., p. 99. Emphasis Flamsteed's.
94. Ibid., pp. 98, 229.
95. I.N. Corres., V:224–25.
96. R.S.W. Biog., p. 693.
97. F.B., p. 305.
98. Ibid., p. 314.
99. I.N. Corres., VI:255–56.
100. F.B., p. 101.
101. Ibid., p. 305.

Chapter Nineteen: "Second Inventors Count for Nothing"

1. I.B.C. Intro., p. 189.
2. I.N. Corres., III:311–12, 338.
3. W.G.H., p. 41.
4. I.B.C. Intro., p. 199.
5. Keynes MS. 130.6, Book 2.
6. I.B.C. Intro., p. 200.
7. I.N. Principia, p. xxxiii.
8. R. J. Waite, Dr. Bentley: A Study in Academic Scarlet (London, 1965).
9. I.N. Corres., IV:518–19.
10. W.W., pp. 133, 136.
11. J.E., pp. xvi, 1.
12. I.N. Corres., V:3.
13. Ibid., V:5, 7.
14. Ibid., V:7.
15. Keynes MSS. 130.6, Book 2; 130.7, sheet 1.
16. I.N. Corres., V:24, 51, 56.
17. R.S.W. Biog., p. 696.
18. A.R.H., M.B.H., IX:493–94.
19. A.R.H., pp. 54–55; Joseph E. Hofmann, Leibniz in Paris, 1672–1676 (Cambridge, England, 1974), pp. 26–27.
20. A.R.H., M.B.H., IX:566.
21. See Hofmann, Leibniz in Paris, especially Chapter 13.
22. I.N. Corres., II:32.
23. Ibid., II:39.
24. Ibid., II:65.

25. Hofmann, *Leibniz in Paris*, Chapter 20.
26. I.N. *Corres.*, II:109.
27. *Ibid.*, II:130, 163.
28. *Ibid.*, II:134, 153, fn. 25.
29. *Ibid.*, II:219, 234.
30. *Ibid.*, II:235, 110.
31. D.T.W., IV:415–16; I.N. *Corres.*, II:396.
32. I.N. *Corres.*, II:179.
33. D.T.W., IV:565–76.
34. Gottfried Wilhelm Leibniz, *Sämtliche Schriften und Briefe*, ser. I (Berlin, 1923), IV:475–77.
35. I.N. *Principia*, p. 655.
36. A.R.H., p. 5.
37. Quoted in *Ibid.*, p. 77.
38. U.L.C. Add. MS. 3968.29, f. 422^v.
39. I.N. *Corres.*, III:170.
40. *Ibid.*, III:338.
41. Christiaan Huygens, *Oeuvres complètes*, pub. by *Société hollandaise des sciences* (The Hague, 1888–1950), X:214.
42. *Ibid.*, X:257–58.
43. I.N. *Corres.*, III:258.
44. *Ibid.*, III:286.
45. For fragmentary drafts of this correspondence see *Ibid.*, III:219–22.
46. Huygens, *Oeuvres complètes*, X:675.
47. I.N. *Corres.*, III:387, fn. 1.
48. *Ibid.*, IV:100, 101, fn. 7. Although the volume was published in 1695, two years after the revised *Algebra*, it was numbered volume I. The *Algebra* was numbered volume II.
49. Quoted in A.R.H., pp. 118, 105.
50. Nicolas Fatio de Duillier, *Lineae brevissimi descensus investigatio geometrica duplex* (London, 1699), p. 18.
51. W.G.H., pp. 26–27.
52. Quoted in A.R.H.; p. 133.
53. Newton, *Opticks*, 1st ed., p. 166.
54. "Isacci Newtoni tractatus duo, de speciebus & magnitudine figurarum curvilinearum," *Acta Eruditorum* (January 1705), p. 35. Emphasis Leibniz's.
55. Pierre Des Maizeaux, *Recueil des pièces, sur la philosophie, la religion naturelle, l'historie, les mathématiques, etc. Par Mrs. Leibniz, Clarke, Newton, & autres autheurs cèlèbres* (Amsterdam, 1720), II: 49.
56. John Keill, "Epistolia . . . de Legibus Virum Centripetarum," *Philosophical Transactions*, 26 (1708): 185.

57. I.N. *Corres.*, V:115, and J.B.C.R.S., X: 195.
58. I.N. *Corres.*, V:95.
59. *Ibid.*, V:97.
60. J.B.C.R.S., X:264.
61. I.N. *Corres.*, V:115, 117.
62. J.B.C.R.S., pp. 266–67.
63. *Ibid.*, p. 270.
64. I.N. *Corres.*, V:142–52.
65. J.B.C.R.S., X:290–91; U.L.C. Add. MS. 3968.30, f. 422^v; and I.N. *Corres.*, V:132.
66. I.N. *Corres.*, V:142, 149.
67. *Ibid.*, V:207–8.
68. *Ibid.*, V:213.
69. J.B.C.R.S., X:369, 375, 377, 386, and I.N. "Account," p. 211.
70. J.B.C.R.S., X:391.
71. I.N. "Account," p. 194.
72. U.L.C. Add. MS. 3968.37, f. 539^r. For details of the *Commercium* itself see Add. MS. 3968.8.
73. William Whiston, *Historical Memoirs of the Life of Dr. Samuel Clarke* (London, 1730), p. 132.
74. I.N. *Corres.*, V:349.
75. Quoted in A.R.H., p. 153. Emphasis Leibniz's.
76. I.N. *Corres.*, V:203.
77. I.N. *Principia*, pp. 398–99, 543.
78. Richard S. Westfall, "Newton and the Fudge Factor," *Science*, 179, no. 4075 (February 23, 1973): 751–58.
79. I.N. *Corres.*, V:226, 278.
80. *Ibid.*, V:361.
81. I.N. *Principia*, pp. 543–46.
82. *Ibid.*, p. 547.
83. *Ibid.*
84. I.N. *Corres.*, V:389.
85. *Ibid.*, V:391, 400.
86. I.N. *Principia*, p. xx.
87. I.N. *Corres.*, V:413–14, 417.
88. J.E., p. lxxvii.

Chapter Twenty: "They Could Not Get Me to Yield"

1. I.N. *Corres.*, VI:3–5.
2. *Ibid.*, VI:8–9.
3. *Ibid.*, VI:18–19.
4. *Ibid.*, VI:80, 106, 142, 158.

5. I.B.C. *Intro.*, pp. 252–53, 256–57.
6. *Ibid.*, pp. 254–56.
7. I.N. *Corres.*, VI:71, 105–6.
8. *Ibid.*, VI:140.
9. *Ibid.*, VI:153, 185, 71, 158.
10. *Ibid.*, VII:83.
11. I.N. "Account," pp. 194, 206.
12. *Ibid.*, pp. 215, 218, 221.
13. F.M. *Portrait*, p. 339.
14. I.N. "Account," p. 224.
15. A.R.H. p. 231.
16. I.N. *Corres.*, V:xlv–xlvi.
17. Wilhelm Gottfried Leibniz, Die *philosophischen Striften*, ed. by C. J. Gerhardt (Berlin, 1875–90), III:655.
18. I.N. *Corres.*, VI:222, 239, 250.
19. *Ibid.*, VI:204.
20. *Ibid.*, VI:288, fn. 1.
21. *Ibid.*, VI:458, fn. 1.
22. *Ibid.*, VI:275.
23. *Die Werke von Leibniz*, ed. by Onno Klopp (Hanover, 1888), XI:115, 90.
24. I.N. *Corres.*, VI:288, 285–86.
25. *Ibid.*, VI:339–40, 342.
26. H. G. Alexander, *The Leibniz–Clarke Correspondence* (New York, 1956), p. 190.
27. *Ibid.*, pp. 11–12.
28. Quoted in L.T.M., p. 601.
29. Alexander, *The Leibniz–Clarke Correspondence*, pp. 193–94.
30. I.N. *Corres.*, VI:376.

Chapter Twenty-One: Infinity

1. J.B.C.R.S., XI:190.
2. I.N. *Corres.*, VII:74.
3. *Ibid.*, VII:17.
4. U.L.C. Add. MSS. 3968.26, ff. 372–80; 3968.12, ff. 146–72; 3968.13, ff. 173–218, 223–35.
5. I.N. *Corres.*, VI:385.
6. *Ibid.*, VI:387.
7. *Ibid.*, VI:383.
8. *Ibid.*, VI:435–39.
9. *Ibid.*, VII:45–46.
10. *Ibid.*, VII:64, 70. Emphasis Newton's.
11. *Ibid.*, VII:73, 130.
12. *Ibid.*, VII:77–79.
13. *Ibid.*, VII:149.
14. *Ibid.*, VII:185.
15. *Ibid.*, VII:221.
16. *Ibid.*, VII:72.
17. I.N. *Opticks*, pp. 350–51.
18. R.S.W. *Biog.*, p. 794.
19. I.N. *Corres.*, VII:114.
20. *Ibid.*, VII:16.
21. *Ibid.*, VII:205.
22. H.P., Preface, unpaged.
23. W.S., p. 17.
24. I.B.C. *Intro.*, pp. 265–72.
25. H.P., Preface.
26. I.N. *Principia*, p. 400.
27. I.N. *Corres.*, VII:302.
28. J.E., p. lxxviii, fn. 186.
29. James Wilson, biographical preface to Henry Pemberton, *A Course of Chemistry, divided into twenty-four lectures* (London, 1771).
30. *The Autobiography of Benjamin Franklin*, ed. by Leonard W. Labaree, Ralph L. Ketcham, Helen C. Boatfield, and Helene H. Fineman (New Haven and London, 1964), p. 97.
31. I.N. *Corres.*, VII:279, 311, 322.
32. See Frank Manuel *Isaac Newton, Historian* (Cambridge, Mass., 1963), pp. 21–36, 89–102.
33. Isaac Newton, "Remarks upon the Observations made upon a Chronological Index . . . published at Paris," *Philosophical Transactions*, 33, no. 389 (July and August 1725):315–21.
34. *The Lives of Dr. Edward Pocock by Dr. Twells; of Dr. Zachary Pearce and Dr. Thomas Newton by themselves; and of the Rev. Philip Skelton by Mr. Burdy* (London, 1816), I:434–35.
35. Keynes MS. 136; W.S., pp. 62–63.
36. I.N. *Corres.*, VII:382.
37. W.S., pp. 12–13.
38. Keynes MS. 3, pp. 1–27.
39. I.N. *Corres.*, III:89.
40. W.S., p. 70.
41. *Ibid.*, pp. 65–68.
42. E.T., p. 165.
43. Francis Atterbury, *The Epistolary Correspondence, Visitation Charges, Speeches, and Miscellanies* (London, 1783) I:180.
44. Thomas Hearne, *Remarks and Collections*, ed. by C. E. Doble, 11 vols. (Oxford, 1885–1921), 11:100.
45. W.S., p. 68.
46. I.N. *Corres.*, VII:289.

47. W.S., pp. 13–14, 66.
48. R.V., p. 10.
49. Keynes MS. 130 (6).
50. W.S., pp. 14–15.
51. *Ibid.*, p. 13.
52. I.N. *Corres.*, VII:210.
53. Joseph Spence, *Anecdotes, Observations, and Characters, of Books and Men*, ed. by Samuel S. Singer (London, 1820), p. 368.
54. E.T., p. 164.
55. W.S., p. 68.
56. I.N. *Corres.*, V:201, 345.
57. Keynes MS. 129A, p. 29.
58. I.N. *Corres.*, V:251–52, 353.
59. W.S., pp. 68–69.
60. Thomas Dunham Whitaker, *The History and Antiquities of the Deanery of Craven* . . . . 3d ed. (London, 1878), p. 542; John Nichols, *Illustrations of the Literary History of the Eighteenth Century*, 8 vols. (London, 1817–58), 4:34.
61. E.T., p. 165.
62. *Ibid.*, pp. 165–66; Keynes MSS. 130 (6), Book 2, 130 (7), sheet 2.
63. E.T., pp. 172–73.
64. W.S., pp. 71–77.
65. E.T., p. 173, fn. 2.
66. W.S., pp. 19, 22.
67. I.N. *Corres.*, VII:355–56.
68. E.T., p. 166, and W.S., pp. 82–83.
69. Keynes MS. 130 (6), Book 1.
70. E.T., p. 166.
71. J.B.C.R.S., XIII:62.
72. I.N. *Corres.*, VII:355.

Acknowledgments

The author of any work of nonfiction, especially one that has taken the better part of six years to complete, inevitably becomes indebted to many individuals and institutions. This writer is certainly no exception. Newton once observed that if he had seen farther than other men it was because he stood on the shoulders of giants. Whether or not I have helped bring Newton into sharper focus I leave to the judgment of my readers, but there can be no doubt that some of our most gifted scholars have provided the requisite tools with which to do so. I thank them, one and all, for allowing me to stand on their reassuring shoulders: I. Bernard Cohen, Betty Jo Teeter Dobbs, A. R. Hall, Marie Boas Hall, John Herivel, David Kubrin, Frank Manuel, J. E. McGuire, P. M. Rattansi, and D. T. Whiteside. One among their number has done even more. Having written copiously and with great insight on Newton over the past two decades, Richard S. Westfall mercifully befriended this innocent abroad, shared unstintingly of his vast knowledge and rich personal archives, tolerated my endless intrusions on his always demanding schedule, and somehow held enough compassion in reserve to encourage me every step of the way. Thanks, Sam, it was infinitely more than I deserved or can ever hope to repay. To colleague and friend Rebecca Shepherd Shoemaker, who read and helped edit the entire work while in manuscript, goes a special gesture of tribute for what must have often seemed an endless and thankless task. And to those who rashly assert that the days of the skilled and sympathetic editor are no more, I can only say that they haven't had the privilege of working with Joyce Seltzer at The Free Press. Special thanks are also due to Celia Knight and Norman

Sloan for their expert handling of the more technical aspects of editorial production. Lest any faults in this work be attributed to one or more of the above persons, let me simply add: *mea culpa*.

I also wish to thank the following for granting me access to Newton materials and for the right to reproduce certain of them: Babson College (Grace K. Babson Collection); the Warden and Fellows of the New College, Oxford; the Syndics of the Cambridge University Library (Portsmouth Papers); Lord Portsmouth and the Trustees of the Portsmouth Estates; Bibliothèque et Universitaire de Genève; the National Portrait Gallery; the Syndics of the Fitzwilliam Museum; the Herzog Anton Ulrich-Museum, Braunschweig; the Graduate Library and Swain Hall Science Library, Indiana University; Cunningham Memorial Library, Indiana State University; the Jewish National and University Library (Yahuda Manuscript Collection); the Provost and Scholars of King's College, Cambridge (Keynes Manuscript Collection); Controller of H.M. Stationery Office (Crown-copyright records in the Public Record Office); Lincoln Cathedral Archives; The Royal Society; the Master and Fellows of Trinity College, Cambridge; University of California Press; Cambridge University Press; Dover Publications, Inc.; Harvard University Press; and The University of Wisconsin Press. Again, I thank them all.

Index

Gale E. Christianson received his Doctor of Arts degree in History from Carnegie-Mellon University. He joined the faculty of Indiana State University in Terre Haute in 1971, where he is now Professor of History. He is the author of *This Wild Abyss: The Story of the Men Who Made Modern Astronomy* and has published numerous articles in both scholarly and popular journals.